Centrifugal Compressors in Hydrocarbon Processing Industries

Neetin Ghaisas

Centrifugal Compressors in Hydrocarbon Processing Industries

A Comprehensive Engineering Overview

Springer

Neetin Ghaisas
Calgary, AB, Canada

ISBN 978-3-032-03686-5 ISBN 978-3-032-03687-2 (eBook)
https://doi.org/10.1007/978-3-032-03687-2

This Springer imprint is published by the registered company Springer Nature Switzerland AG
The registered company address is: Gewerbestrasse 11, 6330 Cham, Switzerland

Preface

This book aims to bridge the gap between theoretical concepts and practical applications, providing readers with a comprehensive understanding of the subject matter. It presents a unique perspective on the thermodynamic, aerodynamic, and mechanical design features of centrifugal compressors. Their auxiliary systems are elaborated for the benefit of engineering professionals involved with the specification, application, and selection of compression strings and packages.

Throughout this book, I have endeavored to present the material in a clear and accessible manner. Each chapter builds on the previous one, gradually increasing in complexity to ensure a smooth learning curve. I have also included numerous images and graphics to enhance the reader's understanding of the subject.

This book comprises a detailed review of centrifugal compressor casings, their internals, rotor design, and impeller thermodynamics and aerodynamics. The chapters on surge detection and surge control, flexible-element couplings, dry gas seals, seal support systems, and lubricating oil systems provide comprehensive insight into the relevant theory, principles of operation, and design processes.

The chapters on lateral rotordynamics, torsional analysis, shop testing, special engineering analyses, and condition monitoring and diagnostics address key knowledge-based engineering parameters that are significant to acceptable designs, optimum performance, and safety and long-term reliability in service of the equipment.

In addition to a description of conventional concrete foundations in the chapter on grouting, the subject matter on machinery modularization includes novel considerations on installation by moving a significant portion of the site work to module yards, reducing material movement at job sites, and decreasing the costs associated with construction and many precommissioning activities.

Comprehensive and elaborate technical specifications for inquiry and purchasing aim at eliminating guesswork, misinterpretation, and lost scope. Errors or omissions in purchase orders can be cost prohibitive or challenging to mitigate during project execution. The chapter on compressor specifications, proposal evaluation, and design coordination presents this subject with the help of real-world examples.

Considering the ongoing global initiatives and efforts to reduce greenhouse gas emissions, a chapter is dedicated to a narrative on the process of capturing and storing carbon dioxide (CO_2), along with the two compressor types predominantly used for CO_2 compression. This chapter provides an overview of sequestration processes and current challenges, the types of compressors used, new technologies available to extend life of materials used in compressors, and configuration options for improved overall efficiency.

I would like to extend my heartfelt gratitude to all those who have supported me in this endeavor. Special thanks are due to my family, friends, and colleagues for their encouragement and patience. I am also profoundly appreciative of the feedback and suggestions from the technical reviewers, whose expertise has been invaluable in refining the content of this book.

It is my sincere desire that this book serves as a valuable resource to readers.

Calgary, Canada Neetin Ghaisas, P.Eng.(AB)

The original version of the book has been revised. A correction to this book can be found at https://doi.org/10.1007/978-3-032-03687-2_18

Acknowledgments I thank the many contributors to this book, namely, Mitsubishi Heavy Industries Compressor for providing materials and images for the chapters on compressor casing design, rotor design, and impellers, as well as photographs for the book's front cover; Stefano Falomi, Massimo Camatti, and Guido Peano from Baker Hughes and Matthew Gutscher for providing materials and images for the chapter on CO_2 compressors; Krish Ramesh for the chapter on lateral rotordynamics; Malcolm Leader for the chapter on torsional analysis; and Michael Titone for the chapter on condition monitoring and diagnostics. This book could not have assumed its distinct form without their support.

I offer a special note of thanks to the staff at my publisher, Springer Nature, for making this book writing and publication experience truly gratifying. Their professionalism has ensured the successful publication of this book. Their efforts in editing, designing, and marketing have been exceptional.

I extend my earnest gratitude to my family for their unwavering support and encouragement throughout this journey. Their patience and understanding have been invaluable.

Finally, I am indebted to the readers of this book and to the technical community for their continuous support and interest in my work. It is their enthusiasm and feedback that continue to inspire me to strive for excellence.

Competing Interests The author has no competing interests to declare that are relevant to the content of this manuscript.

Contents

About the Author

Neetin Ghaisas, P.Eng.(AB) is a registered practicing professional engineer with expertise in process plant machinery. With a proven track record in all aspects of rotating equipment, he brings extensive experience across petrochemical manufacturing facilities—including ethylene, olefins, and plastics—as well as diverse petroleum projects such as refinaries, hydrogen compression, upgraders, oil sands developments, and ventures in the natural gas sector.

Mr. Ghaisas has contributed to the development of American Petroleum Institute (API) standards and process industry practices for many years and currently serves positions at the API as Chairman of API Standard 612, Chairman of API Standard 672, as well as a member of the Subcommittee on Mechanical Equipment. He is also a member of the American Society of Mechanical Engineers (ASME). Mr. Ghaisas has authored and co-authored numerous papers, many of which have been featured in prominent international magazines and showcased at global conferences. In addition, he holds a U. S. patent on machinery modularization, with a second patent currently pending.

Chapter 1
Compressor Casings

Abstract A centrifugal compressor casing houses both stationary and rotating internal components. Stationary elements include the inlet volute, guide vanes, discharge volute, diaphragms, return channels, sealing systems, and bearings. The dynamic assembly—referred to as the rotor—comprises the shaft, impellers, and balance drum (or piston). Process nozzles are either welded onto the casing or integrally cast as part of its structure, and provisions are also made for auxiliary piping connections. This chapter offers an in-depth exploration of casing joint types and their pressure-handling capabilities, various casing configurations, shear ring designs specific to radially split compressors, and the methods used in casing fabrication. Additional topics covered include the interface between the casing and its mounting pedestal, design considerations for discharge volutes and nozzle configurations, and the role of water or oil injection in compressor operation.

1.1 Types of Joint Splits, Pressure Limits

Centrifugal compressors typically feature two distinct casing split configurations: axial split and radial split. Figure 1.1 illustrates a common classification method employed by one manufacturer to determine the appropriate casing split type, based on inlet volumetric flow rate (measured in actual cubic feet per minute [ACFM]) and discharge pressure (expressed in pounds per square inch absolute [psia]).

The axially split casing, also referred to as the horizontally split casing, is composed of two halves—top and bottom—with each having a flange in the horizontal plane. These flanges are bolted together at the split line to secure the casing, prevent atmospheric leakage from internal compressor components, and support both the rotor and stationary internal elements. In addition to the split line, radial end covers serve as another critical sealing point. Integral end covers are preferred over separable covers because they help eliminate additional leakage paths and reduce manufacturing and machining complexities. This casing style has pressure limitations, particularly at the joint, from the consideration of gas leakage. Horizontally split compressors are suitable for handling high-molecular-weight gases in low-pressure applications, typically up to 725 psig (5000 kPag). Figure 1.2 illustrates a typical horizontally split casing. This configuration, which accommodates sideloads,

corrected publication 2025
N. Ghaisas, *Centrifugal Compressors in Hydrocarbon Processing Industries*,
https://doi.org/10.1007/978-3-032-03687-2_1

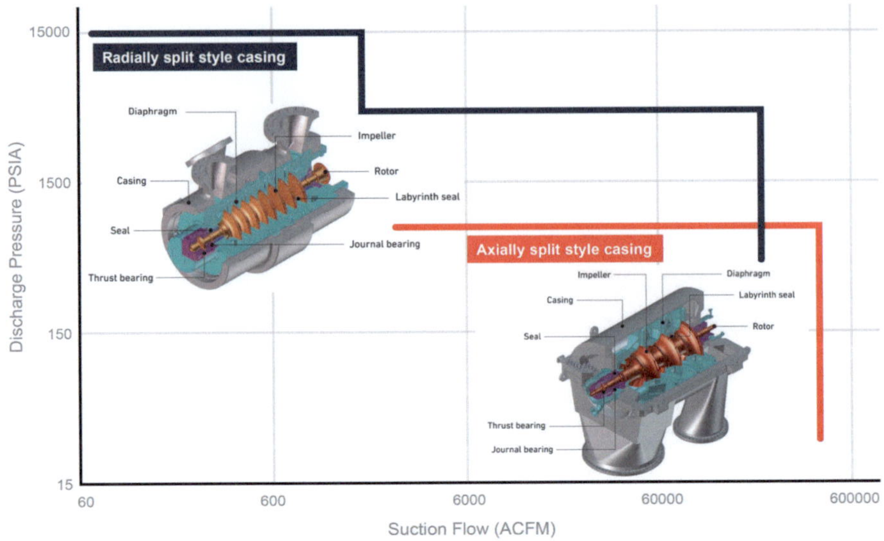

Fig. 1.1 Casing split styles (Image courtesy of Mitsubishi Heavy Industries Compressor, Hiroshima, Japan)

is commonly used in refrigeration systems involving hydrocarbons such as propane, propylene, and ethylene—especially where multiple pressure stages are required.

The radially split casing—also known as the vertically split or barrel-style casing—features a cylindrical body with end plates or covers mounted on both sides of the centrifugal compressor. In this configuration, the rotor and stationary internal components are integrated into a single structural unit called the inner barrel assembly, which is inserted into the outer casing like a cartridge. The radial clearance between the inner barrel and the outer casing is exposed to discharge pressure, enabling an effective and reliable seal between the two halves of the inner barrel. This design is ideal for handling low-molecular-weight gases and is capable of withstanding high sealing pressures ranging from 10,000 to 14,500 psig (69–100 MPag). American Petroleum Institute (API) Standard 617 [1] stipulates that a radially split compressor is required if the partial pressure of hydrogen in the process gas exceeds 200 psig (1380 kPag) at the maximum allowable casing pressure. Figure 1.3 depicts a typical radially split casing.

Horizontally split compressors with upward-facing nozzles necessitate the removal of piping for maintenance access, making this configuration uncommon. In contrast, the more prevalent design in process plants features downward-oriented nozzles, which facilitate easier maintenance and overhauls. Radially split compressors offer the advantage of allowing the inner bundle to be extracted from one end without disconnecting process piping. However, this design requires extra axial space to accommodate bundle removal, posing a disadvantage in multicasing compressor string arrangements.

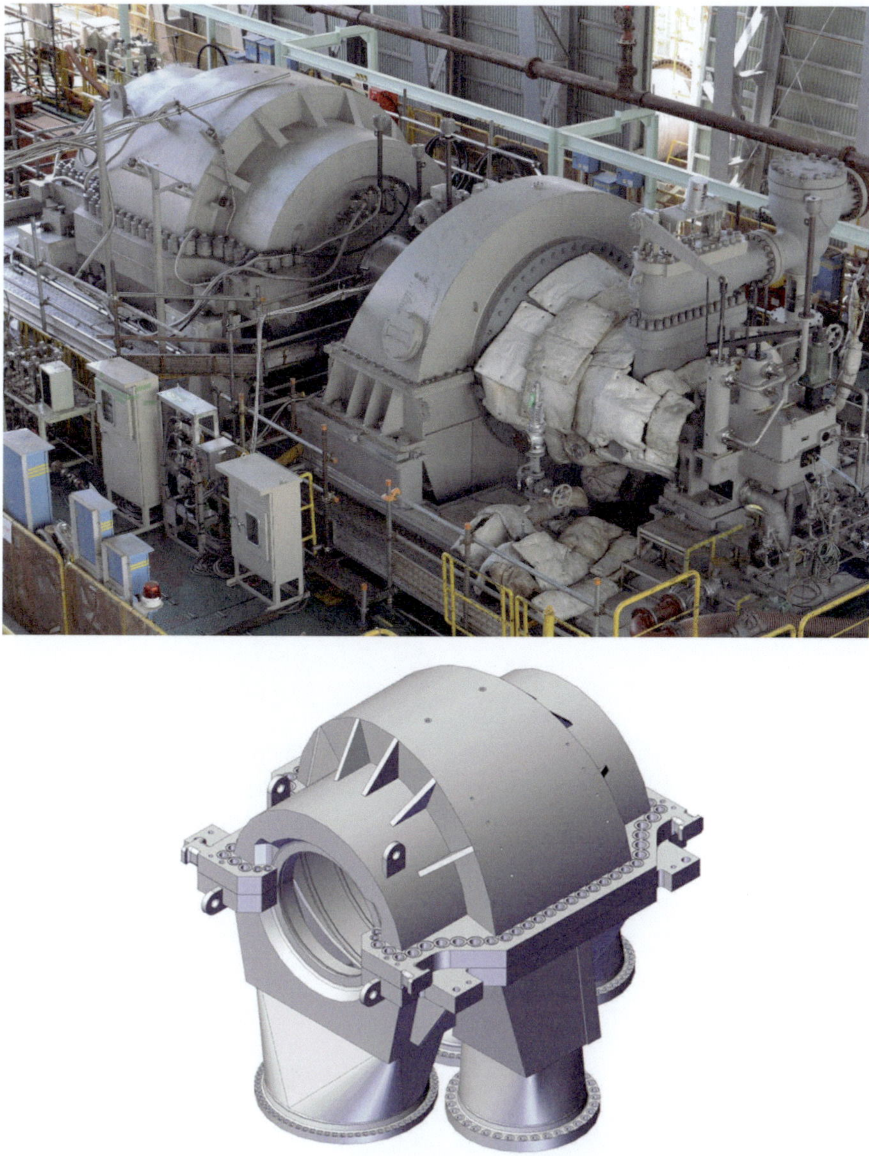

Fig. 1.2 Axially split type casing (Image courtesy of Mitsubishi Heavy Industries Compressor, Hiroshima, Japan)

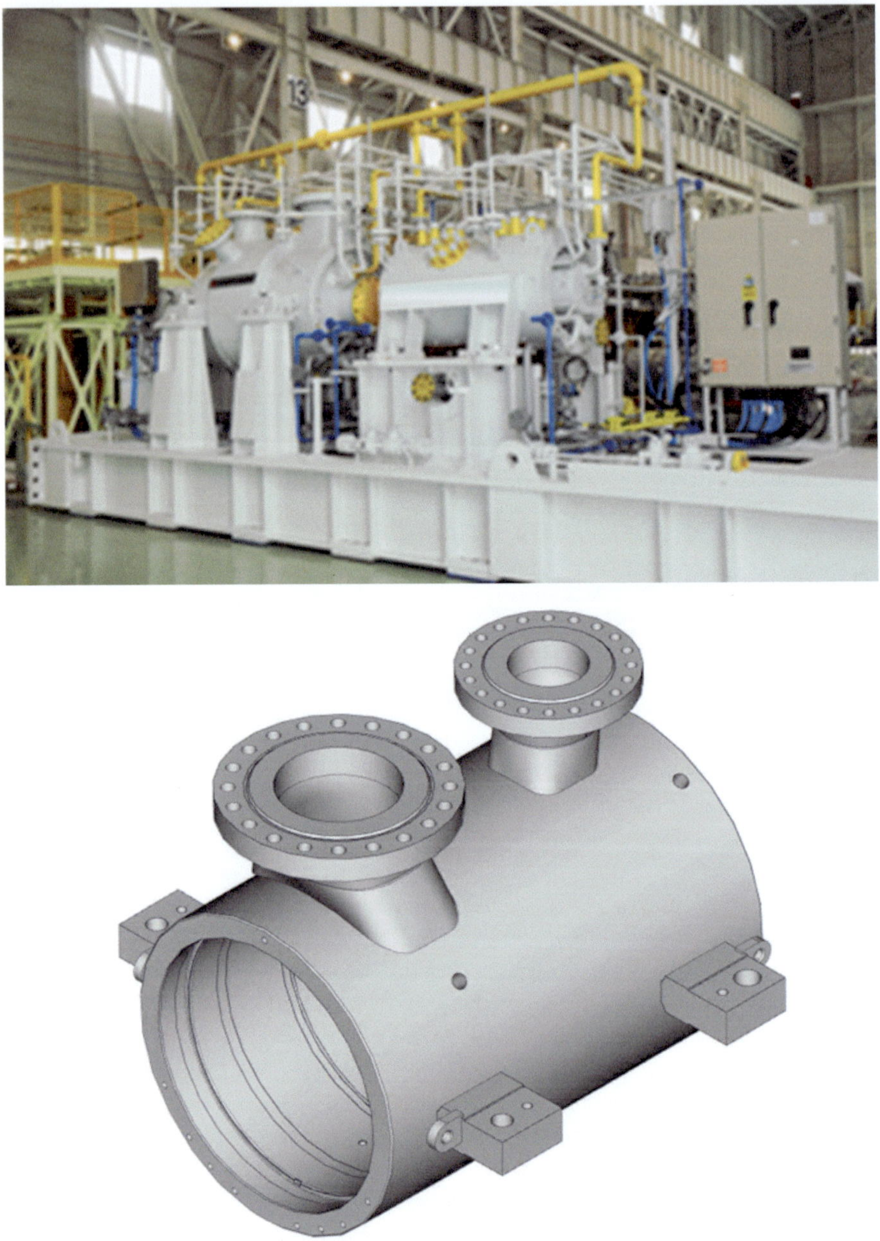

Fig. 1.3 Radially split type casing (Image courtesy of Mitsubishi Heavy Industries Compressor, Hiroshima, Japan)

1.2 Casing Configurations

This section outlines several compressor design configurations—equalized, unequalized, double-flow, and balanced opposed with center seal—along with their typical applications (refer to Chap. 3 for details on impeller thrust and balance piston arrangements).

Centrifugal compressor casings can be configured in various ways, each featuring distinct internal flow passage designs. The most straightforward is the straight-through flow path, illustrated in Fig. 1.4. This design consists of a single process stage with impellers arranged in series from inlet to discharge, and axial thrust is counteracted by a balance piston. The casing includes two primary nozzles—one for gas inlet and one for discharge—making this configuration well-suited for standard, uncomplicated compression applications.

Compound and sidestream (or sideload) compressor configurations also feature straight-through impellers oriented in the same direction within the casing. However, unlike single-stage designs, these configurations incorporate multiple process stages within a single casing. As a result, additional nozzles are included beyond the primary inlet and final discharge to accommodate intermediate flows.

A compound centrifugal compressor has a division wall between the sections. The two sections are arranged in one casing and have an external discharge volute (Fig. 1.5). A sidestream centrifugal compressor also has several process stages in a single casing, but each stage serves as a continuous process gas passage, with side-in and/or side-out gas passages, as shown in Fig. 1.6. Both compound and sidestream compressors are commonly employed in systems requiring multiple pressure levels, such as refrigeration and synthesis gas applications.

A balanced opposed centrifugal compressor can be distinguished as either equalized or unequalized. Both types of centrifugal compressors have two sections in a single casing, with the sections positioned in the opposite orientations. An equalized-type balanced opposed centrifugal compressor, as shown in Fig. 1.7, is called a double-flow centrifugal compressor or a dual-flow centrifugal compressor. In a

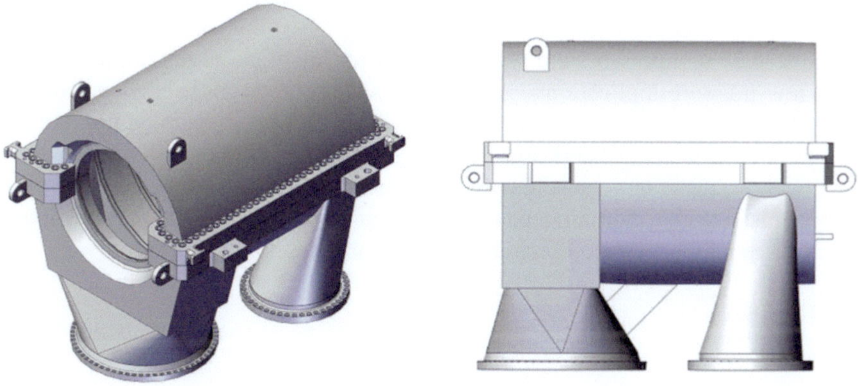

Fig. 1.4 Straight-through type compressor configuration (Image courtesy of Mitsubishi Heavy Industries Compressor, Hiroshima, Japan)

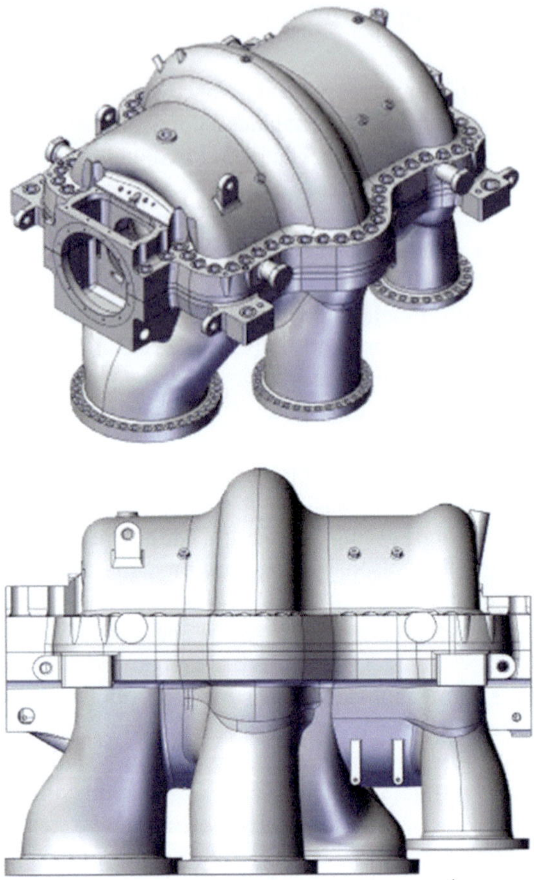

Fig. 1.5 Compound type compressor configuration (Image courtesy of Mitsubishi Heavy Industries Compressor, Hiroshima, Japan)

double-flow centrifugal compressor, the two sections are aerodynamically identical with back-to-back impeller stages, resulting in cancelation of the net axial thrust force generated by the impellers in each section. The double-flow configuration has three total nozzles on the casing, namely, the two identically shaped nozzles for the inlet and a common single nozzle for discharge.

Double-flow compressors are typically used for handling large volumetric flow rates in low-pressure applications—such as cracked or charge gas in ethylene production—where casing dimensions can be substantial, and the assembled unit weight may exceed 500,000 lb (227,000 kg). If transportation constraints limit the allowable size or weight, these large casings often require partial disassembly for shipment. Components like the mounting pedestals, soleplates, rotor, and upper and lower casing halves may be shipped separately to the job site.

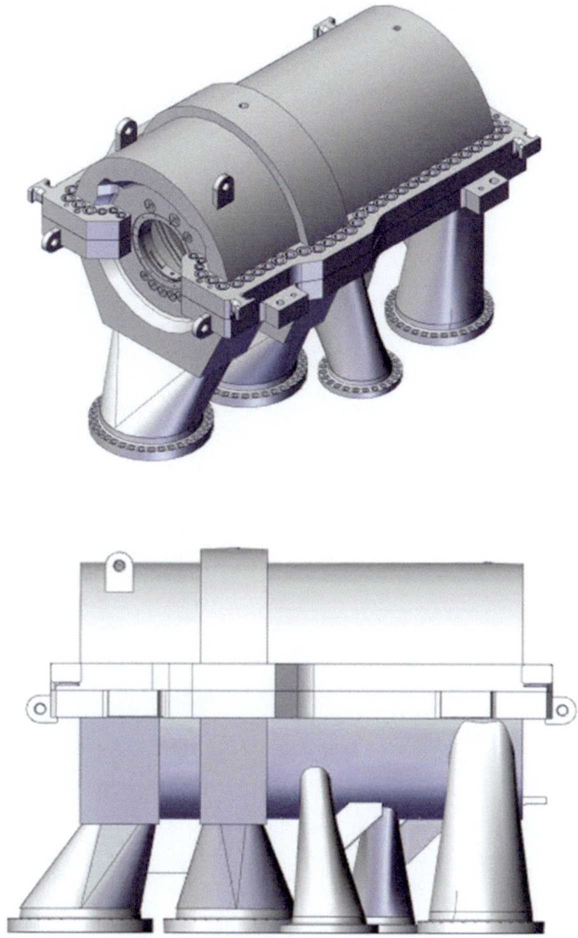

Fig. 1.6 Sidestream type compressor configuration (Image courtesy of Mitsubishi Heavy Industries Compressor, Hiroshima, Japan)

The need to disassemble for shipment offsets the advantages of conducting mechanical run and gas leakage tests at the manufacturing facility. Reassembling the compressor on-site can be complex due to limitations in infrastructure, equipment, and environmental conditions. These challenges must be carefully weighed when determining the appropriate casing size for high-flow compressors. Ultimately, it's a trade-off—such as opting for two casings instead of three or consolidating compression duties into a single train rather than using two parallel trains. Each configuration presents its own set of benefits and limitations. The overall dimensions and structural design of compressors are additionally shaped by a variety of factors, including the following:

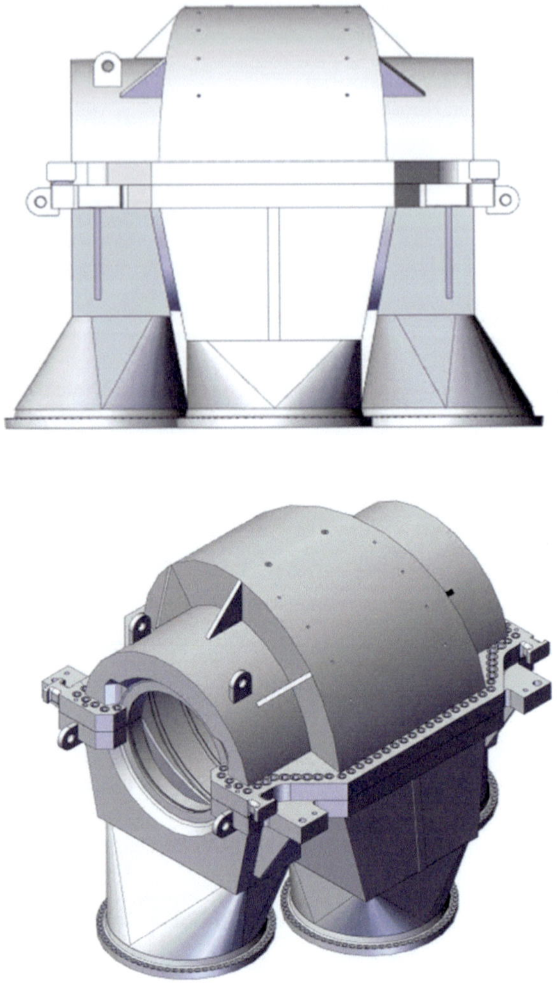

Fig. 1.7 Double-flow type compressor configuration (Image courtesy of Mitsubishi Heavy Industries Compressor, Hiroshima, Japan)

- Thermodynamic Considerations
  - Total pressure ratio across the compressor from inlet to outlet flanges.
  - Volume flow rate at the inlet.
  - Presence and magnitude of any sideloads.
- Aerodynamic Factors
  - Maximum impeller diameter, driven by tip speed and flow area requirements.
  - Required head, which determines the number of stages within a single casing.
  - Impeller design type—whether two-dimensional, three-dimensional, or a combination.
  - Spacing between stages.

- Rotordynamic and Mechanical Design
  - Influence of bearing span on shaft stiffness and critical speeds.
  - Shaft diameter selection.
  - Shaft-end diameter, based on peak torque during startup and operation.
  - Impact of stiffness ratio (the ratio of the bearing stiffness to shaft stiffness) on rotor mode shapes.

These parameters are interconnected and must be evaluated collectively to achieve optimal compressor casing and rotor design. Balancing them effectively ensures mechanical integrity, performance efficiency, and manufacturability.

An unequalized compressor—commonly referred to as a back-to-back centrifugal compressor, as illustrated in Fig. 1.8—features a distinct sectional arrangement designed to manage axial thrust. In this configuration, opposing thrust forces are counteracted either by a thrust bearing or by incorporating a balance piston between the two sections. If axial loads in each section are excessive and in the proximity of maximum load capacity of the thrust bearing, an additional balance piston can be needed to absorb the thrust.

This back-to-back layout includes four process nozzles: inlet and discharge for both the first and second sections. Such compressors are particularly well-suited for high-pressure applications requiring substantial pressure ratios, such as process gas compression and gas injection services.

The diaphragms are the stationary aerodynamic components in the compressor casing. They include diffusers, return bends, and return channels. Diffusers are designed with an increasing cross-sectional area to transform the kinetic energy of the gas exiting the impellers into pressure energy. Return bends and return channels serve to redirect the flow—guiding the gas from the diffuser outlet of one stage to the inlet of the next stage impeller, as illustrated in Fig. 1.9.

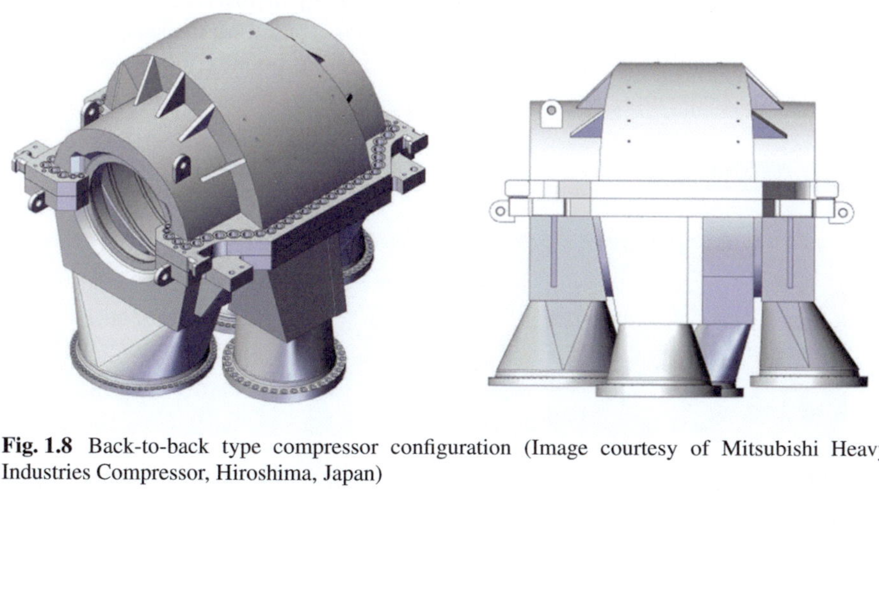

Fig. 1.8 Back-to-back type compressor configuration (Image courtesy of Mitsubishi Heavy Industries Compressor, Hiroshima, Japan)

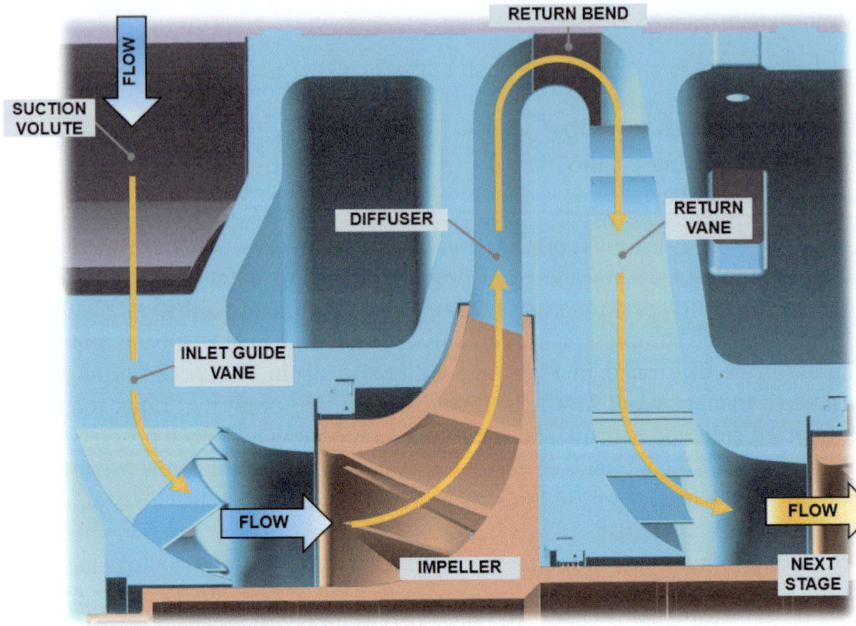

Fig. 1.9 Typical casing internals (Image courtesy of Mitsubishi Heavy Industries Compressor, Hiroshima, Japan)

In radially split casings, diaphragms are fastened using bolts or welds (Fig. 1.10). Axially split casings accommodate diaphragms within precisely machined grooves or stepped recesses (Fig. 1.11).

Figure 1.12 illustrates typical pressure increase within the stationary elements of a centrifugal compressor. Enhancing diffuser efficiency involves refining its cross-sectional geometry—whether conical or rectangular—based on the specific performance requirements of the application and considering boundary layer separation effects on flow behavior.

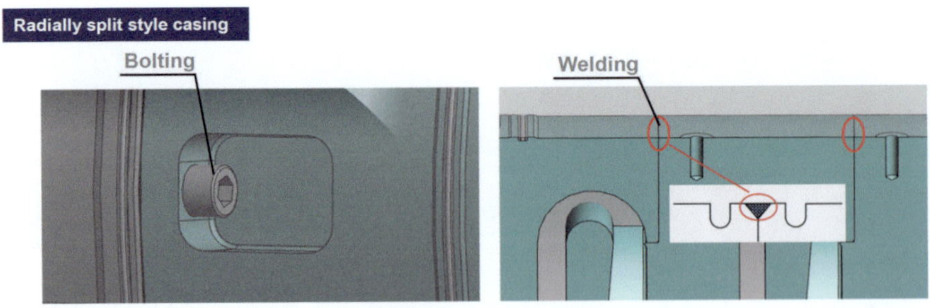

Fig. 1.10 Diaphragm fixation in the inner barrel of a radially split type compressor (Image courtesy of Mitsubishi Heavy Industries Compressor, Hiroshima, Japan)

Fig. 1.11 Diaphragm fixation in the casing of an axially split type compressor (Image courtesy of Mitsubishi Heavy Industries Compressor, Hiroshima, Japan)

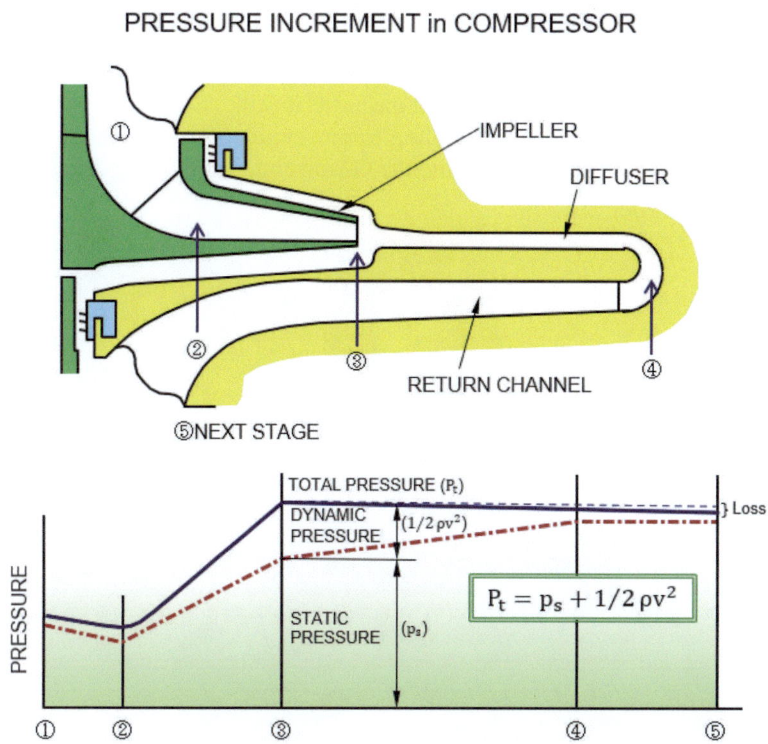

Fig. 1.12 Velocity and pressure profiles in an impeller and diffuser (Image courtesy of Mitsubishi Heavy Industries Compressor, Hiroshima, Japan)

Efficiency is also influenced by the selection of interstage seals and the aerodynamic design of the impellers. Further details on interstage seals are provided in Chap. 2, and impeller design considerations are covered in Chap. 3.

1.3 Radially Split Design: Inner Barrel Design and Sealing Features

The radially split casing design incorporates an assembled inner barrel that contains both stationary and rotating internal components. This inner assembly is inserted axially into the compressor casing, as illustrated in Fig. 1.13, necessitating adequate maintenance clearance at the nondrive end to facilitate removal and reinstallation of the inner barrel.

Considering the structural features of the radially split style casing design, sealing against gas leakage between the casing and the barrel and between the casing and the head is critical. The radially split style casing design is usually employed in higher-pressure applications such as those involving refinery recycle gas and synthesis gas.

O-rings at the joints are critical to sealing. O-rings on the barrel slide on the inner diameter of the casing during insertion of the barrel into the casing. This step demands precise execution and specialized tooling to prevent damage. Several factors must be carefully evaluated before finalizing the O-ring and back-up ring design:

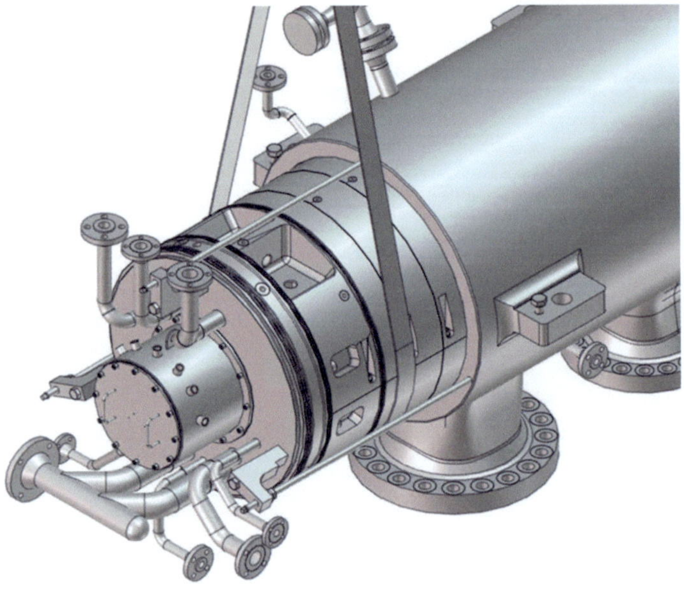

Fig. 1.13 Inner bundle insertion into a radially split type compressor casing (Image courtesy of Mitsubishi Heavy Industries Compressor, Hiroshima, Japan)

- Proper compression of the O-rings.
- Use of back-up rings to prevent extrusion or overhang.
- Quantity and placement of O-rings.
- Material compatibility with process gas and suitability for operating conditions (temperature, pressure, and chemical exposure).
- Acceptable decompression rate following casing depressurization.

1.4 Shear Ring Design

The fastening method for a radially split compressor casing varies according to the manufacturer's design philosophy. Typically, a flanged end plate secured with studs and shear rings is used to attach the inner barrel to the casing. This configuration maintains internal pressurization during operation and prevents process gas leakage at the interface between the end cover (or end wall) and the casing.

Shear rings are positioned within a groove on the inner side of the nondrive end of the casing, as illustrated in Fig. 1.14. These rings are held in place by four locking rings, which are further secured by locking plates mounted on the casing. Locking studs tighten the end cover against the shear rings, eliminating any clearance within the groove.

This assembly method ensures precise radial and axial alignment of the end cover with both the casing and the bearing housing, regardless of whether the compressor is in a pressurized or unpressurized state. Under operating pressure, axial forces are transmitted through the head and inner barrel directly to the shear rings.

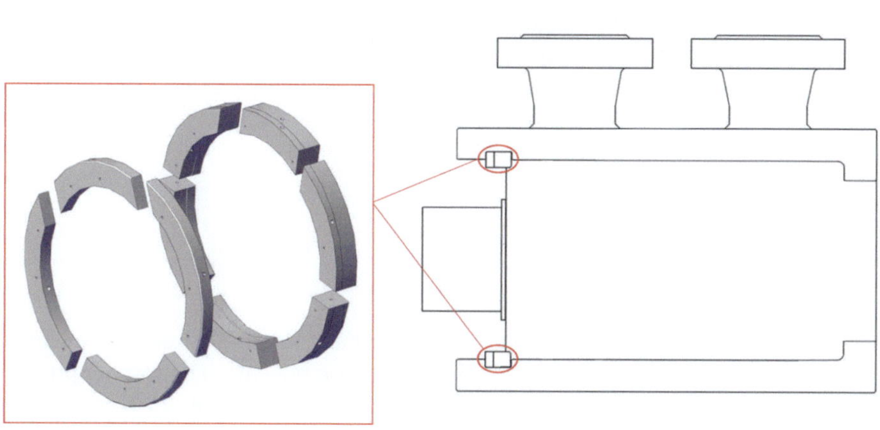

Fig. 1.14 Shear ring (Image courtesy of Mitsubishi Heavy Industries Compressor, Hiroshima, Japan)

1.5 Varieties of Casing Manufacturing

Casing materials are selected based on several key factors: the compressor's design temperature range (both minimum and maximum), the nature of the process gas, and the specific casing construction method. Carbon steel is the predominant choice for compressor casings, and it may be utilized in cast, forged, fabricated plate, or hybrid forms. For applications involving extreme temperatures or corrosive environments, alloy steels and stainless steels are preferred.

This section explores three primary casing manufacturing methods—casting, forging, and fabricating. While these methods are independent of the casing joint configuration (whether radially or axially split), each possesses distinct characteristics that can influence the optimal choice for a given application.

1.5.1 Cast Casing

Casting enables the production of complex, versatile, and architecturally intricate compressor casings—provided that a suitable mold can be fabricated and the molten metal flows smoothly throughout the casting process. Process gas connection nozzles are typically integrated into the mold design itself. If multiple casings of the same design are needed, casting becomes cost-effective compared to other manufacturing methods, as the mold can be reused repeatedly. Figure 1.15 illustrates an example of a cast compressor casing.

However, casting is not without its challenges. Defects—such as those shown in Fig. 1.16—can occur during the process and are difficult to eliminate entirely. Surface or subsurface imperfections may require corrective welding using certified procedures and documented qualification records. These repairs can be labor-intensive and time-consuming. In some cases, even after repair, the casting may fail to meet quality standards and must be scrapped, leading to increased production cycle times.

1.5.2 Forged Casing

Forged materials are generally recognized for their superior structural integrity and minimal defects compared to traditional cast components. The forging process enhances the metallurgical quality and consistency of the material, regardless of its thickness. As a result, forged casings are typically selected for high-pressure compressor applications. An example of a forged compressor casing is shown in Fig. 1.17.

However, forged casings are predominantly limited to cylindrical geometries, which restricts the ability to incorporate complex shapes—such as those illustrated in Fig. 1.12—particularly in areas like the inlet volute and discharge scroll. To

Fig. 1.15 Cast casing (Image courtesy of Mitsubishi Heavy Industries Compressor, Hiroshima, Japan)

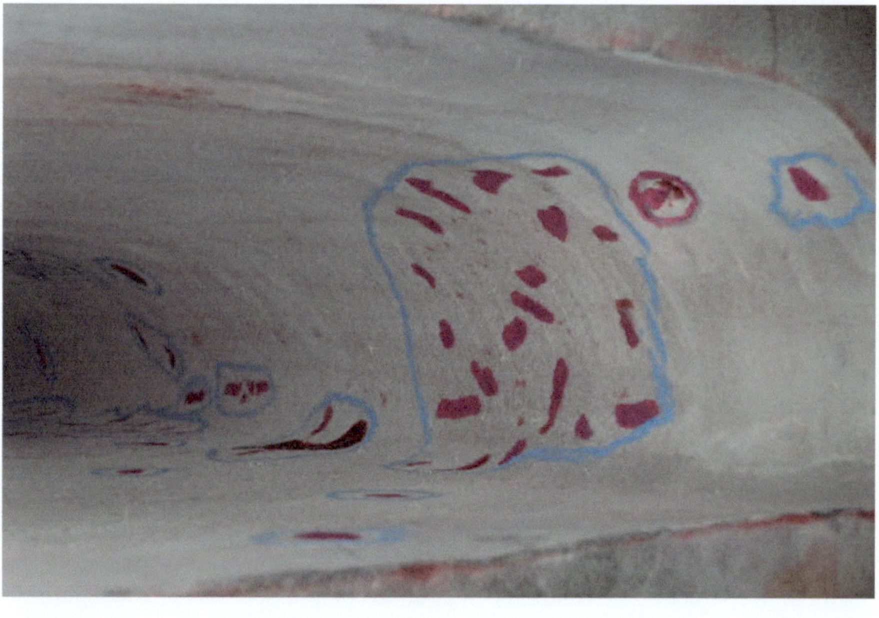

Fig. 1.16 Detection of casting defects by nondestructive testing (Image courtesy of Mitsubishi Heavy Industries Compressor, Hiroshima, Japan)

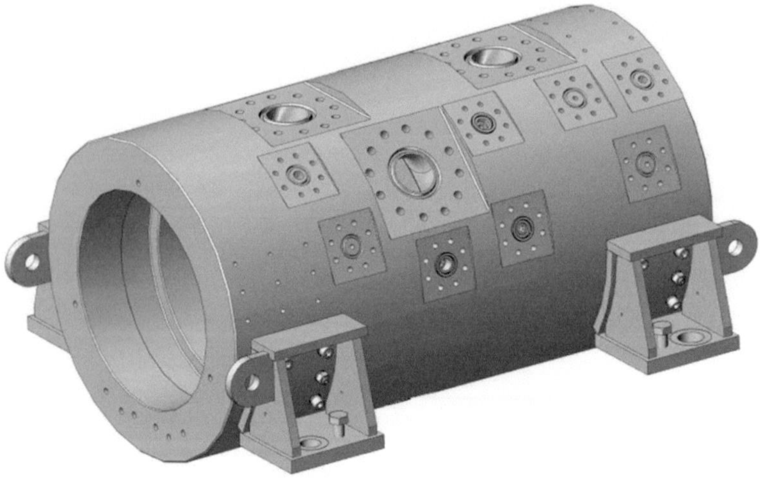

Fig. 1.17 Forged compressor casing (Image courtesy of Mitsubishi Heavy Industries Compressor, Hiroshima, Japan)

accommodate these design features, process gas nozzles are fabricated separately and subsequently welded onto the casing.

1.5.3 Fabricated Casing

Fabricated compressor casings are constructed by joining multiple steel plates, which are shaped—typically bent into circular or semicircular forms—to create the outer shell of the vessel-style casing. Figure 1.18 illustrates an example of this construction method.

However, this construction method comes with certain constraints. The thickness of the steel plates and their permissible bending radius are governed by the internal pressure requirements of the casing. Additionally, since fabrication relies heavily on welding, there remains an inherent risk of welding defects. Even under rigorous quality control protocols, such imperfections cannot be entirely eliminated during the manufacturing process.

1.6 Casing-to-Mounting Pedestal Design

The mounting structure for centerline-supported compressors serves a dual purpose: it not only bears the full weight of the centrifugal compressor but also preserves precise alignment tolerances. This is essential to account for factors such as lateral and longitudinal thermal expansion of the casing and forces or movements at the

Fig. 1.18 Fabricated compressor casing (Image courtesy of Mitsubishi Heavy Industries Compressor, Hiroshima, Japan)

nozzles, which can lead to misalignment within the coupled machinery train. To ensure stability and performance, casing supports or pedestals must be engineered with consideration for the rotordynamic behavior of the compressor, as outlined in Sect. 1.6.2.

1.6.1 Structure of Casing Mounting, Aspects of Thermal Movement

A centrifugal compressor is typically supported at four distinct points, as illustrated in Fig. 1.19. These support elements—commonly referred to as "feet," lugs, or portions of the horizontal split flanges—are secured using bolts and nuts. To accommodate factors that influence machinery string alignment, such as thermal expansion and nozzle-induced forces, not all support points should be rigidly fixed. Select points must remain "free" to permit movement of the casing in the horizontal plane.

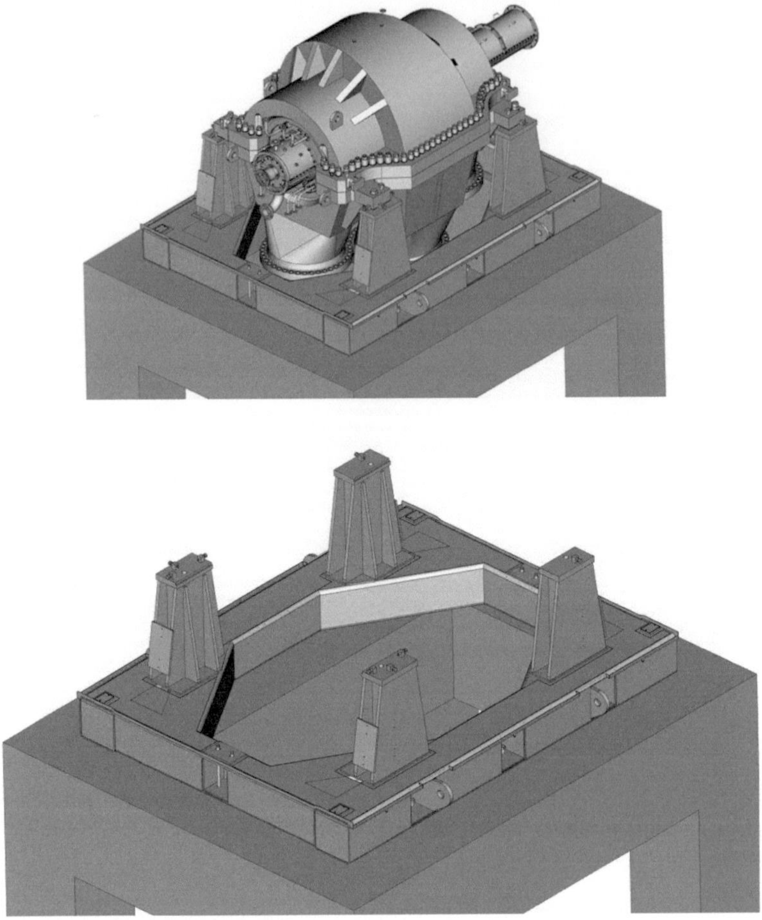

Fig. 1.19 Structure of a casing mounting, with and without the casing (Image courtesy of Mitsubishi Heavy Industries Compressor, Hiroshima, Japan)

Maintaining a uniform elevation across all mounting surfaces is essential to ensure proper train alignment within the tolerances specified by the manufacturer or API Standard 617 [1]. Additionally, the mounting pedestals must possess sufficient structural rigidity to prevent deflection under the compressor's weight.

Flatness of each machined mounting surface is defined as the difference between the highest and lowest measured points, determined using a dial indicator. This measurement is taken by traversing the indicator across the four edges and diagonals of the surface. Parallelism is assessed by calculating the difference between [(maximum height – datum height) – (minimum height – datum height)], also using a dial indicator.

1.6.2 Location of Casing Fixed and Flexible Supports

The positioning of fixed and flexible support points on a compressor casing is not governed by any formal codes or standards. Instead, these locations are determined by the design philosophy of each individual compressor manufacturer.

Fixed support points are typically selected based on combinations of the following configurations:

- Suction side versus discharge side.
- Drive-end versus nondrive-end.
- Thrust-bearing side versus nonthrust-bearing side.

Each configuration presents its own set of advantages and limitations. Manufacturers determine the optimal support arrangement by assessing several critical factors such as:

- Thermal expansion and deformation of the casing.
- Potential misalignment of process nozzles.
- Shaft and coupling alignment considerations.
- Structural rigidity to withstand piping-induced forces.
- Pedestal stiffness and damping characteristics relevant to rotordynamic behavior.

1.7 Discharge Volute Types and Design

In a centrifugal compressor, the compressed gas exits through the discharge scroll—also referred to as the discharge volute or exit scroll. Its primary function is to convert the radial velocity head of the gas leaving the diffuser into pressure head, maximizing energy recovery before the gas reaches the discharge nozzle. This conversion process preserves the angular momentum of the gas by transforming kinetic energy into pressure energy.

The geometry of the discharge volute is influenced by the casing's manufacturing method—whether cast, forged, or fabricated. Its configuration is typically classified based on:

- Radial location on the casing's periphery (internal, external, or intermediate).
- Cross-sectional shape (circular, rectangular, or oval).

Figure 1.20 illustrates a representative discharge volute design. In radially split forged casings, the volute is generally positioned internally along the casing's periphery. Cast or fabricated casings offer greater design flexibility, enabling optimization of the volute shape to reduce dynamic pressure losses.

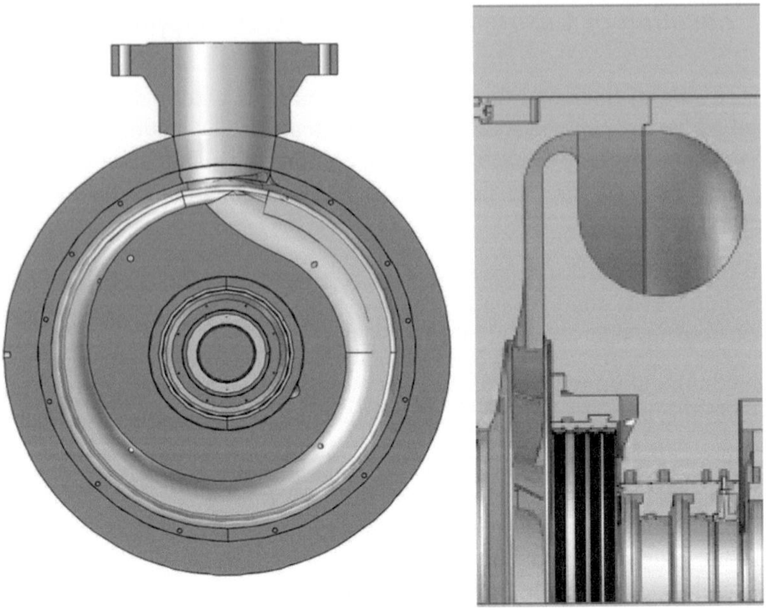

Fig. 1.20 Representation of a discharge volute design (Image courtesy of Mitsubishi Heavy Industries Compressor, Hiroshima, Japan)

1.8 Nozzle Design

The fabrication method used for centrifugal compressor nozzles typically mirrors that of the casing itself. Several key factors influence the selection of the nozzle fabrication approach:

- Nozzle orientation (upward or downward).
- Quantity and placement of nozzles.
- Interface type between nozzle and casing (integrally cast or welded).
- Gas flow velocity limitations in the inlet and discharge nozzles.

Process nozzle orientation should be determined in conjunction with the installation strategy at the jobsite, considering the pipe rack layout, auxiliary system configuration, and foundation design—such as tabletop foundations. Axially split casings offer enhanced maintainability due to the placement of process nozzles on the lower half of the casing (Fig. 1.21). Similarly, radially split compressors mounted on tabletop foundations typically feature downward-facing nozzles. However, upward-facing nozzles may be suitable for radially split casings if layout at the installation site provides sufficient overhead clearance and crane accessibility for maintenance operations (Fig. 1.22).

Fig. 1.21 Vertically downward-oriented compressor process nozzles (Image courtesy of Mitsubishi Heavy Industries Compressor, Hiroshima, Japan)

The configuration of process nozzles can be limited by several factors, including the total number of nozzles, their positioning on the casing, and the feasibility of integrating them with the casing structure. If nozzles are welded to the casing, additional spacing between them is necessary to accommodate welding operations.

To minimize dynamic pressure losses, manufacturers establish specific guidelines for gas flow velocities within suction and discharge nozzles. As a general rule, flow velocities should not exceed 150 ft/s (46 m/s).

1.9 Injection Nozzles on Casings

In processes where the gas stream contains heavy hydrocarbons and is prone to fouling, liquid injection into the internal flow path of a centrifugal compressor can be employed as a mitigation strategy. Typically, clean water and/or oil is introduced into the compressor. The specific advantages associated with each type of liquid injection are outlined in the following section.

Fig. 1.22 Vertically upward oriented compressor process nozzles (Image courtesy of Mitsubishi Heavy Industries Compressor, Hiroshima, Japan)

1.9.1 Water Injection

Boiler feedwater is commonly injected into the gas flow path of a centrifugal compressor to reduce the temperature of the process gas through latent heat absorption. This temperature reduction is primarily intended to prevent the formation of polymeric compounds derived from heavy hydrocarbons during operation. Such polymers tend to accumulate on internal surfaces—including impellers, diffusers, return channels, suction lines, and discharge volutes (Fig. 1.23)—leading to reduced compressor efficiency and overall performance degradation.

1.9.2 Oil Injection

Oil injection is employed to clean the internal surfaces of the gas flow path in centrifugal compressors, particularly when performance degradation is observed due to extended continuous operation or polymer buildup. Online liquid injection—typically using oil or water—supports sustained productivity and helps maintain plant throughput. However, even when operated within manufacturer-recommended flow rates, prolonged exposure to liquid injection can lead to erosion of internal compressor components due to repeated impact of liquid particles. This potential for erosion must be carefully weighed against the operational advantage of maintaining uninterrupted production.

Fig. 1.23 Polymer deposits on compressor casing internals (Image courtesy of Mitsubishi Heavy Industries Compressor, Hiroshima, Japan)

An example of a typical injection nozzle configuration is illustrated in Fig. 1.24.

1.10 Application Areas for Milled (Machined) Flats for SAE Connections

To achieve a more compact design, some manufacturers incorporate milled flats for Society of Automotive Engineers (SAE) flanged connections on components such as equalizing lines, bearing housings, casings, and stage drains. This approach offers space-saving advantages over traditional American Society of Mechanical Engineers (ASME)/American National Standards Institute (ANSI) standard flanges, which are typically rated for higher pressures. However, the use of milled flats for SAE connections may require prior approval from the purchaser.

Fig. 1.24 Typical configuration of injection nozzles on a compressor casing (Image courtesy of Mitsubishi Heavy Industries Compressor, Hiroshima, Japan)

Acknowledgements Mitsubishi Heavy Industries Compressor, Hiroshima, Japan, has contributed most of the material and all of the images/graphics/figures used in this chapter.

Reference

1. API Standard 617 (2022) Axial and centrifugal compressors and expander-compressors, 9th edn. American Petroleum Institute, Washington, DC

Chapter 2
Rotor Design

Abstract This chapter provides a comprehensive overview of rotor design principles for centrifugal compressors. The rotating assembly comprises several key components, including the shaft, impellers, sleeves, balance piston(s), thrust disk, and coupling hubs. Shaft design is guided by both lateral and torsional rotordynamic behavior, as well as its role in transmitting torque to the impellers and any additional casings located at the nondrive end. Material selection for the rotor assembly is influenced by factors such as process gas composition, operating and design parameters, and the installation environment. This chapter outlines a systematic approach to rotor design, examining how shaft length and diameter affect rotordynamic performance, and comparing stiff versus flexible shaft configurations along with their respective applications. It also covers shaft manufacturing techniques and commonly used materials. Additional topics explored in detail include impeller thrust forces, sizing criteria for balance pistons, the types of interstage and balance piston seals, interference fits on the shaft, and protective surface treatments for rotor components.

2.1 Compressor Rotor Design Features

Rotor design for a centrifugal compressor encompasses various elements, including shaft geometry, step diameters, end configurations, torque transmission requirements, stress analysis, attachment fits and tolerances, and material selection. The individual rotor components are illustrated in Fig. 2.1.

2.1.1 Shaft Design: Factors to Consider When Calculating the Shaft-End Diameter for Torque and Power

For a given set of process requirements, selecting an appropriate impeller diameter and rotational speed is closely tied to the shaft design. The shaft must possess sufficient rigidity to transmit both steady-state and anticipated transient torques, while

N. Ghaisas, *Centrifugal Compressors in Hydrocarbon Processing Industries*,
https://doi.org/10.1007/978-3-032-03687-2_2

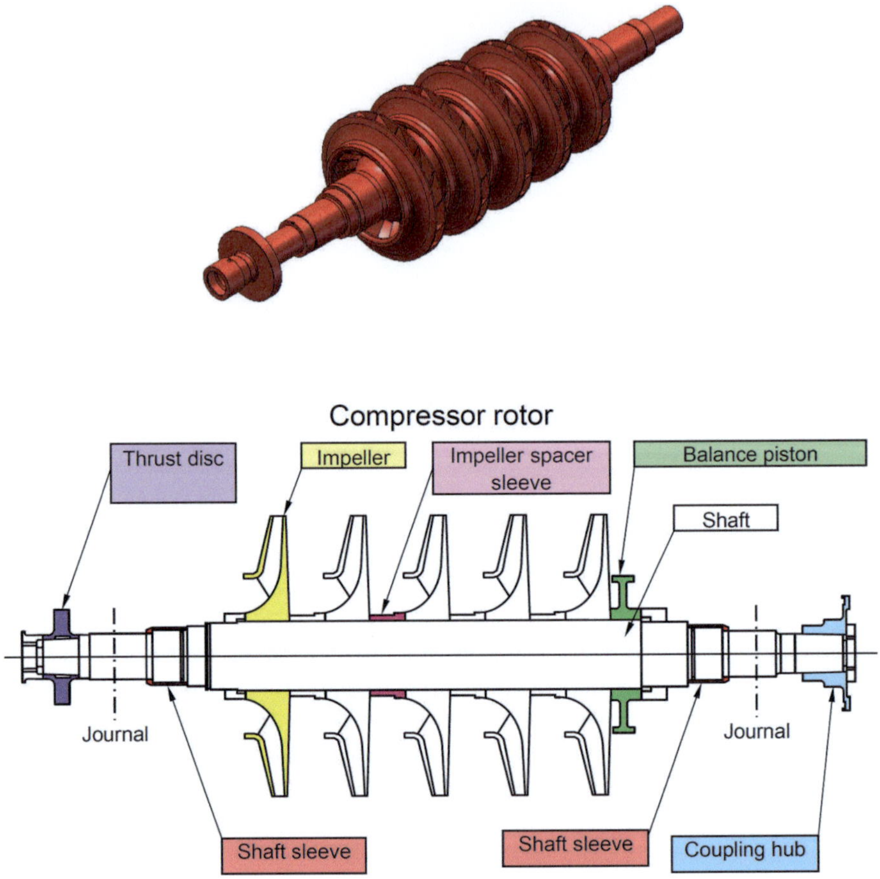

Fig. 2.1 Centrifugal compressor rotor (Image courtesy of Mitsubishi Heavy Industries Compressor, Hiroshima, Japan)

ensuring that the rotor's natural frequency maintains the separation margins specified in API Standard 617 [1] relative to the operating speed range. At the impeller fit locations, the shaft typically features stepped diameters arranged in ascending order, with the largest step positioned near the midspan to facilitate assembly.

Designing a suitable shaft structure involves compliance with several engineering guidelines. Key design checks include:

- Sizing of shaft end(s) based on transmitted power and torque.
- Journal bearing dimensions.
- Impeller tip velocity constraints.
- Minimum impeller bore diameter.
- Shaft diameters at impeller fit locations, accounting for gripping force and loss of interference fit due to centrifugal and operational stresses.

- Interference fits at shaft sleeves and at the hydraulically fitted coupling hub interface.
- Progressive shaft step configuration from bearing to seal to impeller locations.
- Thermal expansion considerations at shaft ends.
- Preliminary rotordynamic evaluations, including:
  - Diameter-to-length squared (D/L^2) ratio (refer to Sect. 2.1.2).
  - Bearing stiffness relative to shaft bending stiffness and its influence on critical speed.
  - Mass distribution analysis.
  - Undamped critical speed map.
- Thrust force assessment.
- Additional checks, such as shaft manufacturability and assembly requirements for impellers, seals, and coupling(s).

It's important to note that some of these evaluations are typically performed during the detailed engineering phase following compressor procurement, and specific criteria may vary across manufacturers.

2.1.2 *Influence of Shaft Length and Diameter on Rotordynamics*

As highlighted in the previous section, lateral rotordynamics plays a pivotal role in compressor rotor design, ensuring that vibration levels remain within the acceptable limits defined by API Standard 617 [1]. Excessive vibration can lead to operational instability, unplanned shutdowns, and reduced equipment reliability and availability.

Accurately assessing the dynamic behavior of rotating machinery requires modeling the rotor system with the added mass of impellers and gas seals, along with the stiffness and damping properties of hydrodynamic bearings and their supports. These elements are typically represented as a spring-mass-damper system, as illustrated in Fig. 2.2.

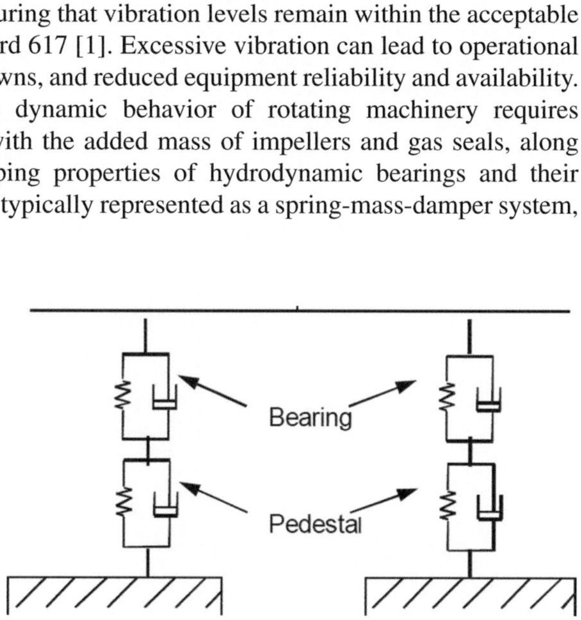

Fig. 2.2 Spring-mass-damper model (Image courtesy of Mitsubishi Heavy Industries Compressor, Hiroshima, Japan)

Rotordynamic performance is influenced by several factors, including bearing geometry and load, tilting-pad preload, and the pressure and temperature of the lubricating oil. While extensive literature exists on rotordynamics and vibration analysis, the scope in this section is limited to preliminary design checks.

One of the most frequent sources of vibration issues is rotating unbalance in the shaft and impellers. Resonance occurs when the frequency of an external forcing function aligns with the rotor's natural frequency, producing synchronous vibrations. The amplitude of these vibrations is governed by the amount of damping or the system's damping capacity. API 617 [1] outlines specific criteria for separation margins, amplification factors, and related parameters.

During early design phases, full-scale rotor modeling using advanced analytical tools may not be practical. Instead, simplified methods are employed, such as evaluating the shaft length-to-diameter ratio (L/D) and comparing the D/L^2 value—where D refers to the shaft's midspan diameter—against the first natural frequency. These preliminary assessments, supported by manufacturer experience and theoretical analysis, are effective in predicting whether the rotor system will meet more rigorous rotordynamic requirements. Refer to Chap. 12 for additional insights.

The D/L^2 method itself is derived from the foundational spring-mass-damper model, which serves as the basis for simulating real-world dynamic systems.

This approach involves correlating the D/L^2 value with the rotor's first critical speed (Fig. 2.3), guided by manufacturer-specific recommendations. Standard benchmarks for D/L^2 relative to the maximum continuous operating speed are illustrated in Fig. 2.4. Equation 2.1 defines the relationship between the shaft's natural frequency and its material properties—specifically Young's modulus—as well as its diameter, length, and vibrational mode count.

Another commonly evaluated parameter is the shaft's length-to-diameter (L/D) ratio. Based on industry experience, this ratio is typically kept below 10:1. However, several reputable compressor manufacturers have successfully implemented designs with L/D ratios exceeding this threshold. As such, the 10:1 ratio should be viewed as a conservative design guideline rather than a strict limit.

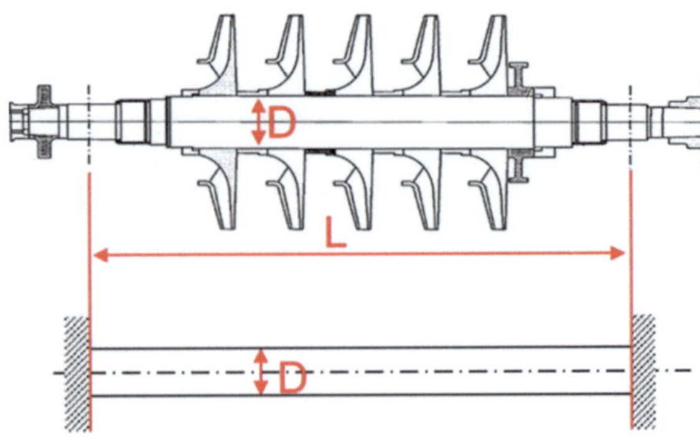

Fig. 2.3 Diagram of the rotor shaft for use in the D/L^2 method

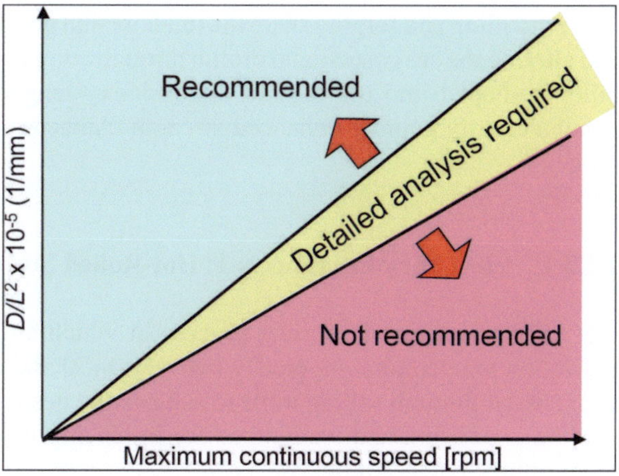

Fig. 2.4 Typical guidelines for rotor rigidity (Image courtesy of Mitsubishi Heavy Industries Compressor, Hiroshima, Japan)

$$f = \frac{\lambda_{\mathrm{n}}}{2\pi L^2} \frac{D}{4} \sqrt{E} \propto \frac{D}{L^2} \tag{2.1}$$

where D is the shaft diameter, L is the shaft length, f is the natural frequency, λ_{n} is the constant number of modes of the natural frequency, and E is Young's modulus.

2.1.3 Rolled Shafts Versus Forged Solid Shafts

According to API Standard 617 [1], compressor shafts without through-bolts must be constructed from one-piece, heat-treated forged steel if the finished diameter exceeds 8 in. (20.3 cm). For shafts with finished diameters of 8 in. (20.3 cm) or less, either forged steel or hot-rolled bar stock is acceptable. Forged steel shafts are produced through a process that shapes the metal using localized compressive forces—typically by repeated hammering or pressing. This method enhances the mechanical properties of the steel, resulting in superior strength, improved ductility, and enhanced fatigue resistance, making forged shafts particularly well-suited for compressor applications.

The most commonly used materials for forged shafts are alloy steels designated by the American Iron and Steel Institute (AISI) and the American Society for Testing and Materials (ASTM), specifically grades 4140 and 4340.

When selecting steel for a given application, design engineers must understand the distinctions between various steel grades and manufacturing techniques. This section focuses on two primary forming methods used in compressor shaft manufacture: forging and rolling.

Both rolling and forging share the fundamental goal of reducing material thickness and altering the cross-sectional profile through compressive force. In essence, a metal billet is shaped into the desired dimensions—length, diameter, and form—while simultaneously gaining enhanced strength characteristics through the deformation process.

2.1.3.1 Steel Forming Process 1: Hot-Rolled Steel

Hot rolling is a metal forming process in which steel is heated above its recrystallization temperature—typically between 1450 and 2100 °F (788–1150 °C)—and then passed through rolling mills to achieve the desired shape and dimensions. This method is efficient and well-suited for producing both individual components and high-volume batches quickly.

However, a key limitation of hot rolling is its relatively poor dimensional accuracy. As the steel cools, it undergoes thermal contraction, which can result in uneven thickness across the material and a rough, scale-covered surface finish.

2.1.3.2 Steel Forming Process 2: Cold-Rolled Steel

Cold rolling follows a process similar to hot rolling but includes an additional deformation step carried out at temperatures below the steel's recrystallization point—typically near room temperature. Because this step occurs after recrystallization, it introduces strain hardening into the steel, enhancing its mechanical properties. As a result, cold-rolled steel offers improved surface finish and tighter dimensional tolerances, making it more suitable for applications requiring precision.

2.1.3.3 Steel Forming Process 3: Forged Steel

Similar to hot rolling, the forging process begins with heating an ingot until it reaches a malleable state. However, instead of shaping the bar through rolling, forging achieves the desired rough dimensions by hammering or applying hydraulic pressure. This method enhances the uniformity of the material's composition and internal structure, resulting in superior metallurgical integrity and mechanical performance. Forged components exhibit a high strength-to-weight ratio and exceptional toughness, particularly beneficial for cryogenic, low-temperature environments—making forging the preferred method for manufacturing most shafts in centrifugal compressors.

From a manufacturing perspective, rolled shafts are typically restricted to diameters under 4 in. (10.2 cm) due to their mass production for automotive parts, sheet metal, tubing, and general machinery. In contrast, forged shafts are custom-fabricated and well-suited for centrifugal compressors, which often demand significantly larger diameters—up to 24 in. (61 cm) in the largest configurations.

2.1.4 Stiff Shaft and Flexible Shaft Designs

A rotor whose entire operating speed range remains below its first critical speed is classified as a rigid (or stiff) shaft. The critical speed refers to the rotational speed at which a distinct lateral vibration response peak occurs [2]. For low-speed systems—such as 2-pole (3600 rpm) and 4-pole (1800 rpm) motors driven by adjustable-frequency drives (AFDs), also known as variable-frequency drives (VFDs)—it is essential that the first critical speed exceeds the typical operating range of 1300–3800 rpm. Alternatively, the rotor system must be critically damped near or below the operating speed to avoid excluding desirable speed ranges, thereby ensuring a rigid shaft design.

Conversely, a rotor that operates above its first critical speed is referred to as a flexible shaft. This configuration is commonly used in centrifugal compressors due to several design factors, including the aerodynamic characteristics of the impeller and containment of the gas pressure within the casing. Unlike turbines and axial compressors, which benefit from larger blade diameters for optimal efficiency, centrifugal compressors achieve higher efficiency with smaller impeller inlet eye diameters. This design requirement results in a longer, more slender rotor that naturally operates in the flexible shaft region [2].

Operating in this region requires the manufacturer to carefully process the rotor during fabrication to avoid balance shifts during operation. Prior to assembly, each flexible shaft rotor—comprising the shaft, balance piston, and impellers—is sequentially dynamically balanced across multiple axial planes. These components are balanced to stringent tolerances defined by ISO 21940 Quality Grade G0.67 [3] or the American Petroleum Institute—API $4W/N_{mc}$ [1, 2]. The balancing process includes both low-speed balancing of individual components and the complete rotor, as well as high-speed (operating speed) balancing if specified by the manufacturer's internal criteria or purchaser requirements. Final verification of vibration behavior at operating speed is conducted during the compressor's mechanical running test.

2.2 Shaft-End Design and Coupling Hub Installation

Coupling attachment to the shaft has historically posed significant challenges across various industrial applications. As a result, the shaft end of a rotor is a critical focus during the preliminary design phase. It must be capable of transmitting the maximum anticipated torque from the driver to the rotating elements along the shaft and to any casings located at the compressor's nondrive end. The dimensions of the shaft end and the overhung moment introduced by the coupling play a pivotal role in shaping the rotor's dynamic behavior—particularly influencing the coupling mode and vibrational response, as illustrated in Fig. 2.5.

Selecting the appropriate coupling hub-to-shaft connection warrants careful evaluation of several design considerations, including:

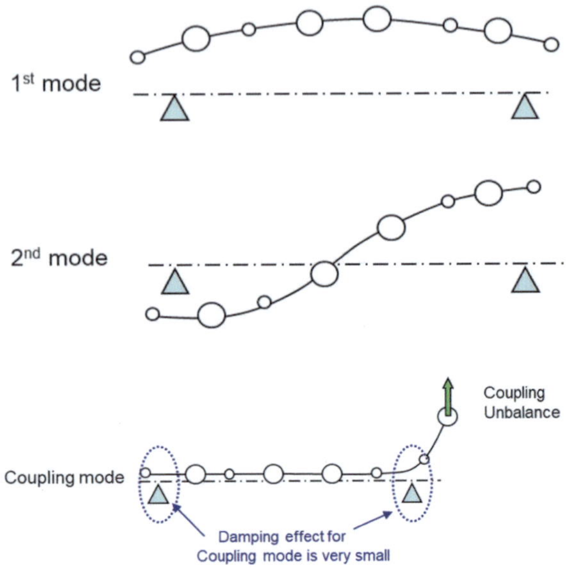

Fig. 2.5 Typical mode shapes (Image courtesy of Mitsubishi Heavy Industries Compressor, Hiroshima, Japan)

- Ease of assembly and disassembly.
- Connection reliability.
- Accessibility for shaft end seals and/or impeller removal.
- Capability to transmit both steady-state and peak transient torques (with the shaft-to-coupling interface rated for at least 115% of the maximum transient torque).

This discussion focuses exclusively on hydraulically fitted coupling hubs used in centrifugal compressor shafts, excluding integral flanged shaft ends and other coupling configurations. Historically, keyed hubs were commonly used; however, the industry has transitioned to keyless hubs mounted on tapered shaft ends. These hubs rely on an interference fit, and their design requires analysis of three critical parameters:

- The degree of interference necessary to transmit the required torque.
- The hydraulic pressure needed for hub installation and removal.
- The stresses induced in the hub during mounting, dismounting, and operational conditions.

When selecting a coupling hub-to-shaft connection, particular attention must be given to transient operating scenarios that may demand torque levels exceeding nominal ratings. For instance, during air-run conditions in propane dehydrogenation plants, torque requirements can reach up to twice the normal process torque.

In applications involving synchronous motors and adjustable-frequency drives (AFDs), the shaft end must also accommodate noncontinuous peak torques, short-circuit events, and inverter-induced torque pulsations.

2.3 Typical Materials

Alloy steels such as AISI/ASTM 4140 and AISI/ASTM 4340, when subjected to specialized heat treatments, deliver the necessary mechanical strength and toughness for demanding applications. These materials have demonstrated reliable performance at temperatures as low as – 175 °F (– 115 °C), offering tensile strengths of up to 128,000 psi (880 MPa) and minimum yield strengths of 99,000 psi (685 MPa).

In corrosive environments, material selection must account for protection against wire wooling—a wear phenomenon occurring in narrow clearances between rotating and stationary components, such as labyrinth seals. This issue is often linked to the formation of chromium carbide in materials like 17-4PH and 400 series stainless steels. The embedded carbide particles in stationary surfaces act as abrasive agents against the rotor. One effective mitigation strategy is the use of hard chromium plating on the susceptible surfaces.

API 617 [1] outlines standard materials commonly used for shafts and compressor components. Beyond these recommendations, compressor manufacturers often rely on field-proven alternatives or tailor material choices to meet specific application requirements.

2.4 Balance Piston Sizing

In a multistage compressor, the rotor incorporates a set of impellers to generate the pressure head required for the intended application. As gas flows through each impeller stage, a pressure differential is created, resulting in a net axial thrust directed toward the suction side of the impeller. To alleviate the load on the thrust bearing, balance pistons are employed in multistage dynamic compressors—both centrifugal and axial types. These pistons feature a cross-sectional area exposed to discharge pressure on one side and interstage or suction pressure on the opposite side, effectively counteracting the axial forces produced by the impellers.

Balance pistons are sized to minimize total thrust across all operating conditions, thereby protecting the thrust bearings from excessive loads. As illustrated in Fig. 2.6, the piston dimensions are optimized to reduce thrust loading. However, excessive flow through the balance line can lead to undesirable recirculation, which not only wastes energy but also reduces compressor capacity—a clear performance drawback. Improperly designed balance pistons may result in elevated thrust forces and potential thrust bearing failure.

To safeguard against such issues, API 617 [1] mandates that the balance line must be sized to accommodate gas leakage from the balance piston at twice the initial design balance piston seal clearance—without exceeding the thrust bearing's load rating.

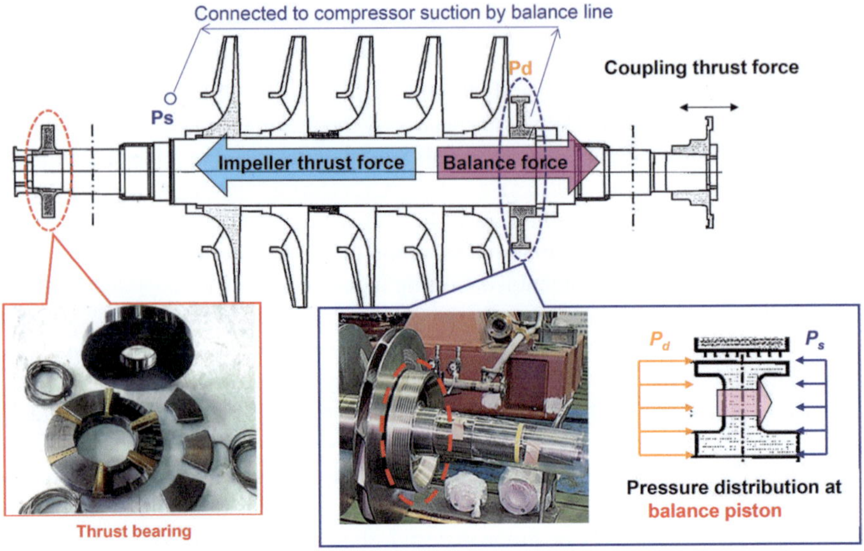

Fig. 2.6 Concept of thrust force balancing (Image courtesy of Mitsubishi Heavy Industries Compressor, Hiroshima, Japan)

2.5 Balance Piston Arrangements

Compression trains can be configured in either series or parallel arrangements, with casing designs varying to meet the specific demands of each application. The location and configuration of the balance piston are influenced by both the casing geometry and the impeller arrangement.

The most basic compressor setup is the straight-through configuration, which features one or more impellers oriented in the same direction, along with a single suction and discharge nozzle. In this design, the balance piston is positioned on the discharge side, downstream of the final impeller stage. Its rear side is connected to the suction via a balance line, as illustrated in Fig. 2.7.

The second compressor configuration is the back-to-back design, characterized by impeller sections arranged in opposing directions. This setup incorporates an additional seal known as the division wall seal (Fig. 2.8), which eliminates the need for a balance piston. The division wall seal serves two critical functions: it prevents gas leakage from the high-pressure side to the low-pressure side of the compressor, and it counteracts the net axial thrust caused by pressure differentials between these sections.

In high-speed, high-pressure back-to-back compressors, the tight clearance division wall seal is typically positioned at the rotor's midspan—where rotor sag is most pronounced. This configuration is prone to cross-coupling effects, which can destabilize the rotor. These destabilizing forces, often induced by entry swirl within the

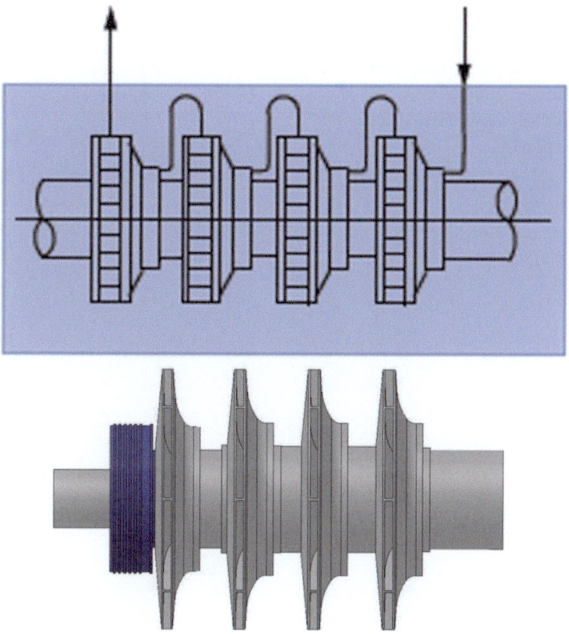

Fig. 2.7 Straight-through design (Image courtesy of Mitsubishi Heavy Industries Compressor, Hiroshima, Japan)

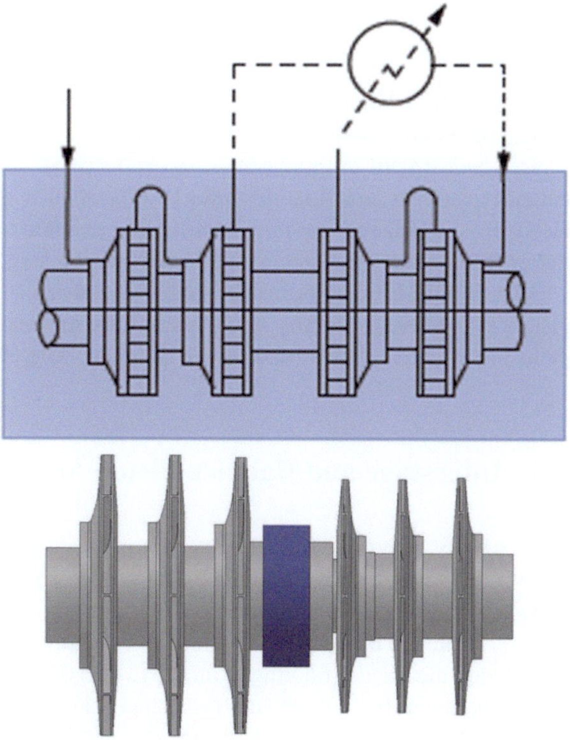

Fig. 2.8 Back-to-back configuration (Image courtesy of Mitsubishi Heavy Industries Compressor, Hiroshima, Japan)

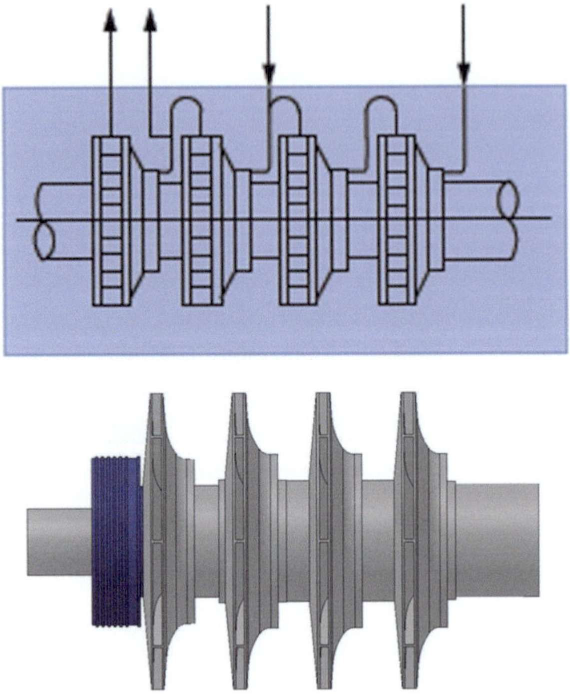

Fig. 2.9 Sidestream configuration (Image courtesy of Mitsubishi Heavy Industries Compressor, Hiroshima, Japan)

division wall or balance piston seals, can be mitigated through the use of damper seals, honeycomb seals, or shunt lines [2].

In a sidestream compressor—also referred to as a sideload compressor—the balance piston is positioned similarly to that in a straight-through configuration. However, the balance line may be routed to an interstage suction or discharge point, rather than being connected to the initial suction stage (Fig. 2.9).

In a double-flow compressor with symmetrically designed sections, a balance piston is not required. Any slight variations in axial thrust between the two flow paths are effectively absorbed by the thrust bearing (Fig. 2.10).

2.6 Interstage and Balance Piston Seals

Traditionally, labyrinth shaft seals—featuring a combination of annular orifices and a tooth-on-stator design (Fig. 2.11)—have been constructed from metallic materials such as aluminum or stainless steel. The amount of leakage through these seals is directly related to their operating clearance. Aluminum labyrinth seals typically have installation clearances ranging from 0.012 to 0.040 in. (0.30–1.0 mm), depending on the fin diameter. In contrast, stainless steel seals require larger clearances due to the differing thermal expansion characteristics of the material compared to aluminum.

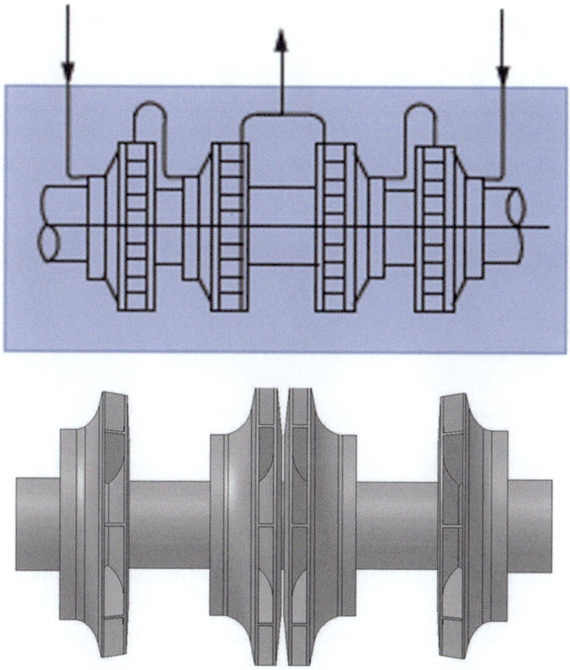

Fig. 2.10 Double-flow configuration (Image courtesy of Mitsubishi Heavy Industries Compressor, Hiroshima, Japan)

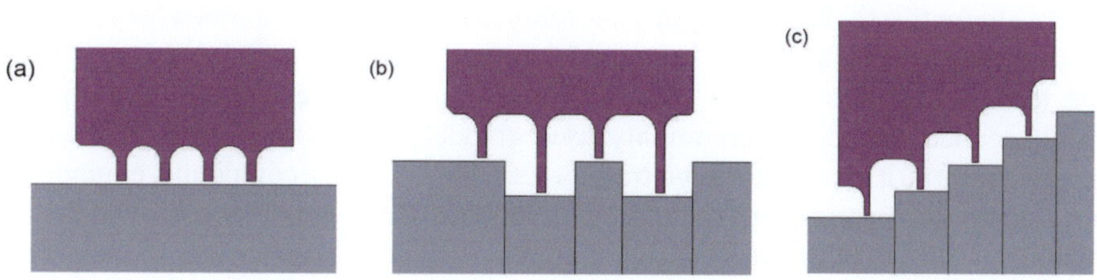

Fig. 2.11 Tooth-on-stator labyrinth seal types: **a** straight labyrinth, **b** labyrinth with different diameters, **c** stepped labyrinth (Image courtesy of Mitsubishi Heavy Industries Compressor, Hiroshima, Japan)

Labyrinth seals positioned at critical locations—such as the impeller eye, interstage sections, balance piston, and bearing housing ends—are vulnerable to rotor rub if their running clearances are not optimized (Fig. 2.12). Several factors influence these clearances, including:

- Differential thermal expansion between the rotor and seal materials.
- Oil film thickness in radial bearings.
- Rotor vibrations.
- Temperature gradients between bearing pedestals and seal locations.

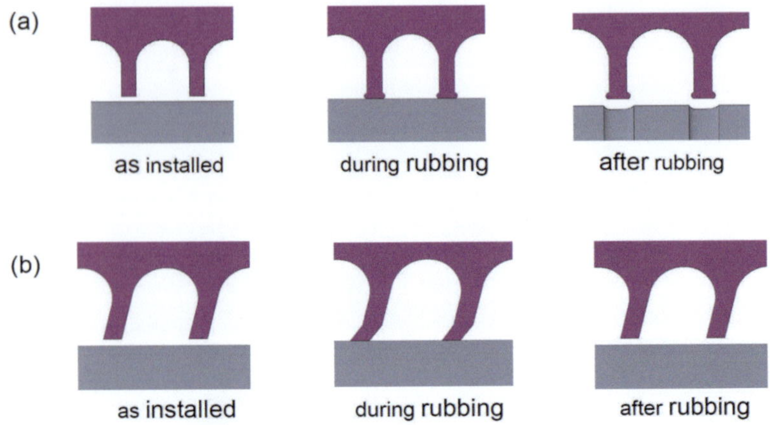

Fig. 2.12 Labyrinth seal rubbing: **a** aluminum or stainless material and **b** PAI, PEI, or PEEK material (Image courtesy of Mitsubishi Heavy Industries Compressor, Hiroshima, Japan)

To enhance rub tolerance, high-performance thermoplastics like polyamide-imide (PAI), polyetherimide (PEI), and polyether ether ketone (PEEK) are increasingly used for seal fins. These materials are softer than traditional metals such as aluminum and stainless steel, allowing them to deform during transient events—such as passing through critical speed—and then recover their original geometry post-rub. This resilience enables tighter-clearance designs, which significantly reduce leakage during operation and thereby improve compressor efficiency.

An alternative sealing approach is the tooth-on-rotor configuration, where rotating teeth interface with a smooth, abradable surface (Fig. 2.13). This design enables the teeth to cut into the soft material, establishing tight clearances that significantly reduce leakage. A commonly used abradable material for this application is Fluorosint®, a modified polytetrafluoroethylene (PTFE) compound engineered specifically for bearing and sealing environments. It features low tensile strength, making it ideal for controlled material removal during operation. Figure 2.14 illustrates an abradable seal installed at the impeller eye location.

2.7 Impeller Thrust

The total thrust force exerted on the rotor is calculated by summing the axial forces generated at each impeller, shaft end, and the balance piston. As illustrated in Fig. 2.15, the thrust loads on the impellers arise from the pressure differential between their front and rear faces. These axial forces can be further categorized into static force components, including:

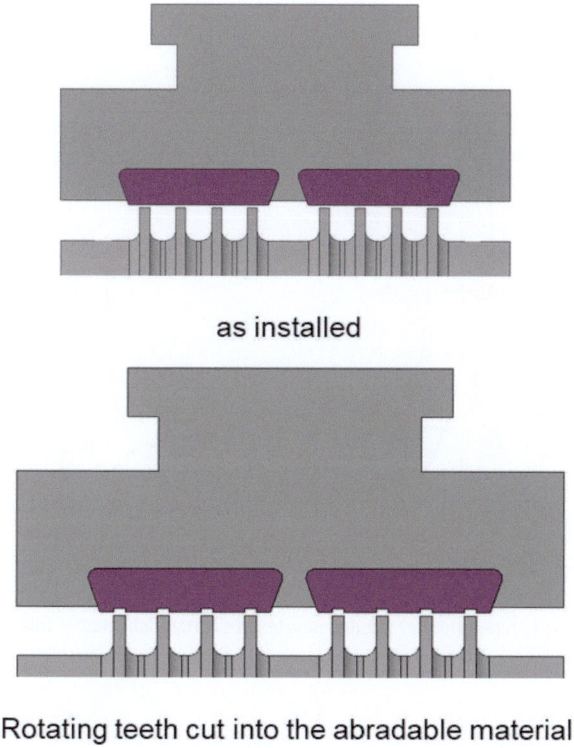

Fig. 2.13 Abradable seal: tooth-on-rotor configuration (Image courtesy of Mitsubishi Heavy Industries Compressor, Hiroshima, Japan)

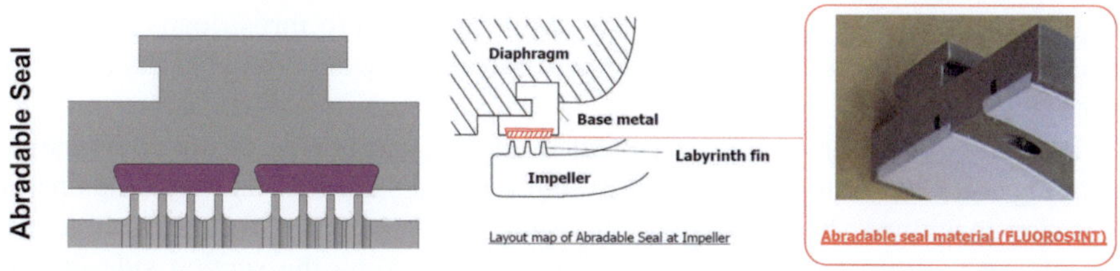

Fig. 2.14 Abradable seal at the impeller eye (Image courtesy of Mitsubishi Heavy Industries Compressor, Hiroshima, Japan)

- F_i: force due to the stage inlet pressure.
- F_s: force from the discharge pressure acting on the shroud.
- F_h: force from the discharge pressure acting on the hub

These forces are influenced by factors such as labyrinth seal clearances, shaft diameters, and the overall static pressure distribution.

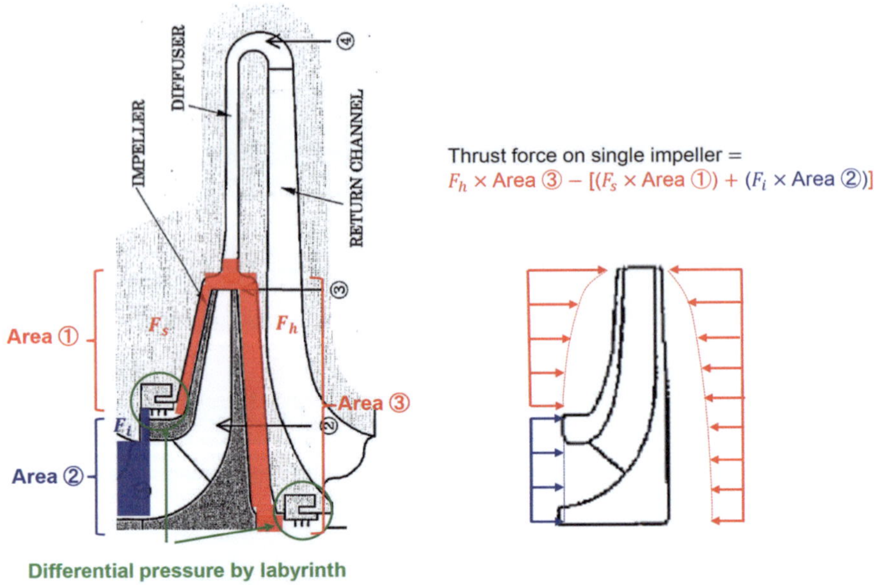

Fig. 2.15 Thrust load on an impeller (Image courtesy of Mitsubishi Heavy Industries Compressor, Hiroshima, Japan)

Thrust force variation resulting from changes in seal clearances can be substantial in high-pressure compressors, often necessitating the use of both analytical methods and computational fluid dynamics (CFD) models to accurately predict axial loads.

To minimize axial instability, it's advisable to design the impeller thrust loading to act in a single direction across all operating conditions and clearance scenarios (e.g., Cl × 1.0, Cl × 2.0). Designs that result in near-zero thrust loads should be avoided, as they can lead to axial fluctuations caused by directional shifts in thrust force (Fig. 2.16).

Typically, the thrust bearing is mounted on the suction side of the compressor, allowing rotor thrust to be absorbed by the thrust disk, which is seated against a precision-machined shaft shoulder. This setup prevents thrust forces from acting directly on the thrust disk locknut. Additionally, locating the suction side on the nondrive end simplifies the assembly and disassembly of both the thrust bearing and thrust disk. This configuration should be evaluated during the preliminary engineering phase, in coordination with the purchaser's process piping layout, to ensure an optimized compressor string arrangement.

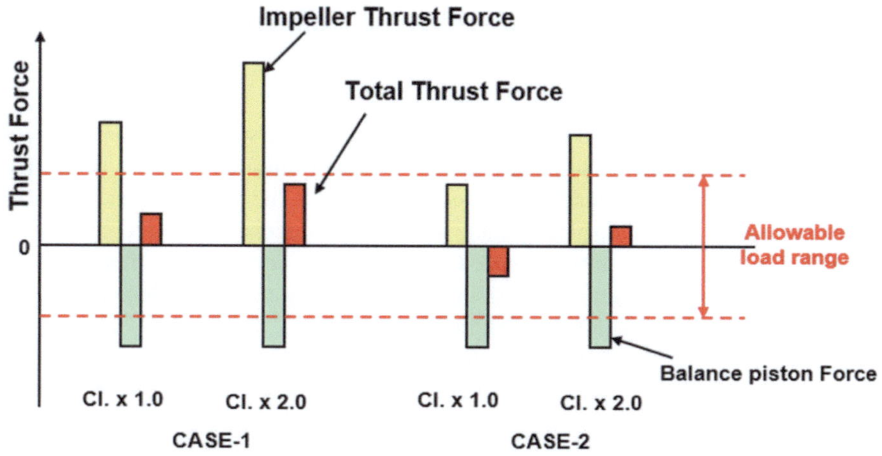

Fig. 2.16 Overall thrust evaluation (Image courtesy of Mitsubishi Heavy Industries Compressor, Hiroshima, Japan)

2.8 Typical Interference Fits on Shaft

Nominal interference fits for shrunk-on impellers typically range between 0.001 and 0.0015 in. per inch of shaft diameter (or 0.01 and 0.015 mm per centimeter of shaft diameter) under the impeller. The specific interference fit selected for a given application is influenced by several key factors, including:

- Maximum stage power and torque transmission requirements.
- Centrifugal and thermal stresses encountered during operation.
- The required design safety margin between the tangential stress at the impeller hub-to-shaft interface and the minimum yield strength of the impeller and shaft materials.

Additionally, the interference fit must prevent surface damage or distortion during impeller installation and rotor disassembly.

For shaft sleeves, nominal interference fits generally range between 0.0007 and 0.001 in. per inch of shaft diameter (or 0.007 and 0.01 mm per centimeter of shaft diameter) under the sleeve. Components such as the balance piston and hydraulically fitted coupling hub typically require a tighter interference fit of 0.002 in. per inch of shaft diameter (or 0.02 mm per centimeter).

Compressor manufacturers determine the appropriate shrink fits for these and other components—such as the hydraulically fitted thrust disk—based on the specific application and design criteria. API TR 684-1 [2] mentions the effects of shaft sleeves on rotordynamic behavior. To ease rotor assembly and disassembly, relieved shaft sleeves are often used. In contrast, nonrelieved shrunk-on sleeves generate higher internal friction, which can adversely impact rotor stability.

2.9 Influence of the Type of Driver on Compressor Shaft Sizing

Process compressor train drivers are selected based on the specific requirements of the application. For variable-speed operations, steam turbines or gas turbines are traditionally employed, depending on process and project needs. However, advancements in inverter technology have made adjustable-frequency drive (AFD) motors a practical alternative for variable-speed control. While conventional motor drivers are well-suited for constant-speed applications, AFD systems introduce additional rotordynamic complexity into the design.

Regardless of the driver type, careful consideration must be given to the operating map, particularly transient conditions and their impact on the shaft end. In systems utilizing synchronous motors or AFD drives, the shaft end must be capable of handling peak torque loads during short-circuit events and inverter-induced torque pulsations. Additionally, torque fluctuations during startup and trip scenarios in motor-driven configurations demand higher safety margins for both the coupling and shaft end compared to turbine-driven systems.

A detailed analysis of synchronous motor behavior and its interaction with electrical systems is provided in Chap. 13.

2.10 Coatings

Chapter 1 covers the topics of oil and water injection. To mitigate operational degradation, protective coatings are employed to shield components from corrosion, erosion caused by solid or liquid contaminants in the gas stream, elevated temperatures, mechanical wear, and fouling. Corrosion and erosion progressively remove material, increasing the risk of unexpected shutdowns for component repair or replacement (Fig. 2.17).

Gas streams containing wet chlorides or hydrogen sulfide are particularly corrosive. Additionally, processes such as gas cracking are susceptible to fouling, often due to polymer formation. Fouling within aerodynamic elements—including impellers, diffusers, return bends, and return channels—can significantly impair compressor performance by increasing pressure drop and reducing both throughput and efficiency.

A wide range of industrial coating materials is available, including metal-ceramic-polymer composites and PTFE-based coatings (commonly known as Teflon™). Among these, metallic coatings demonstrate superior durability, particularly in resisting erosion caused by wash oil or water injection. Standard coating thicknesses typically range between 0.001 and 0.004 in. (0.025 and 0.10 mm).

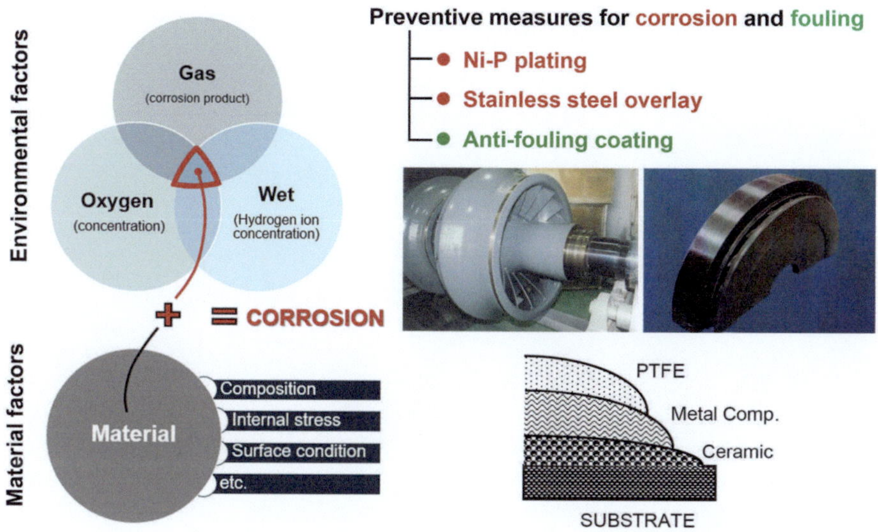

Fig. 2.17 Corrosion-promoting factors and preventive measures (Image courtesy of Mitsubishi Heavy Industries Compressor, Hiroshima, Japan)

Metallic-ceramic and metallic-ceramic-polymer coatings can achieve a smooth surface finish of up to 32 Ra μin. (0.81 Ra μm), where Ra denotes the average surface roughness. The choice for the type of coating should consider factors such as bond strength, ductility, abrasion resistance, surface finish quality (critical for maintaining compressor performance and efficiency), and the coating's ability to protect against hydrogen embrittlement, corrosion, erosion, and high-temperature exposure.

It is standard practice to conduct a final dynamic balance check following the application of coatings on rotor components. Figure 2.18 illustrates a protective coating applied to a compressor impeller.

Fig. 2.18 Protective coating on an impeller (Image courtesy of Mitsubishi Heavy Industries Compressor, Hiroshima, Japan)

Acknowledgements Mitsubishi Heavy Industries Compressor, Hiroshima, Japan, has contributed most of the material and all of the images/graphics/figures used in this chapter.

References

1. API Standard 617 (2022) Axial and centrifugal compressors and expander-compressors, 9th edn. American Petroleum Institute, Washington, DC
2. API TR 684-1 (2019) API standard paragraphs rotordynamic tutorial: lateral critical speeds, unbalance response, stability, train torsionals, and rotor balancing, 1st edn. American Petroleum Institute, Washington, DC
3. ISO 21940-11:216 (2016, confirmed 2021) Mechanical vibration—rotor balancing—part 11: procedures and tolerances for rotors with rigid behaviour, 1st edn. International Organization for Standardization, Geneva, Switzerland

Chapter 3
Impellers

Abstract This chapter presents its content in a list-based format, focusing on key concepts related to thermodynamics, aerodynamics, stage design, curve profiles, performance maps, and specific speed as they pertain to centrifugal compressor impellers. These topics are supported by illustrative figures to aid in the explanation and comprehension of fundamental principles. While quantitative data such as typical ranges and values for impeller aerodynamic design can be found extensively in existing literature, this chapter opts for a qualitative approach to the subject matter. Additionally, it offers a detailed overview of how inlet condition variations, surge and stonewall phenomena, impeller manufacturing techniques, standard materials, material selection criteria, and operational stresses influence overall compressor performance.

Nomenclature

a	Acoustic velocity
C_{p}	Specific heat at constant pressure
C_{v}	Specific heat at constant volume
D	Impeller outer diameter
H_{p}	Polytropic head
H_{th}	Theoretical head
K	Specific heat
MW	Molecular weight
n	Polytropic volume exponent
N_{mc}	Maximum continuous speed
P	Pressure
Q	Inlet volume flow
$\hat{R}$	Universal gas constant
T	Temperature
v	Specific volume
Z	Compressibility factor

N. Ghaisas, *Centrifugal Compressors in Hydrocarbon Processing Industries*,
https://doi.org/10.1007/978-3-032-03687-2_3

ζ Slip factor
η_p Polytropic efficiency
Φ_D Flow coefficient

3.1 Thermodynamics

- Figure 3.1 shows the basic thermodynamic equations for the acoustic velocity, polytropic exponent, polytropic head, polytropic efficiency, and gas power of a compressor impeller.
- Variations in inlet conditions of a centrifugal compressor have a direct impact on its performance. For instance, a drop in suction pressure (P_s) leads to a corresponding decrease in discharge pressure (P_d) and a reduction in absorbed shaft power, as illustrated in Figs. 3.2 and 3.3.
- A decrease in suction temperature (T_s) leads to an increase in both discharge pressure and absorbed shaft power in the compressor. Lower suction temperatures result in higher gas density, which in turn reduces the specific volume. Consequently, for a fixed weight flow rate, the volumetric flow rate decreases, as illustrated in Figs. 3.4 and 3.5.
- An increase in the gas molecular weight (MW) leads to a corresponding rise in the compressor's discharge pressure and absorbed shaft power, as illustrated in Figs. 3.6 and 3.7.
- At the same rotational speed, centrifugal compressors develop a higher head when compressing high-molecular-weight gases relative to lighter or low-molecular weight gases. The degree of this variation increases with increasing inlet volumetric flow rates. As a result, for a given degree of impeller backward lean, the characteristic head–flow (H–Q) performance curve of compressor is less steep for high-molecular-weight gas compared to low-molecular-weight gas.
- The head required is directly proportional to the inlet compressibility factor.
- The head produced by a centrifugal compressor is determined solely by the impeller tip speed and is independent of inlet conditions.
- An increase in the specific heat capacity ratio (K) results in higher discharge temperature, discharge pressure, absorbed shaft power, and rotational speed. Conversely, a decrease in K decreases the discharge temperature.
- According to the affinity laws, changes in compressor speed affect performance as follows:
  - Inlet volumetric flow varies linearly with the speed ratio.
  - Generated head varies with the square of the speed ratio.
  - Absorbed shaft power varies with the cube of the speed ratio.
- Impellers should be sized based on the highest specified inlet temperature, lowest gas molecular weight, and maximum pressure ratio for the intended service.

- The impeller head is influenced by the velocity triangle at the impeller tip, which is governed by the blade angle.
- Entropy of gas changes during a polytropic process, but remains constant in an adiabatic process.
- Compressor inlet flow can be specified as either mass flow (weight flow) or standard volumetric flow, as per API Standard 617 [1].
- Weight flow is independent of inlet conditions and gas composition, making it the preferred metric—especially for sideload flow—and is derived from the process heat and mass balance sheet.
- Manufacturers typically use inlet volumetric flow for sizing. As compression progresses, the specific volume of the gas decreases, leading to a reduction in volumetric flow from inlet to discharge, while the mass flow rate remains constant.

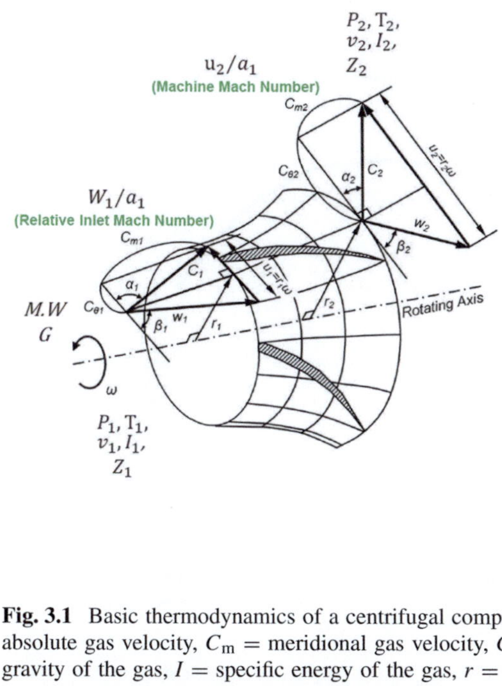

Fig. 3.1 Basic thermodynamics of a centrifugal compressor impeller. Additional variables: C = absolute gas velocity, C_m = meridional gas velocity, C_θ = tangential gas velocity, G = specific gravity of the gas, I = specific energy of the gas, r = impeller radius, u = impeller tip velocity, w = gas velocity relative to the impeller blade, α = angle between the absolute gas velocity and the tangent to the blade, β = angle between the blade inlet and the tangent to the inlet radius, ω = angular velocity. Subscripts: 1 = at the impeller inlet, 2 = at the impeller outlet (Image courtesy of Mitsubishi Heavy Industries Compressor, Hiroshima, Japan)

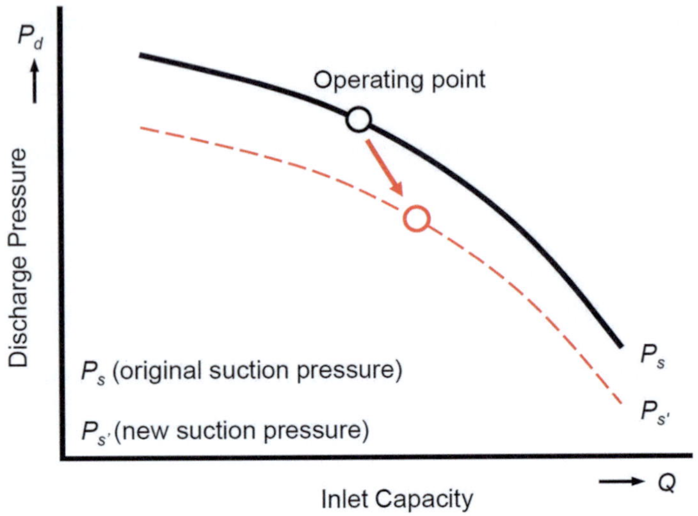

Fig. 3.2 Effect of a decrease in suction pressure ($P_{S'} < P_S$) on the compressor discharge pressure (Image courtesy of Mitsubishi Heavy Industries Compressor, Hiroshima, Japan)

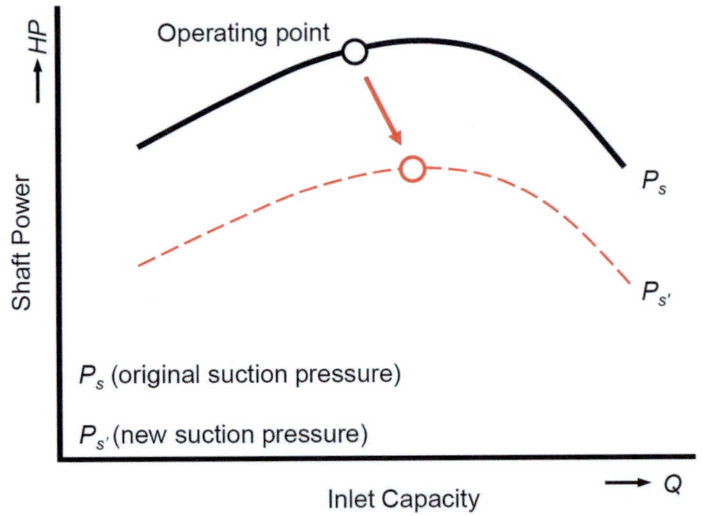

Fig. 3.3 Effect of a decrease in suction pressure ($P_{S'} < P_S$) on the compressor shaft power (Image courtesy of Mitsubishi Heavy Industries Compressor, Hiroshima, Japan)

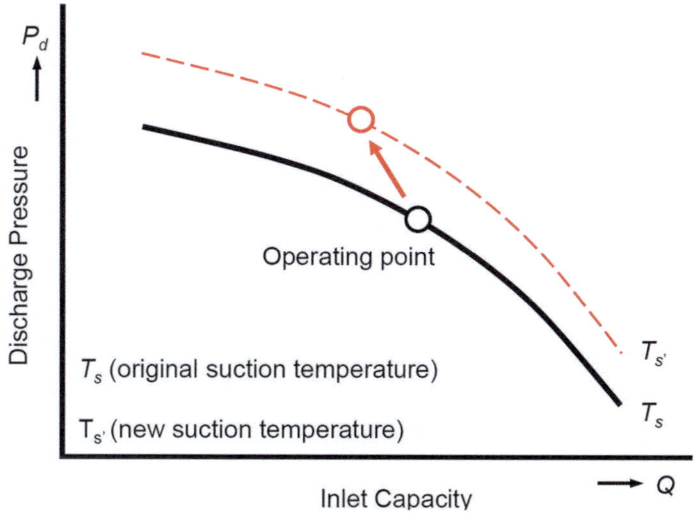

Fig. 3.4 Effect of a decrease in suction temperature ($T_{s'} < T_s$) on the compressor discharge pressure (Image courtesy of Mitsubishi Heavy Industries Compressor, Hiroshima, Japan)

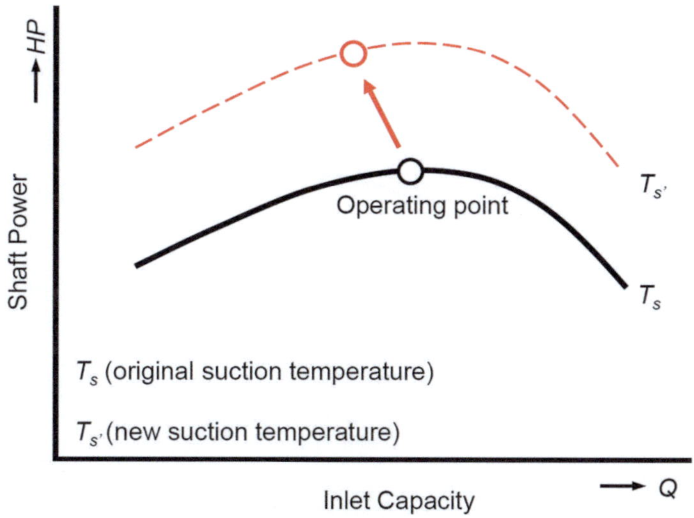

Fig. 3.5 Effect of a decrease in suction temperature ($T_{s'} < T_s$) on compressor shaft power (Image courtesy of Mitsubishi Heavy Industries Compressor, Hiroshima, Japan)

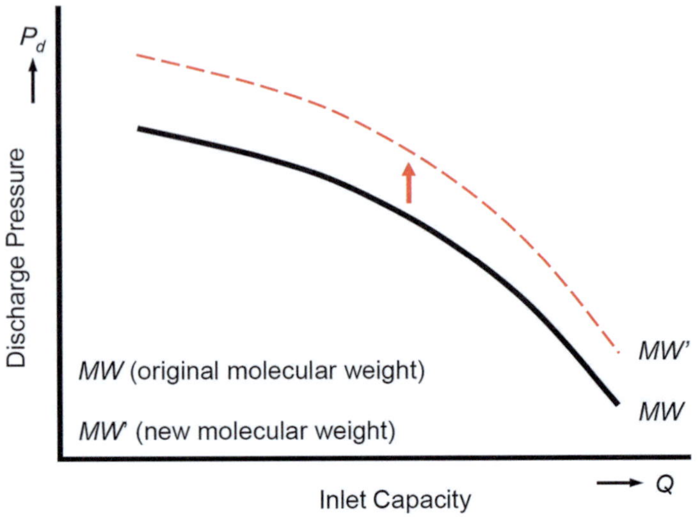

Fig. 3.6 Effect of an increase in gas molecular weight (MW′ > MW) on the compressor discharge pressure (Image courtesy of Mitsubishi Heavy Industries Compressor, Hiroshima, Japan)

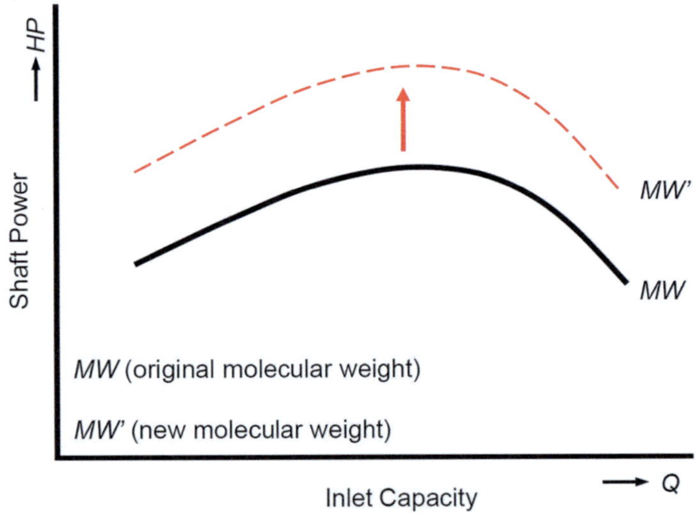

Fig. 3.7 Effect of an increase in gas molecular weight (MW′ > MW) on the compressor shaft power (Image courtesy of Mitsubishi Heavy Industries Compressor, Hiroshima, Japan)

3.2 Aerodynamics

- Blade thickness is calculated based on the maximum operational stress levels encountered during compressor operation.
- The blade backward lean angle typically ranges between 40 and 60°, depending on the specific application requirements.
- In modern centrifugal compressors, impeller flow coefficients (Eq. 3.1) range between 0.04 and 0.19, depending on impeller design:
  - Radial-bladed impellers exhibit the lowest flow coefficients
  - Two-dimensional (2D) impellers fall in the midrange
  - Three-dimensional (3D) impellers have higher values, typically between 0.15 and 0.19

$$\text{Flow coefficient}\, \text{Ø} = \frac{Q}{ND_{\text{out}}^3} \tag{3.1}$$

  where Ø is the flow coefficient, Q is the inlet volumetric flow rate, N is the operating speed, and D_{out} is the impeller outlet diameter.

- Higher flow coefficients correlate with larger gas flow paths within the impeller, as illustrated in Fig. 3.12.
- The impeller head (or pressure) coefficient (Eq. 3.2) generally ranges between 0.5 and 0.6.

$$\text{Head coefficient}\, \mu = \frac{H_{\text{p}} g_{\text{c}}}{u^2} \tag{3.2}$$

  where μ is the head or pressure coefficient, H_{p} is the impeller polytropic head, g_{c} is the gravitational constant, and u is the impeller tip speed.
- As outlined in Sect. 3.1, lighter gases yield lower head coefficients compared to heavier gases.
- The head coefficient decreases as the flow rate through the impeller increases.
- Radial-bladed impellers cannot produce a rising head-versus-capacity (H–Q) curve, which is critical for maintaining controllability and operational stability in centrifugal compressors. An impeller with a greater degree of backward lean yields a higher head rise at surge conditions while exhibiting a lower-pressure coefficient. The influence of backward lean diminishes at low-flow rates, where its impact is less pronounced. At minimum flow and constant tip speed, a high-lean impeller generates a head comparable to that of a low-lean design.
- The number of blades, their lean angles, and the blade profiles are engineered to minimize sudden velocity shifts and prevent boundary layer separation. Additionally, high-performance seals at the impeller eye and interstage locations help reduce leakage losses, contributing to improved polytropic efficiency.

- Figure 3.8 highlights how a higher boss ratio influences rotordynamics and how a shorter-span impeller enhances stage efficiency. Increasing the bore diameter (D_1) relative to the impeller tip diameter (D_2) results in a stiffer shaft, which improves mechanical integrity. Typical boss ratios for centrifugal compressors range from 0.44 to 0.50. In high-density gas applications, impellers with larger boss ratios are preferred to counteract destabilizing forces that can induce rotor vibrations. Figure 3.9 demonstrates how reducing the number of blades or vanes and optimizing the vane profile can further enhance stage efficiency.
- The speed of sound in a gas decreases as its molecular weight increases, due to its inverse relationship with the square root of molecular weight. Consequently, heavier gases have lower sonic velocities than lighter ones. This causes sonic conditions—marked by higher Mach numbers—to occur at lower volumetric flow rates. As a result, compressors operating with high-molecular-weight gases have a narrower stable operating range, reaching surge and stonewall limits more quickly.
- The head developed by an impeller is directly proportional to the product of the blade tip speed U_2 and the tangential component of the gas velocity V_T relative to the blade tip $V_{T.rel}$. In other words, at a given tip speed, an increase or decrease in the inlet volumetric flow will decrease or increase V_T, respectively, and thus, result in a corresponding decrease or increase in the head developed. This effect on the head-versus-capacity (H–Q) curve is demonstrated in Figs. 3.1 and 3.10.
- Aerodynamic losses in impellers are represented in Fig. 3.11. They result in increased frictional losses; thus, the heat generated is realized in a higher discharge temperature. A greater amount of energy input into the gas reduces the compressor

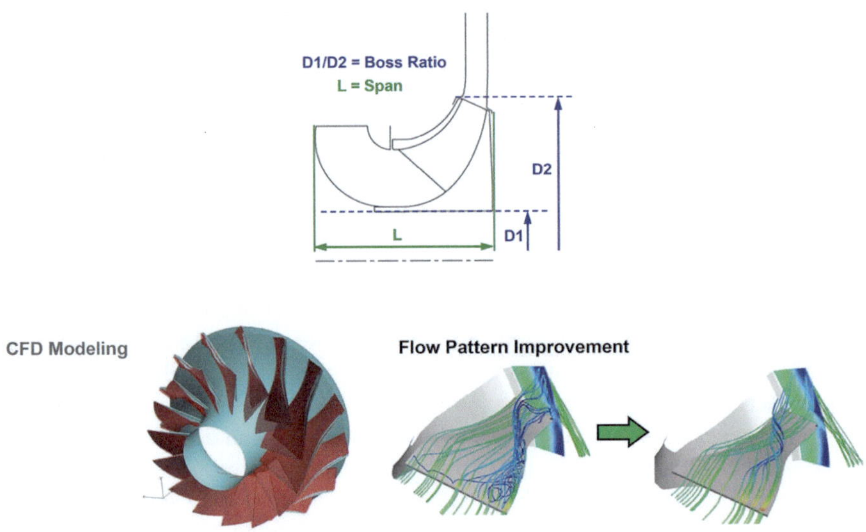

Fig. 3.8 Computational fluid dynamics (CFD) analysis for improving impeller stage efficiency with a higher boss ratio and shorter span (Image courtesy of Mitsubishi Heavy Industries Compressor, Hiroshima, Japan)

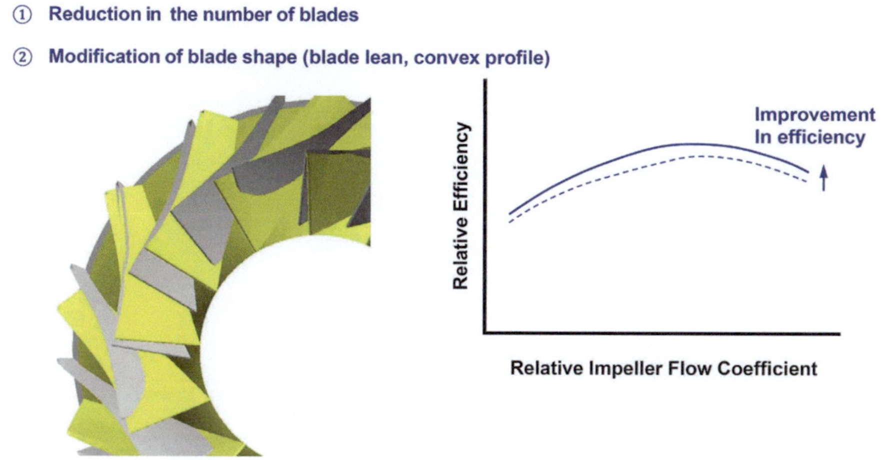

Fig. 3.9 Improvement in impeller stage efficiency using enhanced blade profile and number of blades (Image courtesy of Mitsubishi Heavy Industries Compressor, Hiroshima, Japan)

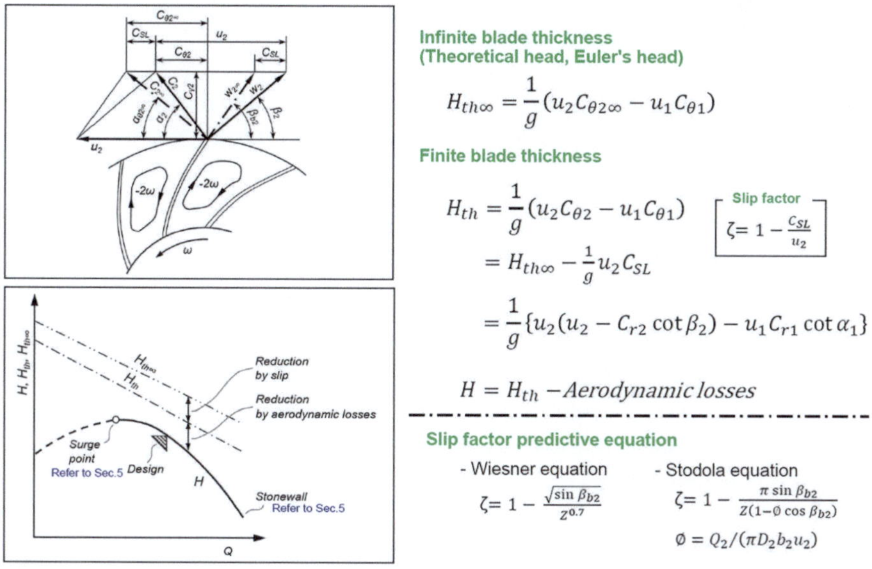

Fig. 3.10 Basic aerodynamics of a centrifugal compressor impeller. Additional variables: C = absolute gas velocity, g = conversion factor, ω = angular velocity. Subscript: 2 = at the impeller outlet. For definitions of C_{SL}, C_{γ}, C_{∞}, $C_{\theta\infty}$, $H_{th\infty}$, α, α_{θ}, β, and β_b, please refer to Euler's equation (Image courtesy of Mitsubishi Heavy Industries Compressor, Hiroshima, Japan)

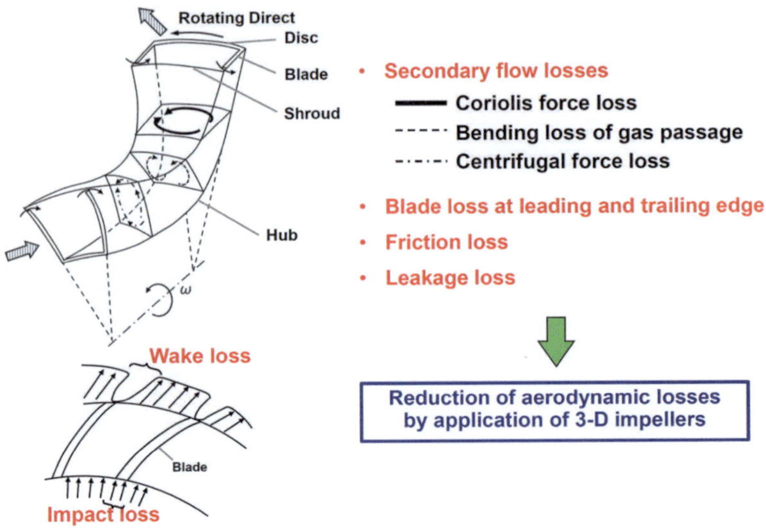

Fig. 3.11 Aerodynamic losses in an impeller (Image courtesy of Mitsubishi Heavy Industries Compressor, Hiroshima, Japan)

efficiency, as suggested in Fig. 3.1. Equation 3.3 shows the relationship between power, polytropic head, mass flow, and polytropic efficiency for values measured in US customary units

$$\text{Power}\, P = \frac{(H_\text{p})(m)}{(33{,}000)(n_\text{p})} \tag{3.3}$$

where H_p is the polytropic head (ft-lbf/lbm), m is the mass flow (lbm/min), n_p is the polytropic efficiency (%), P is the horsepower (hp), and 33,000 is the conversion constant (in US customary units).

Equation 3.4 shows the corresponding relationship for values measured in Système international d'unités (SI) units

$$\text{Power}\, P = \frac{(H_\text{p})(m)}{(3600)(n_\text{p})} \tag{3.4}$$

where H_p is the polytropic head (kN m/kg), m is the mass flow (kg/h), n_p is the polytropic efficiency (%), P is the horsepower (kW), and 3600 is the conversion constant (in SI units).

3.3 Three-Dimensional Impellers

- The optimal flow coefficient range for three-dimensional (3D) impellers lies between 0.15 and 0.19, while the ideal head coefficient range is from 0.5 to 0.6, depending on the inlet volumetric flow rate. Figure 3.12 compares the flow coefficients and corresponding polytropic efficiencies of radial, mixed-flow, and 3D impellers. Operating outside the recommended flow coefficient range either too low or too high can negatively impact stage efficiency (refer to Sect. 3.5).
- Enhanced machining precision and a high-quality aerodynamic surface finish contribute significantly to improved compressor efficiency.
- 3D impellers are typically fabricated using five-axis milling machines from a single block, and especially selected for applications requiring large flow areas.
- For impeller designs with narrow gas passages, electrical discharge machining (EDM) or weldless manufacturing technique is preferred to achieve the necessary dimensional accuracy and structural integrity.

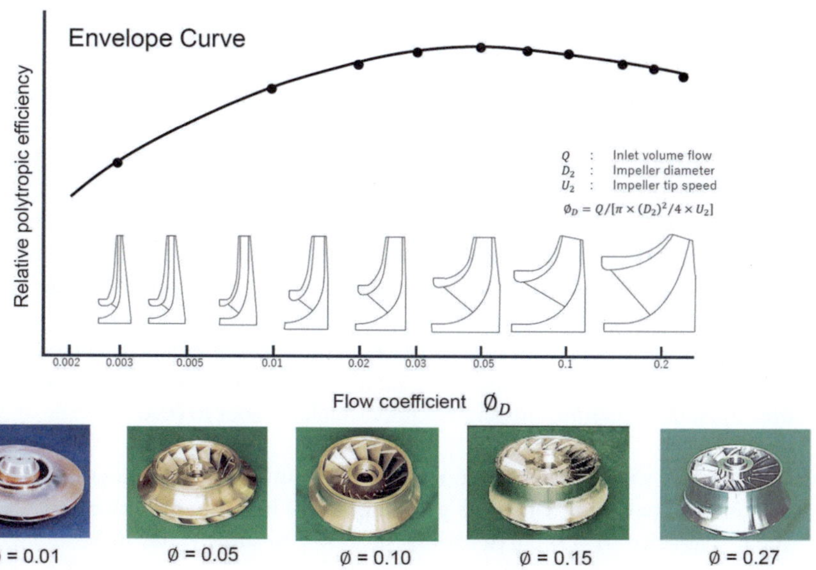

Fig. 3.12 Application of high-performance three-dimensional impellers for all stages (Image courtesy of Mitsubishi Heavy Industries Compressor, Hiroshima, Japan)

3.4 Surge and Stonewall Effects

- Figure 3.13 illustrates the behavior of stable and unstable flows governed by converging and diverging flow principles. The stable operating range of a compressor is constrained by the minimum and maximum allowable speeds, as well as the surge and overload boundaries shown on the compressor performance map in Fig. 3.14.
- As outlined in Chap. 4, surge events may occur under several conditions: during startup or shutdown, when operating at low-flow rates near the surge limit, or at high flow rates when sudden process disturbances such as pressure fluctuations, gas composition changes, fouled aftercoolers, clogged inlet filters or strainers, or loss of driver power disrupt normal operation. Surge cycles typically last between 0.3 and 3 s per cycle. The surge and stonewall (overload) regions are depicted on the operating envelope of a centrifugal compressor in Fig. 3.14.
- The consequence of surge includes rapid fluctuations in flow and pressure, a severity-dependent exponential rise in gas temperature, process instability, potential machine or process shutdowns, and mechanical damage to components such as seals, bearings, impellers, and shafts, depending on the energy intensity and compressor design.
- The machine Mach number is defined as the ratio of the gas velocity at the tip of the largest-diameter impeller to the acoustic velocity of the gas.
- The relative inlet Mach number refers to the ratio of the gas velocity relative to the blade to the acoustic velocity at the inlet. As this value approaches Mach 1, flow separation and reduced flow area near the impeller blades may occur, contributing to stonewall conditions. Figure 3.1 presents the mathematical expressions for both the relative inlet Mach number at the blade leading edge and the machine Mach number.
- Prolonged operation in stonewall subjects the leading edges of impeller blades to elevated mechanical stresses, potentially leading to fatigue failure. To mitigate this, most designs limit the relative inlet Mach number to 0.90 at the design point flow.
- The velocity triangle (vector geometry) at the impeller inlet plays a critical role in stonewall behavior, while the geometry at the impeller tip influences the head-versus-capacity (*H*–*Q*) curve and the head rise to surge (refer to Sect. 3.2).

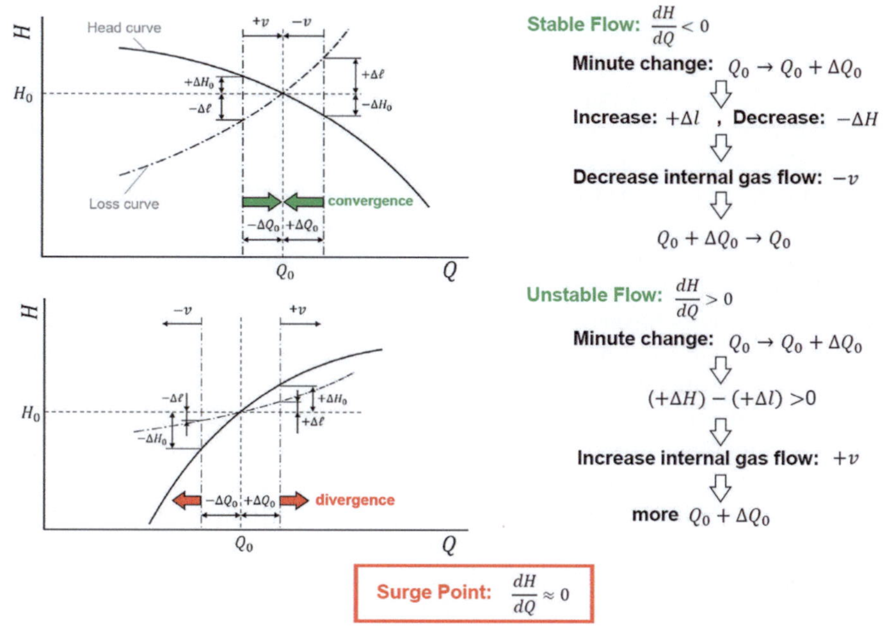

Fig. 3.13 Effects of surge and stonewall (Image courtesy of Mitsubishi Heavy Industries Compressor, Hiroshima, Japan)

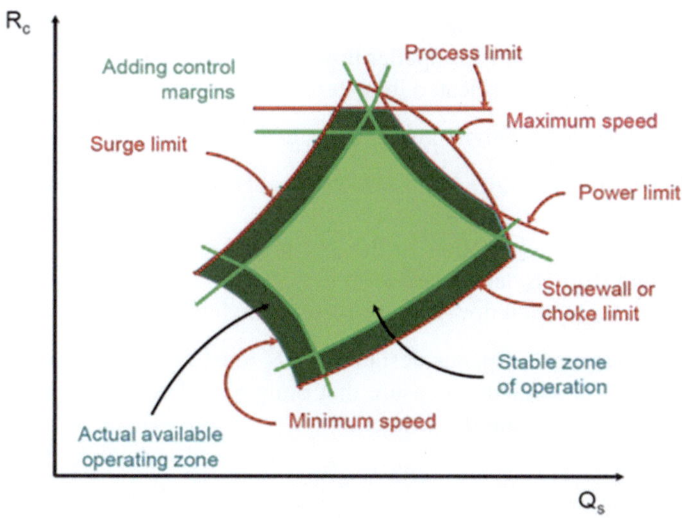

Fig. 3.14 Performance map and typical operating envelope of a centrifugal compressor (duplicated from Chap. 4) (Image courtesy of Compressor Controls Corporation I Honeywell, Des Moines, IA)

3.5 Specific Speed [2]

- Figure 3.15 illustrates the concept of specific speed, which guides the design of impeller flow paths for specific applications by scaling standard impeller stages according to the laws of similarity.
- Specific speed is a dimensionless parameter derived from the flow and head coefficients (refer to Eqs. 3.1 and 3.2). While specific speed provides a general classification, the flow and head coefficients are more directly tied to the aerodynamic performance of the impeller.
- Radial-flow, two-dimensional impellers typically exhibit low specific speeds. Mixed-flow and axial-flow impellers have higher specific speeds due to their larger flow coefficients.
- The base impeller geometry can be scaled up or down to design larger or smaller impellers for standardized applications. With advancements in computational fluid dynamics (CFD), simulation accuracy has greatly improved, enabling precise predictions of blade and flow path performance and facilitating the optimization of compressor characteristic curves. Enhanced impeller profiles can yield higher polytropic efficiencies, as shown in Figs. 3.8 and 3.9 (refer to Sect. 3.2).
- Very low or high specific speeds tend to reduce stage efficiency, and the flow coefficient similarly influences both capacity and polytropic efficiency.
- Additional factors affecting capacity and stage efficiency include:
  - Blade lean angle.
  - Blade height and thickness.
  - Surface roughness of the impeller flow path.
  - Weld distortions in fabricated impellers.
- Stage performance sensitivity also extends to:
  - Blade width at the inlet and tip, which impacts head coefficient and polytropic efficiency.
  - Excessive weld penetration in the gas passage, which can lower the head coefficient from its design target.
  - Eye area and internal flow path geometry, which influence stage performance.
- Therefore, analytical impeller design models must be validated against actual manufactured components to ensure that dimensional deviations remain within the manufacturer's required tolerances.

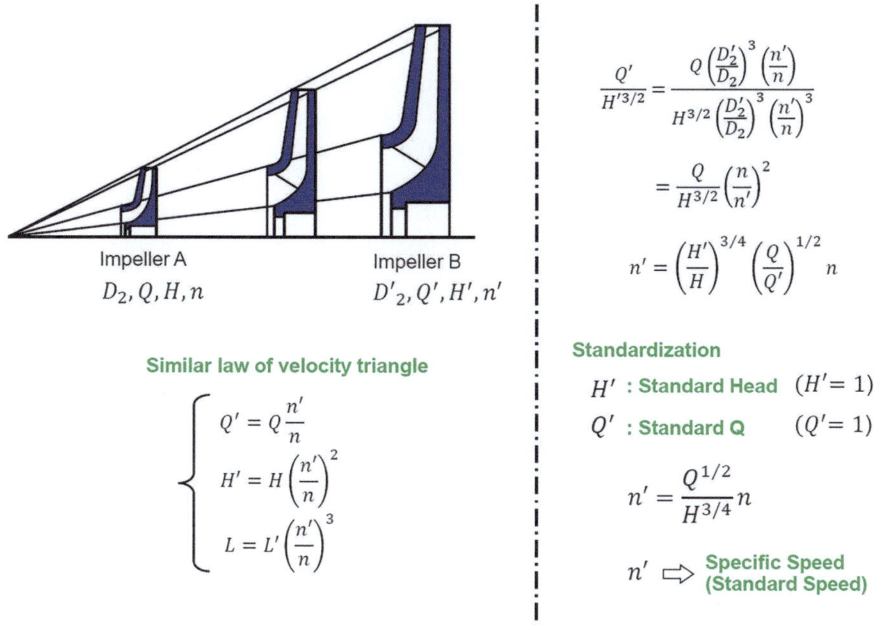

Fig. 3.15 Impeller specific speed and similarity laws. Additional variables: D_2 = impeller outer diameter at the impeller outlet, L = characteristic length (chord length or blade length) (Image courtesy of Mitsubishi Heavy Industries Compressor, Hiroshima, Japan)

3.6 Impeller Fabrication Methods [2]

- Figure 3.16 illustrates standard fabrication techniques for one-piece, two-piece, and three-piece impellers. Among these, one-piece and two-piece impellers produced by milling offer superior machining precision and surface finish, both of which contribute to enhanced stage efficiency.
- Impeller blades are typically cast or die-formed. In fabricated impeller designs, blade fillet welds must be full-penetration and continuous to ensure structural integrity. Details on inspection protocols and nondestructive testing are provided in Chap. 16.
- In certain two-piece configurations, blades are machined into the cover and then welded to the disk or hub, either from the inside or outside, depending on the blade tip width. Alternatively, blades are machined on the disk and welded to the cover. This method is favored for impellers with large diameters or high rotational speeds, as it subjects the blade fillet welds to comparatively lower dynamic stresses (refer to Sect. 3.9).
- For impellers with narrow tip widths ($\leq$ 0.50 in. [12.7 mm]), vacuum brazing serves as an alternative to electrical discharge machining (EDM). EDM is particularly effective for duplex stainless steels, including austenitic types 304 and 304L

and ferritic type 430, which are challenging to weld and prone to deformation during stress relief.

- The EDM process involves melting the surface material, which then resolidifies to form a recast layer. Beneath this layer, a heat-affected zone (HAZ) develops due to elevated temperatures. These surface alterations can negatively impact impeller efficiency and corrosion resistance.
- To mitigate these effects and enhance surface quality, electrochemical machining (ECM) is employed. ECM removes approximately 1970 μin. (50 μm) from the surface, achieving a finish finer than 236 Ra μin. (6 Ra μm).
- Figure 3.17 showcases one-piece or weldless impellers fabricated from solid forgings using 3D milling and EDM techniques.
- The EDM fabrication sequence for one-piece or weldless impellers includes:
  - Heat treatment for stress relief of the forging.
  - Liquid penetrant or magnetic particle testing before and after rough machining.
  - EDM followed by ECM to eliminate the recast layer and HAZ, and improve surface finish.
  - Dimensional verification.
  - Final nondestructive testing before and after finish machining.
  - Overspeed testing, followed by comprehensive dimensional and nondestructive inspections.

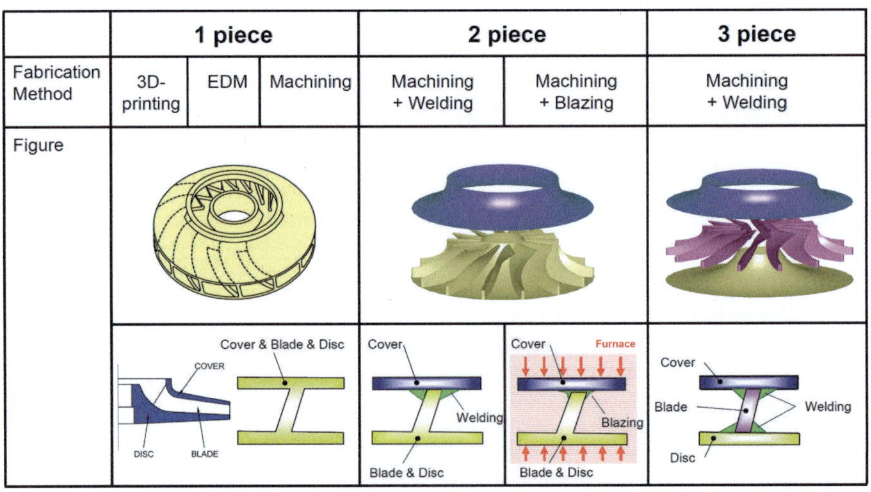

Fig. 3.16 Comparison of impeller fabrication methods (Image courtesy of Mitsubishi Heavy Industries Compressor, Hiroshima, Japan)

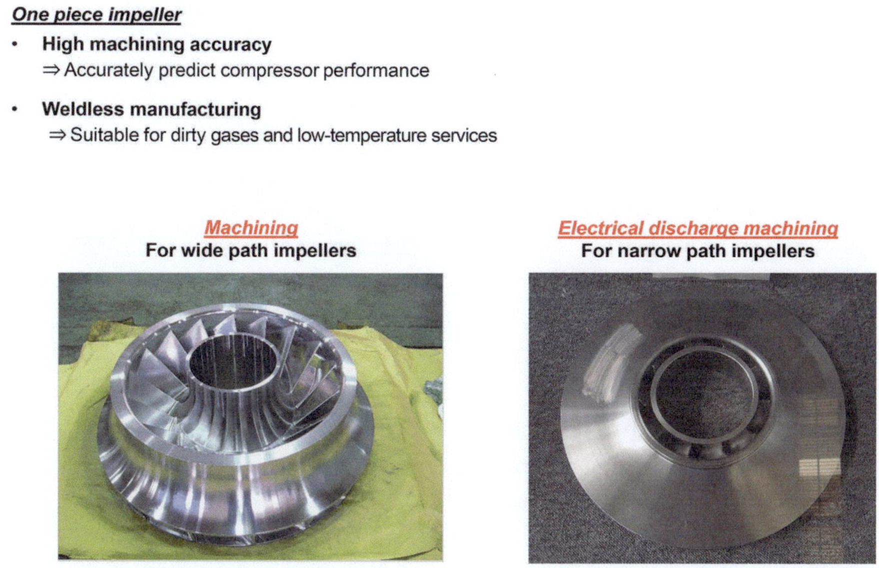

Fig. 3.17 One-piece or weldless impellers (Image courtesy of Mitsubishi Heavy Industries Compressor, Hiroshima, Japan)

3.7 Impeller Attachment to Shaft

- Figure 3.18 illustrates the two commonly employed methods for securing impellers to shafts in centrifugal compressors:
  - Interference fit only.
  - Interference fit combined with a key.
- The choice between these fixation techniques is primarily influenced by factors such as stage torque, stage thrust, and the anticipated start-stop modes of the compressor.
- For shrunk-on impellers, the typical interference fit falls within the range of 0.001–0.0015 in. per inch of shaft diameter (or 0.01–0.015 mm per centimeter of shaft diameter) under the impeller. For further details, refer to Chap. 2.

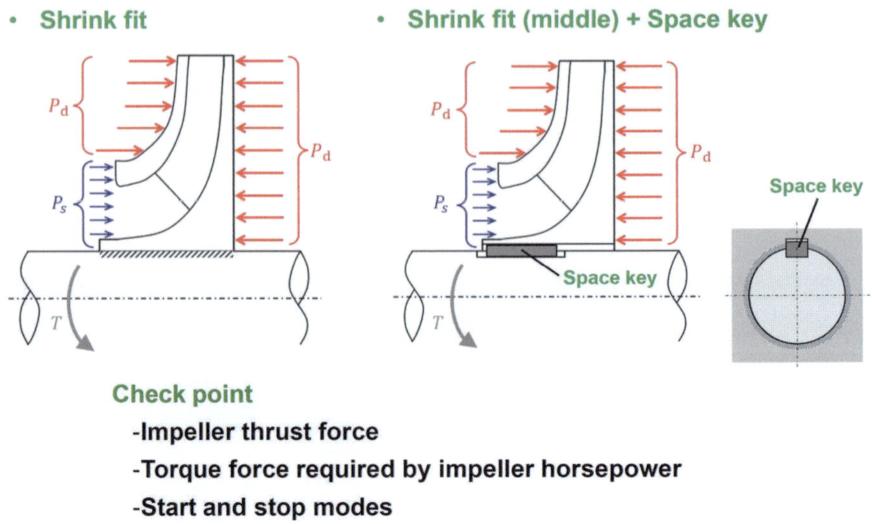

Fig. 3.18 Methods of impeller fixation or attachment to the shaft (Image courtesy of Mitsubishi Heavy Industries Compressor, Hiroshima, Japan)

3.8 Standard Impeller Materials [1, 3]

- API Standard 617 [1] outlines commonly used materials for compressor components. The selection of materials for impellers and other parts exposed to process gases is influenced by several factors, including:
  - The type of gas (e.g., hazardous, toxic, acidic).
  - The gas composition, such as the partial pressures and presence of hydrogen sulfide, carbon dioxide, mercury, ammonia, cyanide, and water vapor.
  - The installation environment.
  - The corrosion resistance and impact strength of both base materials and weld metals, particularly in low-temperature refrigeration applications involving gases like ethylene and propylene.
- For fabricated impellers, the chemical composition and thermal, physical, and mechanical properties of the selected materials must be evaluated against the expected aerodynamic loads, rotordynamic forces, and operational stresses (refer to Sect. 3.9). Key properties requiring thorough assessment include:
  - Coefficient of thermal expansion
  - Yield strength
  - Weldability
  - Hardenability
  - Corrosion resistance
  - Tensile strength.

- American Iron and Steel Institute (AISI) type 410 stainless steel and precipitation-hardened grades such as 17-4PH and 15-5PH provide enhanced corrosion resistance compared to conventional alloy steels.
- Austenitic stainless steels including 304, 304L, 316, and 316L are particularly resistant to corrosion caused by gases containing sulfur dioxide. These materials are suitable for low-tip-speed impellers due to their relatively low yield strength, approximately 30,000 psi (207 MPa).
- API 617 [1] enforces strict requirements for weld qualification, hardness limits, and maximum yield strengths for materials exposed to wet hydrogen sulfide or chloride-containing gases. Specifically, it prohibits the use of materials with yield strengths exceeding 120,000 psi (827 MPa) or hardness values above Rockwell C 34 in hydrogen gas environments where the hydrogen partial pressure exceeds 100 psig (690 kPag) or the hydrogen concentration exceeds 90 mol % at any pressure (Fig. 3.19).
- The positive material identification (PMI) process is employed to confirm the chemical composition of alloys and metals, as illustrated in Fig. 3.20.
- The implications of site-specific conditions and process gas characteristics on material selection are summarized in Fig. 3.21.

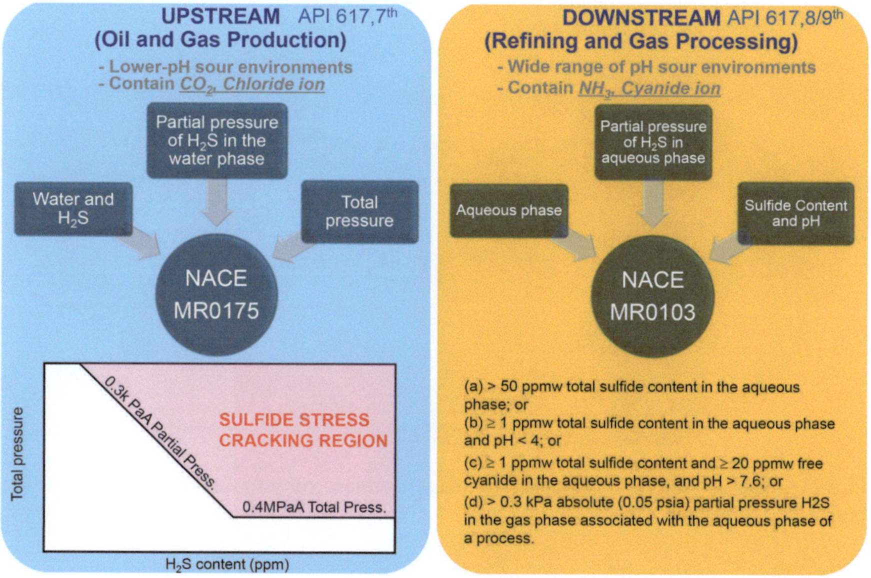

Fig. 3.19 Implications of National Association of Corrosion Engineers (NACE) standards NACE MR0103 [4] and NACE MR0175 [3], as specified in API Standard 617 (Image courtesy of Mitsubishi Heavy Industries Compressor, Hiroshima, Japan)

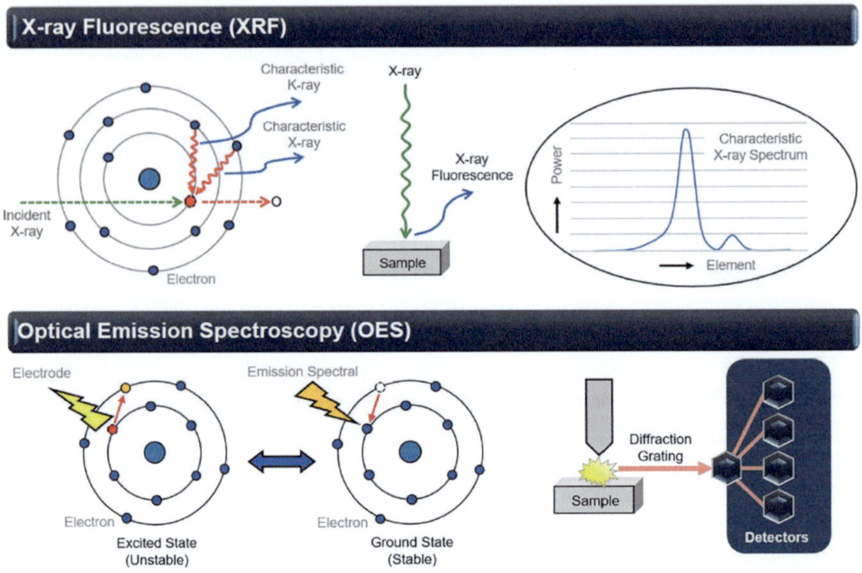

Fig. 3.20 Positive material identification (PMI) process (Image courtesy of Mitsubishi Heavy Industries Compressor, Hiroshima, Japan)

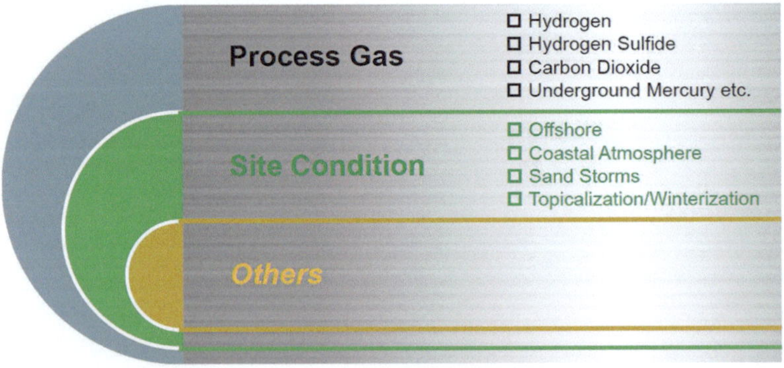

Fig. 3.21 Material selection influences (Image courtesy of Mitsubishi Heavy Industries Compressor, Hiroshima, Japan)

3.9 Impeller Stresses

- Impeller stress during operation increases proportionally with the square of the rotational speed. In a centrifugal compressor, the largest-diameter impeller experiences the highest centrifugal stress, making it the limiting factor for the machine's maximum speed.

- Assuming that the tip stress approaches the minimum yield strength (YS) of the impeller material at overspeed conditions (defined as 115% of the maximum continuous speed, N_{mc} or MCS), the resulting speed–stress profile for variable-speed compressors is outlined in Table 3.1 [5].
- It is essential to evaluate all relevant stresses during the impeller design process, including centrifugal stress, tangential and tensile stress, compressive and thermal stress, and vibratory stress to ensure a sufficient safety margin relative to the material's minimum yield strength.
- Additional insights into stress behavior at the impeller hub-to-shaft interface and finite-element analysis (FEA) of impellers are provided in Chaps. 2 and 9, respectively.
- In welded impeller designs, blade fillet welds are subject to elevated stress levels that may exceed their material strength under overspeed conditions, as indicated in Table 3.1.
- To ensure structural integrity, fillet weld geometry and dimensions must be rigorously assessed so that the calculated stresses remain well below the minimum yield strength of the weld material.
- Figure 3.22 depicts a blade that is machined into the impeller cover and internally welded to the disk, highlighting the stress vectors acting on both the fillet welds and the blade structure.
- Figures 3.23 and 3.24 present Goodman diagrams, which are used for estimating fatigue life under specific stress ratios, considering the material properties, geometry, and surface condition of the component. Fatigue-related impeller failures are typically caused by persistent operational stresses. The fatigue performance of a material based on its ultimate tensile strength and endurance limit can be evaluated using the applied mean stress and alternating stress. If the mean stress surpasses the ultimate tensile strength, or the alternating stress exceeds the endurance limit, the material is at high risk of failure.

Table 3.1 Centrifugal stresses at the impeller tip at different speeds

Speed step	Speed (%)	Stress (%)	% of YS
Design (rated)	100	100	68
MCS (1.05 × design)	105	110	75
Trip (1.10 × N_{mc})	115	132	90
Overspeed (1.15 × N_{mc})	121	146	100

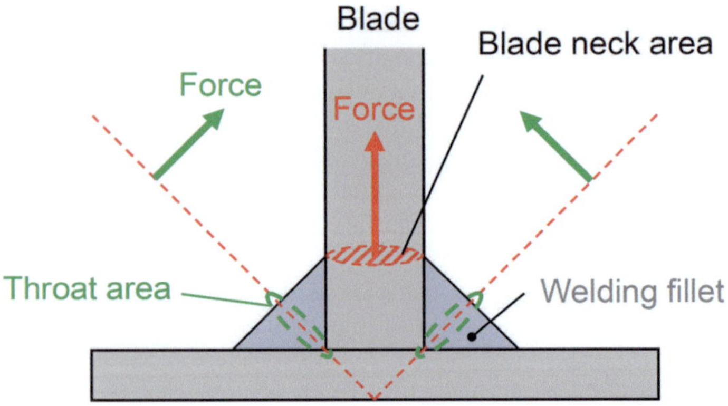

Fig. 3.22 Stresses in the impeller blade and weld filets (Image courtesy of Mitsubishi Heavy Industries Compressor, Hiroshima, Japan)

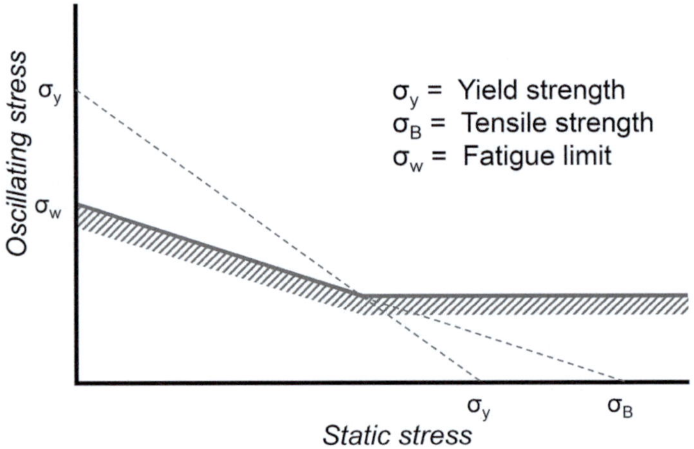

Fig. 3.23 Representative Goodman diagram (Image courtesy of Mitsubishi Heavy Industries Compressor, Hiroshima, Japan)

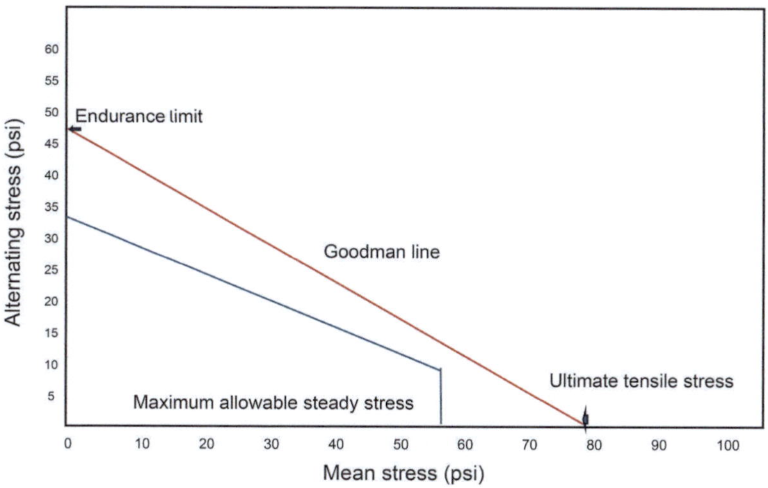

Fig. 3.24 Goodman diagram

Acknowledgements Mitsubishi Heavy Industries Compressor, Hiroshima, Japan, has contributed all of the images/graphics/figures used in this chapter, except for the Goodman diagram in Fig. 3.24.

References

1. API Standard 617 (2022) Axial and centrifugal compressors and expander-compressors, 9th edn. American Petroleum Institute, Washington, DC
2. Pash N, Hayles GC Jr, Hirata D, Kiuchi D (2012) A comparison of manufacturing methods, accuracy, quality control and testing methods as they relate to high head low flow impeller efficiency and overall compressor performance. Paper presented at the 41st turbomachinery symposium, Houston, TX, 24–27 Sept 2012
3. NACE MR0175-2021/ISO 15156:2020 Petroleum and natural gas industries—Materials for use in H_2S-containing Environments in oil and gas production—Part 1: General principles for selection of cracking-resistant materials. NACE International/ISO, Houston, TX
4. ANSI/NACE MR0103-2015/ISO 17945:2015 (R2023) Petroleum, petrochemical and natural gas industries—Metallic materials resistant to sulfide stress cracking in corrosive petroleum refining environments. NACE International, Houston, TX
5. Cameron JA, Danowski FM Jr (1973) Some metallurgical considerations in centrifugal compressors. In: Proceedings of the second turbomachinery symposium, Gas Turbine Laboratories, Texas A&M University, College Station, TX, Oct 1973

Chapter 4
Antisurge and Performance Control Systems

Abstract Surge detection and control in centrifugal compressors are inherently complex functions that demand specialized knowledge and expertise in machinery dynamics, control systems, and process behavior. Achieving effective control involves the development of sophisticated algorithms, robust fallback mechanisms, and custom-engineered systems capable of executing high-speed computations for optimal turbomachinery performance. This chapter outlines the foundational concepts, including the dynamic compressor map, operating envelope, surge cycle characteristics, and methods for surge detection. It further explores the roles of antisurge and performance controllers within a comprehensive control strategy, detailing antisurge response mechanisms and advanced control functionalities such as loop decoupling, load sharing, performance override, limiting loops, and fallback operations. Additionally, the chapter addresses key design and operational considerations for recycle piping and valve sizing, hot-gas bypass valve specifications, and instrumentation requirements critical to reliable surge management.

4.1 Dynamic Compressor Map [2]

Centrifugal compressors transform the energy supplied by the driver into both potential and kinetic energy within the compressed gas. Figure 4.1 presents a typical dynamic compressor map, depicting this energy transformation across two dimensions. The map's horizontal axis represents the volumetric flow rate at the inlet Q_S, and the vertical coordinate of the map represents the pressure ratio R_c.

Figure 4.2 displays the typical operating envelope of a centrifugal compressor, highlighting how the stable operating region is defined. This zone is established by incorporating control margins within the limits of surge, stonewall, and the compressor's minimum and maximum speeds, along with any other constraints imposed by the process or compressor train configuration.

N. Ghaisas, *Centrifugal Compressors in Hydrocarbon Processing Industries*,
https://doi.org/10.1007/978-3-032-03687-2_4

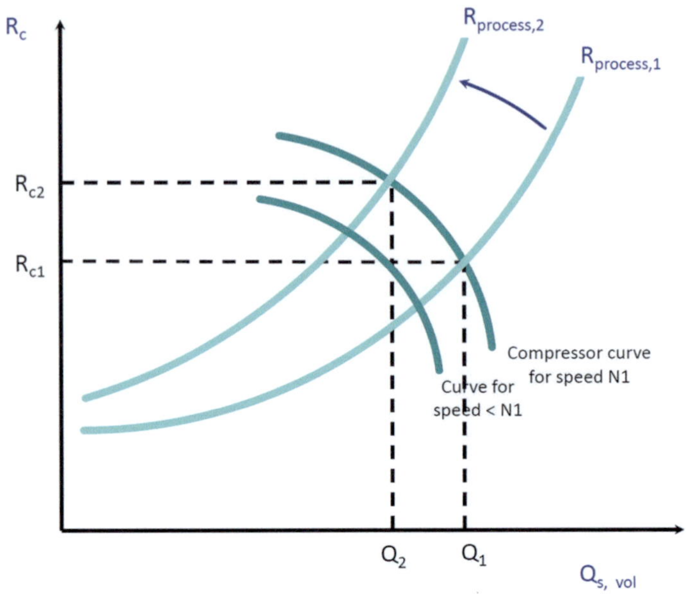

Fig. 4.1 Dynamic compressor map showing the inlet volumetric capacity versus pressure ratio (Image courtesy of Compressor Controls Corporation I Honeywell, Des Moines, IA)

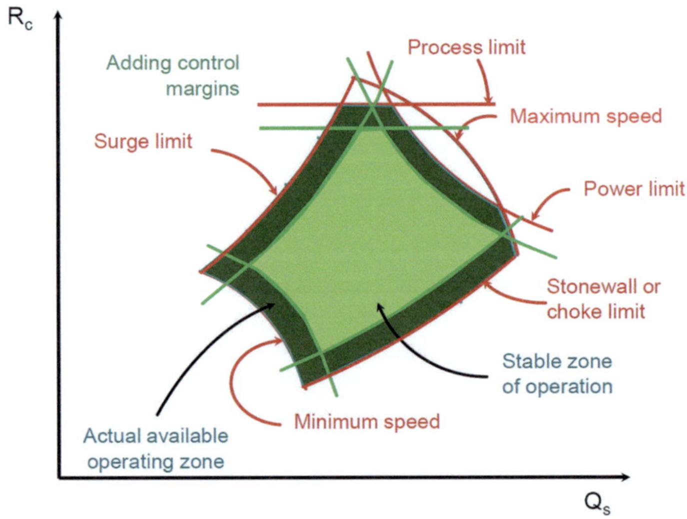

Fig. 4.2 Typical operating envelope of a centrifugal compressor (Image courtesy of Compressor Controls Corporation I Honeywell, Des Moines, IA)

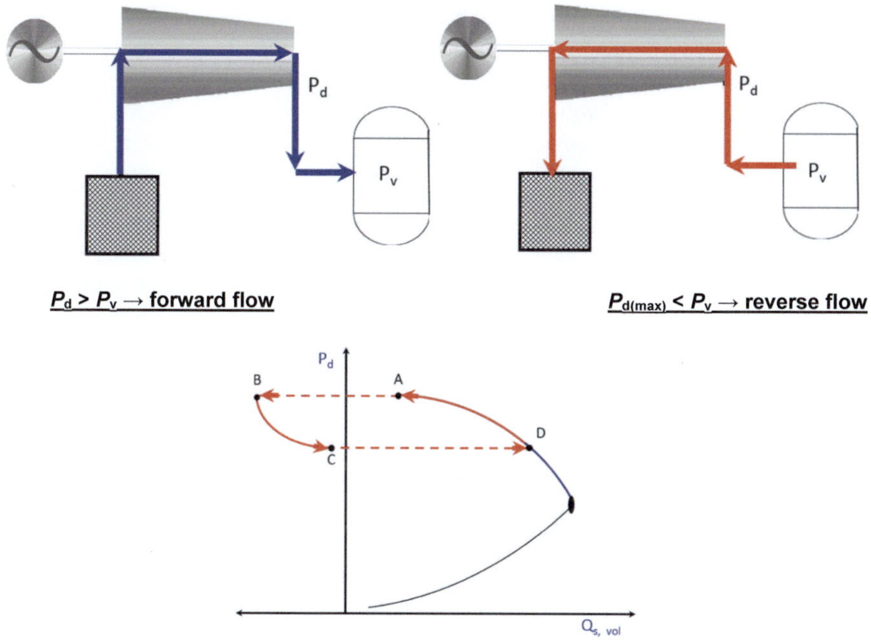

Fig. 4.3 Development of a surge cycle (Image courtesy of Compressor Controls Corporation I Honeywell, Des Moines, IA)

4.2 Surge Cycle and Surge Detection

Figure 4.3 illustrates the progression of a surge cycle. As pressure accumulates within the system, resistance increases, causing the operating point to trace along the performance curve. At point A, the potential energy in the discharge vessel exceeds the capability of the compressor. When the vessel pressure (P_v) exceeds the compressor's maximum discharge pressure ($P_{d(max)}$), that is, $P_v > P_{d(max)}$—the flow reverses. This reversal leads to a drop in pressure downstream of the compressor, enabling it to resume forward flow. If no corrective measures are taken, the surge cycle will continue to repeat.

Surge events can occur under several conditions, including:

- During compressor startup or shutdown.
- At low-flow rates, especially near the surge limit line.
- At high flow rates in centrifugal compressors when sudden process changes occur—such as shifts in pressure or gas composition.
- In cases of equipment malfunction, like fouled aftercoolers, clogged inlet filters or strainers, or loss of driver power.

Typical surge cycle durations range from 0.3 to 3 s per cycle. The effects of surge can be severe, involving:

- Rapid fluctuations in flow and pressure.
- A sharp, severity-dependent rise in gas temperature.
- Instability in the overall process.
- Potential shutdowns of equipment or processes.
- Possible damage to compressor components such as seals, bearings, impellers, and shafts—depending on the surge energy and compressor design.

Surge detection is a fundamental component of the surge control system. Its primary role is to notify plant operators, activate control mechanisms to take corrective measures—such as increasing the safety margin to avoid surge—and, if necessary, initiate a compressor shutdown to prevent equipment damage.

Detection methods rely on monitoring key indicators, which can include:

- Sudden changes in pressure.
- Rapid variations in flow rate.
- Temperature spikes relative to upstream gas temperature.
- Fluctuations in driver power.
- Or a combination of these parameters.

Figure 4.4 illustrates a surge detection event, triggered by a sharp drop in compressor discharge pressure (P_d) and a significant shift in differential pressure across a flow orifice ΔP_{o}.

4.3 Compressor Map with Invariant Coordinates

Compressor inlet volumetric flow (Q_{s}) and polytropic head (H_{p}) are standard parameters used as map coordinates, and their values are defined by the following equations (Eq. 4.1):

$$Q_{\mathrm{s}} = A\sqrt{\frac{\mathrm{d}P_{\mathrm{o}}T_{\mathrm{s}}Z_{\mathrm{s}}}{P_{\mathrm{s}}\mathrm{MW}}}, \quad H_{\mathrm{p}} = \frac{ZRT_{\mathrm{s}}}{\mathrm{MW}}\left(\frac{Rc^{\sigma}-1}{\sigma}\right) \tag{4.1}$$

where

$$Z = (Z_{\mathrm{s}} + Z_{\mathrm{d}})/2$$
$$R_{\mathrm{c}} = P_{\mathrm{d}}/P_{\mathrm{s}}$$
$$\sigma = (k-1)/(k\eta)$$
$$k = C_{\mathrm{p}}/C_{\mathrm{v}}$$

A is the inlet flow orifice coefficient, C_{p} is the gas specific heat at constant pressure, C_{v} is the gas specific heat at constant volume, $\mathrm{d}P_{\mathrm{o}}$ is the differential pressure flow measurement signal, MW is the gas molecular weight, P_{d} is the absolute discharge pressure, P_{s} is the absolute suction pressure, R is the gas constant, R_{c} is the pressure

- The nature, magnitude, and frequency of process disturbances.
- The durability of the compressor design.
- The importance of the compressor's role in the overall operation.

Even after accounting for these factors, it's not possible to pinpoint an exact surge margin for a given system without site-specific testing and real-world operational data. Since recycling gas consumes energy unnecessarily, it's essential to carefully optimize the surge margin to improve efficiency and lower operating costs.

4.5 Antisurge Controller Structure [2]

An antisurge controller must be equipped to handle a wide array of transient conditions while also maintaining precise control during steady-state operation—regardless of whether continuous recycling is present. In a typical compressor setup, a sudden drop in feed gas can trigger a surge if the controller fails to react swiftly. When low-flow conditions persist, the controller must stabilize the system along the surge control line, making fine adjustments to the valve position to ensure the compressor maintains its minimum required flow. Due to the broad spectrum of operating scenarios and potential load disturbances, the antisurge controller must feature a sophisticated architecture (as illustrated in Fig. 4.8) and deliver exceptional speed and accuracy in its performance.

4.6 Levels of Surge Protection Within the Antisurge Controller

Four basic types of control responses are commonly used in surge prevention, as follows:

- "Predictive" response: Proportional plus integral (PI).
- Open-loop response: Recycle Trip™.
- Adaptive surge control line response: Safety On®.
- Derivative response.

Figure 4.9 shows a basic recycle loop with flow and pressure measurement for a variable-speed centrifugal compressor.

Figures 4.10, 4.11 and 4.12 illustrate the various levels of protection offered by antisurge controllers. When the operating point crosses the surge control line (SCL), the predictive control mechanism (PI) begins to gradually open the recycle valve. This action helps maintain the minimum required inlet volumetric flow through the compressor.

However, the PI response is often too slow to prevent surge effectively. To address this, a second, more assertive control layer is activated when the operating point

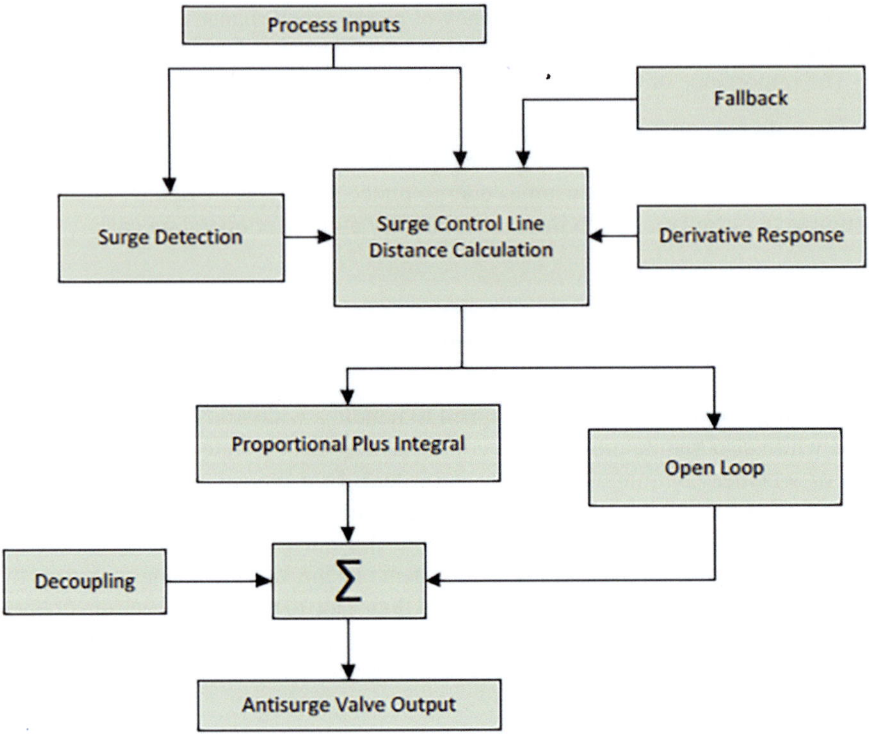

Fig. 4.8 General structure of an antisurge controller (Image courtesy of Compressor Controls Corporation I Honeywell, Des Moines, IA)

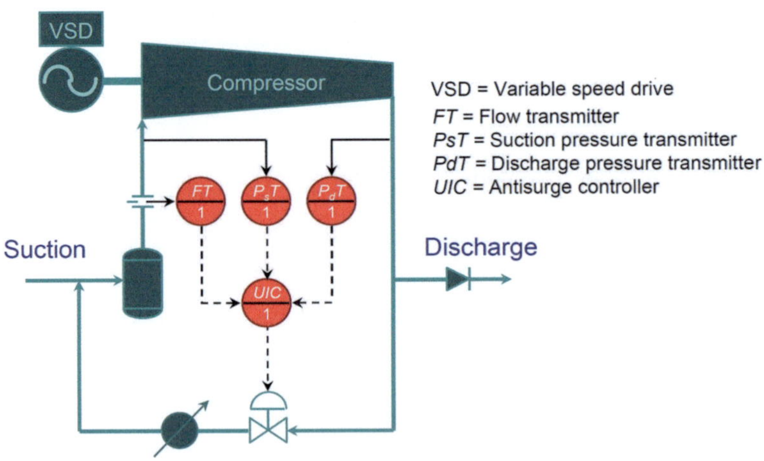

Fig. 4.9 Recycle loop of a variable-speed centrifugal compressor (Image courtesy of Compressor Controls Corporation I Honeywell, Des Moines, IA)

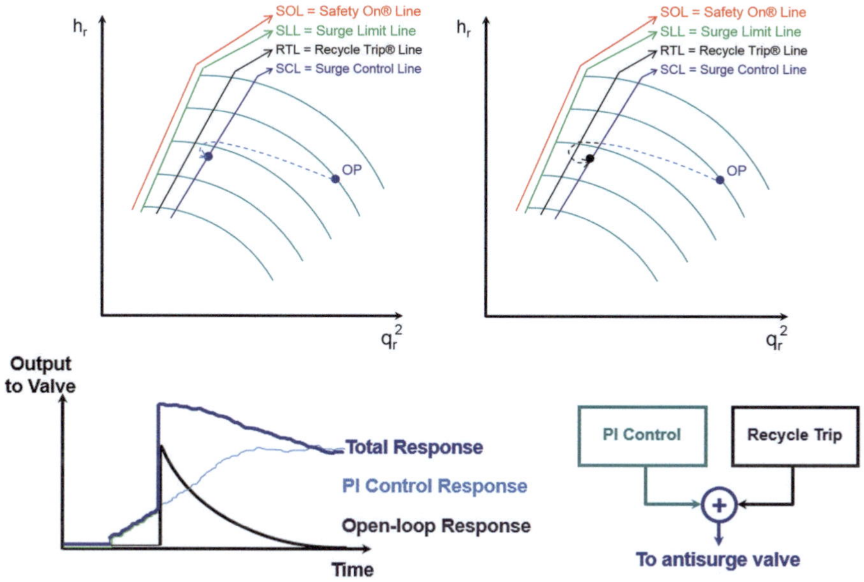

Fig. 4.10 PI and Recycle Trip™ (open-loop) responses (Image courtesy of Compressor Controls Corporation I Honeywell, Des Moines, IA)

crosses the Recycle Trip™ Line (RTL). At this stage, an open-loop response supplements the PI control. The magnitude of this open-loop action is determined by the rate at which the operating point approaches surge (rate of change dS_S/dt). This method significantly enhances surge prevention.

If a surge event does occur, a third level of control is triggered to break the surge cycle. Crossing the Safety On® Line (SOL) initiates a system alarm and introduces an additional surge margin. Once normal operating conditions are restored, the antisurge controller must be manually reset to clear the Safety On® alarm and reinstate the normal surge margin.

In the fourth level of control, a rapid shift in the operating point—specifically when the rate of change (dS_S/dt) surpasses a defined threshold—activates a derivative response. This response dynamically adjusts the surge control line, shifting it closer to the operating point. The result is an increased surge margin and earlier engagement of the predictive control (PI).

The key benefit of this approach lies in its selectivity: the surge margin is enhanced only during significant transient conditions, minimizing disruption to the process. Compared to open-loop responses, the PI control remains more stable and less intrusive.

Once triggered, the derivative response gradually tapers off, resetting itself until the next transient event occurs.

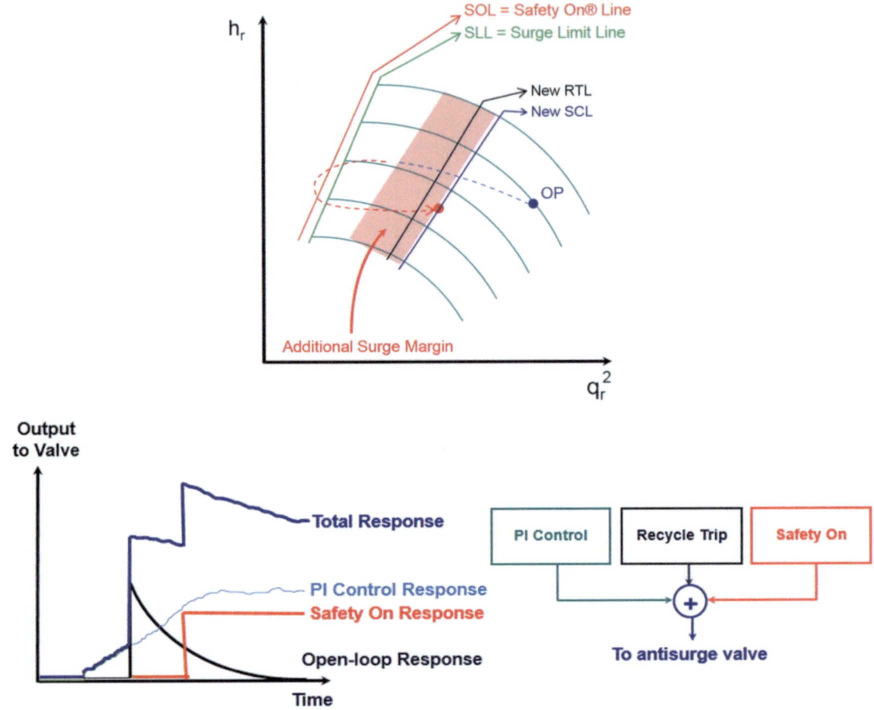

Fig. 4.11 Detection of a surge, triggering a Safety On® response (Image courtesy of Compressor Controls Corporation I Honeywell, Des Moines, IA)

Fig. 4.12 Derivative response, moving the surge control line toward the operating point (Image courtesy of Compressor Controls Corporation I Honeywell, Des Moines, IA)

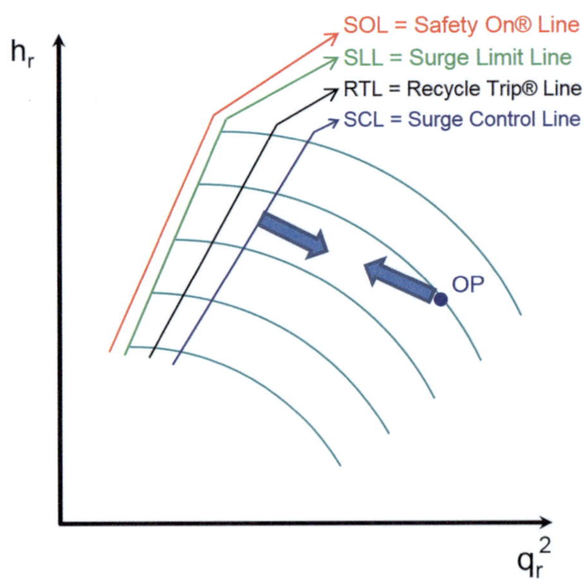

4.7 Performance Control

The primary function of a performance controller is to regulate a process variable, keeping it aligned with its designated set point. The controller's output determines the set point for compressor throughput, enabling precise control that can significantly enhance plant productivity.

Common process variables include:

- Suction pressure.
- Discharge pressure.
- Compression ratio.
- Volumetric or mass flow rate.

Figure 4.13 illustrates several methods for managing throughput, such as suction throttle valves, inlet guide vanes, and variable-speed drives. Selecting the appropriate method requires careful consideration of factors like

- Electrical energy costs.
- Availability and cost of steam or fuel gas.
- Overall process efficiency.

While constant-speed electric motor drives paired with throttling are the most straightforward option for startup and basic operation, they tend to be less energy-efficient compared to more advanced alternatives.

The throughput control method has a significant impact on the antisurge control system. For instance, if inlet guide vanes are used, it may be necessary to feed the vane position signal back into the antisurge system to implement a tailored characterizer

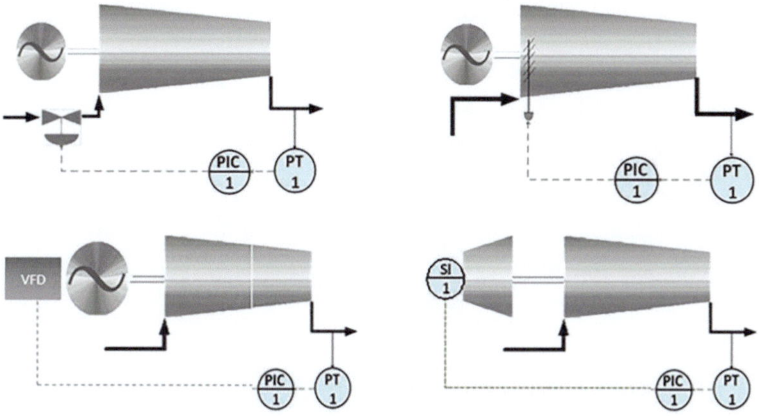

Fig. 4.13 Compressor throughput options to maintain a desired discharge pressure. Abbreviations: PIC = pressure-indicating controller, PT = pressure transmitter, SI = turbine drive, VFD = variable-frequency drive or adjustable-frequency drive (Image courtesy of Compressor Controls Corporation I Honeywell, Des Moines, IA)

and adjust the surge control line (SCL) accordingly. In cases where turbine speed or a suction throttle valve governs throughput, the antisurge control must be closely coordinated to prevent conflicts between control loops.

Regardless of the approach, throughput and antisurge controls are inherently linked, as they both affect key process variables and ultimately determine the compressor's operating point on its performance map. A tightly integrated control strategy enhances compressor stability, minimizes the risk of exceeding design limits, improves surge protection, sharpens process variable regulation (precise control), and ensures smoother, more consistent automated startup and shutdown sequences or procedures.

4.8 Loop Decoupling

Control loops are considered coupled when the action taken by one loop influences the process variable of another. In multisection compressors with dedicated recycle loops—such as the configuration shown in Fig. 4.14—this coupling is particularly strong and often described as interactive. For example, opening a recycle valve in one section of a two-section compressor may optimize the operating point for that section, but it can simultaneously disrupt the operating point in the other section. This mutual influence can lead to instability between the two recycle loops.

To mitigate this interaction, feed-forward loop decoupling between the recycle controllers (user-defined indicating controllers, or UICs) is critical. By proactively adjusting the other recycle valve, feed-forward control helps maintain balance and significantly enhances the stability of each loop.

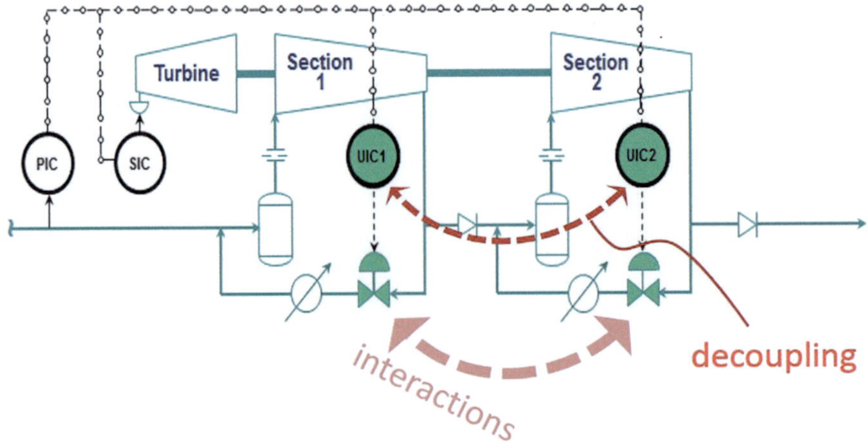

Fig. 4.14 Loop decoupling between recycle controllers. Abbreviations: PIC = performance-indicating controller, SIC = speed-indicating controller, UIC = antisurge controller (Image courtesy of Compressor Controls Corporation I Honeywell, Des Moines, IA)

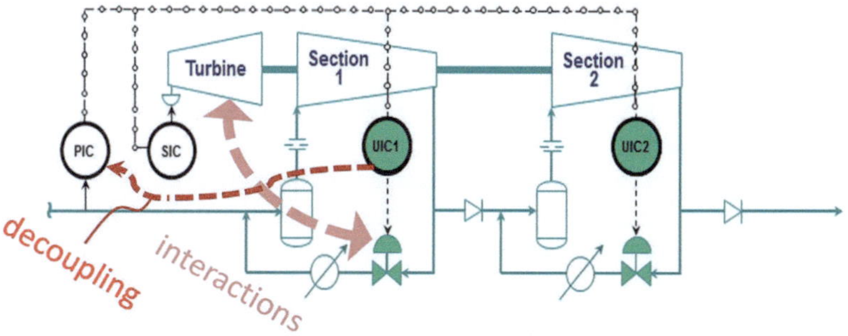

Fig. 4.15 Loop decoupling between the recycle controller and the performance controller. Abbreviations: PIC = performance-indicating controller, SIC = speed-indicating controller, UIC = anti-surge controller (Image courtesy of Compressor Controls Corporation I Honeywell, Des Moines, IA)

Figure 4.15 illustrates the decoupling of control loops between the recycle controller (UIC) and the performance-indicating controller (PIC). As previously discussed, both loops significantly influence the position of the operating point on the compressor map, indicating a strong interdependence. In this example, the performance controller provides a set point to the steam turbine governor—also known as the speed-indicating controller (SIC)—to regulate suction pressure by adjusting turbine speed. The dynamic interaction between the UIC and PIC originates at the surge control line (SCL), where it can induce oscillations and potentially push the operating point into surge conditions before stabilization occurs. Implementing feed-forward control mitigates these oscillations and strengthens the UIC's effectiveness in surge prevention.

4.9 Load Sharing

Parallel compressor operation is commonly employed to meet the varying demands of plant processes. This configuration offers operational flexibility, allowing maintenance on individual units without disrupting overall production or compromising the ability to handle wide process flow ranges. Strategic cycling of compressors—bringing units online or offline as needed—can yield substantial energy savings.

However, achieving efficient and reliable control of parallel compressors to meet process objectives presents a complex challenge. The key control goals for such networks include:

- Accurate regulation of the primary process variable to ensure maximum production output.
- Ensuring all compressor trains operate within their design limits to maintain equipment integrity.

- Distributing control responsibilities optimally across the compressor network, while simultaneously:
  - Reducing the risk of equipment trips caused by surge, vibration, overspeed, or process disruptions.
  - Lowering overall energy consumption across the network.
  - Preventing process disturbances during transitions when compressors are brought online or taken offline.

A variety of control strategies exist for managing parallel compressors, though their effectiveness in meeting these objectives can vary significantly. Common approaches include:

- Base-load and swing machine configuration: Utilizing both base-load and swing machines can lead to substantial gas recycling through the swing machine. In this setup, only the swing machine responds to process disturbances, which heightens the risk of operational trips and accelerates wear and maintenance requirements for the swing unit.
- Independent performance controllers per train: Assigning separate performance controllers to each compressor train can cause flow oscillations when multiple controllers attempt to regulate a shared process parameter. To mitigate these oscillations, controller tuning is often dampened or control loops are placed in manual mode. However, these adjustments significantly weaken control over the primary process variable and increase the likelihood of trips during disturbances.
- Master controller for load balancing: A master controller can be implemented to regulate the primary process variable by setting targets for each train's performance controller (also known as load-sharing controllers). As illustrated in Fig. 4.16, this control structure ensures that both compressor trains operate in unison toward the main process objective. However, this method assumes identical compressor performance maps, which is rarely the case due to variations in compressor design i.e., as-built dimensions are not identical and differences in suction/discharge piping. Consequently, flow distribution becomes uneven, and under low-flow conditions, some compressors may enter recycle mode while others do not.
- Master controller for distance-to-surge balancing: An alternative control strategy involves using a master controller that balances load based on each compressor's distance to the surge control line (SCL), as determined by its antisurge controller. Figure 4.17 presents a block diagram of this coordinated, deviation-based load-sharing approach involving the master, load-sharing, and antisurge controllers. By equalizing the relative surge margin across all compressors, this method ensures optimal flow distribution, minimizes total recycle, and enhances overall efficiency. It also enhances antisurge protection, as all compressors begin to open their recycle valves simultaneously during a disturbance. This approach effectively accounts for differences in compressor capacities and performance characteristics, as shown in Fig. 4.18.

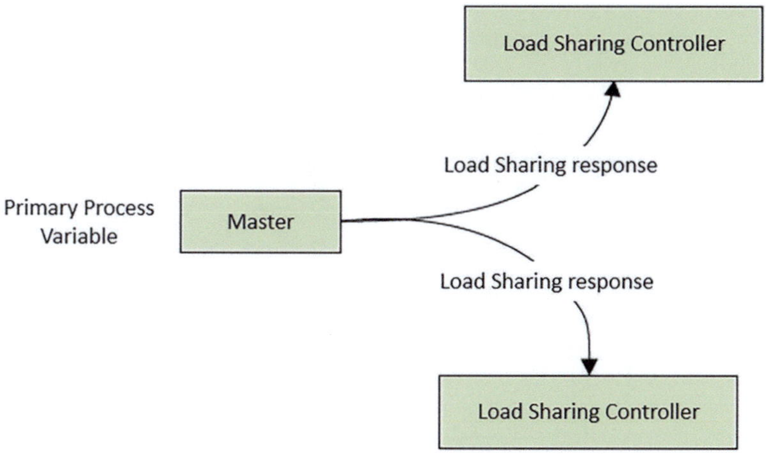

Fig. 4.16 Load-sharing response (Image courtesy of Compressor Controls Corporation I Honeywell, Des Moines, IA)

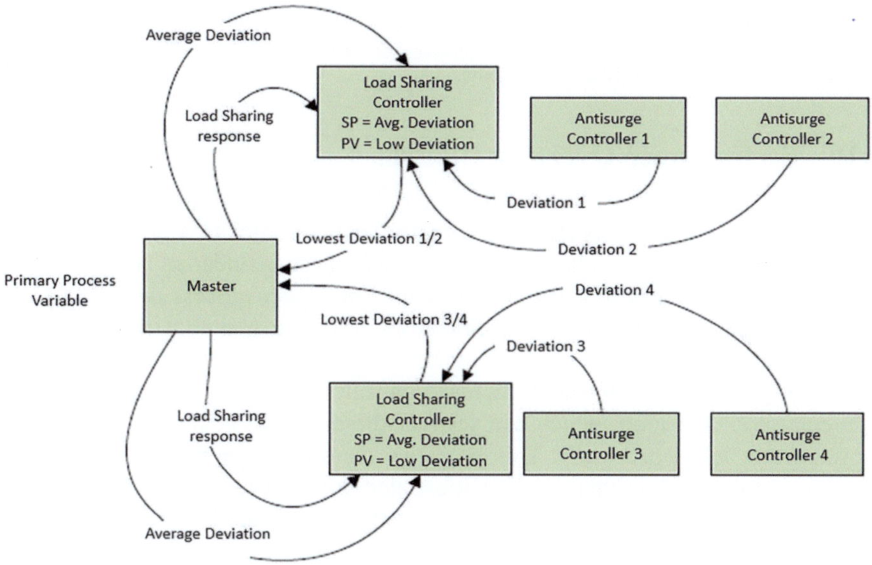

Fig. 4.17 Load-sharing response based on distance to surge (Image courtesy of Compressor Controls Corporation I Honeywell, Des Moines, IA)

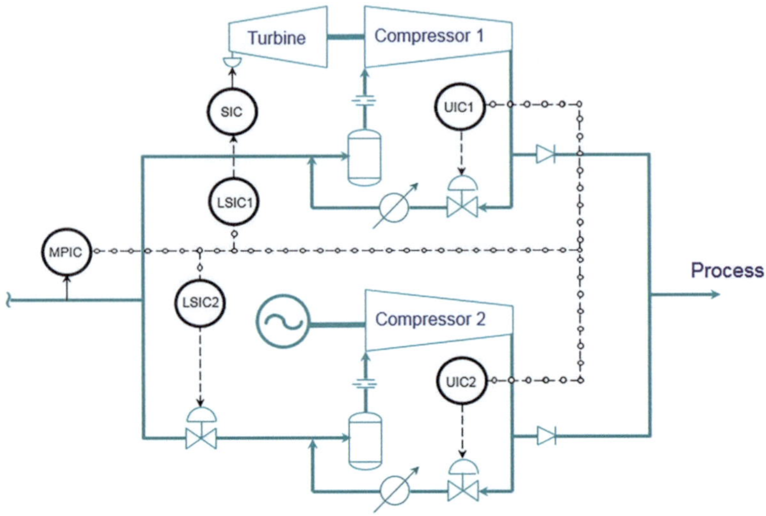

Fig. 4.18 Deviation-based load sharing for compressors of different types. Abbreviations: LSIC = load-sharing compressor performance controller, MPIC = master pressure/flow integral controller, SIC = speed-indicating controller, UIC = antisurge controller (Image courtesy of Compressor Controls Corporation I Honeywell, Des Moines, IA)

4.10 Performance Override Control

The main function of the performance controller is to regulate a process variable within tightly defined boundaries. Operating near these boundaries helps optimize production output. However, surpassing them can trigger process disruptions and push equipment beyond its design specifications, leading to potential trips and flaring events.

To mitigate this risk, the performance controller includes a performance override control (POC) loop, which can instruct the antisurge controller to open the recycle valve when necessary. This proactive measure helps prevent the process variable from exceeding its limits. As illustrated in Fig. 4.19, the POC mechanism enables rapid engagement of the antisurge controller, facilitating quicker recovery during process disturbances.

By enhancing the system's responsiveness to sudden process fluctuations, the POC loop significantly reduces the likelihood of excursions and trips, ensuring safer and more stable operation even under demanding conditions.

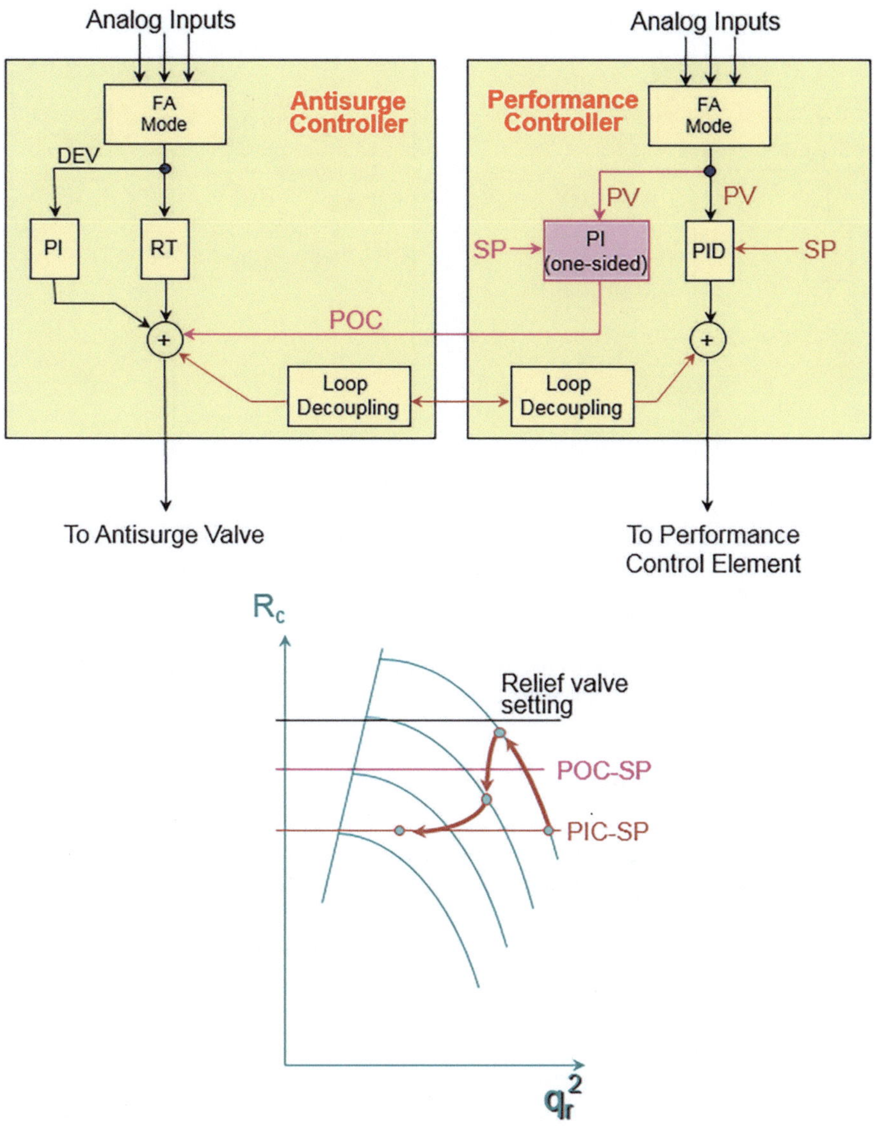

Fig. 4.19 Performance override control. Abbreviations: PIC-SP = performance-indicating controller set point, POC-SP = performance override control set point (Image courtesy of Compressor Controls Corporation I Honeywell, Des Moines, IA)

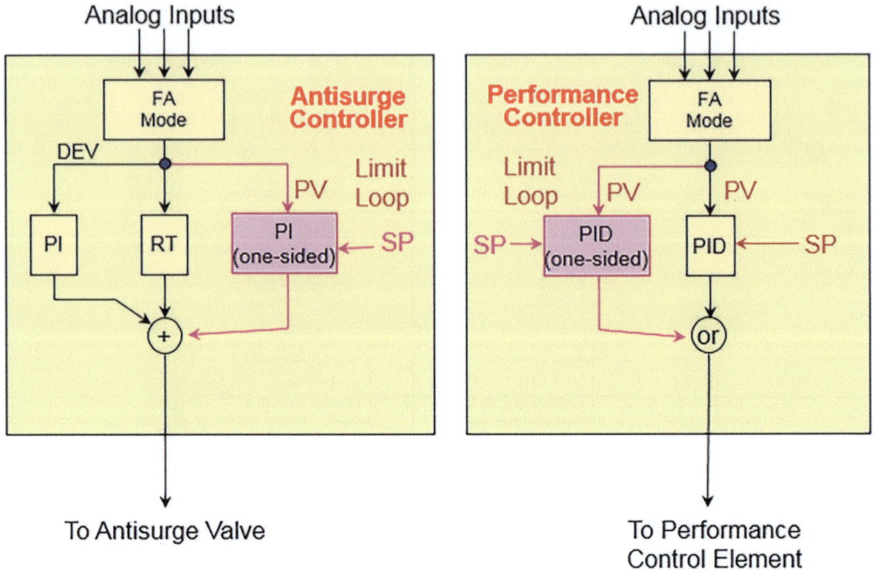

Fig. 4.20 Limiting loops in antisurge and performance controllers. Abbreviations: DEV = deviation, FA mode = fault-avoidance mode in performance override control, PI = proportional–integral response, PID = proportional–integral–derivative response, PV = low deviation, SP = average deviation (Image courtesy of Compressor Controls Corporation I Honeywell, Des Moines, IA)

4.11 Limiting Loops

In addition to their core control functions, both antisurge and performance controllers can incorporate limiting control loops to manage their analog input signals. These loops typically utilize a standard proportional–integral (PI) algorithm to correct any deviations from the set point. In the case of the antisurge controller, it can adjust the recycle or blowoff rates to mitigate excessive discharge pressure or insufficient suction pressure. Similarly, the performance controller's limiting loop can override its output to constrain key process variables such as mass flow rate, suction pressure, discharge pressure, and compression ratio. Additionally, it can be configured to limit the compressor driver's power. Figure 4.20 illustrates the integration of limiting loops within both controller types.

4.12 Fallback Modes

Fallback modes are designed to improve system reliability by preventing unnecessary shutdowns. When specific data become unavailable, these modes offer alternative methods for calculating critical variables, ensuring continued operation. Typically, fallback modes are activated by transmitter failures and enable safe, conservative control until the issue is resolved.

In performance controller applications, a transmitter failure may lead to either a default output value or a fixed output state. For antisurge controllers, the calculation of reduced polytropic head—used to determine the distance to surge—relies on flow, temperature, and pressure transmitters. If any of these transmitters fail, an alarm is triggered, and the system switches to a fallback mode based on the following logic:

- Temperature transmitter failure: Estimate surge proximity using only pressure and flow data.
- Pressure transmitter failure: Use a predefined minimum flow to assess surge proximity.
- Flow transmitter failure: Adjust the antisurge valve to either a configured minimum opening or maintain its current position—whichever is open the most.

4.13 Recycle Piping Considerations

To ensure optimal performance and protection of the compressor, the surrounding piping system should be engineered to minimize the volume within the discharge line—specifically between the compressor discharge flange, the antisurge valve inlet, and the check (nonreturn) valve. Excessive gas volume in this section acts as an energy reservoir, which delays the recycle response and heightens the potential for compressor damage during surge events.

The extent of damage incurred during a surge is often directly related to the volume of gas present in the discharge line. Simply put, the larger the volume, the longer it takes for the gas to be recycled through the compressor, increasing the likelihood of harm to its internal components.

The most effective strategy to mitigate or halt a surge cycle is to quickly lower the discharge pressure. With that principle in mind, the following guidelines are recommended for configuring a recycle piping system:

- To reduce the volume in the compressor discharge line, position the recycle takeoff as near as possible to the compressor discharge flange. Install a check valve immediately downstream of the recycle takeoff, and place the recycle valve in close proximity to this point. Ideally, the recycle valve should be situated within five pipe diameters of the takeoff location. Figure 4.21 illustrates the section of discharge piping where volume minimization is critical.
- Evaluate the noise and turbulence of gas flow at the recycle valve before finalizing its placement within the recycle loop to ensure stable operation.
- Use damped-opening check valves to prevent abrupt movements. If a check valve is positioned upstream of the recycle valve takeoff, it should be of the nonslam type to avoid pressure shocks.
- To reduce discharge volume, it's preferable to install recycle coolers downstream of the antisurge valve. This setup supports continuous recycle operation when

needed. While some facilities opt to use compressor discharge coolers to cut costs, this approach significantly increases discharge volume and compromises surge protection.

- Prior to installing a cooler upstream of the antisurge valve, verify hydrocarbon and water dew points to avoid the risk of two-phase flow through the valve, which can erode the valve seat.
- Liquid condensate must be removed before reaching the compressor suction. This is achieved by installing a suction knockout drum between the recycle cooler and compressor suction. The drum should be properly sized to accommodate both process flow and the combined cold and hot recycle streams, especially if a hot-gas bypass valve (HGBV) is present.
- As illustrated in Fig. 4.22, interstage coolers are occasionally repurposed as recycle coolers in multisection compressor systems to reduce capital and installation costs. However, this configuration can severely impair surge protection for the low-pressure section—particularly when the high-pressure stream enters recycle mode and causes a rapid rise in discharge pressure on the low-pressure side. Such setups may result in adverse control interactions.
- For multisection compressors and compressor trains with multiple casings, it's recommended to implement separate recycle loops for each section. This approach enhances surge protection and operational control. An optimized piping configuration is illustrated in Fig. 4.23.
- The recycle stream should be introduced downstream of the suction throttle valve, inside the suction block valve, and positioned upstream of both the inlet flow element and the suction knockout drum. For compressors driven by electric motors, it may be beneficial to allow recycle flow to enter upstream of the suction throttle valve to help reduce startup load.
- If the suction throttle valve is part of the recycle loop, it must include a minimum stop setting to guarantee sufficient flow through the compressor during recycle operation.
- Any restrictions in the recycle line—such as those illustrated in Fig. 4.24—can impair the effectiveness of antisurge control and should be carefully evaluated during design.

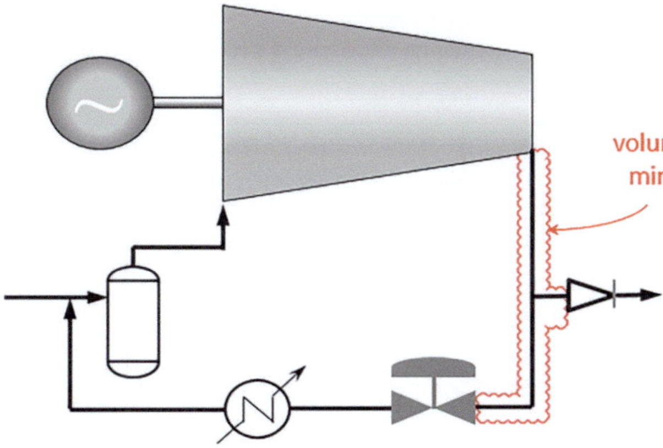

Fig. 4.21 Discharge volume (Image courtesy of Compressor Controls Corporation I Honeywell, Des Moines, IA)

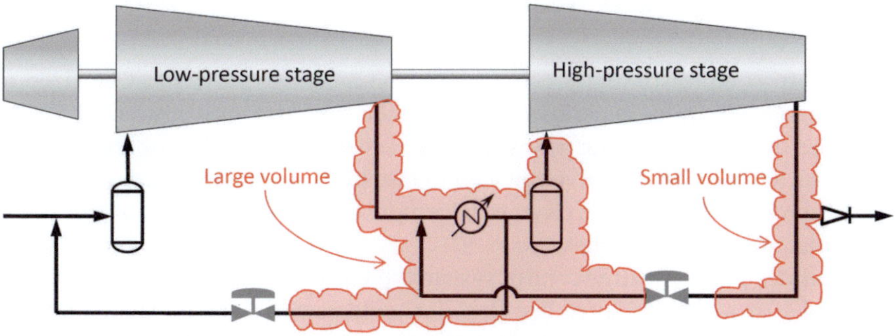

Fig. 4.22 Recycle cooling using a common interstage cooler (Image courtesy of Compressor Controls Corporation I Honeywell, Des Moines, IA)

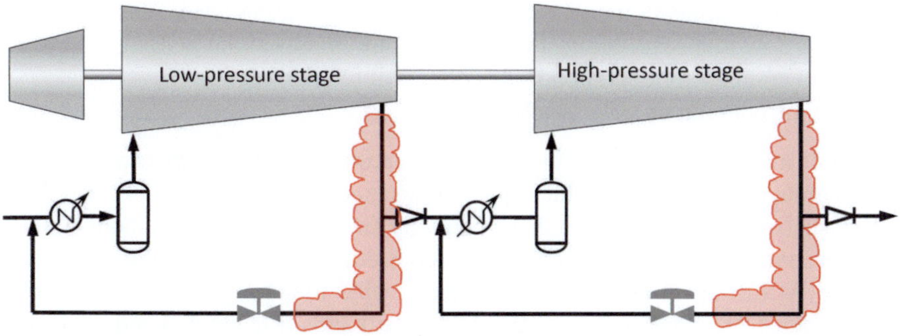

Fig. 4.23 Improved piping design (Image courtesy of Compressor Controls Corporation I Honeywell, Des Moines, IA)

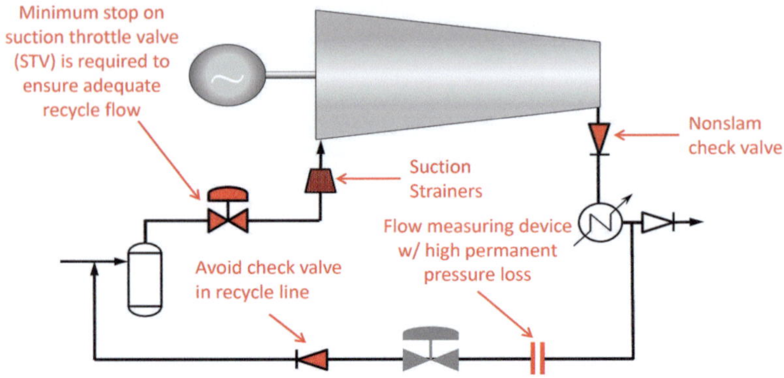

Fig. 4.24 Recycle path restrictions (Image courtesy of Compressor Controls Corporation I Honeywell, Des Moines, IA)

4.14 Antisurge Valve Requirements

To maintain compressor and process safety, the antisurge valve must be sized to prevent surge—even in scenarios where all other flow paths are obstructed. At the same time, the valve must offer precise and responsive control under normal operating conditions. Oversizing the valve can lead to instability, as small control inputs may cause disproportionately large changes in compressor flow. Therefore, proper sizing is essential to avoid such issues.

A well-designed antisurge control system typically operates with a surge margin of around 10% above the surge line flow. It should ensure smooth regulation at the surge control line (SCL) and have the additional capacity to rapidly open the valve in response to sudden flow reductions caused by process disturbances.

Key sizing objectives for the antisurge valve are:

- Ensure surge protection under the most extreme operating conditions.
- Deliver flow rates exceeding steady-state requirements at the SCL.
- Prevent compressor operation in the stonewall (choke) region.
- Avoid introducing process instability.

In addition to proper sizing, the valve's actuation system must be capable of fast and accurate response to meet these performance goals.

Industry experience indicates that a correctly sized antisurge valve should deliver 100% of the compressor's minimum flow requirement at approximately 50% valve opening. At full opening, the valve should be able to supply twice the minimum steady-state flow. Applying a $\pm$ 10% tolerance yields a recommended maximum valve flow capacity (C_{Vmax}) between 1.8 and 2.2 times the flow at the surge point.

Figure 4.25 shows valve C_{V} curves for different values of C_{V}. A valve with a C_{V} value of 50 would provide the minimum amount of flow. For effective antisurge control, the recommended C_{V} should lie within the range of 90–110, to ensure sufficient flow flexibility and responsiveness.

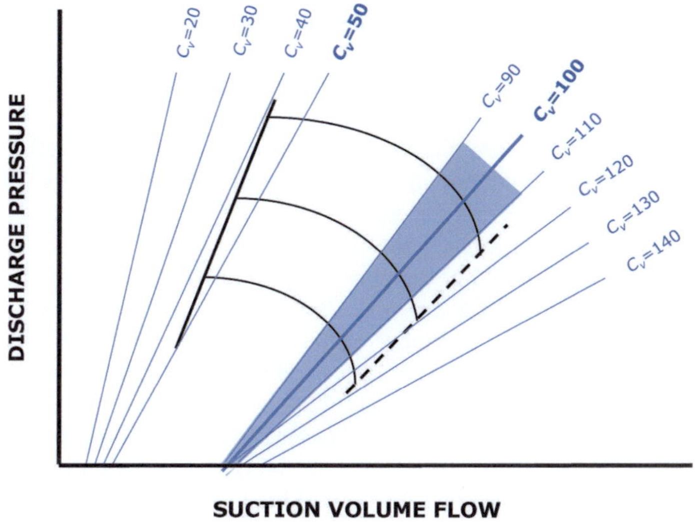

Fig. 4.25 Antisurge valve C_V curves compared with a conventional compressor map (Image courtesy of Compressor Controls Corporation I Honeywell, Des Moines, IA)

Figure 4.26 illustrates the sizing limits for the antisurge valve, with capacity demands at sub-minimum speeds established through dynamic simulation.

Linear trim characteristics are generally favored for antisurge valves. While larger valves may enhance pressure equalization due to their faster response, they often lack precise control during throttling operations in process control. When the recycle valve

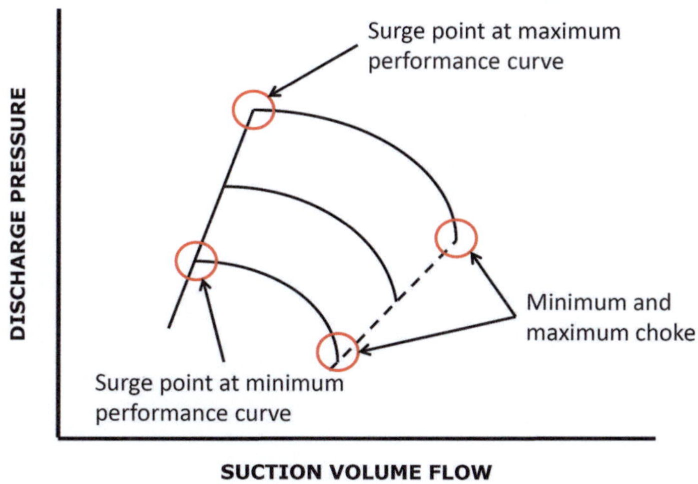

Fig. 4.26 Antisurge valve sizing boundaries (Image courtesy of Compressor Controls Corporation I Honeywell, Des Moines, IA)

is required to function in throttling mode, incorporating a smaller parallel valve may be necessary to maintain accurate control at low recycle flow rates. In such configurations, the antisurge controller can be configured with split-range output to manage both valves effectively.

Figure 4.27 illustrates an antisurge control system designed to deliver smooth and accurate valve positioning in alignment with the controller's output. To ensure stable operation, valve overshoot should remain within approximately 10% of the controller's step change, whether in the opening or closing direction.

The antisurge valve should be capable of opening from fully closed to at least 95% open within 2 s or less. Closing time should not exceed 2–3 times the opening duration. Valve overshoot must remain within 10% of the controller's step change in both directions, and directional hysteresis should be kept below 0.2 s. These performance criteria are illustrated in Figs. 4.28 and 4.29.

To prevent mechanical impact at the end of the stroke, a "cushioning" mechanism should be incorporated into the actuator system—typically within the final 5% of travel—to avoid damage from the actuator striking its mechanical stops.

Noise levels from the antisurge valve should not exceed 85 dBA at a distance of 3 ft (1 m). In cases where noise reduction is necessary, noise abatement trims may be employed. Whisper trims are effective but require clean process gas to prevent

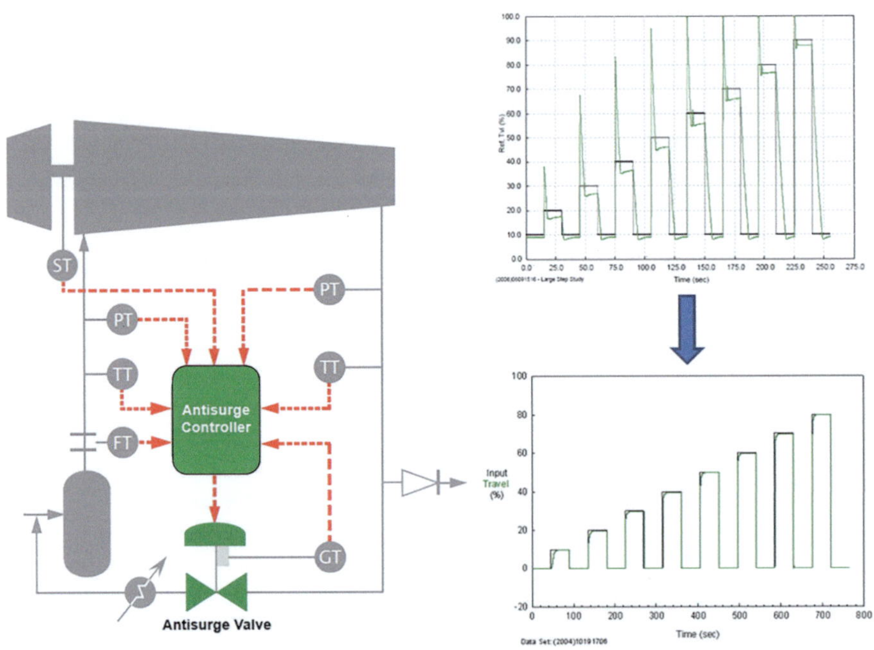

Fig. 4.27 Recycle valve response. Abbreviations: FT = flow transmitter, GT = a certain "greater than" parameter in the recycle valve response, PT = pressure transmitter, Ref. T_M = step change, ST = suction temperature, TT = additional temperature transmitter (Image courtesy of Compressor Controls Corporation I Honeywell, Des Moines, IA)

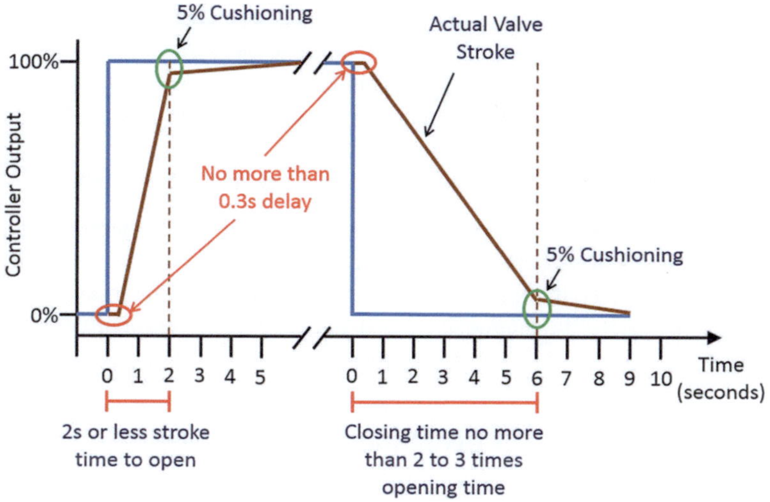

Fig. 4.28 Antisurge valve dynamic requirements (Image courtesy of Compressor Controls Corporation I Honeywell, Des Moines, IA)

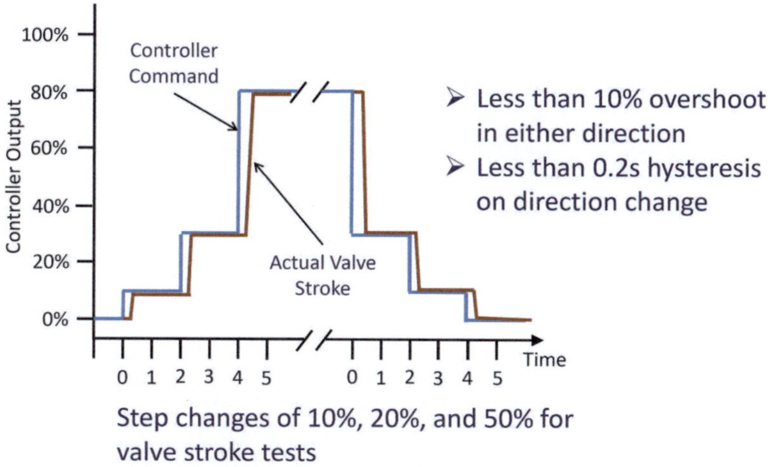

Fig. 4.29 Antisurge valve stroking accuracy requirements (Image courtesy of Compressor Controls Corporation I Honeywell, Des Moines, IA)

clogging. If the gas contains contaminants, downstream diffusion trim or startup trash trim should be considered as alternatives [1].

The actuation system for the antisurge control valve typically comprises the following key components:

- Digital positioner capable of handling both slow and rapid command signals from the antisurge controller.

- Volume boosters that enhance the responsiveness of the actuator's motive fluid, facilitating swift valve movement in both opening and closing directions.
- Quick-release mechanism (such as a solenoid valve) designed to rapidly open the antisurge valve upon receiving an emergency shutdown (ESD) signal, even if the signal originates outside the antisurge controller.
- Valve position feedback, which plays a crucial role in system reliability.

Valve position feedback is especially vital, as it enables the generation of alarms when discrepancies arise between the valve's actual position and the controller's command. Additionally, tracking valve position relative to controller output provides valuable insight during the diagnosis of critical events or system malfunctions.

4.15 Antisurge Control Instrumentation

To prevent compressor surge, the control system must continuously and accurately assess how close the operating conditions are to surge in real time. This requires a minimum set of process measurements, including compressor flow rate, suction pressure, and discharge pressure. Additionally, monitoring changes in gas composition, as well as suction and discharge temperatures, is essential.

If inlet guide vanes are used to regulate compressor throughput, their position must be tracked to reflect changes in flow geometry resulting from vane angle adjustments. Measurements such as power or torque and compressor speed are often incorporated into the antisurge controller's logic during startup and shutdown sequences. For diagnostic purposes, especially in high-speed trending and data recording, compressor speed and antisurge valve position feedback are critical parameters.

4.15.1 Flow Element

A flow measurement device inherently causes a permanent pressure drop within the system. Because of this, venturi flow meters—known for their low-pressure loss—are generally favored over orifice plates. Orifice plates typically result in a permanent loss of approximately 50% of the flow signal, whereas venturi elements lose only about 5%.

When selecting a flow meter, several key factors should be considered: response time (ideally between 100 and 150 ms), repeatability, signal-to-noise ratio, pipe diameter, and the availability of straight piping runs. In situations where straight runs are limited, a multihole orifice plate may be used to mitigate flow disturbances. Reputable compressor manufacturers recommend a minimum of eight pipe diameters upstream and two diameters downstream for venturi meters. In contrast, single-hole orifice plates require more extensive straight runs—at least 20 diameters upstream and eight diameters downstream.

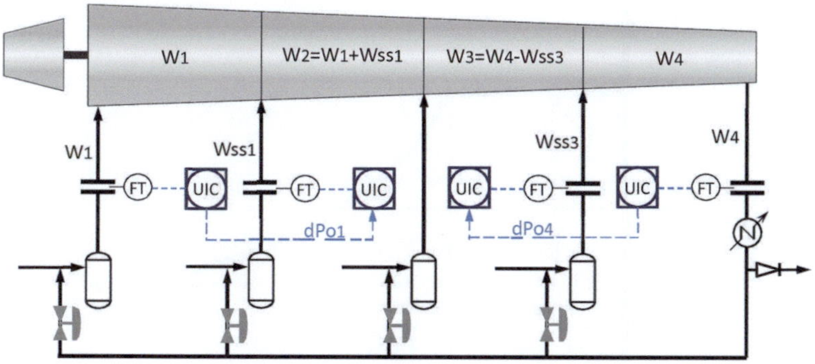

Fig. 4.30 Flow element location. Abbreviations: FT = flow transmitter, UIC = antisurge controller (Image courtesy of Compressor Controls Corporation I Honeywell, Des Moines, IA)

Beyond piping layout, the flow element's accuracy, linearity, and noise performance are critical. Incorrect installation can compromise the antisurge control system. The flow element must capture the total flow rate of the compressor, meaning no interstage headers or flare-line takeoffs should be placed between the compressor and the flow meter.

The flow meter must be properly sized to cover the compressor's full operating range. Under maximum flow conditions, the differential pressure should exceed 10 in. water column WC (0.36 psi or 2.5 kPA).

Direct measurement of the flow is not always possible, as in the case of the multisection compressor shown in Fig. 4.30. (Note that, in this figure, W_1–W_4 represent the mass flow rates of the corresponding compressor sections; W_{ss1} and W_{ss3} are the mass flow rates of the sidestreams between sections 1 and 2 and sections 3 and 4, respectively; and dP_{o1} and dP_{o4} are the differential pressures between sections 1 and 2 and sections 3 and 4, respectively.) In these situations, the preferred method involves combining the flow from adjacent compressor sections with any sidestream contributions and then measuring the flow at both the suction and discharge points. This approach assumes uniform gas composition across all inlet streams. When pressure and temperature conditions are consistent, the flows can be directly summed. However, if there are variations in pressure or temperature, compensation must be made to ensure accurate flow calculations.

4.15.2 Transmitters

In selecting transmitters for antisurge control systems, response speed and reliability are key considerations. Pressure transmitters are expected to respond within 400 ms or less, while differential pressure transmitters should achieve response times of 150 ms or less. Figure 4.31 illustrates these response times as plotted along the transmitter output signal curve.

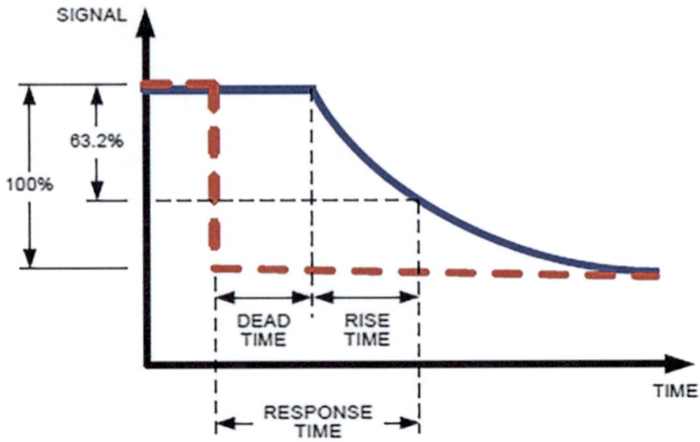

Fig. 4.31 Transmitter response times (Image courtesy of Compressor Controls Corporation I Honeywell, Des Moines, IA)

Flow measurement using differential pressure transmitters requires a range of 0–100 in. of water column WC (3.6 psi or 25 kPA), with a minimum detectable signal of 20 in. WC (0.72 psi or 5.0 kPA) at the lowest surge flow condition. If the flow transmitter is positioned upstream of the throttle valves, an additional pressure transmitter must be installed to accurately adjust the antisurge flow calculation (refer to Fig. 4.32). If the flow transmitter is placed downstream of a discharge cooler, a supplementary temperature measurement is necessary to compensate for variations in the flow calculation (refer to Fig. 4.33).

4.16 Hot-Gas Bypass Valve

A hot-gas bypass valve (HGBV) should be considered in compressor systems where the volume between the discharge flange, discharge check valve, and recycle valve is insufficient to prevent surge during an emergency shutdown (ESD). In such cases, the system's reaction time may be too short for the standard recycle system to respond effectively. Dynamic simulation studies or operational experience often reveal that abrupt changes—such as a rapid drop in compressor speed or sudden shifts in pressure ratio or differential pressure—can exceed the response capabilities of conventional recycle controls.

For ESD scenarios, the HGBV must feature a stroke time of 0.5 s or less and exhibit a linear opening characteristic. Valve sizing should be determined through process simulation.

HGBVs are particularly essential for two-section compressors with back-to-back impeller configurations, where they help prevent thrust overload during ESD and restart. In these setups, the rate at which the compressor casing depressurizes is

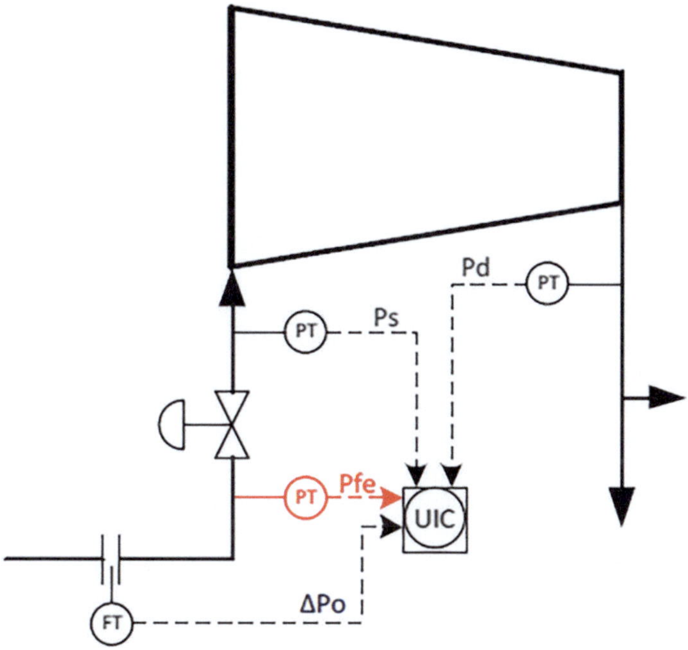

Fig. 4.32 Additional pressure transmitter for differential pressure compensation. Abbreviations: FT = flow transmitter, ΔP_o = differential pressure across a flow orifice, P_d = discharge pressure, Pfe = differential pressure compensation provided by the additional pressure transmitter, P_s = suction pressure, PT = pressure transmitter, UIC = antisurge controller (Image courtesy of Compressor Controls Corporation I Honeywell, Des Moines, IA)

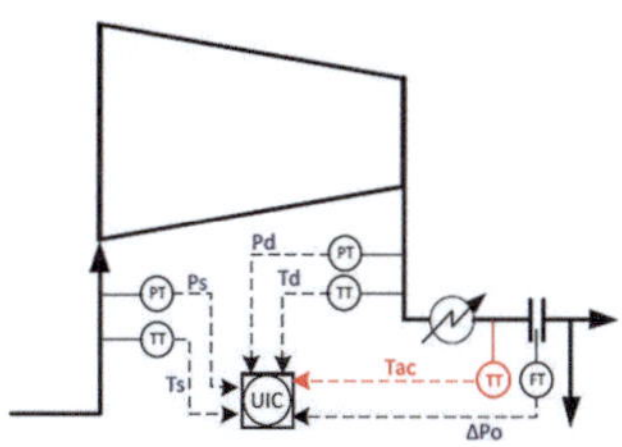

Fig. 4.33 Additional temperature transmitter for flow measurement compensation. Abbreviations: FT = flow transmitter, ΔP_o = differential pressure across a flow orifice, P_d = discharge pressure, P_s = suction pressure, PT = pressure transmitter, Tac = flow measurement compensation provided by the additional temperature transmitter, T_d = discharge temperature, T_s = suction temperature, TT = temperature transmitter, UIC = antisurge controller (Image courtesy of Compressor Controls Corporation I Honeywell, Des Moines, IA)

critical to avoiding excessive thrust forces during shutdown from full or partial load, as well as during surge events.

Beyond ESD protection, HGBVs are commonly integrated into antisurge control strategies. A secondary benefit is their ability to prevent liquid formation at the compressor suction caused by frequent or continuous cold recycle operation. If both the HGBV and cold recycle valve are used in tandem to mitigate surge, their combined maximum flow coefficient combined C_{Vmax} must be considered during sizing. Oversizing these valves increases the risk of operating the compressor under choke flow conditions.

The HGBV return line should be routed downstream of the suction throttle valve, within the suction block valve, and upstream of both the inlet flow element and the suction knockout drum.

For initial design, the HGBV loop should be sized to handle 100% of the recycle flow. Final sizing of the line and valve is typically confirmed after completing the dynamic simulation study. To avoid costly redesigns, the HGBV line should be incorporated early in the engineering phase and included in the piping model. Final verification of line size and valve dimensions must follow the simulation results.

HGBVs used in general antisurge control are modulating in nature, while those intended for ESD applications are strictly quick-acting—either fully open or fully closed.

Figure 4.34 illustrates the typical layout of a cold recycle loop combined with an HGBV loop for a centrifugal compressor system.

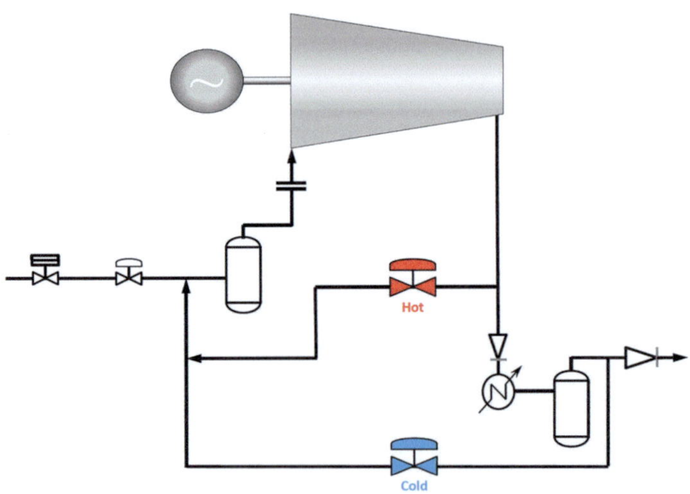

Fig. 4.34 Cold recycle loop and HGBV loop for a centrifugal compressor (Image courtesy of Compressor Controls Corporation I Honeywell, Des Moines, IA)

4.17 Antisurge Systems for Multisection Compressors

The schematics in Figs. 4.35, 4.36 and 4.37 illustrate simplified antisurge systems commonly used in critical compressor trains in ethylene and LNG facilities. In the compressor train shown in Fig. 4.35, maintaining accurate suction pressure control is vital for optimizing furnace output and preventing process disruptions. To achieve this, a coordinated approach involving antisurge, performance, and speed control is necessary to keep suction pressure within its target range.

Compressor stages 1, 4, and 5 are particularly vulnerable to surge due to frequent furnace trips and process fouling. Therefore, surge protection is essential for these stages at a minimum. Stages 1 and 4 utilize a shared recycle valve managed by their respective antisurge controllers, with valve priority given to the stage operating nearest to surge conditions. Meanwhile, stages 2 and 3 are part of compressor section 2.

Figure 4.36 illustrates propane and mixed refrigerant trains operating under a balanced power configuration for LNG production. The use of identical gas turbine drivers is feasible in this setup. To optimize LNG output, precise control of suction pressure and compressor speed is critical. Moreover, integrated management of antisurge systems, performance parameters, and speed regulation is vital to prevent cascading failures that could result in a complete plant shutdown.

Figure 4.37 depicts two parallel end flash gas compressor trains equipped with load-sharing control systems. A master integrated controller (MIC) regulates suction header pressure by coordinating load distribution across the individual load-share controllers (LSICs). Each LSIC adjusts the speed of its respective compressor train according to its proximity to surge conditions. Surge margin data are provided by the antisurge controllers (UICs) at each compressor stage, and the minimum surge distance from each train is used to balance the system. This strategy enhances process efficiency by reducing recycle flow and enables both trains to effectively respond to fluctuations in process demand.

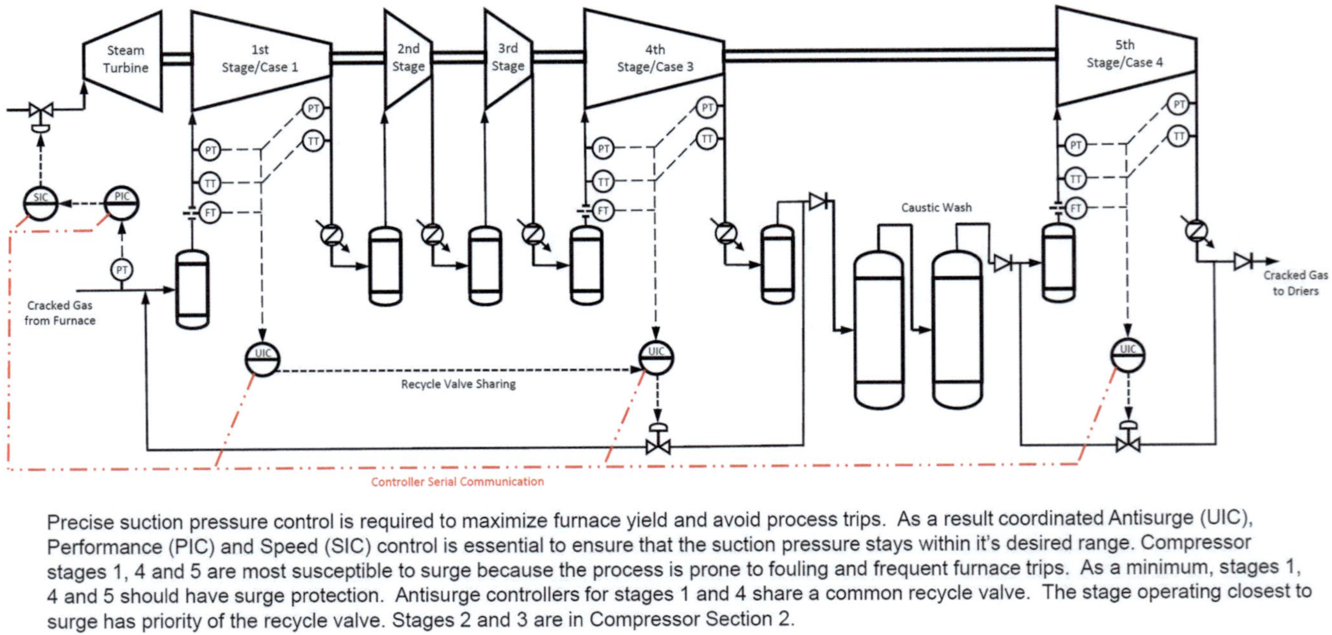

Fig. 4.35 Antisurge system for cracked gas compressor train. Abbreviations: FT = flow transmitter, PIC = performance-indicating controller, PT = pressure transmitter, SIC = speed-indicating controller, TT = temperature transmitter, UIC = antisurge controller (Image courtesy of Compressor Controls Corporation I Honeywell, Des Moines, IA)

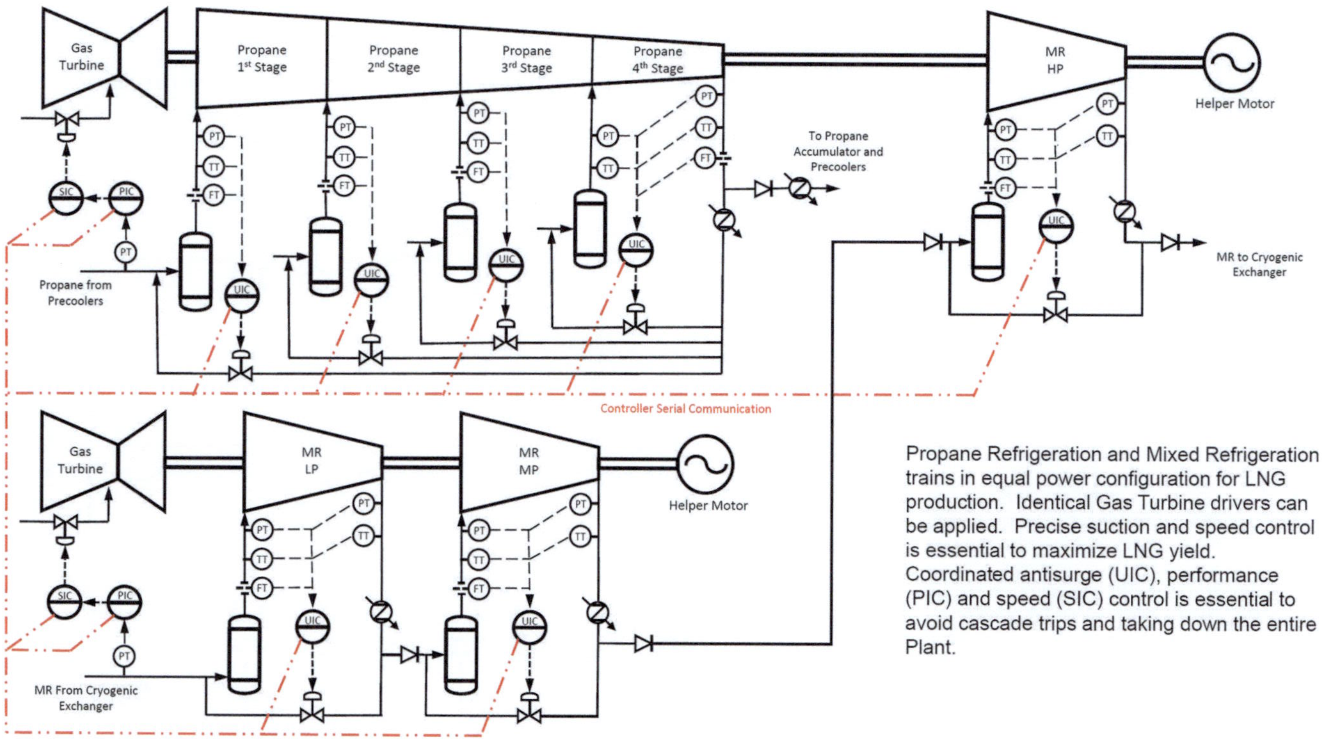

Fig. 4.36 Antisurge system for a propane and mixed refrigerant (MR) compressor train. Abbreviations: FT = flow transmitter, MR LP = mixed refrigerant low-pressure casing, MR HP = mixed refrigerant high-pressure casing, PIC = performance-indicating controller, PT = pressure transmitter, SIC = speed-indicating controller, TT = temperature transmitter, UIC = antisurge controller (Image courtesy of Compressor Controls Corporation I Honeywell, Des Moines, IA)

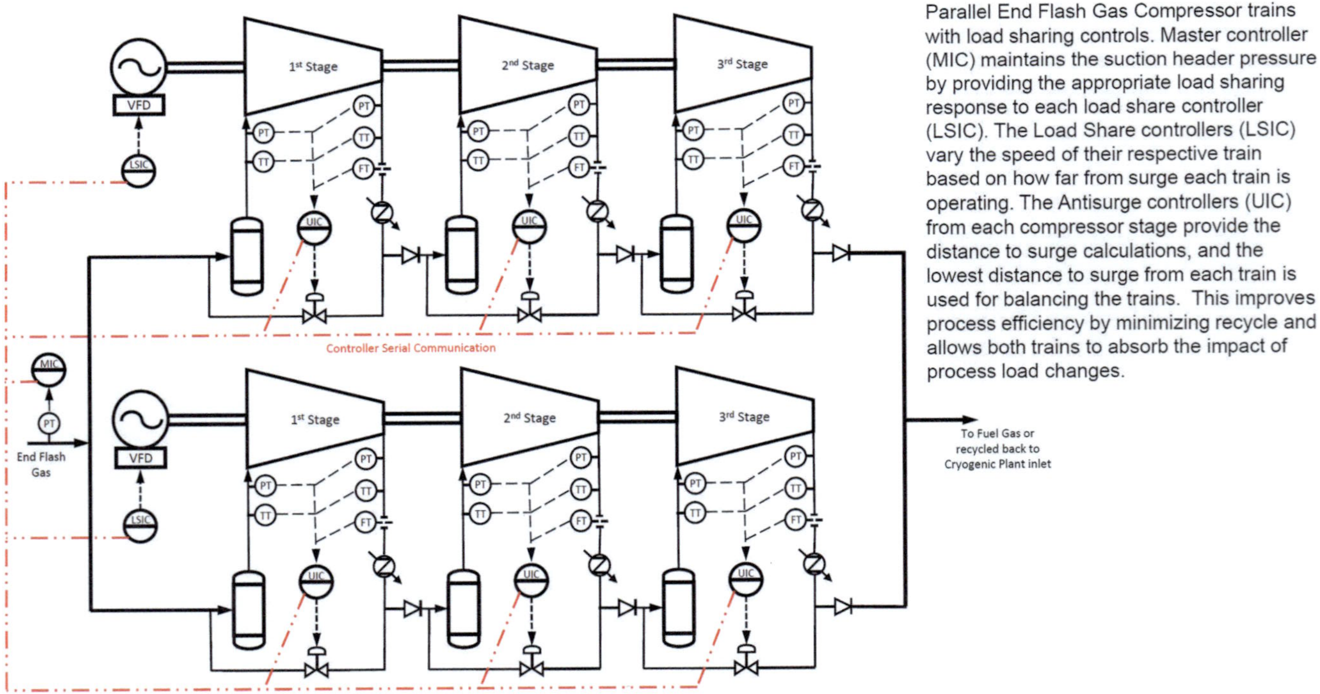

Fig. 4.37 Antisurge system for parallel end flash compressor trains. Abbreviations: FT = flow transmitter, LSIC = load-sharing compressor performance controller, MIC = motor control for regulation of compressor flow and speed, PIC = performance-indicating controller, PT = pressure transmitter, TT = temperature transmitter, UIC = antisurge controller, VFD = variable-frequency drive (Image courtesy of Compressor Controls Corporation I Honeywell, Des Moines, IA)

Acknowledgements All of the images/graphics/figures and some of the material in this chapter were provided by Compressor Controls Corporation (now part of Honeywell Process Solutions).

This material was reprinted with permission of the publisher from Ghaisas N, Reitsma T (2016) Surge detection and surge control systems for centrifugal compressors—Part 1. CompressorTECH^2 Magazine (July 2016), pp 64–67. Ghaisas N, Reitsma T (2016) Surge detection and surge control systems for centrifugal compressors—Part 2. CompressorTECH^2 Magazine (Aug/Sept 2016), pp 62–66.

References

1. Brun K, McBain M, Kurz R (2008) Application guideline for centrifugal compressor surge control systems, release version 4.3, Apr 2008. Gas Machinery Research Council, Southwest Research Institute, San Antonio, TX
2. Mirsky S, Jacobson W, Tiscornia D, McWhirter J, Zaghloul M (2012) Development and design of antisurge and performance control systems for centrifugal compressors. In: Proceedings of the forty-second turbomachinery symposium, Turbomachinery Laboratory, Texas A&M Engineering Experiment Station, Oct 2012

Chapter 5
Dry Gas Seals

Abstract Dry gas seals play a critical role in the performance of modern centrifugal compressors. However, specifying engineers often overlook the importance of clearly defining the scope and carefully selecting the most suitable seal type, its control system, and associated auxiliaries for the intended application. Instead, this responsibility is frequently deferred to compressor manufacturers, who make these selections based on their own standards. To ensure optimal design and performance, several key factors must be thoroughly evaluated—these include the gas composition, the thermodynamic properties of the seal or buffer gas, the choice of barrier gas, operational scenarios such as slow roll conditions, compatibility with bidirectional rotation, and overall system rating. Ideally, this evaluation should be a collaborative effort involving the equipment vendor, the seal manufacturer, and machinery engineers representing both the owner and the engineering contractor. Choosing the correct dry gas seal and its support system is essential for ensuring the long-term reliability and availability of centrifugal compressor trains in process plants. This chapter explores several important aspects, including the operating principles of dry gas seal faces, techniques for controlling seal gas, procedures for slow rolling, and strategies for managing demanding sealing environments.

5.1 Dry Gas Seal Faces

A dry gas seal features a rotating mating ring with a hard surface, incorporating a circumferential spiral groove pattern—either machined or etched—and a stationary primary ring made from a softer material. While the primary ring is fixed radially, it retains the ability to move axially. In static, depressurized conditions, springs positioned behind the primary ring press the seal faces together, keeping them in contact. When static but pressurized, the sealed gas enters the groove tip and flows between the seal faces. A sealing dam ensures even pressure distribution across the faces, contributing to hydrostatic lift. This lift minimizes startup torque, reduces heat generation, and lowers parasitic power losses.

N. Ghaisas, *Centrifugal Compressors in Hydrocarbon Processing Industries*,
https://doi.org/10.1007/978-3-032-03687-2_5

During dynamic operation, the gradually shallower spiral grooves channel the sealed gas toward the central dam, generating a gas film pressure that separates the seal faces, allowing them to operate without contact. The groove geometry is designed to produce a consistent hydrostatic and hydrodynamic pressure profile, maintaining an optimal gap between the faces for gas film stiffness. These design elements enable rapid liftoff—even at low speeds (as low as 10 ft/s or 3 m/s under pressures below 50 psig or 345 kPa)—and allow the seal faces to respond quickly to changing process conditions.

The configuration of the spiral grooves determines whether the seal is unidirectional (Fig. 5.1) or bidirectional (Fig. 5.2). Bidirectional seals offer protection against reverse rotation, such as in the event of a discharge check valve failure, and eliminate the need for spare seal cartridges on both ends of a between-bearings centrifugal compressor. A spare cartridge can be installed on either end. Figures 5.1 and 5.2 illustrate the sealing dam and land, while Fig. 5.3 presents typical properties of the materials used in seal faces.

Figure 5.4 depicts the pressure distribution across the face of a dry gas seal during dynamic operation. It also shows the typical gas film thickness that forms between the seal faces, highlighting its impact on leakage flow. The seal ring, also known as the primary ring, is designed to float and move axially, allowing its position relative to the mating ring to be governed by the forces acting upon it. Initially, a light axial spring force—approximately 1 psi (6.9 kPa)—keeps the seal and mating ring faces aligned until pressurized gas is introduced.

Once in operation, the gas pressure force (calculated as pressure × area) becomes the dominant influence on the seal ring. The geometry of the seal face and the placement of the dynamic secondary sealing element (typically an O-ring) enable a natural hydraulic balance. A pressure gradient forms across the interface between the mating and seal rings, extending from the outer to the inner diameter. This gradient

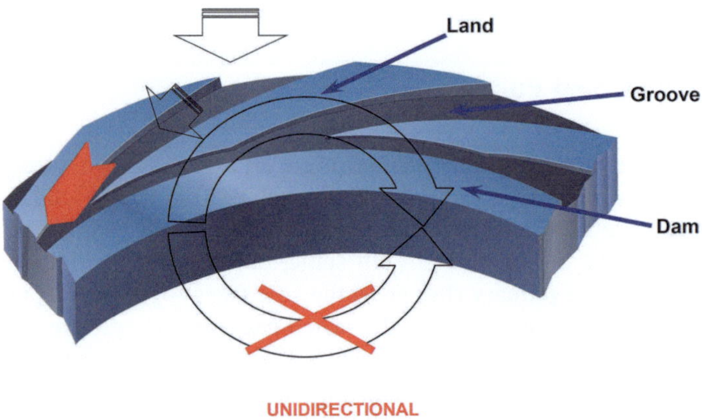

Fig. 5.1 Unidirectional spiral groove pattern on the mating ring of a gas seal (Image courtesy of Flowserve Corporation)

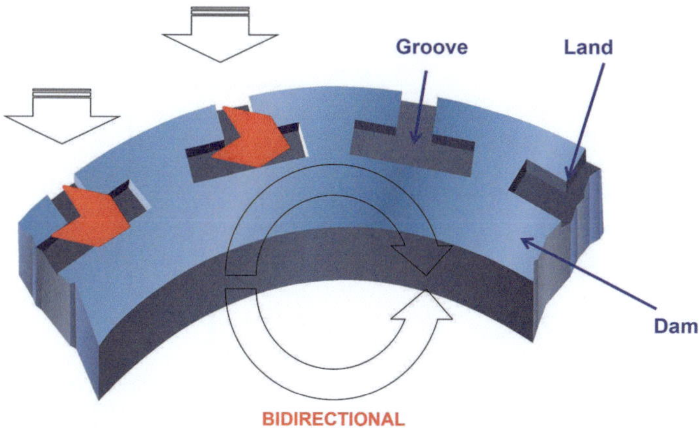

Fig. 5.2 Bidirectional spiral groove pattern on the mating ring of a gas seal (Image courtesy of Flowserve Corporation)

Material property	Tungsten carbide	Silicone carbide	Silicon nitride	Desirable level
HA (Kg/mm2)	1,870	2,800	1,800	High
ME (E6 psi)	90	51	45	High
SW (Ratio)	5	10	19	High
TE (E-6 in/in°F)	3	2	2	Low
HC (BTU-in/ft^2-hr°F)	568	830	208	High

- Hardness (HA)
- Modulus of Elasticity (ME)
- Tensile Strength to Specific Weight Ratio (SW)
- Coefficient of Thermal Expansion (TE)
- Heat Conductivity (HC)
- Desired values are in green highlight

Fig. 5.3 Typical properties of common seal face materials (Image courtesy of Flowserve Corporation)

generates an opening force (pressure × interface area) that tends to push the seal ring away from the mating ring. At the same time, the full pressure differential acts on the rear side of the seal ring, producing a closing force (pressure × back-side area) that pushes the seal ring toward the mating ring. Figure 5.5 provides details on commonly used O-ring materials for various process gas applications, along with the typical properties of secondary gaskets.

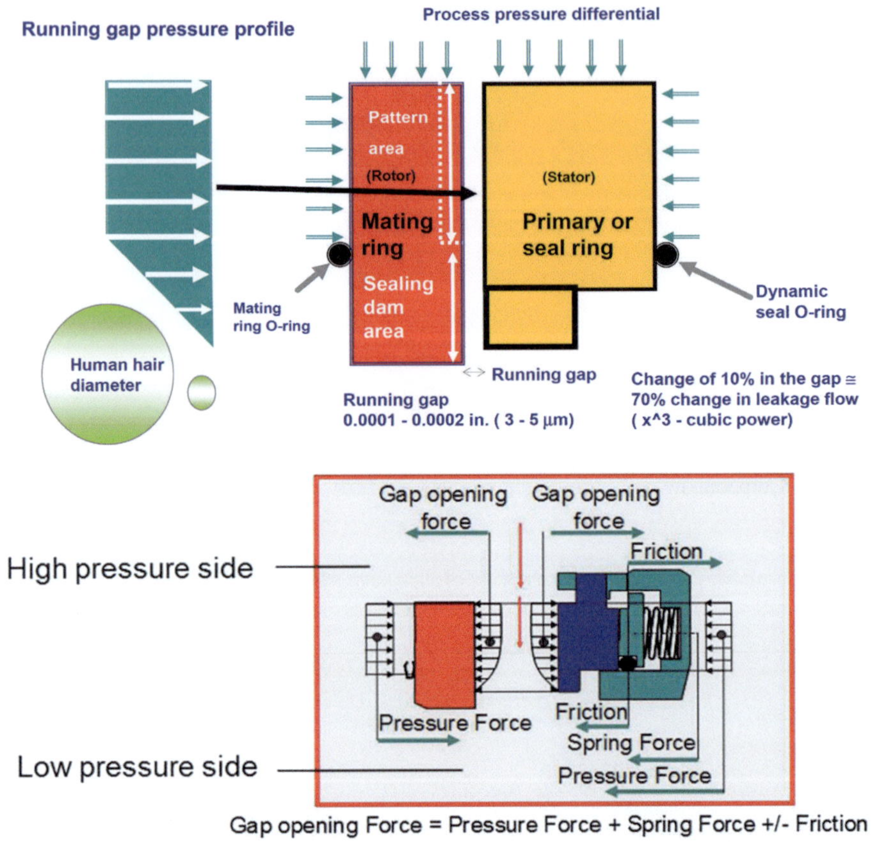

Fig. 5.4 Dry gas seal faces during dynamic operation (Image courtesy of Flowserve Corporation)

5.2 Slow-Speed Operation of Seal Faces (Slow Roll)

During standstill conditions, rotors in between-bearings centrifugal compressors, gas turbines, and steam turbines undergo elastic deflection or bending. This flexure is caused by external forces acting perpendicular to the rotor axis—primarily gravity, or a combination of gravity and thermal gradients depending on the operating environment.

To alleviate rotor bowing before startup, a slow roll of the compressor and steam turbine train is performed at rotational speeds ranging from 2 to 500 rpm determined by factors such as thermal gradients, rotor mass, anticipated shaft deflection, and the span between bearings. This includes coastdown or ratchet-style slow rolls at approximately one-eighth turn per minute (typically not exceeding 50 rpm). For this process to be effective, dry gas seals must be engineered to maintain face separation during slow roll conditions, whether the seal chamber is pressurized or not.

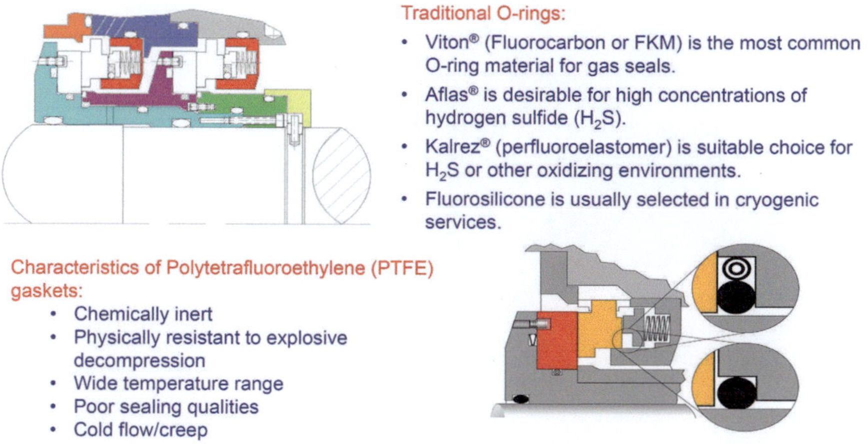

Fig. 5.5 Materials used in secondary seals and their characteristics (Image courtesy of Flowserve Corporation)

Seal face contact may occur during slow roll if hydrostatic or hydrodynamic forces are insufficient to lift the seal faces apart. The durability of the seal faces under such contact depends on several factors: rotor speed, gas dew point, seal face material properties and combinations, surface finish, contact load, and the duration and frequency of contact events.

In noncontacting slow roll scenarios, the stiffness of the gas film is lower than during normal operation, increasing the risk of seal face rubbing—especially if the sealed gas contains entrained liquids. Additional challenges include axial shaft displacement and reverse pressurization of the inner seal. Partial contact between seal faces can also happen during coastdown at speeds below the liftoff threshold.

To ensure reliable performance during slow roll, dry gas seals are designed with a balanced combination of seal face loading, spring force, and friction characteristics. They can accommodate sealed pressures from 3 to 450 psig (21–3100 kPag) and axial displacements up to ± 0.06 in. (± 1.5 mm).

5.3 Types of Dry Gas Seals

A dry gas seal functions as a mechanical seal composed of two main components: a rotating mating ring and a stationary seal ring. Figure 5.6 illustrates a typical dry gas seal assembly along with a diagram of its basic components, and Fig. 5.7 outlines commonly used seal face materials and their properties. The three primary types of dry gas seals used in turbomachinery applications are detailed in the following sections.

On the inboard side of the primary seal, a process gas labyrinth is installed. To prevent reverse flow of process gas into the seal cavity, two control strategies are

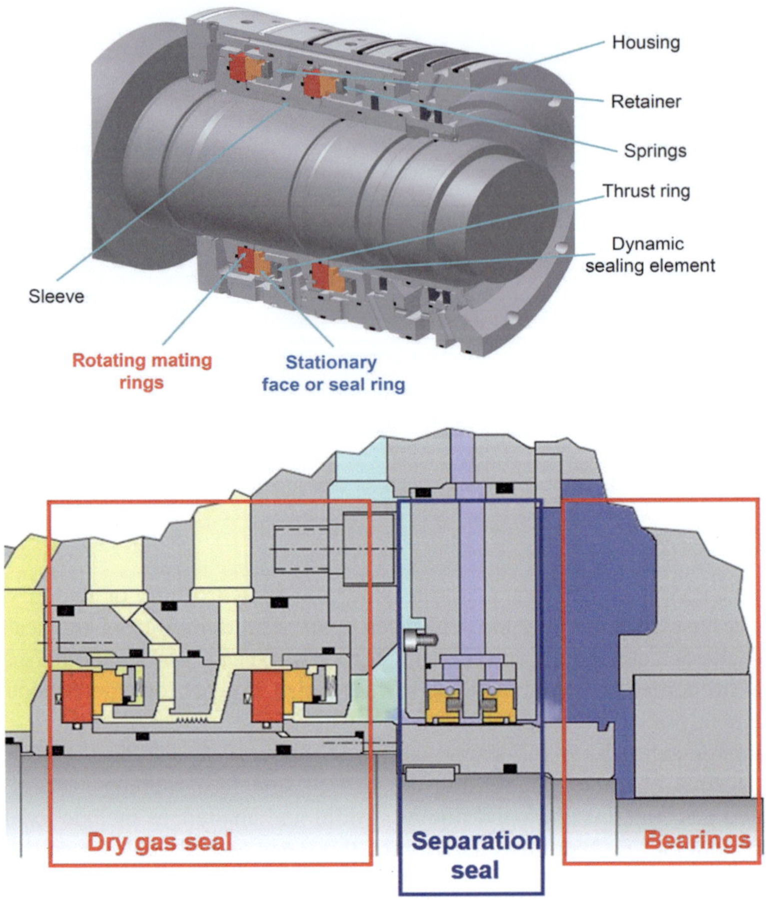

Fig. 5.6 Seal cartridge layout (Image courtesy of Flowserve Corporation)

employed: the seal gas flow control method and the differential pressure control method. Both approaches ensure a consistent, positive flow of seal gas across the labyrinth.

Seal gas is typically sourced from clean process gas, inert gas, or sweet gas—provided it is compatible with the process gas. Maintaining a continuous and stable supply of seal gas is critical for the seal's performance. The seal gas pressure should exceed the highest sealed pressure by at least 50 psi (345 kPa).

It is common practice to draw seal gas from the compressor's discharge. However, during operational disturbances, such as startup, shutdown, or process upsets, the pressure rise across the compressor may be insufficient to maintain positive seal gas flow. In such cases, auxiliary systems like seal gas boosters can be employed to elevate the seal gas pressure and prevent contamination at the primary seal faces. Seal gas booster systems are further discussed in Chap. 6.

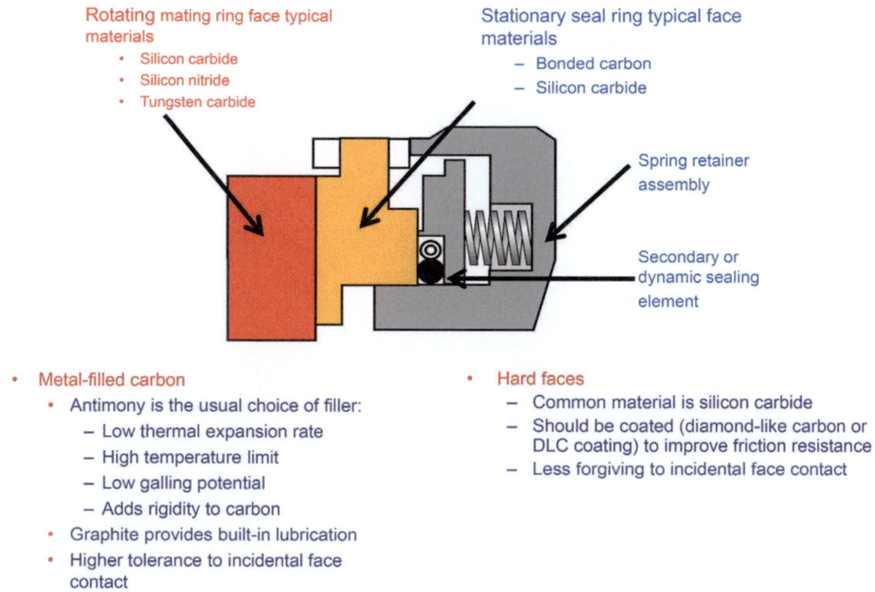

Fig. 5.7 Seal face materials (Image courtesy of Flowserve Corporation)

5.3.1 Single Seal

Single-seal configurations are typically employed in compressors that handle non-toxic, non-flammable gases which pose no environmental risk, and where the sealed pressure does not exceed 300 psig (2068 kPag). Common examples include carbon dioxide, air, and nitrogen. To mitigate gas leakage in the event of seal failure, a labyrinth-type seal may be added to the single-seal arrangement.

As described in American Petroleum Institute (API) Standard 692 [1], the majority of the seal gas introduced between the seal faces and the process-side seal flows into the compressor from the process side. Any seal leakage is vented to the atmosphere, as illustrated in Fig. 5.8.

5.3.2 Double Seal

A double seal configuration (illustrated in Fig. 5.9) features two opposing seals with a buffer gas introduced between them. This buffer gas is maintained at a pressure at least 50 psi (345 kPa) higher than the sealed process gas pressure, but it does not pass through the seal faces. Any leakage of seal gas across the secondary seal is typically vented to the atmosphere.

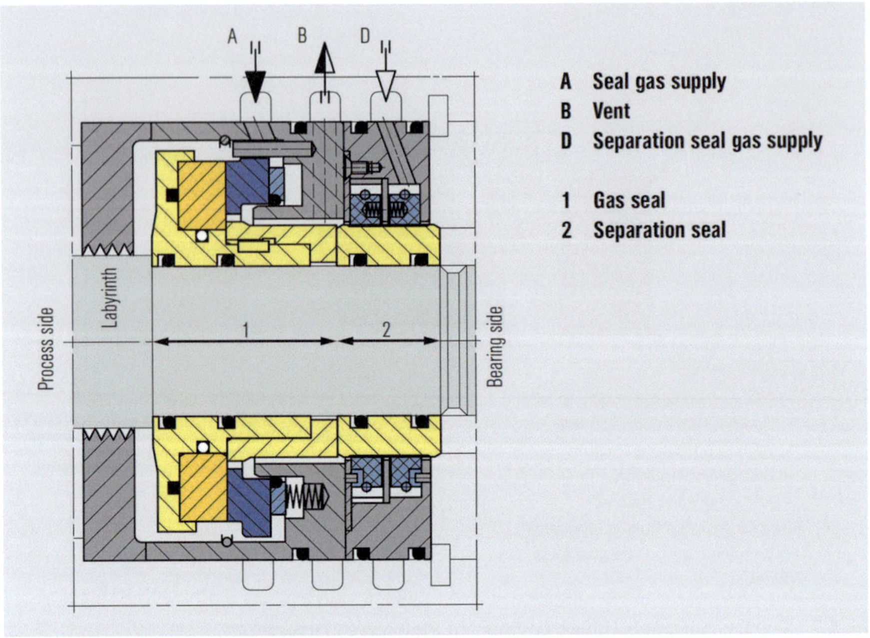

Fig. 5.8 Single seal (Image courtesy of Eagle Burgmann, Houston, TX)

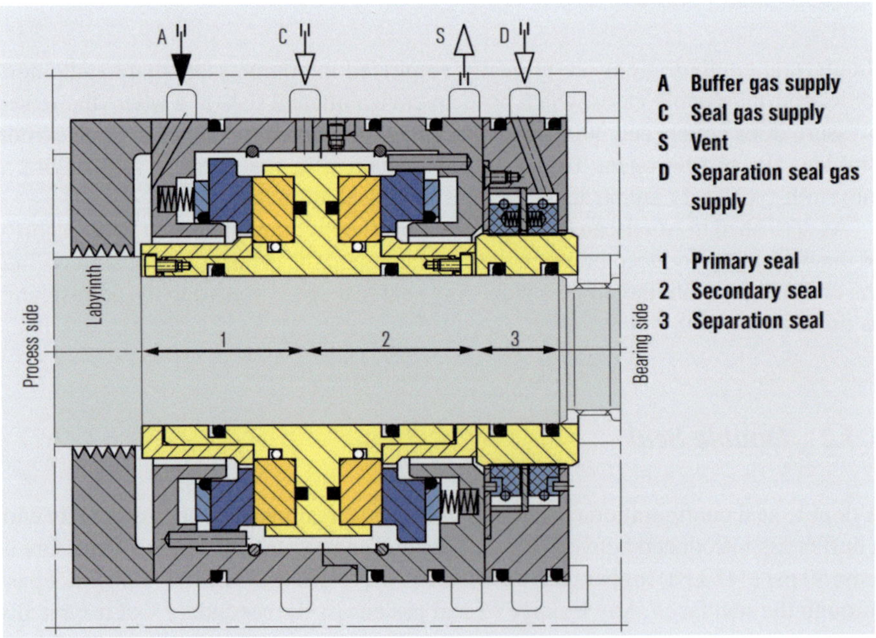

Fig. 5.9 Double seal (Image courtesy of Eagle Burgmann, Houston, TX)

Double seals are ideal for applications involving poisonous, dirty or hazardous gases—such as hydrogen sulfide—where external leakage must be strictly avoided or where minimizing seal gas consumption is critical. This arrangement is also beneficial if axial space for the seal cartridge is limited or if the compressor's suction pressure is close to the pressure of the vent system. In such cases, the seal system must be balanced to an intermediate pressure above the vent system pressure.

Double seals are generally restricted to applications with sealed pressures not exceeding 200 psig (1380 kPag).

5.3.3 *Tandem Seal*

In a tandem-seal configuration, the full system pressure is absorbed across the primary seal faces. The secondary (or outboard) seal typically operates at a lower pressure than the primary (inner) seal, but it is engineered to withstand the full system pressure in the event of a primary seal failure. The primary seal's vent is directed to the plant's flare system, while the secondary seal's vent is generally discharged to the atmosphere. Depending on the specific seal design and the number of seal rings used, tandem seals can handle sealed pressures ranging from 1200 to 1800 psig (8273–12,410 kPag). Figure 5.10 illustrates a typical tandem seal arrangement.

To prevent process gas from escaping into the atmosphere, a tandem seal equipped with an intermediate labyrinth is chosen. A clean secondary seal gas is injected at this labyrinth at a pressure slightly above that of the primary vent, establishing a differential pressure that inhibits the process gas from reaching the secondary seal faces. This setup also facilitates the sweeping of any seal gas leakage from the primary vent and separation gas leakage from the secondary vent (refer to Fig. 5.11). Tandem-seal configurations are more commonly implemented than single or double seals, particularly in hydrocarbon and other critical process applications.

5.3.4 *Separation Seal*

The main role of a separation seal is to prevent bearing housing lubricant from entering the dry gas seal cavity, while also inhibiting process gas from contaminating the bearing oil. Common separation seal types include labyrinth seals (Fig. 5.12) and segmented dual carbon ring bushings. Carbon ring bushings, impregnated with graphite, may be designed for either contacting or noncontacting operation. The noncontacting variant is generally preferred, as it minimizes localized heat buildup and reduces the risk of premature wear. An effective separation seal must support bidirectional shaft rotation and remain stable when exposed to bone-dry nitrogen. Secondary sealing components are typically constructed from fluorocarbon materials such as Viton®. Figure 5.13 illustrates the configuration of a carbon ring separation seal within a dry gas seal cartridge.

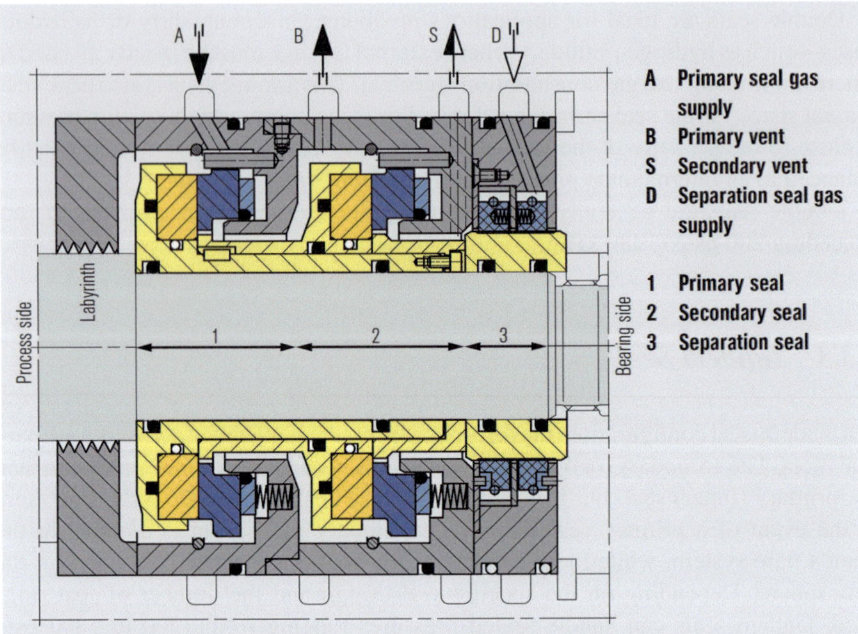

Fig. 5.10 Tandem seal without an intermediate labyrinth (Image courtesy of Eagle Burgmann, Houston, TX)

To prevent lubricant from the outboard bearing housing from migrating into the seal cavity, a clean and dry separation gas is continuously supplied to the separation seal at a pressure higher than that of both the bearing housing and the seal vent cavity. This separation gas must be regulated using a fixed-pressure control system—maintaining pressure above the secondary vent—and safeguarded with dual check valves on the secondary gas supply module to prevent reverse pressurization. According to API 692 [1], filters used for separation gas in noncontacting bushing seals must achieve a minimum efficiency of 99.9% for particles ≤ 10 μm (394 μin.) with a beta ratio of $\beta_{10} \geq 1000$. For contacting bushing seals, the required minimum efficiency is 99.9% for particles ≥ 1 μm (39.4 μin.), with a beta ratio of $\beta_1 \geq 1000$. This requirement is equally applicable to both the seal gas supply and the secondary seal gas supply.

5.3.5 Process-Side Seal

Process-side seals are typically categorized as labyrinth, noncontacting bushing, or abradable types. Among these, labyrinth seals are widely employed across various applications. Compared to traditional labyrinth seals, noncontacting carbon ring

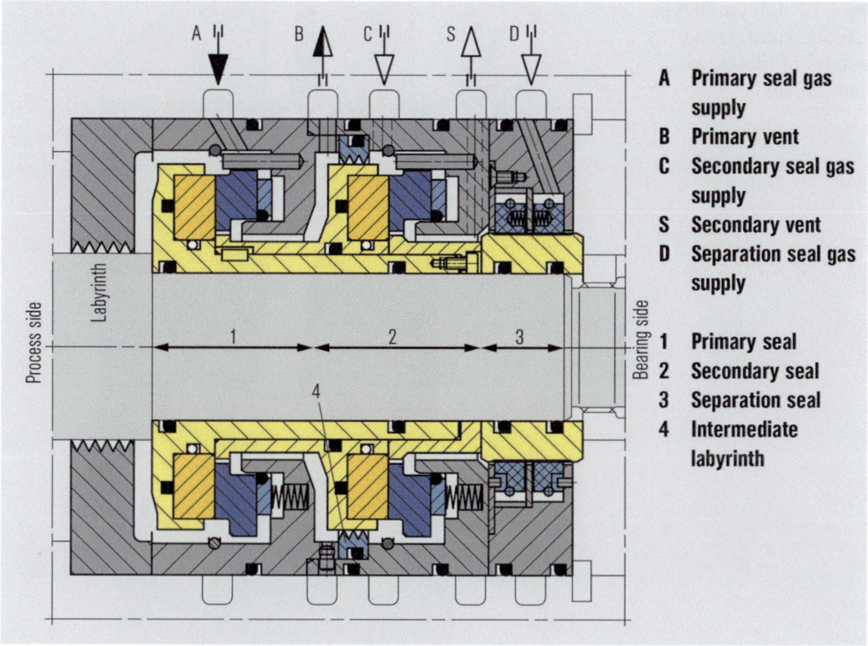

Fig. 5.11 Flow through a tandem dry gas seal with an intermediate labyrinth (Image courtesy of Eagle Burgmann, Houston, TX)

(single or dual) and abradable designs offer reduced gas leakage due to their narrower rotor-to-stator clearances. The functional principles governing abradable and labyrinth seals—outlined in Sect. 2.6 of Chap. 2—are equally relevant to their use as the process-side seals.

5.4 Separation Gas

Nitrogen or air is typically used as the separation gas in dry gas seal systems. Due to its inert nature, nitrogen is preferred for safety reasons. If nitrogen is not readily available on-site, installing nitrogen generation equipment should be considered. Most of the flow through the secondary vent of a dry gas seal consists of the separation gas, with only a minimal amount of seal gas present.

Using air as the separation gas introduces the risk of forming an explosive atmosphere in the secondary vent, especially when it mixes with combustible process gases. Combustion may occur if the air-to-process gas ratio falls within explosive limits and an ignition source is present.

If air must be used, the seal control system must be designed to ensure the secondary vent environment is either lean or rich—both conditions help prevent

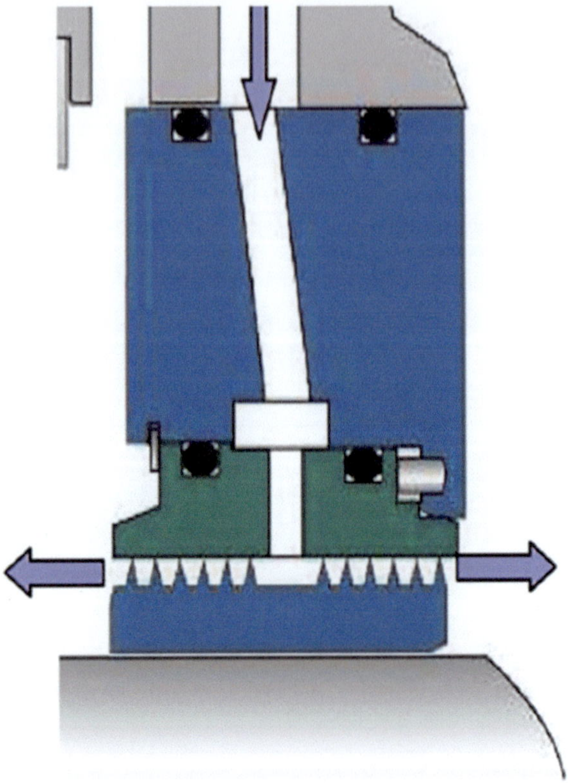

Fig. 5.12 Labyrinth type separation seal (Image courtesy of Flowserve Corporation)

the formation of an explosive mixture. A lean mixture is achieved by injecting air into the vent, resulting in a hydrocarbon/air composition below the lower explosive limit (LEL). Conversely, a rich mixture is created by injecting process gas, producing a composition above the upper explosive limit (UEL).

Explosive mixtures typically contain 5–15% methane (CH_4) in air or 17–54% CH_4 in oxygen (O_2). Mixtures with less than 5–17% fuel are considered lean (excess oxidizer), while those with more than 15–54% fuel are rich (excess fuel).

If a rich mixture is present in the secondary vent, it must be directed to the plant's flare system. The required amount of air or process gas injection should be determined through a comprehensive engineering study that accounts for all operating scenarios, including potential primary seal failure.

Figure 5.14 illustrates the relationship between hydrocarbon concentration and molecular weight in hydrocarbon/air mixtures. The graph identifies rich mixtures above the UEL line, lean mixtures below the LEL line, and explosive mixtures in the region between the two.

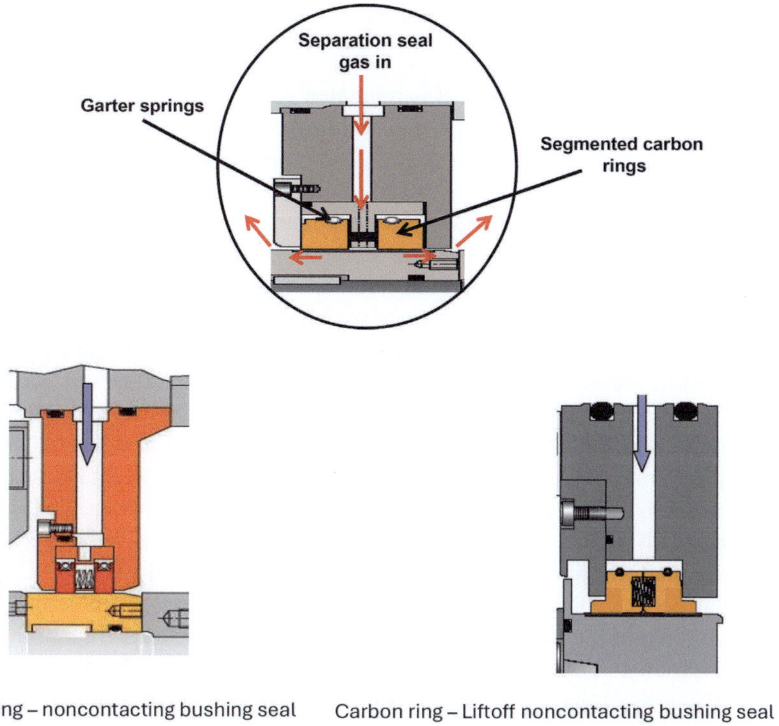

Fig. 5.13 Segmented dual carbon ring bushing separation seal (Image courtesy of Flowserve Corporation)

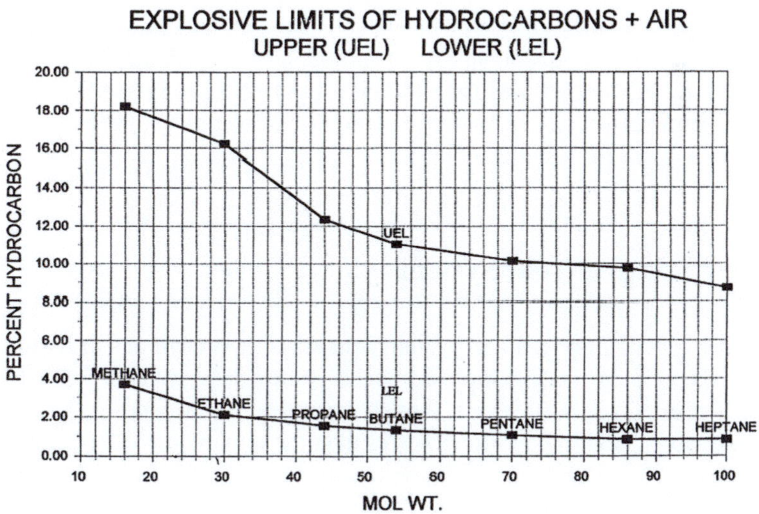

Fig. 5.14 Upper explosive limits (UELs) and lower explosive limits (LELs) for hydrocarbon/air mixtures (Image courtesy of Flowserve Corporation)

5.5 Seal Gas Control Methods

Seal gas flow to the primary seal is regulated using either a flow control system or a differential pressure control system. Both approaches aim to ensure a consistent sweep of seal gas across the process gas labyrinth, preventing backflow of process gas into the dry gas seal.

In a flow control system, seal gas is regulated through an orifice positioned upstream of each seal. A remote flow controller monitors the flow rate and adjusts the control valve to maintain a minimum velocity of 16 ft/s (5 m/s) across the inner process labyrinth. This velocity is determined using the "low-select reference pressure" measured downstream of the orifices. According to API 692 [1], minimum gas velocities must be maintained across all labyrinth and clearance seals at twice the maximum design clearance, accounting for seal eccentricity during slow roll, sub-slow roll, and higher speeds.

A differential pressure control system, on the other hand, maintains seal gas pressure at a fixed differential—typically 10 psid (69 kPa)—above the sealed reference pressure. This setup includes a differential pressure control valve with a remote controller and a bypass line equipped with manual block valves.

Regardless of the control method, the dry gas seal system must be capable of sustaining a minimum gas velocity of 16 ft/s (5 m/s) across the inner labyrinth and throughout the full range of clearance values, up to twice the maximum design clearance. This is essential to prevent contamination of the seal faces and to keep seal face temperatures within safe operating limits. Achieving the right balance between seal gas flow rate and velocity is crucial. Excessive flow increases energy consumption, reduces system efficiency, and necessitates larger components.

Flow control systems are generally preferred due to their ability to minimize seal gas usage while maintaining adequate velocity. They are especially suitable for most seal types and applications, except in double-seal configurations or systems with very low sealed pressures (below 100 psig or 690 kPag). Additionally, flow control systems do not rely on reference pressure measurements, which are critical in differential pressure systems.

5.6 Dry Gas Seal Vent System

If the sealed gas in a dry gas seal system consists of hydrocarbons or hydrocarbon mixtures, the primary vent is directed to the plant's flare system. Typically, flare pressure is near atmospheric levels, though it can reach up to 50 psig (345 kPag). If the compressor's suction pressure is lower than the maximum vent pressure of the primary seal, the seal chambers must be equalized to an intermediate pressure that exceeds the suction pressure.

To ensure safety, bypass relief valves on the primary seal vent lines should be positioned as close to the dry gas seals as possible. Additionally, the piping and venting system must be properly sized to prevent overpressurization of the bearing housings in the event of a seal failure.

The secondary vent is usually discharged to the atmosphere—unless the hydrocarbon/air mixture exceeds the upper explosive limit. As outlined in Sect. 5.4, if a rich mixture is present in the secondary vent cavity, it must be redirected to the plant's flare system to mitigate explosion risks.

5.7 Dry Gas Seals in Cryogenic Applications

The selection of materials for dry gas seal faces, secondary sealing elements, and metallic components must align with the full range of operating conditions and the specific gas composition of the application. For centrifugal compressors used in liquefied natural gas (LNG) service, the following considerations are particularly important:

- Employ a bidirectional silicon nitride rotating face paired with a silicon carbide/carbon stationary face, selected based on operating pressure, seal size, and rotational speed.
- Apply an amorphous diamond-like carbon (DLC) coating to the silicon carbide stationary face to enhance hardness, reduce friction, and improve chemical and thermal resistance.
- Utilize specialized Teflon™ and spring-energized Teflon™ gaskets and O-rings designed for cryogenic environments.
- Incorporate low-speed liftoff face technology for tertiary sealing applications.
- Ensure the seal is rated for bone-dry nitrogen service at − 94 °F (− 70 °C).
- Adopt a flow control strategy for managing seal gas delivery.
- Install flow meters on each individual seal gas supply line for precise monitoring.
- Include a seal gas conditioning system equipped with a booster to maintain optimal gas quality and pressure.

5.8 Challenging Environment: Sulfur Contamination

Sulfur can be present in process gas compositions, commonly in the form of hydrogen sulfide (H_2S). Elemental sulfur has a melting point between 235 and 246 °F (113–119 °C), but its solidification can occur at varying temperatures and pressures depending on its concentration in a saturated gas stream. In systems where H_2S and/or sulfur vapor are present, solid sulfur deposits can form due to chemical reactions influenced by catalysts and fluctuations in gas pressure and temperature.

One notable reaction involves H_2S interacting with trace levels of oxygen (typically 3–5%) found in nitrogen, which is often used as the separation gas. This reaction can lead to sulfur separation and deposition. Potential areas within a dry gas seal

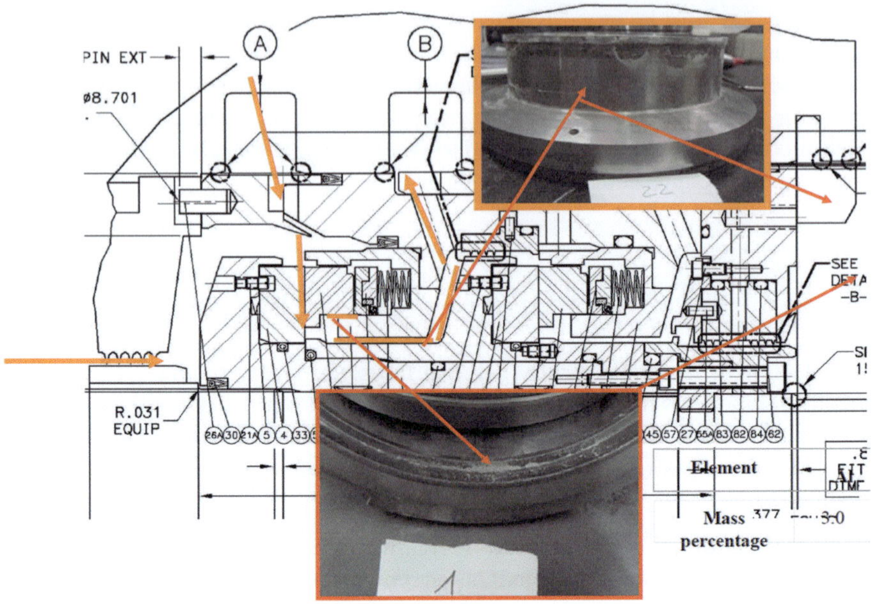

Fig. 5.15 Possible locations of sulfur deposition in a dry gas seal passage (Image courtesy of Flowserve Corporation)

passage where sulfur buildup may occur are illustrated in Fig. 5.15, and Fig. 5.16 shows examples of seal components affected by sulfur contamination.

For centrifugal compressors operating with hydrogen sulfide (H_2S) gas, it's advisable to incorporate gas conditioning components into the dry gas seal control system. Two proven strategies for preventing sulfur solidification and buildup include using a coalescing prefilter to purify the seal gas and installing a seal heater to elevate the seal gas temperature to 250 °F (121 °C). Additionally, integrating a booster unit helps prevent the intrusion of contaminated gas into the seal zones during transient operating conditions.

5.9 Storage and Protection of Dry Gas Seals

This guidance applies to seals that are either stored in appropriate packaging or installed in equipment for extended periods without being operated. Over time, critical components of the seal may undergo changes such as aging or deformation, which can compromise their performance. To ensure proper preservation and functionality, the following protective measures are recommended:

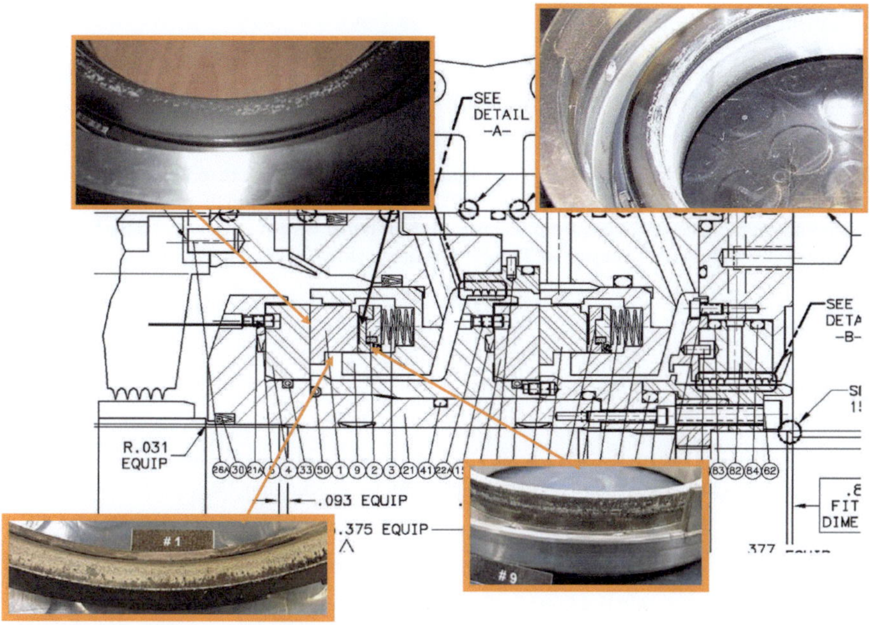

Fig. 5.16 Dry gas seal components contaminated by sulfur deposition (Image courtesy of Flowserve Corporation)

Storage Conditions:

- Store seals in a dry environment with relative humidity below 65% and minimal dust.
- Maintain ambient temperature between 68 °F ± 10 °F (20 °C ± 5 °C).
- Avoid exposure to direct sunlight, heat sources, steam, ozone, and radiation.
- Use specialized storage containers with nitrogen blanketing to safeguard the seals.

Chemical Exposure:

- Do not apply corrosion inhibitors, as they may negatively affect seal performance.

Requalification Protocol:

- For seals stored for 36 months or longer, requalification is advised before installation.
- Replace elastomeric secondary sealing elements.
- Inspect seal faces for any signs of degradation that could impair functionality.

Spare Component Handling:

- Keep spare gaskets and O-rings away from heat sources, including sunlight, UV-emitting lights, and appliances that generate heat.
- Avoid storing these components at temperatures below 50 °F (10 °C).

Radiation Protection:

- Shield dry gas seals from all forms of ionizing radiation, which can damage both the seals and their internal components.

Acknowledgements The images/graphics/figures in this chapter were provided by Flowserve Corporation and Eagle Burgmann (Houston, TX). Some of the material in this chapter was reprinted with permission of the publisher from Ghaisas N, Majumdar S (2017) Dry Gas Seal Systems for Centrifugal Compressors. CompressorTECH^2 magazine (June 2017), pp 34, 36–42.

Reference

1. API Standard 692 (2018) Dry gas sealing systems for axial, centrifugal, rotary screw compressors and expanders, 1st edn. American Petroleum Institute, Washington, DC

Chapter 6
Seal Gas Treatment

Abstract Dry gas seals serve as a highly effective, dependable, and resilient sealing solution for compressors operating across a wide range of applications in the petrochemical, natural gas, and petroleum sectors. Their optimal performance hinges on a continuous, uninterrupted supply of clean and dry seal gas delivered within the specified flow and pressure parameters. This ensures proper lift-off of the seal faces, minimizing wear and maximizing efficiency. Both pressurized and unpressurized configurations rely on clean, dry seal gas introduced upstream of the seal to form a protective barrier against contamination from the process stream. Seal gas treatment involves the removal of particulates and aerosols to maintain purity. Dedicated seal gas support systems are engineered to deliver the required flow rate, temperature, and pressure, with these variables actively monitored and controlled. This chapter provides a qualitative overview of critical subsystems—such as dew-point management and pressure boosting—that contribute to the reliability and performance of dry gas seals.

6.1 Seal Gas Quality

The filtration specifications for the seal gas supply, secondary seal gas supply, and separation gas supply are outlined in Sect. 5.3.4 of Chap. 5. These rigorous standards are designed to prevent contamination and ensure optimal performance of the dry gas seals. Additionally, maintaining a dew-point margin (superheat) of no less than 35 °F (20 °C) across the entire dry gas seal system is critical to avoid the formation of liquid. To confirm this margin, a phase-map simulation of the system—from the primary seal gas supply point to the primary vent—must be conducted to assess any risk of condensation. It is important to measure the seal gas temperature at the point of entry into the seal itself, rather than at the supply source. A typical phase map curve is illustrated in Fig. 6.1.

To ensure the desired quality of seal gas, it is often necessary to integrate the seal gas treatment system with the broader dry gas control system. The seal gas conditioning equipment includes components designed to deliver clean, dry gas.

N. Ghaisas, *Centrifugal Compressors in Hydrocarbon Processing Industries*, https://doi.org/10.1007/978-3-032-03687-2_6

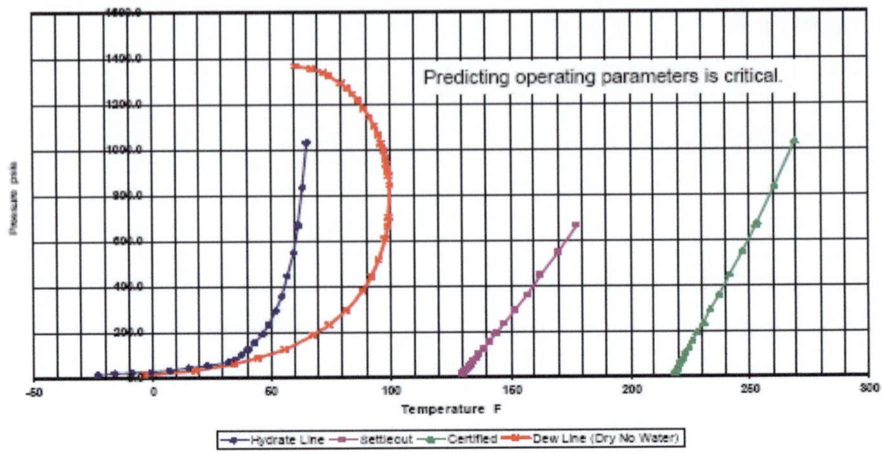

Fig. 6.1 Dew-point analysis or gas phase map (Image courtesy of Flowserve Corporation)

These typically consist of coolers, wet gas prefilters or separators, and—if required—a seal gas heater to achieve the desired dryness. For purification, wet gas demisters and dual-stage filters are employed.

In many compressor applications, a pressure-boosting unit is essential to maintain a sufficient positive differential pressure, preventing seal contamination during transient operating conditions such as startup, shutdown, process transitions, slow roll, recycle, and settle-out. To protect the primary seal, the seal gas supply pressure should exceed the sealed pressure by at least 50 psi (345 kPa).

According to API Standard 692 [1], separators must be sized to handle three times the normal seal gas flow and achieve 98% removal efficiency for solid particles $\geq$ 10 μm (394 μin.) and liquid particles down to 50 ppmw. However, designing separators for maximum efficiency at triple flow rates may lead to reduced performance under normal flow conditions. Figure 6.2 illustrates a typical cross-sectional view of a seal gas separator featuring an impingement plate and a demister.

6.2 Seal Gas Heaters

The direct-type seal gas heater employs an electric immersion heating coil, as illustrated in Fig. 6.3. In contrast, the indirect-type heater utilizes either an oil-immersed coil or a pressurized seal gas routed through a spiral coil housed within a cast aluminum shell—typically made from Alloy 356—which contains the electrical heating elements. This configuration is particularly advantageous if the compressor's seal gas has a high temperature or elevated dew point. The aluminum shell facilitates heat transfer from the heating elements to the seal gas without direct contact.

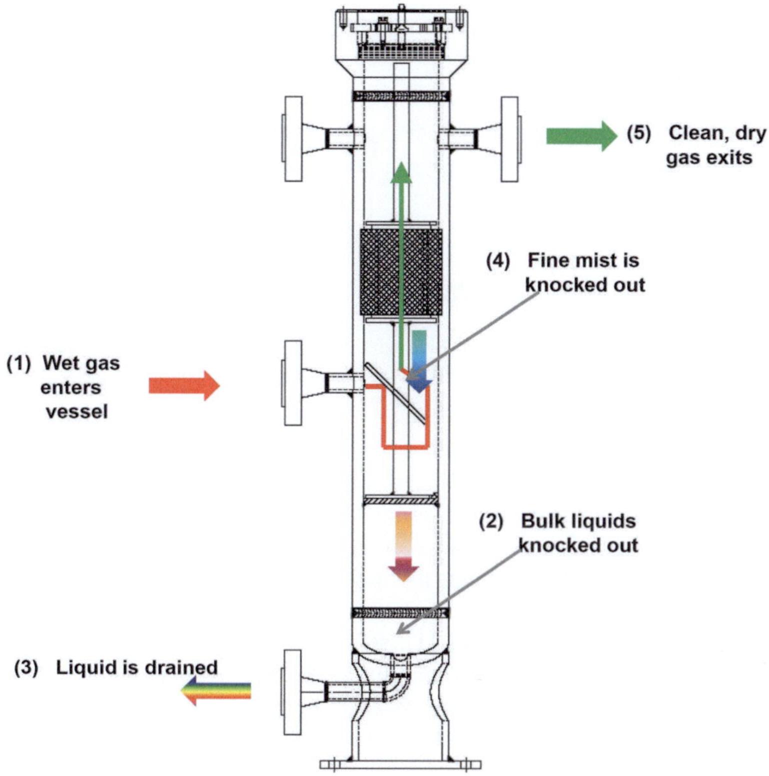

Fig. 6.2 Seal gas flow through a separator (Image courtesy of Flowserve Corporation)

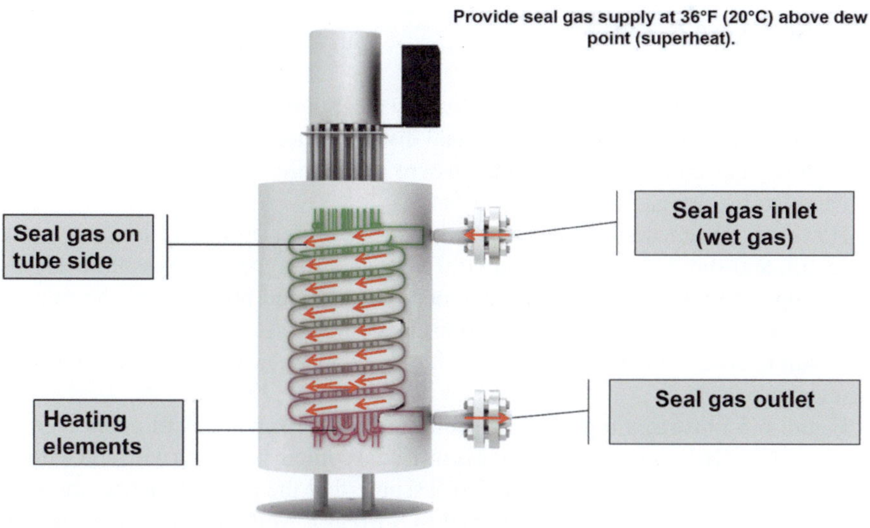

Fig. 6.3 Seal gas heater (Image courtesy of Flowserve Corporation)

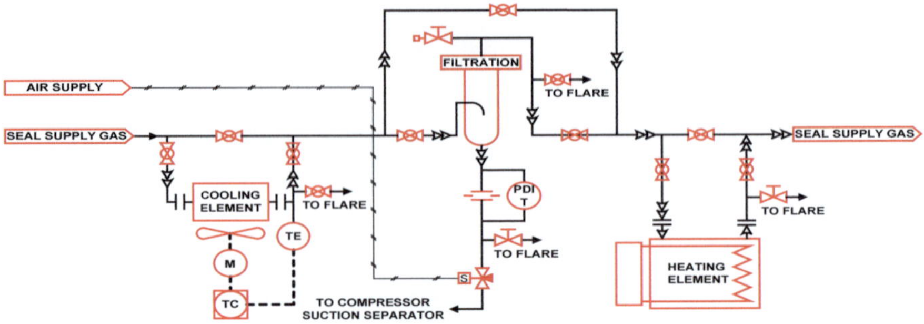

Fig. 6.4 Dew-point management system. Abbreviations: M = motor, PDIT = pressure differential indicator transmitter, S = three-way solenoid valve, TC = temperature controller, TE = temperature element (Image courtesy of Flowserve Corporation)

Temperature sensors are incorporated to monitor the heater casting, internal enclosure high limits, and cascade control settings (including both high and low temperature thresholds). Pyrogel® insulation is commonly used for the heater shell to enhance thermal efficiency. Figure 6.4 depicts a seal gas heater integrated into a dry gas seal dew-point management system.

According to API Standard 692 [1], seal gas heaters must be sized based on three times the normal flow rate, the minimum required temperature, the maximum allowable seal gas pressure, and any user- or supplier-defined limits on seal gas contact temperature. These stringent criteria can often lead to frequent cycling of the heaters during operation.

6.3 Seal Gas Boosters

During normal operation, compressors supply a continuous stream of seal gas from a high-pressure point—most often the final discharge nozzle. This seal gas is routed through dry gas conditioning systems, where it is filtered and treated before being delivered to the dry gas seals.

However, during transitional phases such as startup, pressurization, recycling, standstill, and shutdown, the pressure differential across the compressor decreases, disrupting the seal gas flow. When this occurs, unfiltered process gas can migrate into the seal area, leading to contamination and compromised seal performance. This can result in unsafe operating conditions, increased machine downtime, and elevated maintenance costs.

Figure 6.5 illustrates typical flow profiles (measured in standard cubic feet per minute [SCFM]) during the startup of a centrifugal compressor, both with and without a seal gas pressure boost. These profiles assume that the pressure across the seals matches the suction pressure of 700 psig (4.8 MPag), while the discharge pressure is 880 psig (6.1 MPag).

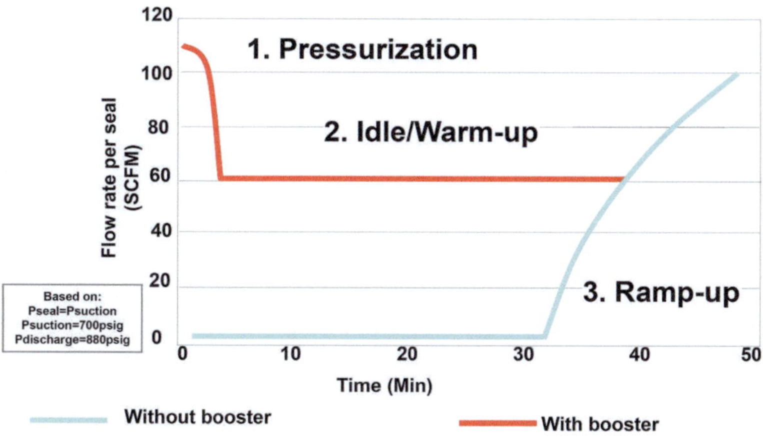

Fig. 6.5 Seal gas flow with and without a pressure boosting device during the startup of a centrifugal compressor (Image courtesy of Flowserve Corporation)

During steady-state operation of a centrifugal compressor, a pressure differential is established, allowing clean, dry seal gas from the compressor's final discharge nozzle to be routed through the seal gas system and delivered to the shaft-end dry gas seals (refer to Sect. 6.1).

Figure 6.6 illustrates the projected performance of a seal gas booster operating at an inlet pressure of 725 psig (5 MPa) for a pipeline centrifugal compressor. The figure highlights how variations in the booster's rotational speed influence both the pressure differential and the seal gas flow rate, expressed in actual cubic feet per minute (ACFM) and actual cubic meters per hour (A m^3/h), across the dry gas seal.

As stated in Sect. 6.3, during recycle, pressurization, startup, coastdown, and pressurized standstill (shutdown) conditions, the pressure differential across a centrifugal compressor is insufficient to maintain the necessary seal gas flow rate. As a result, the supply of seal gas to the dry gas seals is interrupted, allowing process gas to migrate into the seal cavity via the process labyrinth and contaminate the seal environment.

Figures 6.7 and 6.8 illustrate how reverse flow of process gas contaminates the dry gas seal environment during recycle mode and pressurized shutdown, respectively. To assess the need for a seal gas booster during startup and coastdown, dynamic simulations are conducted both with and without the booster. These simulations generate plots correlating compressor speed—including maximum slow roll speed—and time durations with flow velocity across the process gas labyrinth, helping to identify the specific operating conditions under which booster activation becomes necessary.

As illustrated in Fig. 6.9, for a compressor operating with an inlet pressure of 1,450 psig (10 MPa), a seal gas flow rate of approximately 3.5 cubic feet per minute (cfm) or 0.1 m^3/min can be maintained at a differential pressure of 56 psid (386 kPa) when running at 3500 rpm. This flow rate is sufficient to prevent contamination within the seal gas cavity.

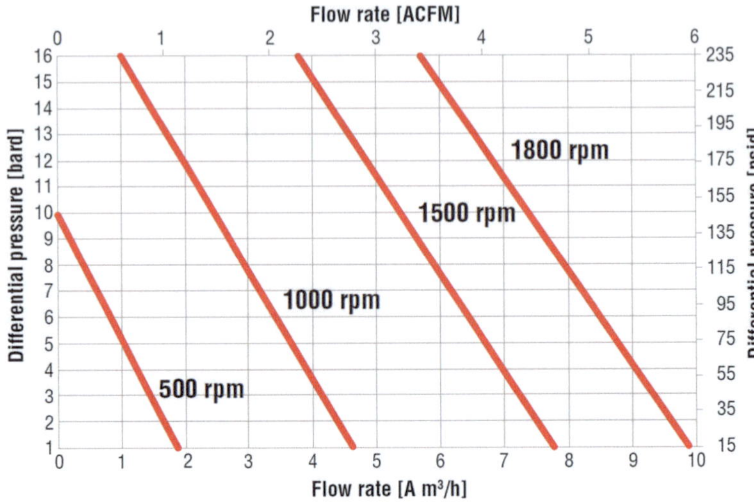

Fig. 6.6 Performance of an electric-motor-driven seal gas booster (Image courtesy of Flowserve Corporation)

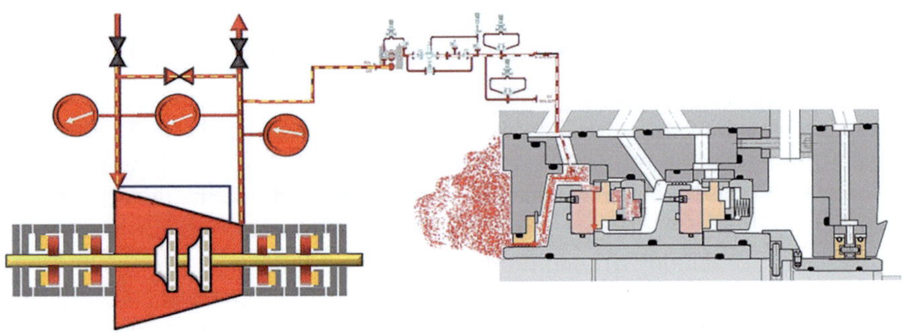

Fig. 6.7 Process gas migration to the dry gas seal cavity during the recycle mode of operation (Image courtesy of Flowserve Corporation)

Modern seal gas boosters are available in two primary configurations: pneumatic and adjustable-frequency electric motor-driven units, typically rated between 7.5 and 25 horsepower (5.6–18.6 kW). These boosters are designed to supply seal gas during periods of low differential pressure, ensuring continuous pressurized protection of dry gas seals across the full operating range of the compressor's performance map.

To prevent contamination of the primary seal, the seal gas supply pressure should exceed the sealed pressure by a minimum of 50 psi (345 kPa). Pneumatic boosters offer pressure boost ratios ranging from 1.2 to 2, depending on a maximum allowable working pressure between 3000 and 6000 psi (21–41 MPa). Their cycle rates typically

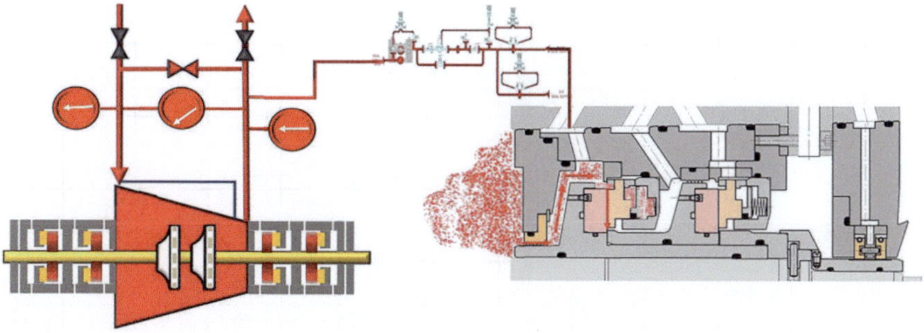

Fig. 6.8 Process gas migration to the dry gas seal cavity during pressurized shutdown (Image courtesy of Flowserve Corporation)

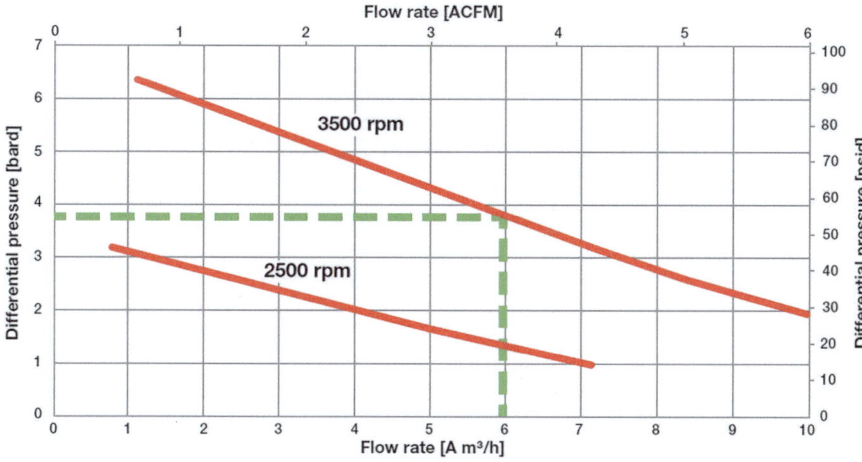

Fig. 6.9 Representative seal gas booster performance curve (Image courtesy of Flowserve Corporation)

range from 60 to 100 cycles per minute, though these figures may vary by model and manufacturer.

Seal gas boosters have become an integral component in modern seal gas treatment systems and are increasingly considered for retrofit applications in hydrocarbon and natural gas processing facilities focused on improving reliability. Figure 6.10 provides a cross-sectional view of a vertical electric motor-driven booster, highlighting the temperature sensor—either a transmitter or a resistance temperature detector (RTD)—mounted on the containment shell. This sensor is usually positioned between the inner and outer magnet rings or between the drive magnet and the casing cover.

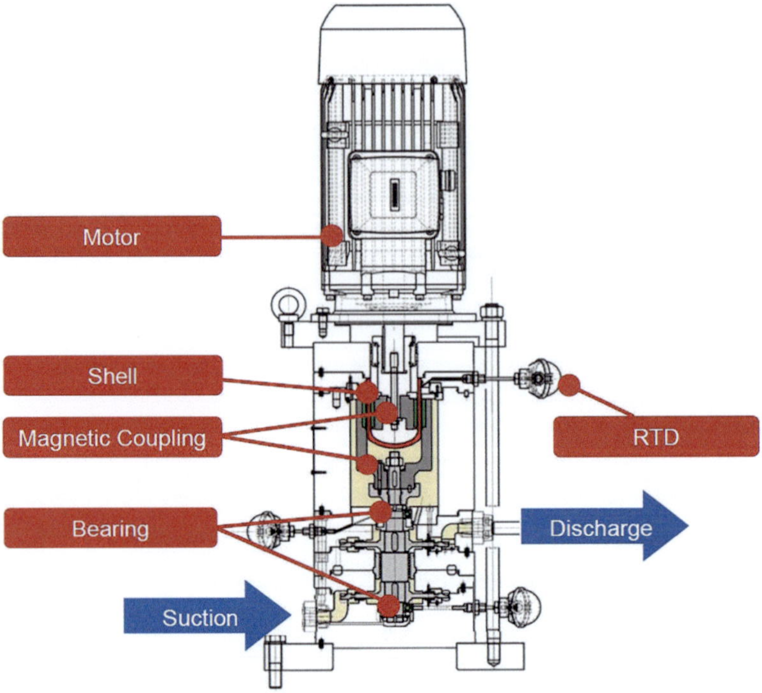

Fig. 6.10 Cross section of an electric-motor-driven seal gas booster (Image courtesy of Flowserve Corporation)

Table 6.1 presents an example of the control logic associated with seal gas boosters, and Fig. 6.11 illustrates a typical seal gas treatment module. For reference, Sect. 16.1.2 of Chap. 16 includes a sample purchase specification for a dry gas seal system.

Table 6.1 Typical control logic for a centrifugal compressor seal gas booster

Event	Action	On/off valve	Booster driver action	Booster mode
Compressor trip and differential pressure across compressor low	Automatic start	Open	Start	Automatic
Booster differential pressure high	Forced stop	Close	Stop	Automatic or manual
Booster casing temperature high	Forced stop	Close	Stop	Automatic or manual
Compressor driver running and differential pressure across compressor not low	Forced stop	Close	Stop	Automatic or manual
On/off valve not open when required to open	Forced stop	Close	Stop	Automatic or manual
On/off valve open during compressor operation	Stop	Open	Stop	Automatic or manual
On/off valve close during booster operation	Forced stop	Close	Stop	Automatic or manual
Booster motor manual start for online test	Manual start	Open	Start	Manual
Booster motor manual stop after online test	Manual stop	Close	Stop	Automatic or manual

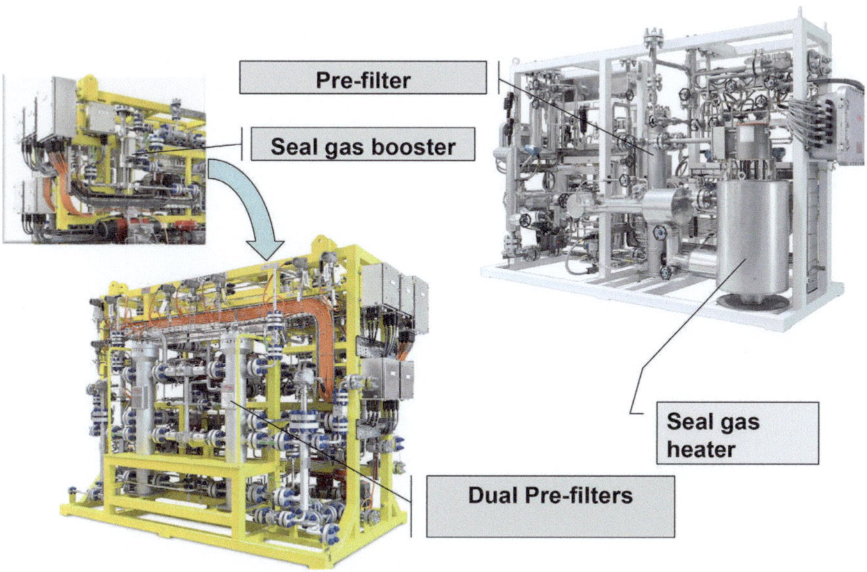

Fig. 6.11 Seal gas treatment system (Image courtesy of Flowserve Corporation)

Acknowledgements All of the images/graphics/figures in this chapter were provided by Flowserve Corporation. Some of the material in this chapter was reprinted with permission of the publisher from Ghaisas N, Majumdar S (2023) Seal gas boosters and heaters in dry gas seal support systems. CompressorTECH^2 magazine (Jan/Feb 2023), pp 36–39.

Reference

1. API Standard 692 (2018) Dry gas sealing systems for axial, centrifugal, rotary screw compressors and expanders, 1st edn. American Petroleum Institute, Washington, DC

Chapter 7
Lubricating Oil Systems

Abstract Lubricating oil systems play a critical role in ensuring the efficient operation, long-term reliability, and overall availability of turbomachinery trains (or strings). Any shortcomings in their design can lead to suboptimal performance and compromise not only the oil system itself but also the reliability and functionality of the entire equipment train it supports. To ensure optimal performance, the scope and configuration of the lubricating oil system for a given application should be collaboratively assessed by the responsible equipment vendor, the oil system designer, and the specifying machinery engineer. A thorough review of all system components is essential, and the design basis should be agreed upon by all parties as early as possible—ideally before the purchase order is issued. This chapter explores the key considerations for selecting and sizing the components that make up a pressurized lubricating oil system for modern turbomachinery strings. The design principles outlined are broadly applicable to oil systems serving a variety of driven and driving equipment types.

7.1 Information Required for Lube Oil System Configuration

To properly configure and design a lube oil system, manufacturers require detailed, application-specific information from the purchaser. Typically, the purchaser is the vendor responsible for the overall compressor-driver string or train.

For the basic or foundational design of a custom-engineered oil system, the following minimum inputs are essential:

- Oil specifications: This includes the type of oil, total flow rate, individual component flow rates, and pressure levels. The most commonly used mineral oils are ISO viscosity grades VG 32 and VG 46. Key properties for design considerations include viscosity across startup and operating temperature ranges, specific gravity or density, specific heat, pour point, and flash point.
- Heat load: The thermal energy that the system must manage.

N. Ghaisas, *Centrifugal Compressors in Hydrocarbon Processing Industries*,
https://doi.org/10.1007/978-3-032-03687-2_7

- Site conditions: Elevation, ambient temperature extremes, and the intended installation location—whether indoors (heated or unheated), outdoors, or exposed (e.g., on a steel module or under a roof).
- Utility and cooling details: Type of cooling medium (water or air) and available utility conditions.
- Electrical requirements: Area classification and any reduced-voltage start specifications.
- Component specifications: Desired features and optional configurations.
- System layout and maintenance: Requirements for accessibility, serviceability, and spatial arrangement.
- Transportation constraints: Limitations such as height restrictions may influence design choices, like orienting a tall reservoir horizontally.
- Equipment elevation and location: The relative position of the oil system to the equipment it serves. Interconnecting piping losses and elevation head differences must be factored into the required outlet pressure from the oil console. In certain cases, even minor elevation differences may necessitate low-profile reservoirs to accommodate sloped oil return lines above the maximum operating level.
- Accessory equipment: Any additional components to be included with the base system, such as oil mist eliminators or oil purifiers.
- End-user specifications: Any specific requirements or standards provided by the final customer.

Figure 7.1 illustrates a typical oil system configuration for a centrifugal compressor string.

Fig. 7.1 Lube oil system for a centrifugal compressor string (Image courtesy of Yoshimine, Etajima, Japan)

7.2 Typical Design Process

Using data provided by the purchaser, the lube oil system designer determines the reservoir size in accordance with the American Petroleum Institute (API) Standard 614 [1]. This includes meeting specifications for oil capacity, oil levels, and ensuring a minimum free surface area of 60 cm^2 per liter per minute (or 0.25 ft^2 per gallon per minute) of normal flow. The reservoir's shape—cylindrical, square, or rectangular—is selected based on available plot space, layout constraints, transportation limitations, structural bracing requirements, and accessibility for maintenance of top-mounted components. For cylindrical reservoirs, orientation is dictated by the space surrounding the lube oil console.

Purchasers typically specify their preferred oil pump type, choosing between positive-displacement rotary-screw or centrifugal pumps. If no preference is indicated, the designer selects the pump type after finalizing the required capacity and maximum discharge pressure.

Purchasers must also specify the heater type for the oil reservoir—either electric or steam. Heater sizing depends on the reservoir's total volume and geometry, the minimum ambient temperature, and whether insulation will be applied. According to API 614 [1], the default requirement is that the heater must raise the oil from the minimum ambient temperature to the minimum equipment startup temperature within 12 h, assuming an uninsulated reservoir and a wind speed of 10 mph (16 kph). Purchasers may request shorter heating durations or additional safety margins.

Pump discharge pressure (differential pressure) is calculated based on the highest pressure needed by the equipment served, such as bearing lubrication systems, hydraulic actuators for steam governors and extraction valves, and inlet guide vane positioners. This calculation also accounts for pressure losses across oil coolers, contaminated filters, piping, fittings, valves, control components, interconnections, and elevation differences.

For positive-displacement pumps, the discharge pressure must also accommodate the relief valve (or pressure-limiting valve) set point plus accumulation. This set point is intentionally set above the maximum required discharge pressure to prevent valve activation during pressure surges, such as when a standby pump is engaged in a dual-pump setup.

Horsepower sizing for steam turbine and electric motor drivers follows the same process, though the actual sizing differs depending on whether the pump is centrifugal or positive-displacement, as detailed later in the chapter.

Once these parameters are defined, components such as oil filters, coolers, transfer valves, relief valves, and pressure-regulating valves (both back-pressure and pressure-reducing) are sized. These results are then validated against the pump's capacity and discharge pressure requirements.

The final design pressure and temperature of the oil system determine the dimensions and specifications of piping and fittings. These are confirmed only after evaluating factors like maximum flow rate, optimal flow velocities in supply and return lines, pressure drops, and ensuring that gravity drain lines remain less than half full under normal operating conditions at peak flow, as per API 614 [1]. If no material preference is stated by the purchaser, piping materials and schedules are selected based on API 614 [1] minimum standards for lubricating and control oil services.

Sizing also includes the rundown tank and lube oil accumulator (for both lubrication and control oil). If applicable, mist eliminators and oil conditioners are selected. API 614 [1] offers standard piping and instrumentation diagrams and modular options for each subsystem. Purchasers must specify key details to enable full customization of the lube oil system, including the location, quantity, and type of flow, as well as instrumentation for pressure, temperature, and oil level.

Further component design specifics are addressed in the following sections.

7.3 Lube Oil Pumps and Drivers

7.3.1 Pumps

Lubrication and control systems may employ either positive-displacement or centrifugal pumps. If using positive-displacement pumps, additional components are necessary—specifically, discharge pressure-limiting valves (PLVs) and a back-pressure or spill-back regulating valve. These elements help manage pressure by redirecting excess pump output back to the reservoir. Figure 7.2 illustrates a back-pressure-regulating valve, with its impulse line—tapped downstream of the oil filters—partially visible in the image.

Centrifugal pump systems offer a simpler design, as they eliminate the need for pressure-limiting valves (PLVs) and back-pressure control regulating valves. However, they are generally less efficient than positive-displacement pumps. The choice between pump types is typically left to the purchaser's preference.

According to API 614 [1], positive-displacement lube oil pumps must be sized based on the lowest expected oil viscosity. Their capacity should be at least 120% of the normal oil flow under any operating condition, plus an additional 50% to accommodate control oil transient flow. This sizing guideline warrants careful evaluation—especially in high-flow systems exceeding 250 gal/min (946 L/min)—as a 20% margin can result in significantly oversized pumps, drivers, and piping. Such oversizing may lead to unnecessary power consumption and instability in oil pressure control. A more appropriate approach would be to define the margin relative to the normal oil flow rate.

Positive-displacement rotary-screw pumps must be capable of continuous operation at rated flow, the set pressure of the pressure-relief or pressure-limiting valve, and the minimum oil viscosity. This minimum viscosity corresponds to the highest oil temperature at the pump suction, plus the temperature rise within the pump. API 614 [1] includes this requirement to ensure pumps can withstand extreme conditions and prevent rotor-to-stator contact.

Fig. 7.2 Back-pressure control valve for positive-displacement oil pumps

For centrifugal pumps, a conventional 10% continuous head rise from the rated point to shutoff can destabilize the lube oil pressure controller due to significant pressure changes from minor flow variations. Therefore, API 614 [1] recommends a lower continuous head rise to shutoff—around 5%—for both single- and dual-pump configurations. This characteristic helps determine the pump's operating point on its performance curve. Additionally, API 614 [1] mandates that centrifugal pumps be sized to accommodate a potential 10% increase in discharge head pressure, should a larger impeller be installed in future.

The maximum required discharge pressure for both pump types is calculated by summing the relief valve accumulation pressure at full flow, static head, pressure losses in piping and fittings, control valve pressure drops, and the pressure drop across a dirty oil filter. API 614 [1] limits the pressure drop across clean filter elements to 5 psi (35 kPa) at 100 °F (40 °C) under normal flow conditions. This limit applies to the entire filter assembly, including the transfer valve and clean elements.

Several discharge pressure scenarios are typically evaluated, with the most demanding case selected for design purposes. The chosen pressure must also be compared against the highest coolant pressure—whether from water or a glycol/water mix—in shell-and-tube heat exchangers. Adjustments are made as needed to comply with API 614 [1], which requires the oil-side pressure to exceed the water-side pressure to prevent contamination in the event of a heat exchanger failure (e.g., tube leakage).

Pump discharge pressure evaluations typically include:

- Clean filter at startup temperature
- Clean filter at operating temperature (lowest pressure case)
- Dirty filter at operating temperature
- Full relief valve accumulation at minimum startup temperature (highest pressure case).

7.3.2 Pump Drivers

Driver Sizing Criteria for Oil Pumps

Steam turbines and electric motors used to drive positive-displacement oil pumps must be sized to meet the most demanding of the following conditions:

- The power required to operate the pump at the cold oil start temperature, considering the pump discharge relief (or pressure-limiting valve, PLV) setting plus accumulation pressure.
- 110% of the power needed to pump the rated flow at the typical reservoir temperature of 150 °F (66 °C), under maximum discharge pressure.
- For centrifugal pumps, driver sizing is also based on the most stringent of these conditions:
- The power required at cold oil start temperature when operating at runout capacity. While runout theoretically represents maximum flow at zero head, manufacturers limit the head-capacity curve just beyond the allowable operating range to prevent overload and excessive vibration. At system startup, centrifugal pumps may briefly operate at runout until flow resistance is established.
- The power needed at the rated head and flow rate, with oil viscosity corresponding to the standard reservoir temperature of 150 °F (66 °C).
- API 614 [1] sets the default cold oil start temperature at 50 °F (10 °C).

Steam Turbine and Motor Driver Considerations:

- Steam turbines are generally reserved for main oil pumps due to their slow startup time. They are not recommended for standby pumps, which require rapid activation.
- These turbines are typically single-stage and conform to API 611 [2].
- Electric motor drivers are typically
  - Designed per NEMA B standards.
  - Have a 1.15 service factor.
  - Selected for full-voltage (across-the-line) startup to minimize pressure dips during automatic standby pump activation.

In some projects, reduced-voltage startup is specified for larger motors based on site-specific electrical conditions. This approach increases startup time and must be accounted for in system design by incorporating accumulators with sufficient capacity.

To avoid inefficient oversizing—which can increase equipment space, weight, and cost—drivers should be evaluated against the following discharge pressure conditions:

- Maximum operating pressure.
- 110% of maximum operating pressure.
- Maximum relief valve setting.
- 110% of relief valve setting.
- Maximum relief valve accumulation pressure (not to exceed 110% of the system's maximum allowable working pressure).
- 110% of the relief valve accumulation pressure.
- API 676 [3] basis for maximum relief valve accumulation pressure.

High-Elevation Adjustments:

At elevations above 3280 ft (1000 m), reduced air density can degrade electric motor performance. Per NEMA MG-1 guidelines, motor ratings may decrease by approximately 3% for every additional 1640 ft (500 m) above this threshold. Higher altitudes also contribute to increased temperature rise, which must be factored into motor selection for high-elevation installations.

7.4 Relief Valves

This section covers the relief valves installed on the pump discharge, the thermal overpressure protection system, and the bypass relief mechanisms.

7.4.1 Pump Discharge Service

Pressure-limiting valves (PLVs) on the discharge side of positive-displacement pumps must be of the modulating type designed for liquid service to avoid valve chatter and excessive blowdown during closure. According to API 614 [1], these valves are not required to comply with ASME code standards outlined in API 520 [4], 521 [5], and 526 [6], and are distinct from conventional "pressure safety" relief valves.

PLVs must be sized to accommodate the full maximum output flow of the pump, ensuring that accumulation pressure does not exceed 110% of the system's design pressure. This aligns with piping and vessel design codes, which allow a temporary operating pressure up to 110% of the rated design pressure during relief events.

Fig. 7.3 Pressure-limiting valves

However, certain non-ASME PLVs may exhibit accumulation pressures exceeding 10%. To remain within code limits, the fully accumulated relief pressure must stay below 10% above the system's design pressure, which may necessitate a lower valve set point.

The highest flow condition for a positive-displacement pump typically occurs during cold start-up, when oil viscosity is at its peak (maximum). Figure 7.3 illustrates the placement of PLVs on oil pump discharge lines.

7.4.2 Thermal Overpressure Protection Service

Standby vessels—such as coolers and filters—that may be isolated from the system can necessitate the use of thermal relief valves. In duplex configurations of coolers and filters, API 614 [1] permits the use of fill equalizing lines with orifices to mitigate oil-side thermal overpressurization, provided the orifice is properly sized for thermal relief and the equalizing valve remains locked in the open position.

Shell-and-tube coolers may also require thermal overpressure protection on the cooling water side if that portion of the system can be blocked in.

It is prudent to consider the volume of fluid trapped and the maximum potential temperature rise under fire exposure conditions in the sizing of thermal relief valves. API 520 [4] and API 521 [5] offer guidance for determining appropriate relief sizing in such scenarios.

7.4.3 Bypass Relief Service

Relief valves may also be necessary to safeguard against overpressure resulting from a malfunction of a pressure-regulating control valve—particularly in scenarios where a lube oil supply pressure-regulating valve fails in the open position.

In such cases, the relief valve must effectively cap the lube oil supply pressure to the unit and be properly sized to redirect the surplus flow back to the reservoir during the failure event. The required relief capacity should be calculated based on the flow rate of the failed-open regulating valve at the designated relief pressure, minus the flow directed to the primary equipment. For this type of service, modulating PLVs are recommended.

7.5 Lube Oil Heat Exchangers

Lube oil heat exchangers (coolers) must be properly sized to ensure complete cooling of the oil. The total cooling duty encompasses both the heat transferred into the oil from the serviced equipment and the thermal energy introduced by the oil pumps.

Depending on purchaser specifications, oil coolers may be either water-cooled or air-cooled. This selection is typically influenced by the availability of a suitable liquid cooling medium, such as cooling water.

7.5.1 Water Coolers

Water coolers are typically designed as either shell-and-tube or plate-and-frame units, depending on the specified requirements. Plate-and-frame coolers are particularly valued for their compact form, which allows for substantial space savings.

According to API 614 [1], a water-side fouling factor of 0.002 h ft^2 °F/Btu (0.352 m^2 K/kW) must be incorporated to maintain effective cooling performance even in the presence of moderate fouling. To help mitigate fouling while ensuring adequate flow, water velocity should be maintained between 5 and 8 ft/s (1.5–2.5 m/s). Although the ideal design aims for a minimum water temperature increase of 20°R (10 K), achieving this target can be challenging when trying to preserve the required flow velocity—which takes precedence.

In critical or high-priority machinery systems, water coolers are arranged in duplex configurations. These setups feature two full-capacity heat exchangers equipped with continuous-flow transfer valves, allowing for seamless online switching or parallel operation.

7.5.1.1 Shell-and-Tube–Type Coolers

According to API 614 [1], shell-and-tube coolers are typically designed to meet TEMA C standards, as defined by the Tubular Exchanger Manufacturers Association. These coolers are generally arranged with oil flowing through the shell side and water through the tube side. To minimize the risk of oil contamination in the event of a heat exchanger failure, the oil pressure is maintained higher than the water pressure. However, purchasers may opt to reverse this pressure differential—allowing water pressure to exceed oil pressure—if environmental considerations warrant it.

Depending on operational experience and specific facility needs, purchasers may also request more robust coolers built to TEMA R specifications, along with additional design requirements. API 614 [1] mandates the use of removable tube bundles and prohibits U-bend tubes unless explicitly approved by the purchaser. A variety of TEMA head and shell configurations are available, as illustrated in Fig. 7.4.

Figure 7.5 illustrates a typical TEMA lube oil exchanger setup known as the AEW configuration, featuring removable channel covers, enclosed stationary heads, a single-pass (once-through) shell-side flow, and externally sealed floating tube sheet return or rear heads.

Use of U-bend tube exchangers requires purchaser approval, as their design can complicate internal tube cleaning. While API 614 [1] outlines the standard materials for cooler construction, alternative materials may be specified at the purchaser's discretion.

7.5.1.2 Plate-and-Frame–Type Coolers

As previously mentioned, plate-and-frame coolers offer a compact design that optimizes space utilization. According to API 614 [1], austenitic stainless steel is the standard material for heat-transfer plates, though purchasers may request alternative materials. These coolers are designed to handle full operational load (100% duty) and must also accommodate future expansion by allowing for the addition of at least 20% more plates. Figure 7.6 presents a cutaway or exploded view of a representative plate-and-frame cooler.

Plate-and-frame coolers consist of a sequence of chevron-patterned heat-transfer plates positioned between two frame plates. Fluid inlet and outlet ports are located at the four corners of the front frame plate and are aligned with corresponding openings in each heat-transfer plate. Oil and cooling water are routed to opposite sides of each plate, generally following a counterflow arrangement to enhance thermal efficiency. Figure 7.7 illustrates the fluid flow path within a typical plate-and-frame cooler.

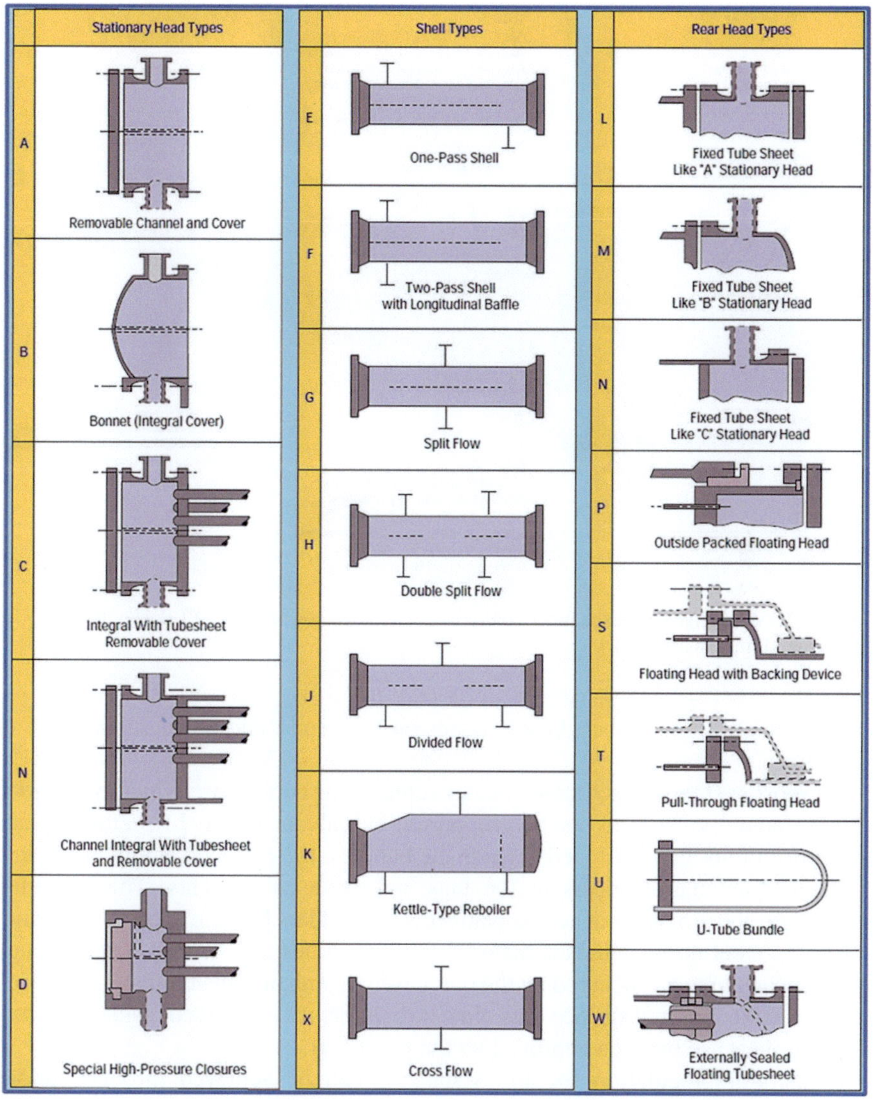

Fig. 7.4 TEMA exchanger head and shell configurations (Image courtesy of G.J. Oliver, Inc., Phillipsburg, NJ)

7.5.2 Air Cooled Heat Exchangers

Air-cooled lube oil heat exchangers are utilized if a suitable cooling fluid system is unavailable. According to API 614 [1], these exchangers, by default, shall include two fans, each capable of handling 100% of the cooling load. The design should

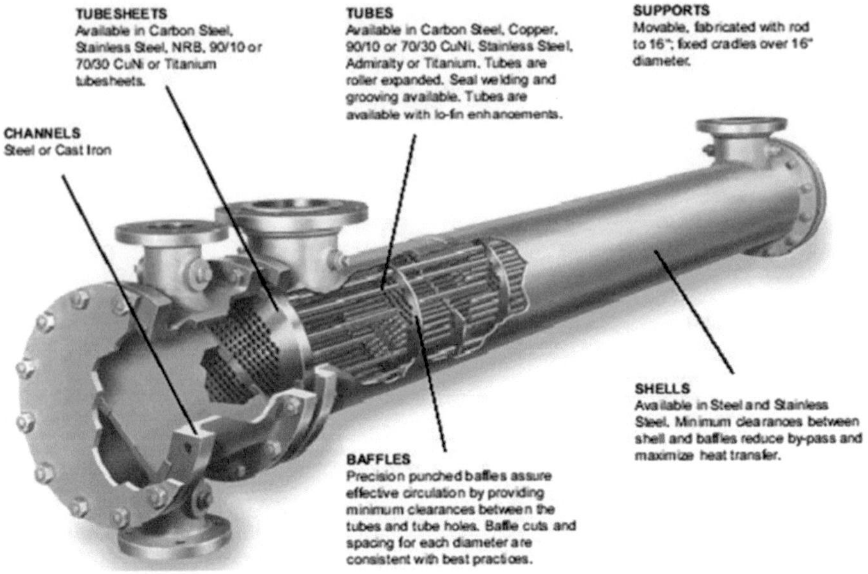

Fig. 7.5 TEMA AEW heat-exchanger configuration (Image courtesy of API Basco, through G.J. Oliver, Inc., Phillipsburg, NJ)

feature straight-finned tubes, header boxes on both ends, and an additional 10% of spare tubes.

For most systems—excluding smaller setups—air coolers are typically installed with horizontally oriented bundles supported by galvanized steel structures. Since forced-draft fans are positioned beneath the bundles, the entire assembly is elevated to allow maintenance access to the fans and drive components. This elevation necessitates the inclusion of access platforms and ladders for safe entry to manways.

Carbon steel is the standard material for both tubes and header boxes, though stainless steel may be specified by the purchaser. To enhance heat transfer efficiency and reduce the required surface area, flow turbulators may be added inside the tubes; however, their use must be approved by the purchaser.

Temperature regulation for air exchangers can be achieved through various methods, including:

- Adjustable-frequency drive (AFD) or variable-frequency drive (VFD) fan motors for speed control
- Louvers to modulate airflow
- Oil bypass systems for thermal control.

To protect the tube fins from damage caused by hail, hail guards are considered essential.

If air-cooled exchangers are employed, the oil volume within the exchanger bundles and associated piping must be factored into the oil reservoir rundown

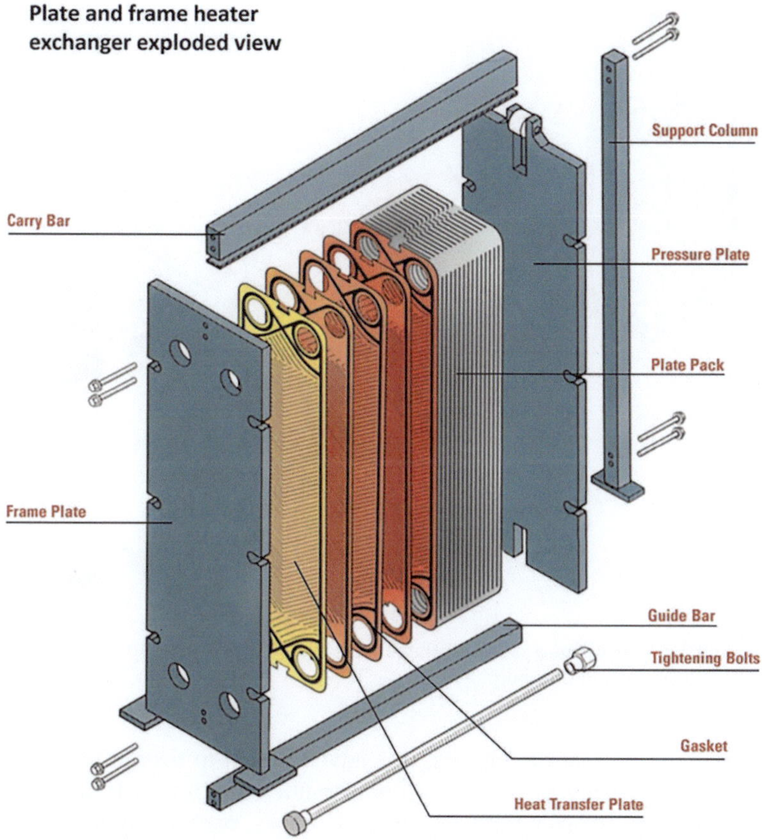

Fig. 7.6 Plate-and-frame cooler arrangement (Image courtesy of Standard-Xchange, through G.J. Oliver, Inc., Phillipsburg, NJ)

capacity. The length and diameter of the interconnecting piping affect not only the total oil volume but also contribute to frictional pressure losses and bundle flow resistance—both of which must be considered when determining the required pump discharge pressure.

As illustrated in Fig. 7.8, space constraints may necessitate mounting the air heat exchanger directly atop the oil console.

7.6 Lube Oil Temperature Control

Oil temperature regulation can be achieved either by routing the entire oil flow through the cooler or by diverting a portion of it around the cooler (bypassing) to maintain the target supply temperature.

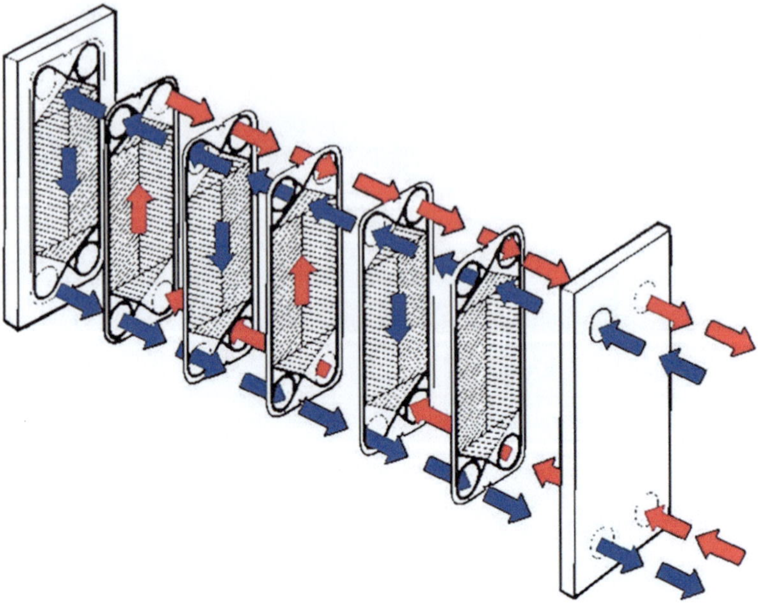

Fig. 7.7 Fluid flow path within a plate-and-frame exchanger (Image courtesy of Standard-Xchange, through G.J. Oliver, Inc., Phillipsburg, NJ)

For water-cooled heat exchangers, it is advisable not to restrict the water flow. Lowering water velocity as a means of temperature control may lead to increased fouling on the heat-transfer surfaces within the tubes, particularly in conventional shell-and-tube configurations.

The following section outlines three primary configurations for oil temperature control valve arrangements.

7.6.1 Thermostatic Temperature Control

The first method of oil temperature control utilizes a self-contained thermostatic three-way mixing valve or regulator. This component features two inlet ports and a single outlet port that delivers oil at a controlled temperature. One inlet receives cooled oil from the heat exchanger's outlet, while the other draws uncooled oil from upstream of the exchanger—commonly referred to as cooler bypass oil.

Inside the valve, a modulating plug element responds to temperature changes by adjusting the flow through the bypass port. This movement is driven by a thermally responsive wax element, which expands or contracts with temperature, thereby shifting the valve plug. The valve blends the hot bypass oil and cooled oil to achieve a preset temperature, determined by the specific thermostatic element installed.

Fig. 7.8 Air heat exchanger mounted on the lube oil system (Image courtesy of G.J. Oliver, Inc., Phillipsburg, NJ)

For added reliability, these valves may include manual override mechanisms that allow the plug to be manually positioned to block the bypass port in case of element failure. Additionally, block and bypass configurations can be integrated to enable online maintenance without interrupting system operation.

7.6.2 Two-Way Cooler Bypass Temperature Control

The second temperature control configuration incorporates a two-way valve that creates a parallel bypass route for hot oil around the heat exchanger. This setup enables the mixing of hot and cooled oil to achieve the target temperature. Typically, the valve is pneumatically actuated and regulated by a temperature controller, with the temperature sensor positioned downstream of the mixing point where bypass and cooled oil converge.

Since the bypass runs parallel to the heat exchanger, it's not possible to redirect the entire flow away from the exchanger. To allow for bypassing roughly 55–65% of the total flow, the valve must be sized so that, when fully open at maximum flow, the pressure drop across the bypass path is between 15 and 25% of the pressure drop across the heat exchanger path under the same conditions.

For ease of maintenance during operation, two-way temperature control valves are generally equipped with block and bypass valves. They may also include manual override hand wheels for added control.

7.6.3 Three-Way Temperature Mixing Control

The most effective temperature regulation setup utilizes a three-way control valve. Functionally, it resembles a self-contained thermostatic valve by blending cooled oil with hot bypass oil to achieve the target temperature. However, unlike the thermostatic variant, this system employs a temperature controller that monitors oil temperature downstream of the mixing point.

Key Advantages:

- Flexible Flow Routing: Oil can be directed entirely through the heat exchanger, entirely through the bypass, or through any proportional mix needed to maintain the desired temperature or conduct diagnostic operations. The controller also allows manual override to position the valve as required.
- Adjustable Set Point: The oil temperature set point can be modified via the controller's input. In certain cases, operating equipment—such as rotor systems—may perform optimally at slightly higher oil viscosity, which can be achieved by lowering the oil supply temperature.
- Bypass Temperature Control: A temperature-control valve is installed on a bypass line around the oil cooler to regulate and stabilize the oil supply temperature. As illustrated in Fig. 7.9, the setup includes a three-way valve with an impulse line sensing the outlet temperature and bypass piping to support valve maintenance.

7.7 Lube Oil Filters

Duplex filters, each rated for 100% capacity, are commonly employed in lubrication and hydraulic oil systems supporting turbomachinery trains. Their primary role is to safeguard equipment bearings by eliminating contaminants, thereby minimizing oil consumption and disposal needs and prolonging lubricant life. Common contaminants found in lubricating oil include:

- Solid particles
- Wear metals
- Acids
- Moisture
- Varnish
- Dissolved gases.

To properly size and select lube oil filters, the following information is essential:

- Type of oil: mineral or synthetic
- Oil viscosity and corresponding minimum/maximum operating temperatures
- Required filtration level
- Nature of contaminants
- System volume or capacity
- Design and peak flow rates
- Operating and design pressures and temperatures
- Filter element type
- Maximum allowable pressure drop and related temperature conditions
- Applicable construction codes (e.g., ASME Section VIII, Division 1).

Fig. 7.9 Three-way oil temperature control valve

According to API 614 [1], filters must be sized to ensure a clean element pressure drop of less than 5 psi (35 kPa) at maximum flow and 100 °F (40 °C). This requirement applies to the entire filter assembly, including the transfer valve, not just the filter element.

When operating under maximum flow with a dirty filter, a differential pressure between 15–25 psi (103–172 kPa) typically signals the need for element replacement. Pressure drop calculations should consider worst-case flow scenarios and cold-start oil viscosity. A safety margin of 3–4 times is recommended between the element's collapse pressure and the maximum pressure drop. API 614 mandates a minimum collapse differential pressure rating of 70 psi (500 kPa) for filter elements.

Flow Direction Considerations:

By default, API 614 specifies that oil should flow from the outside of the filter element toward its center. However, coalescing filter elements—used in oil conditioners and purifiers—typically operate in reverse, flowing from the center outward to enhance coalescing efficiency.

Customization and Configuration Options:

After sizing is finalized, filters can be tailored to meet specific application needs. Custom engineering options include:

- Connection types: flanged or threaded (e.g., Female National Pipe Tapered [FNPT])
- Cover lifters: required for covers over 35 lb (16 kg) per API 614
- Instrumentation: differential pressure gauges or transmitters
- Dedicated control oil filter: if needed
- Valved vents and drains: e.g., ¾-in. ANSI 300# flanged connections
- Code compliance: ASME National Board, Canadian Registration Number (CRN), Pressure Equipment Directive (PED), etc.
- Gasket material: must be compatible with oil type
- Transfer valve options: integral, straight plug, diverter, ball, two-way, six-ported, continuous-flow, with equalizing line or thermal relief valves
- Controlled pleat radius—enhances filtration area, dirt retention, and structural integrity
- Additional specifications: some purchasers may require extra NDE, corrosion allowances, weld neck flanges, and flanged vents/drains.

API 614 [1] establishes a default minimum filtration requirement of 10 μm (394 microinches) with an efficiency of at least 90%, represented by a beta ratio of $\beta_{10} \geq 10$.

Filter performance is quantified using the beta ratio (β_x), which compares the number of particles larger than a specified size x before and after filtration. This ratio is defined by Eq. 7.1.

The filter beta ratio provides a more accurate way to assess filter performance, as 'nominal' ratings can vary significantly between manufacturers due to differing efficiency definitions. By relying on the beta ratio, one can avoid the inconsistencies and potential confusion associated with nominal ratings when comparing filters across various manufacturers and types.

$$\beta_x = \frac{N_u}{N_d} \tag{7.1}$$

where x is the particle diameter (μm), N_u is the number of particles greater than x μm upstream of the filter, and N_d is the number of particles greater than x μm downstream of the filter.

The following two examples illustrate how the beta ratios of filters are calculated:

- If 15,600 particles greater than 6 μm are upstream of the filter and 78 particles greater than 6 μm are downstream of the filter, then $\beta_6 \geq 15{,}600/78$, or $\beta_6 \geq 200$.
- If 15,982 particles greater than 5 μm are upstream of the filter and 213 particles greater than 5 μm are downstream of the filter, then $\beta_5 \geq 15{,}982/213$, or $\beta_5 \geq 75$.

The particle removal efficiency of a filter is determined as $\eta = \{[1 - (1/\beta_x)] \times 100\}$. Table 7.1 lists filter efficiencies for various beta ratios. As a default, API 614 [1] specifies a minimum efficiency of 90% for 10-μm (394 μin.) particles ($\beta_{10} \geq 10$) and a minimum efficiency of 99.5% for 14 μm (551 μin.) particles ($\beta_{14} \geq 200$), when tested in accordance with international standard test method ISO 16889 (Hydraulic fluid power—Filters—Multi-pass method for evaluating filtration performance) to a minimum differential pressure of 50 psi (345 kPa) at the end of a test run.

Cellulose paper cartridge filters are hygroscopic, meaning they absorb moisture, which leads to frequent replacement. As a result, they are rarely used in contemporary oil systems. Operators who prioritize reliability have largely transitioned to synthetic, non-hygroscopic filter elements for lubrication, hydraulic, and control oil applications. As shown in Fig. 7.10, cellulose filters (labeled PL718) exhibit a significantly higher pressure drop under flow conditions compared to three synthetic alternatives (labeled PH718).

Figure 7.11 illustrates a duplex filter configuration featuring an equalizing line and a transfer valve. In Fig. 7.12, the filter elements are shown incorporating a central support tube for added structural integrity.

Static electricity (charges) can accumulate on nonconductive components within a lube oil system—such as filter elements—when the lubricant exhibits low electrical

Table 7.1 Correlation between filter element beta ratio and efficiency (table courtesy of Hilliard Corporation, Elmira, NY)

Beta ratio	Particle removal efficiency (%)
1	0
2	50
4	74
10	90
20	95
75	98.67
100	99
200	99.5
1000	99.9

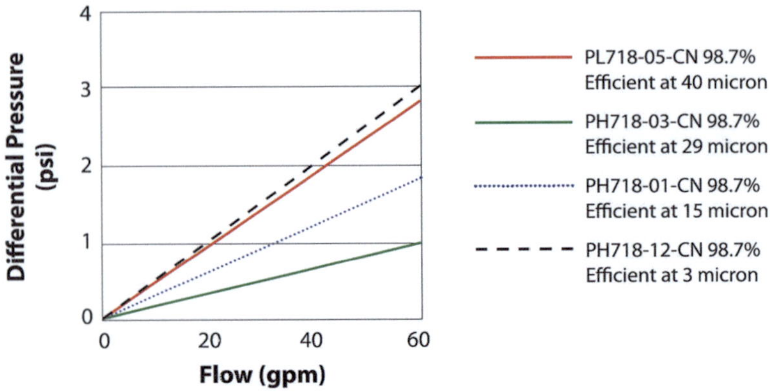

Fig. 7.10 Flow versus pressure drop of cellulose (PL) and synthetic (PH) filters passing 220 SSU oil (where SSU stands for the viscosity unit seconds, Saybolt universal) (Image courtesy of Hilliard Corporation, Elmira, NY)

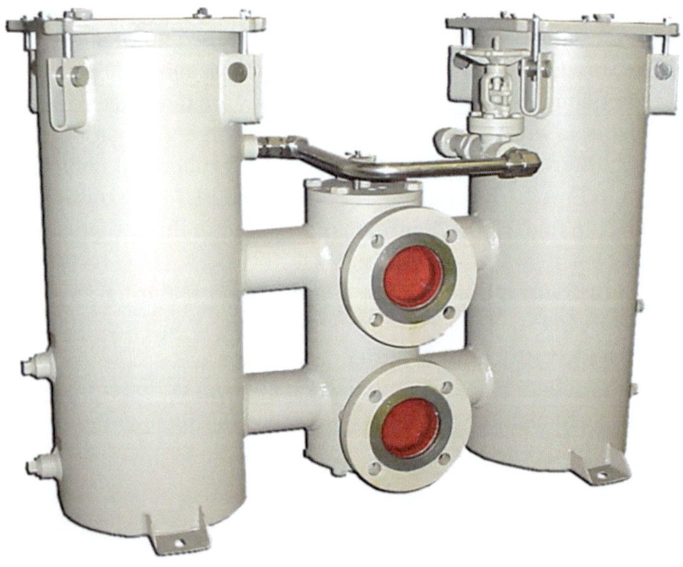

Fig. 7.11 Duplex filter (Image courtesy of Hilliard Corporation, Elmira, NY)

conductivity (below 5×10^{-5} µS/m or 50 pS/m). Once enough charge builds up, a static discharge may occur between areas of differing electrical potential. Over time, these discharges contribute to oil degradation and promote varnish formation throughout the system. Factors such as oil chemistry, localized hot spots, and repeated static discharge events can intensify varnish buildup.

To mitigate this risk, antistatic filters are employed. One effective design incorporates conductive fibers pleated between two layers of microglass media, allowing

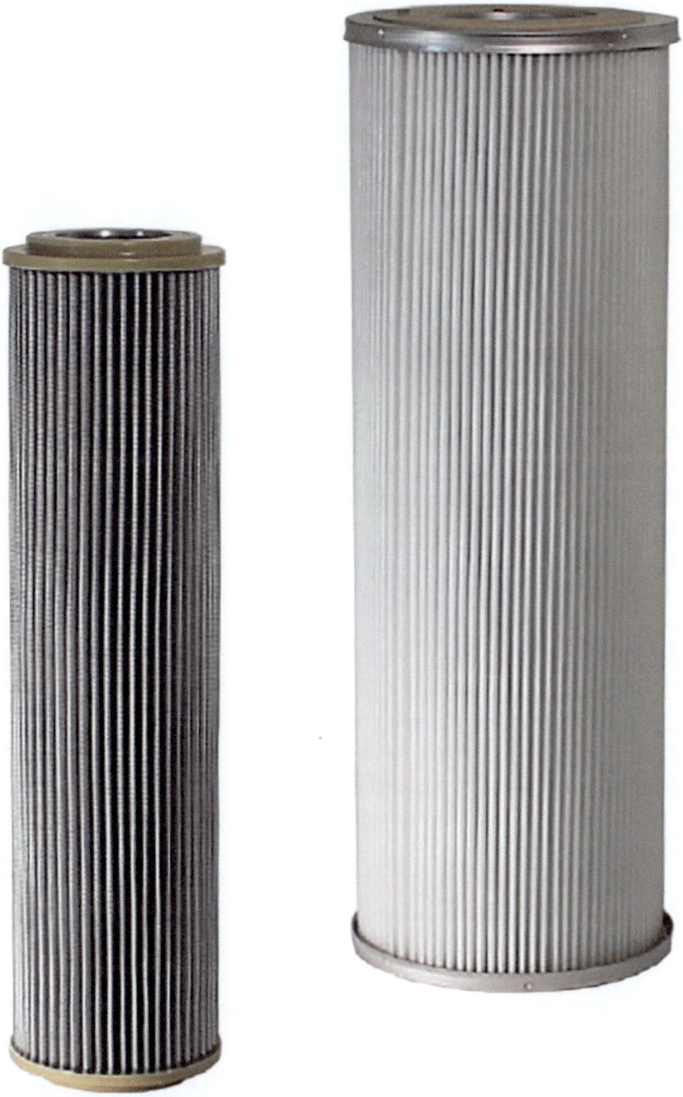

Fig. 7.12 Filter elements (Image courtesy of Hilliard Corporation, Elmira, NY)

static charges to dissipate before discharge can take place. The electrical conductivity of lubricants can be measured using ASTM Standard D2624, which outlines test methods for aviation and distillate fuels.

Both laboratory and field studies have shown that fluid conductivity directly influences the frequency and intensity of static discharge events. Oils with conductivity below 3.5×10^{-5} μS/m (35 pS/m) are susceptible to high-energy discharges, while

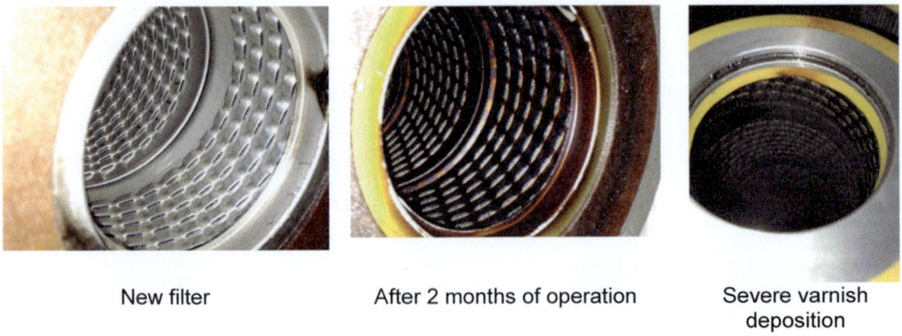

Fig. 7.13 Effects of static discharge on the lube oil filter element (Image courtesy of Hilliard Corporation, Elmira, NY)

those exceeding 2×10^{-4} μS/m (200 pS/m) typically show no discharge-related issues. Figure 7.13 illustrates the progressive impact of static discharge on a lube oil filter element.

7.8 Transfer Valves

Transfer valves are commonly employed in duplex oil filters, as well as in duplex shell-and-tube and plate-and-frame heat exchangers. Modern designs often incorporate equalization ports or pressure equalization lines to ensure seamless flow transition and minimize the torque required when switching from an active unit to a standby component.

Some transfer valves utilize metal-to-metal seals or sealing pads coated with low-friction materials, while others embed elastomeric compounds within the sealing surfaces. Both configurations are engineered to enable low turning torque during operation.

Valves with full-port designs help reduce pressure drop across the system. According to API 614 [1], acceptable configurations include single-body, six-port straight plug-type transfer valves or a pair of three-way ball or plug valves linked by a unified operating lever.

Three-way, six-port valves may feature oil inlet and outlet ports positioned either on the same side or on opposite sides, as depicted in Fig. 7.14. Split-type three-way transfer valves are illustrated in Fig. 7.15, and comparative data on pressure drop versus flow for same-side and split-type configurations is shown in Fig. 7.16.

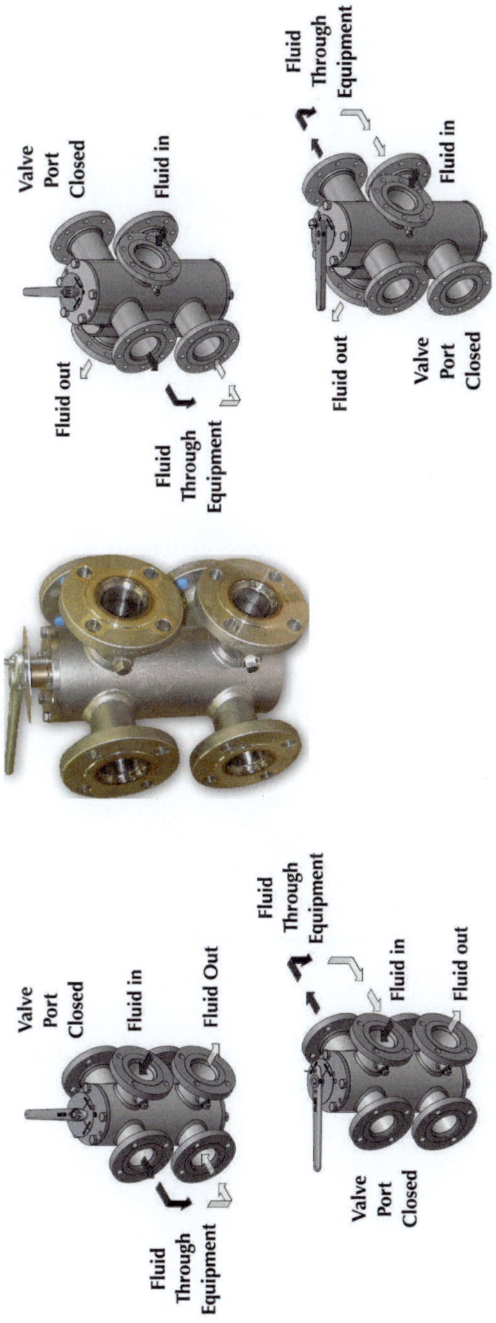

Fig. 7.14 Six-port transfer valves (Image courtesy of Hilliard Corporation, Elmira, NY)

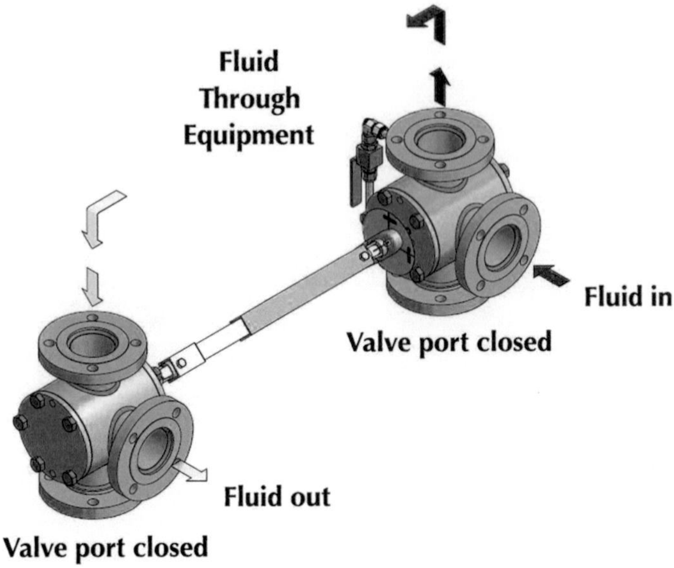

Fig. 7.15 Split three-way transfer valve (Image courtesy of Hilliard Corporation, Elmira, NY)

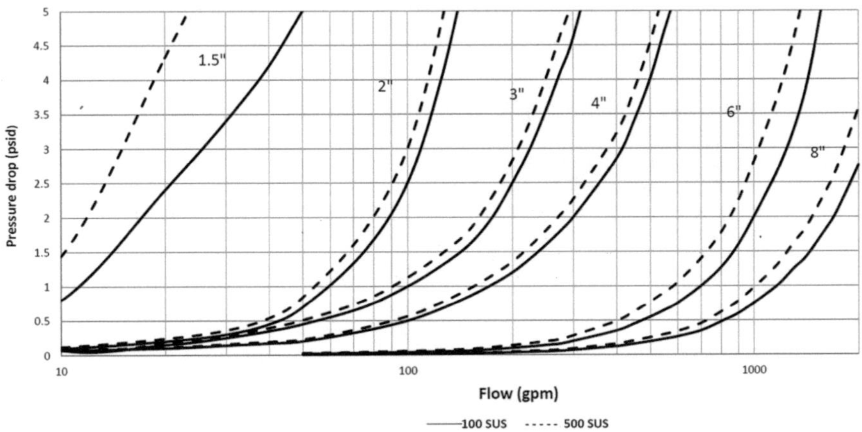

Fig. 7.16 Pressure drop versus flow characteristics for split-type and same-side transfer valves (Image courtesy of Hilliard Corporation, Elmira, NY)

7.9 Pressure Regulators for Lube Oil and Control Oil

This section covers two distinct types of regulators used for lubricating oil and control oil systems.

7.9.1 *Back-Pressure Regulator*

Positive-displacement pumps deliver a constant discharge flow that exceeds the system's required output, as defined by the flow margin criteria in API 614 [1]. Consequently, the surplus flow must be redirected back to the reservoir via a back-pressure regulator valve. These valves must also accommodate the additional flow generated when both the main and standby pumps operate simultaneously—effectively doubling the recirculated volume with the second pump's full discharge.

Back-pressure valves should be of the "fail-close" type and capable of maintaining discharge pressure across all operating conditions, ensuring that pressure-limiting valves (PLVs) do not activate when the standby pump is engaged.

The following example outlines how to determine the required flow coefficient (C_V) range for a back-pressure valve in a system using motor-driven main and standby positive-displacement pumps under these conditions:

- Lube oil reservoir is under atmospheric pressure.
- Fluid: ISO VG 32 with a specific gravity (SG) of 0.87 and a viscosity of 650 s, Saybolt universal (SSU) at 50 °F (10 °C) and 60 SSU at 150 °F 66 °C).
- Lube oil supply flow: 85 gpm at 25 psig (322 L/min at 172 kPa) (including 5 psi [34.5 kPa] combined frictional and head loss to the equipment train).
- Turbine governor control oil: 5 gpm (19 L/min) normal, (15 gpm [57 L/min] transient) at 115 psig (793 kPa).
- Required pump output (based on inclusion of the control oil transient): 120 gpm (454 L/min).
- Actual pump discharge flow: 123 gpm at 150 psig and 150 °F (466 L/min at 1,034 kPa and 66 °C) hot, 125 gpm at 150 psig and 50 °F (473 L/min at 1,034 kPa and 10 °C) cold.
- Pump PLV setting of 175 psig (1,207 kPa).

It's important to recognize that the back-pressure valve experiences a pressure drop equivalent to the full pump discharge pressure, since its outlet returns to an atmospheric reservoir. This condition can potentially lead to flashing in certain fluids. However, most lubricating oils have a vapor pressure close to 0 psia (0 kPa), making flashing unlikely and not a cause for concern.

The minimum C_V value is determined from the minimum recirculation flow, based on one pump running hot with lube oil and control oil at transient flow (Eq. 7.2).

$$Q_{\text{min}} = 123 - (85 + 15) = 23\,\text{gpm} \quad \text{(U.S. Conventional [USC] Units)} \qquad (7.2a)$$

$$Q_{\min} = 466 - (322 + 57) = 87\,\text{L/min} \quad \text{(Système international d'unités [SI] Units)} \tag{7.2b}$$

To be conservative, the minimum flow sizing for C_V is considered at the highest inlet pressure of the back-pressure valve, which is the PLV set pressure of 663 kPa (175 psig).

$$\text{Minimum } C_V = 1.86$$

The maximum C_V value is determined from the maximum recirculation flow based on two pumps running with cold oil with the minimum system output flow of lube oil and normal control oil flow (Eq. 7.3).

$$Q_{\max} = 2 \times 125 - (85 + 15) = 150\,\text{gpm} \quad \text{(USC Units)} \tag{7.3a}$$

$$Q_{\max} = 2 \times 473 - (322 + 19) = 605\,\text{L/min} \quad \text{(SI Units)} \tag{7.3b}$$

To be conservative, the maximum flow sizing for C_V is considered at the normal inlet pressure of the back-pressure valve, which is 150 psig (1,034 kPa).

$$\text{Maximum } C_V = 13.89$$

Note that the C_V calculations presented in Eqs. 7.2 and 7.3 are general computations using the equation

$$C_V = K_f Q \left(\frac{\text{SG}}{P_1 - P_2} \right)^{1/2} \tag{7.4}$$

where C_V is the valve flow coefficient; K_f is a valve factor incorporating the recovery factor, viscosity correction, and so on; Q is the flow rate in gpm (L/min); SG is the specific gravity of the fluid; P_1 is the upstream pressure in psig (kPa); and P_2 is the downstream pressure in psig (kPa).

C_V is typically computed using the valve manufacturer's sizing equations, recovery, and viscosity correction factors and/or software.

The back-pressure valve should be selected such that the required C_V operating range is within the interval between 10 and 90% of the back-pressure valve's available operating C_V range.

Back-pressure valves can be either direct-acting regulator valves or pneumatically operated control valves.

Direct-acting regulator valves offer faster response times than pneumatic control valves, but operate with proportional-only control. This means they require a buildup of pressure before opening further to accommodate increased flow. When designing for a two-pump operation, attention must be paid to the actuator diaphragm and spring sensitivity (expressed in psi per inch of valve travel), as well as the pressure buildup needed to initiate valve movement.

To prevent premature lifting of the pump pressure-limiting valves (PLVs) when the second pump starts, their set pressure should be configured with a margin above the peak pressure observed during dual-pump operation. A transient overshoot in steady-state pressure is expected at startup due to the slower response of the back-pressure valve loop compared to the rapid output of the second pump.

Pressure-sensing lines should be designed to self-purge trapped air—ideally by integrating bleed return line orifices—since air compressibility can delay valve response.

Valve trim characteristics such as quick opening, linear, and equal percentage should be assessed to ensure consistent loop gain and system stability. Some direct-acting back-pressure valves feature actuators with limited stem travel relative to the valve body trim, making quick opening trims advantageous, as only a portion of the valve's full capacity is utilized.

If pneumatic valves are employed, the entire control loop—including the pressure transmitter, controller, valve positioner, and control valve—must deliver rapid yet stable performance. In this setup, a finely tuned proportional-plus-integral (reset) pressure controller is used to optimize both steady-state accuracy and dynamic response.

7.9.2 Pressure-Reducing Valve (Oil Supply Header Pressure Control)

Pressure-reducing valves should be designed as fail-open types and operate within a significantly narrower CV range. This is because both the oil flow rate and outlet pressure remain constant, leaving inlet pressure as the sole variable. Fluctuations in inlet pressure arise from several factors, including pressure loss across the filter depending on element cleanliness, the allowable pressure drop before triggering the standby pump, and the peak pressure generated when the pump runs just below the pressure-limiting valve (PLV) set point.

For the example system described in Sect. 7.9.1, determination of the required valve flow coefficient (C_V) range for the pressure-reducing valve would be performed as follows:

The minimum C_V value is determined from the normal lube oil flow of 85 gpm (322 L/min) with the valve inlet pressure associated with the pump operating near the PLV set pressure of 175 psig (1,207 kPag) with clean filters (5 psid [34 kPa differential]) and normal pressure losses for other parts of the system, which, in this case, amount to 160 psig (1,103 kPa).

$$\text{Minimum } C_V = 7.16$$

The maximum C_V value is determined from the normal lube oil flow of 85 gpm (322 L/min) with the minimum valve inlet pressure associated with the pump operating at 10 psi (60 kPa) below the 150 psig (1034 kPa) normal discharge pressure (for standby pump start), with dirty filters (at 15 psid [103 kPa]) and other normal pressure losses for other parts of the system, which, in this case, correspond to a valve inlet pressure of 115 psig (793 kPa).

$$\text{Maximum } C_V = 8.35$$

Trim characteristics—such as quick opening, linear, or equal percentage—should be selected with careful consideration of the valve's loop gain to maintain system stability. Additionally, it's critical to prevent overpressurization of the equipment train's lube oil supply in the event of a valve failure in the open position. If the valve cannot be configured to avoid excessive pressure under such conditions, it should be equipped with an open travel stop to restrict its output. Alternatively, a pressure-limiting valve (PLV) can be installed downstream of the pressure-reducing valve to divert surplus oil back to the reservoir in case of failure.

Similar to back-pressure control systems, pressure-sensing lines must be designed to purge trapped air effectively. This can be achieved by integrating bleed return line orifices, as air compressibility can delay valve response times. For pneumatic valves, the same dynamic response criteria apply as with back-pressure valves.

Figure 7.17 illustrates a pressure-reducing control valve positioned on an oil supply header, along with its sensing line downstream of the valve.

Fig. 7.17 Pressure-reducing control valve on oil supply header equipment

7.10 Oil System Piping Design

The mechanical design rating for the lubrication and control oil system must be applied to all pressure-containing elements—such as piping, vessels, and components—and must be at least equal to the pump discharge pressure-limiting valve (PLV) set point. The combination of the PLV set point and its accumulation defines the maximum pressure the system must be designed to withstand.

Pressure vessel codes including ASME, Canadian standards, and PED permit operation up to the vessel's design rating during normal conditions, and up to 110% of the system's design rating during relief scenarios. However, certain non-code PLVs may exhibit accumulations exceeding 10% at full relief flow. In such cases, their set points may need to be lowered to ensure that the pressure does not surpass 110% of the vessel's rating during fully accumulated relief conditions.

System piping must be rated to withstand at least the PLV set pressure at the highest operating temperature. While API 614 [1] outlines default minimum pipe schedules (wall thicknesses) for various pipe sizes, these must also meet the wall thickness requirements specified in ASME B31.3 or other applicable piping codes.

Piping design aims to minimize frictional pressure losses, reduce hydrodynamic noise, and maintain flow velocities below 7–10 ft/s (2–3 m/s) to ensure reasonable friction losses. This approach contrasts with high-pressure hydraulic systems, which typically allow much higher flow velocities.

API 614 [1] prohibits the use of certain pipe, fitting, and valve sizes: 1¼ in. (30 mm), 2½ in. (65 mm), 3½ in. (90 mm), 5 in. (125 mm), 7 in. (175 mm), and 9 in. (225 mm).

Flange pressure–temperature ratings are governed by ASME B16.5 and other relevant flange codes. Most API 614 [1] lubrication and control oil systems utilize ASME 150# rated flanges. However, applications requiring higher control oil pressures may necessitate 300# rated flanges.

Note that seal oil systems are excluded from this discussion, as they often demand significantly higher flange ratings. Additionally, control oil systems operating above 300 psig (2068 kPag) are generally not integrated with lubrication systems and are treated as standalone hydraulic systems, typically governed by separate hydraulic standards.

Atmospheric oil drain return lines from primary equipment must be sized to operate half full, based on a minimum continuous slope of ½ in./ft (40 mm/m) toward the reservoir. This half-full design ensures proper ventilation of bearing and gear housings, prevents flooding and leakage, and facilitates the flow of air and nitrogen from separation seals, coupling windage, and breather ingress toward the reservoir for venting—or extraction if blower-assisted demisters are used.

7.11 Accumulators

Bladder, diaphragm, and piston-type accumulators serve a wide range of functions, including:

- Energy storage
- Emergency and safety operations
- Damping of vibrations, pressure fluctuations, and pulsations
- Shock absorption
- Flow stabilization.

API 614 [1] specifically covers bladder and direct-contact-type accumulators. Bladder-type accumulators are most commonly used in lubrication and control oil systems, while direct-contact types are preferred in setups requiring storage of large oil volumes.

Hydraulic accumulators utilize the compressibility of gas to store fluid. Bladder-type accumulators operate on this principle, typically using nitrogen as the compressible medium. These devices consist of two sections—fluid and gas—separated by a bladder that prevents gas and oil from mixing. When oil pressure rises, the bladder compresses the gas as oil enters the accumulator. When pressure drops, the gas expands, pushing the stored oil back into the hydraulic circuit.

Dynamic analysis can help determine the necessity of accumulators in lubrication systems per API 614 [1]. Their roles include:

Lubrication Circuits:

- Act as a buffer or supplementary oil source
- Maintain system pressure above the low-low trip threshold during pump transitions or ramp-up of standby pumps
- Installed on the oil console, downstream of filters on the supply header
- Sized to sustain pressure for at least 4 s during electric motor pump startup

Connections typically include a block valve and bypass line with an orifice or drilled check valve and a drain line routed to the reservoir or return header, equipped with a block valve and restriction orifice.

Control Circuits:

- Support rapid, high-flow oil delivery for short durations
- Allow use of smaller pumps for regular operation
- Maintain control oil header pressure above the minimum required by turbine manufacturers
- Positioned close to hydraulic actuators for quick response during servo motor transients

To properly size an accumulator, the purchaser (i.e., the vendor with overall unit responsibility) should provide:

- Operating Pressure Range: Defined by maximum (P2) and minimum (P1) system pressures

- Required Fluid Volume: Based on the difference between pump output and system demand
- Discharge Rate and Duration: Specifies how much flow is needed, for how long, and how frequently
- Ambient Temperature Range: Influences heat exchange between the accumulator and its surroundings
- Installation Location: Ensures compliance with local pressure vessel regulations
- Connection Types: For both fluid and gas ports

The accumulator designer utilizes all relevant data to determine the appropriate accumulator size and to create performance graphs illustrating pressure and volume relationships. These graphs help demonstrate the accumulator's functional capacity.

As part of the sizing process, the gas precharge pressure is defined. This refers to the pressure of nitrogen gas within the accumulator when the system is inactive. The precharge pressure directly affects the force exerted by the spring mechanism that stores hydraulic energy during operation. A commonly used guideline suggests setting the precharge pressure (P_0) at approximately 90% of the minimum system pressure (P_1). While exceptions may apply depending on specific applications, this serves as a reliable starting point.

Figure 7.18 illustrates how pressure and volume behave within a bladder-type accumulator. In the first image, the accumulator is shown precharged with nitrogen, with the bladder acting as a barrier to isolate the fluid. The middle image indicates that the minimum operating pressure should exceed the precharge pressure to prevent the bladder from repeatedly impacting the fluid port during discharge. The final image demonstrates that when the system reaches its maximum operating pressure, the accumulator provides an effective volume of $\Delta V = V_1 - V_2$.

The typical bladder accumulator features a sealed elastomeric bladder housed within a stainless steel shell, which may be either forged or fabricated. A mechanically actuated antiextrusion valve automatically closes once the oil is discharged, sealing the fluid port. This action traps the bladder inside the shell and safeguards it from being forced out through the discharge port.

In corrosive operating conditions, the shell may require additional protection through internal and/or external coatings, alongside its stainless steel construction, in compliance with API 614 [1]. A cross-sectional illustration of a bladder-type accumulator is provided in Fig. 7.19.

The accumulator bladder is typically installed in a vertical orientation, with the gas port positioned at the top. However, certain bladder accumulators feature pleated bladder designs that allow for horizontal mounting if necessary, though this setup is generally less desirable. In applications demanding high discharge rates or storage capacities beyond the limits of individual accumulators, multiple units can be linked in parallel. Figures 7.20 and 7.21 illustrate parallel accumulator arrangements in horizontal and vertical configurations, respectively.

Table 7.2 provides some typical values for bladder-type accumulators.

Acrylonitrile butadiene rubber—commonly referred to as nitrile rubber or Buna-N—is the standard bladder material for mineral oil applications. For environments with extremely low temperatures, a specialized low-temperature Buna-N variant is

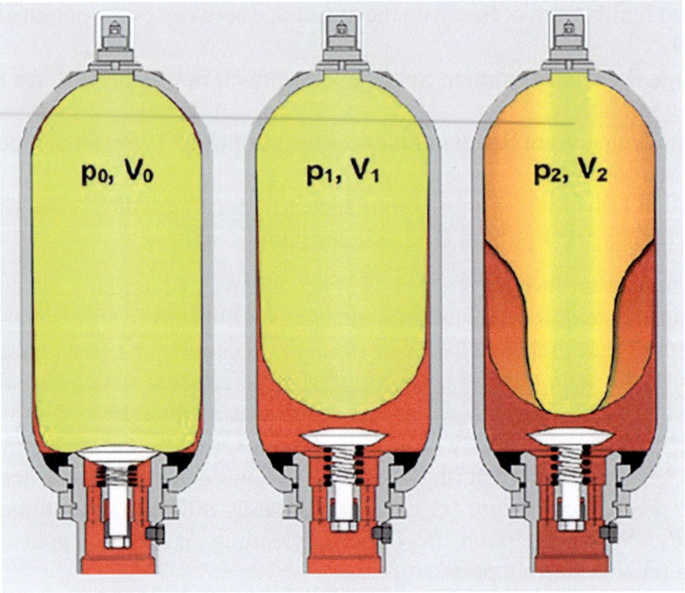

Fig. 7.18 Action of a bladder accumulator (Image courtesy of HYDAC, Welland, Ontario, Canada)

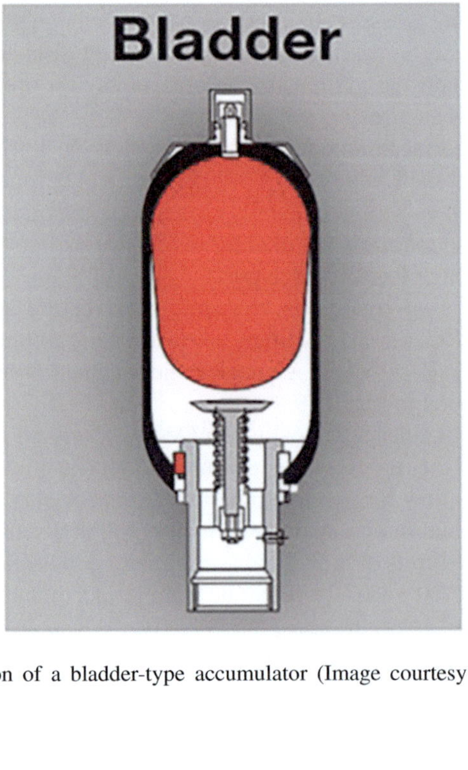

Fig. 7.19 Cross section of a bladder-type accumulator (Image courtesy of HYDAC, Welland, Ontario, Canada)

Fig. 7.20 Multiple parallel horizontal accumulators (Image courtesy of G.J. Oliver, Inc., Phillipsburg, NJ)

Fig. 7.21 Multiple parallel vertical accumulators (Image courtesy: G.J. Oliver, Inc., Phillipsburg, NJ)

Table 7.2 Bladder-type accumulator (table courtesy of HYDAC, Welland, Ontario, Canada)

Nominal volume	Maximum allowable working pressure	Pressure ratio	Flow rate	Preferred installation
1 qt–120 gal (0.95–454 L)	150–5000 psi (1.03–34.47 MPa) up to 10,000 psi (69 MPa)	4:1	Up to 480 gal/min (1817 L/min)	Vertical

employed. If operating across wide temperature ranges, ethylene oxide epichlorohydrin rubber is preferred. Fluorine rubber also serves as a reliable option for both mineral and synthetic oil systems.

In applications where auxiliary oil is used to support pump switching or to accommodate transient flow demands in hydraulic servo systems, accumulator sizing is based on adiabatic gas compression and expansion due to the brief cycle durations. Equation 7.5 provides a method for calculating either the necessary accumulator vessel volume or the quantity of stored oil available, depending on the system's operating parameters and precharge pressure.

$$V_1 = \frac{V_x\left(\frac{P_1}{P_0}\right)^{1/n}}{\left[1 - \left(\frac{P_1}{P_2}\right)^{1/n}\right]} \quad \text{or} \quad V_x = \frac{V_1\left[1 - \left(\frac{P_1}{P_2}\right)^{1/n}\right]}{\left(\frac{P_1}{P_0}\right)^{1/n}} \tag{7.5}$$

where V_1 is the required accumulator vessel volume in in.^{3} (L), V_x is the required store volume in in.^{3} (L), P_0 is the gas precharge pressure in psia (kPa absolute), P_1 is the minimum discharge pressure at the end of the discharge cycle in psia (kPa absolute), P_2 is the maximum (normal) operating pressure in psia (kPa absolute), and n is the ratio of the gas specific heat. (For nitrogen in adiabatic service with a discharge cycle time of less than 6 s and operating pressures of less than 400 psig [2,758 kPa], $n = 1.4$.)

The following example illustrates how an accumulator is sized based on the provided inputs. The actual steps and methods can vary depending on the application and design. In this example, the oil system accumulator is sized for 4-s flows of a 100 gpm (379 L/min) system, which requires a stored volume of 96.7 gal (1,547.7 in.^{3}, 25 L). Atmospheric pressure is assumed to be 14.7 psia (101.3 kPa absolute).

Example service conditions and requirements:

Maximum (normal) system pressure $P_2 = 139.7$ psia (963 kPa absolute)
Minimum system pressure $P_1 = 104.7$ psia (722 kPa absolute)
Gas precharge pressure $P_0 = 94.7$ psia (653 kPa absolute)
Required oil volume $V_x = 6.7$ gal (1,547.7 in.^{3} [25.4 L])
Ambient temperature range $T = 59$–95 °F (15–35 °C)
Duty cycle $D = 4$ s.

These conditions result in a minimum accumulator capacity of 38.7 gal (8,939.7 in.^{3} [146.5 L]) to provide the required 6.7 gal (1,547.7 in.^{3} [25.4 L]) of oil between the given pressures and temperatures. If the capacity falls between a manufacturer's two standard sizes, the next larger size is selected for the application. In this application, a standard 40 gal (150 L) accumulator would be used.

When determining precharge pressure, it is also important to consider changes in the ambient temperature, which impact the accumulator's stored oil volume as a result of the change in gas volume with temperature described by Charles's law ($V_1/T_1 = V_2/T_2$), where the temperatures are the minimum and maximum expressed in absolute temperature units. Keep in mind that, during normal system steady-state operation, the accumulator pressure is maintained constant by the oil system and, hence, the gas volume changes with temperature.

In the previous example, assume the accumulator is precharged at an ambient temperature of 59 °F (15 °C). When the ambient temperature rises to 95 °F (35 °C), the gas inside expands by approximately 6.7%, which reduces the stored oil volume from 6.7 gallons to 6.25 gallons (or from 25.3 to 23.6 L). This decrease in oil volume means the accumulator delivers less than the intended 4 s of oil flow at the higher temperature. Consequently, both accumulator sizing and precharge pressure must account for temperature variations.

Figure 7.22 illustrates this effect: the colored lines on the performance graph represent the accumulator's behavior at the lowest and highest ambient temperatures, assuming precharge occurs at the minimum temperature. The variation between these curves highlights how temperature changes can impact the duration of oil supply.

For installations in North America, accumulators must comply with the ASME Boiler and Pressure Vessel Code and be either registered with the National Board or stamped with a Canadian Registration Number (CRN).

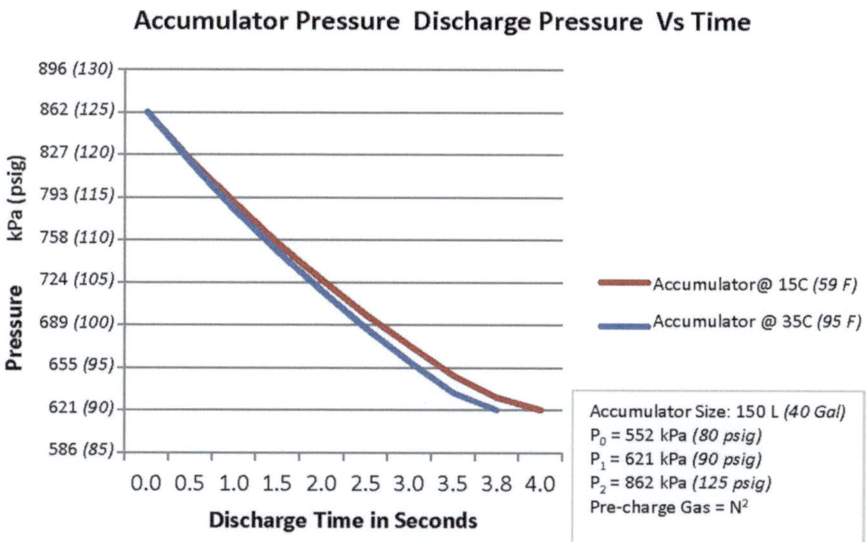

Fig. 7.22 Accumulator system pressure versus time curve. Conditions: accumulator size = 40 gal (150 L), $P_0 = 80$ psig (552 kPa), $P_1 = 90$ psig (621 kPa), $P_2 = 125$ psig (862 kPa), precharge gas = nitrogen (Image courtesy of G.J. Oliver, Phillipsburg, NJ)

In other regions, such as the European Union, pressure vessel designs follow the Pressure Equipment Directive (PED), and specifications may vary depending on the applicable design code. Accumulators designed outside these standards are available but are classified as custom or special designs.

Once the appropriate accumulator has been selected, the designer can determine the necessary accessories to fulfill the system requirements:

- Mounting Configuration:
  - The first decision involves how the accumulator will be mounted. Options vary widely depending on the application. Manufacturer catalogs typically provide multiple mounting choices and guidance on selecting the most suitable configuration.
- Fluid Port Connections:
  - Consider whether the system requires a safety shutoff valve to isolate the accumulator. This is commonly achieved using a single block assembly that integrates a ball valve, relief valve, and drain valve.
- Charging Method:
  - The method for charging the accumulator must be chosen. Manufacturers offer backup nitrogen bottles and various charging and gauging kits to assist maintenance personnel. It's crucial to consider the installation location and the type of standard connection (SI or inch-series) used for the nitrogen source to ensure compatibility.

Loss of precharge pressure over time is typical across all accumulator types. The rate of pressure loss is influenced by factors such as duty cycle, operating temperature, and the elastomer composition of the bladder.

To establish a baseline, the precharge pressure should be checked every three months initially. Based on observed trends, the inspection interval can be adjusted accordingly.

Most accumulators include a permanent pressure gauge port on the gas side for monitoring nitrogen precharge. This pressure should also be verified before restarting lubrication and control oil systems following maintenance or outages.

For bladder-type accumulators, the precharge pressure is measured when the bladder is fully expanded—meaning the oil has been completely drained.

To check the precharge pressure while the oil system is operational:

1. Isolate the accumulator and fully drain it through the return line (this line includes a flow-restriction orifice to regulate drainage).
2. Leave the drain valve open and measure the precharge pressure.
3. Adjust the pressure as needed.

If no pressure is detected and nitrogen addition fails to build pressure, the bladder is likely leaking and must be replaced.

Once the precharge pressure has been verified and adjusted as needed, the precharge gas must be sealed, the drain valve closed, and the accumulator block valve opened slightly. Caution should be taken not to open the block valve too far, as this may redirect excessive flow into the accumulator, potentially leading to low-pressure conditions at the unit. When the precharge pressure matches the system oil pressure, the accumulator is considered fully charged, and the block valve can then be opened completely.

7.12 Oil Console Assembly

Special-purpose lubrication oil systems are typically installed on a dedicated baseplate, separate from the main equipment skid, and commonly referred to as an oil console. However, in space-constrained environments—such as offshore installations—the entire oil system may be integrated directly into the main equipment skid.

Interconnection and Utility Influence:

- The relative positioning of the oil console and the equipment it serves plays a key role in determining optimal locations for oil supply and return connections, ensuring efficient piping arrangements.
- The placement of steam and cooling water utilities significantly impacts the overall layout of the oil system.
- Maintenance access must be carefully planned, with sufficient clearance provided for tasks such as extracting the shell-and-tube cooler bundle and removing the reservoir heater.

Design for Operability and Maintenance:

- The internal arrangement of components within the oil console should prioritize ease of operation and serviceability.
- Adequate spacing around pumps and drivers is essential.
- Block and bypass valves, along with strategically placed flange breaks, should be included to allow equipment servicing without interrupting system operation.
- Instrument gauge panels must be positioned for convenient access during routine operator inspections.

System Components Overview:

- Figures 7.23, 7.24 and 7.25 illustrate a system featuring:
- Motor-driven main and standby horizontal positive-displacement pumps
- Duplex shell-and-tube heat exchangers with cooling water manifold piping
- Duplex oil filters
- An instrument gauge board
- A blower-assisted mist eliminator for the reservoir vent

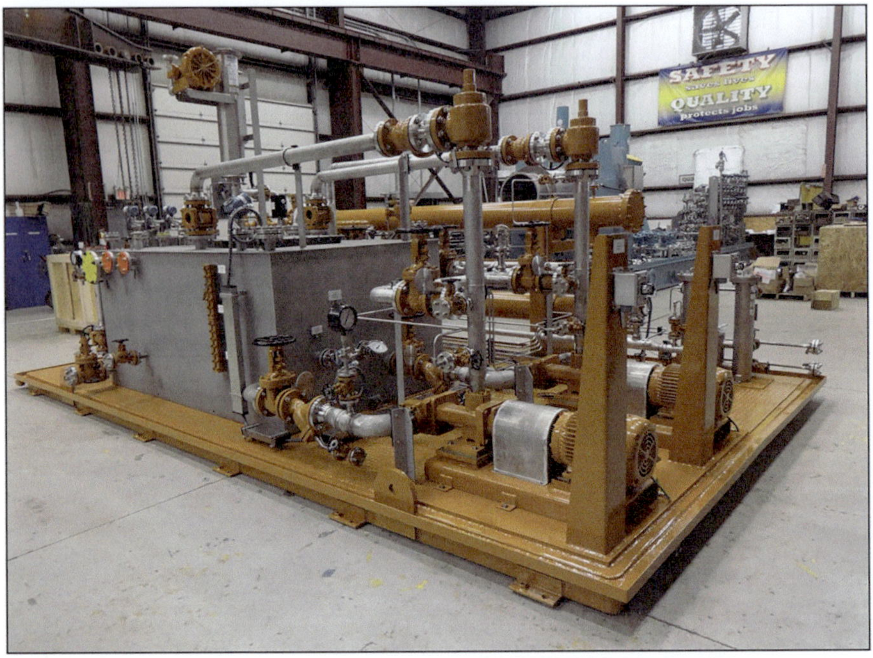

Fig. 7.23 Typical special-purpose lube oil console (Image courtesy of G.J. Oliver, Inc., Phillipsburg, NJ)

Figure 7.26 presents a photograph of an upgraded lube and control oil system, featuring a steam turbine as the primary pump driver, complemented by a motor-driven standby pump. A key design element is the low-profile reservoir, necessitated by the minimal elevation difference between the oil console and the main equipment. To maintain a consistent slope for the oil drain return line—entering the reservoir above its maximum oil level—the reservoir was engineered with reduced height and an expanded surface footprint. Additionally, the oil console includes an accumulator, visible in the upper left corner of the image.

7.13 Structural Steel Welding and Certification

In certain jurisdictions, regulatory and certification bodies classify components such as ladders, handrails, and walkways attached to the oil system baseplate as structural elements. Therefore, it is advisable for purchasers to verify the materials used in these structural steel components prior to issuing a purchase order, ensuring compliance with applicable specifications. Likewise, aspects such as welding procedures, material selection, quality standards, connection design, fabrication and testing methods,

Fig. 7.24 Special-purpose lube oil console reservoir with instruments and piping (Image courtesy of G.J. Oliver, Inc., Phillipsburg, NJ)

welder qualifications, and nondestructive examination must adhere to local certification and authentication requirements. Failure to address these criteria during bid evaluation or proposal review can lead to project delays and increased costs.

7.14 Lube Oil Rundown Tank

API 614 [1] mandates that lube oil rundown tanks be sized to accommodate at least three minutes of normal operating oil flow. This requirement ensures that, even under near-empty conditions, the tank provides sufficient static head pressure to maintain proper bearing lubrication during coastdown following a low-low lube oil pressure trip—typically caused by a loss of supply pressure.

However, the tank's elevation and orientation must be carefully considered. When the tank is full, the resulting static head pressure must remain below the low-pressure shutdown threshold to avoid premature system shutdown. For compressor strings with extended coastdown durations, the standard three-minute sizing criteria should be reevaluated and, if warranted, amended to ensure adequate lubrication throughout the entire deceleration period.

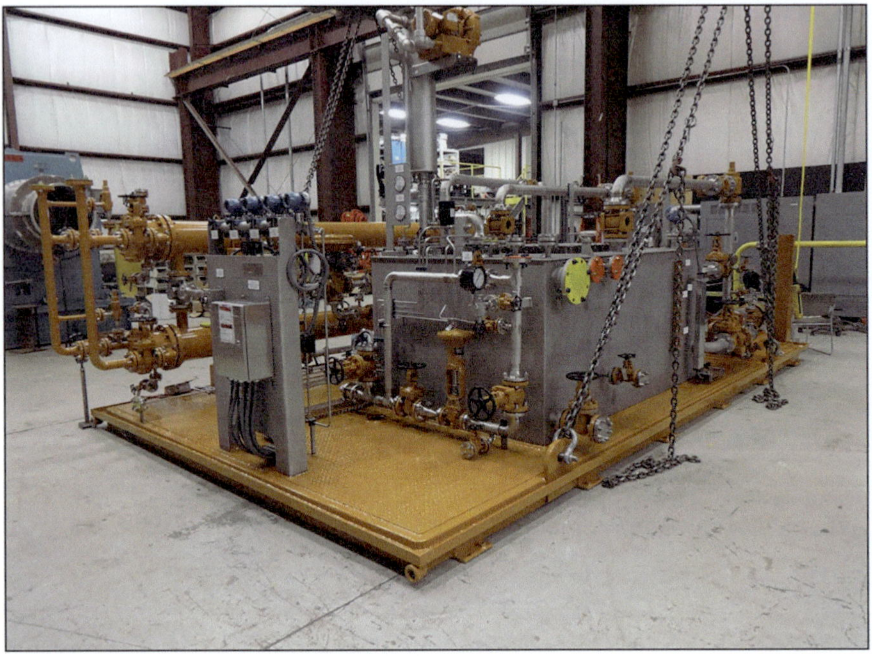

Fig. 7.25 An instrument gauge board on a special-purpose lube oil console (Image courtesy of G.J. Oliver, Inc., Phillipsburg, NJ)

To meet this requirement, the rundown tank should be positioned above the compressor centerline. This placement allows the tank, when nearly empty, to deliver a static head just below the low-low pressure trip setting, ensuring it has already supplied most of the oil needed during coastdown. The remaining oil volume will continue to provide sufficient lubrication until the equipment train comes to a complete stop.

Depending on available space and tank dimensions, rundown tanks may be installed either vertically or horizontally. It is the purchaser's responsibility to specify the available installation space and suitable mounting locations at the required elevation above the unit. In cold climates, additional measures such as heat tracing and thermal insulation may be necessary to maintain oil temperature and prevent viscosity-related issues during low ambient conditions.

Overhead lube oil rundown tanks are available in two configurations: atmospheric and pressurized flow-through types, as illustrated in Figs. 7.27 and 7.28, respectively.

Atmospheric Rundown Tanks:

These are static oil reservoirs that receive a continuous low-volume flow—typically between 1.5 and 4 gallons per minute (6–15 L/min)—to maintain oil temperature. Oil is supplied via a small orifice feed. During startup, tank filling can be expedited

Fig. 7.26 Steam-turbine-driven main pump on lube and control oil console (Image courtesy of G.J. Oliver, Inc., Phillipsburg, NJ)

by opening a quick-fill bypass valve until spillover oil is visibly flowing through the sight glass.

A critical design consideration for atmospheric tanks is the sizing of the overflow line. It must accommodate the maximum flow from a fully open quick-fill valve to prevent oil from spilling out of the tank vent, especially if the valve is mistakenly left open after filling.

To support operational control, atmospheric tanks are equipped with level monitoring instruments. These devices help track oil levels, enable unit-start permissive logic, and trigger low-level alarms.

Pressurized Rundown Tanks:

These tanks are installed in series and designed so that the entire lube oil supply flows through them. They feature floating-ball check valves that purge air during filling. Once filled, the ball float seals the valve, allowing the tank to pressurize to the system's lube oil supply pressure, adjusted for elevation head relative to the shaft centerline of the equipment.

Since the system cannot reach its normal supply header pressure until the pressurized tank is completely filled, a level monitoring instrument for start permissive is not necessary.

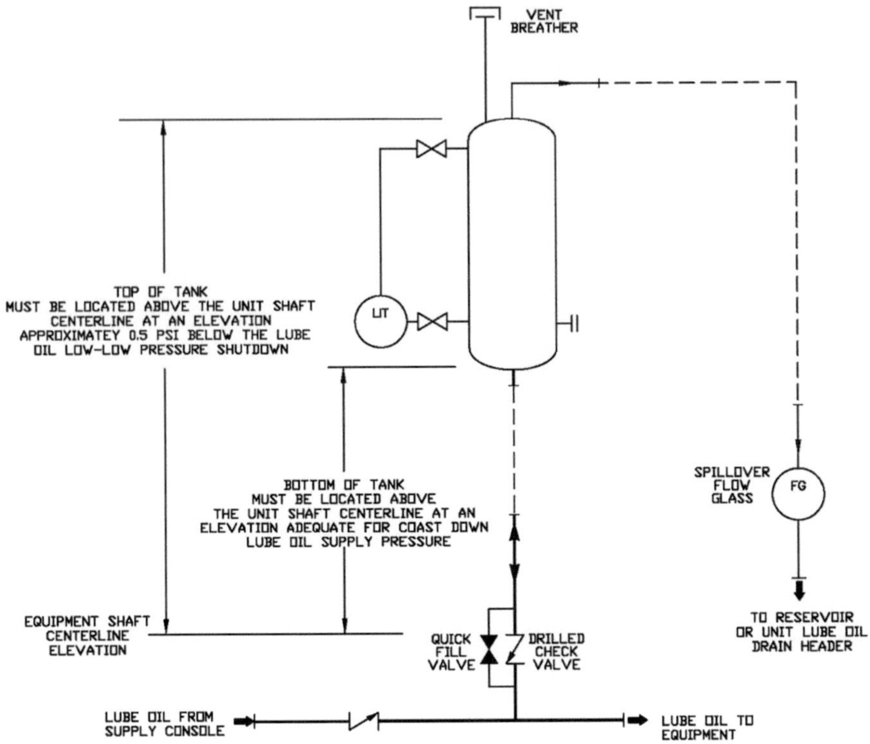

Fig. 7.27 Schematic of an overhead atmospheric lube oil rundown tank. Abbreviations: FG = flow glass, LIT = level indicating transmitter (Image courtesy of G.J. Oliver, Inc., Phillipsburg, NJ)

The total rundown capacity of the oil reservoir must account for the volume of the rundown tank itself as well as all associated piping.

Figure 7.29 illustrates a horizontally mounted, atmospheric-type lube oil rundown tank with an oriented configuration. It features dedicated vent and overflow (spillover) connections, as well as a 12-inch (300 mm) blind flanged inspection cover. For monitoring purposes, a differential-pressure level transmitter is integrated into the design.

Following installation, the tank will be insulated on-site using blanket-type insulation. To facilitate this, insulation studs are provided for securing the insulation material in place.

Figure 7.30 depicts a vertical, pressurized lube oil rundown tank that has been horizontally fastened to a wooden skid for transport.

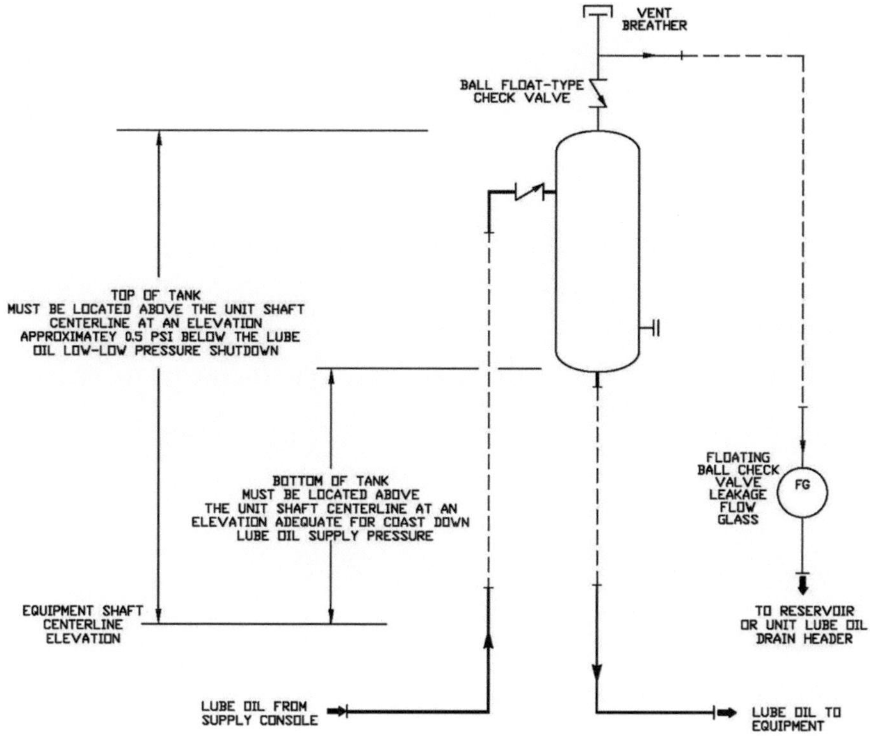

Fig. 7.28 Schematic of an overhead pressure lube oil rundown tank (Image courtesy of G.J. Oliver, Inc., Phillipsburg, NJ)

7.15 Mist Eliminators

Mist eliminators function primarily as coalescers, designed to extract visible oil vapor from the vented air streams of lubricating oil systems. The underlying mechanism—coalescence—is a continuous process where fine aerosol particles interact with the fibers in the filter media. These particles merge with previously captured aerosols, gradually forming larger droplets. Once sufficiently large, the droplets migrate to the downstream side of the media, where they can be drained away by gravity.

By capturing oil mist, these devices prevent environmental contamination of air and soil, aiding facilities in meeting regulatory standards. They also contribute to a safer, cleaner workplace by eliminating oily residues from surfaces such as floors, enclosures, and stairwells. This not only improves indoor and outdoor air quality but also minimizes fire risks. Additionally, mist eliminators help reduce oil loss by recovering and returning it to the reservoir, lowering makeup oil costs.

Mist eliminators are available in several configurations to suit diverse operational needs. Common types include:

Fig. 7.29 Horizontal atmospheric lube oil rundown tank (Image courtesy of G.J. Oliver, Inc., Phillipsburg, NJ)

Fig. 7.30 Vertical pressurized lube oil rundown tank (Image courtesy of G.J. Oliver, Inc., Phillipsburg, NJ)

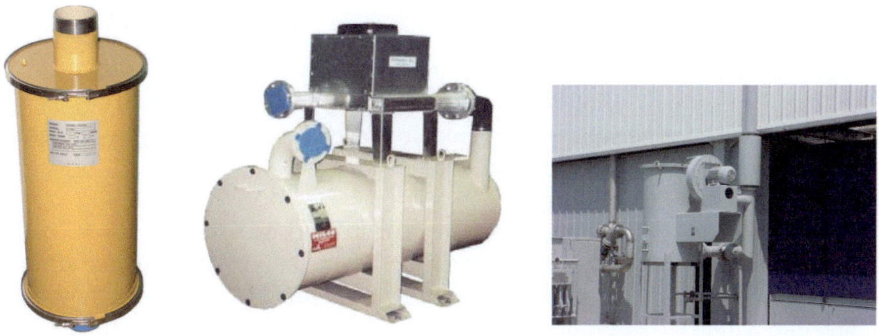

Fig. 7.31 (*Left*) Static or passive mist eliminator, (*center*) mist eliminator with a heat exchanger, and (*right*) blower-assisted mist eliminator (Image courtesy of Hilliard Corporation, Elmira, NY)

- Static units
- Fan- or blower-assisted models
- Systems integrated with heat exchangers

Blowers and fans may be configured as single units or duplex systems (2 × 100% capacity), and can operate at either constant or variable speeds depending on system requirements.

These devices are engineered to remove oil mist with a high efficiency of 99.97%, effectively capturing airborne contaminants across all particle sizes. In fact, API 614 [1] specifies a minimum cartridge efficiency of 99.97% for particles as small as 12 μin. (0.3 μm). Most oil mist particles are under 39 μin. (1 μm), with approximately 60% to 80% measuring less than 20 μin. (0.5 μm).

Mist eliminators are capable of maintaining a vacuum within the oil reservoir ranging from 5 to 10 mbar (2–4 in. of water column) [5–10 cm of water column]. Pressure monitoring systems can be incorporated to ensure proper vacuum levels are maintained.

Before installation, it's important to assess the surrounding environment. Conditions that are humid, dusty, corrosive, or rich in particulates may require additional review to ensure optimal performance and longevity of the mist eliminator.

To properly size a mist eliminator, the following data must be collected:

- Vent flow rate from the oil reservoir
- Acceptable back-pressure or pressure drop
- Connection size on the reservoir
- Operating oil temperature range
- Power rating of the associated equipment train.

The reservoir's vent flow may consist of a combination of sources, including outboard separation seal emissions, bearing buffer air or nitrogen flows, air ingress from bearing and gearbox vent breathers, coupling windage, and nitrogen purge streams.

Mist Eliminator Types and Applications:

- Static or Coalescer-Only Eliminators
  These are suitable if oil system back-pressure is not a concern. They can be sized to meet specific pressure-drop requirements and are also effective in setups where an existing blower aids in oil mist removal.
- Blower-Assisted Eliminators
  Ideal for oil systems that are sensitive to back-pressure or where maintaining low or negative pressure is preferred. These units can incorporate airflow-restriction devices to create a vacuum or be configured to maintain atmospheric pressure within the reservoir.
- Heat Exchanger-Integrated Eliminators
  Recommended for systems where oil temperatures exceed 200 °F (93 °C). Cooling the air stream enhances mist coalescence, thereby improving the efficiency of mist elimination.

Figure 7.31 illustrates these three mist eliminator configurations.

7.16 Lube Oil Conditioners

Water contamination in oil typically appears in two forms: as free water, which can be suspended in either a stable or unstable emulsion, and as dissolved or absorbed water, retained within the oil up to its saturation limit.

Oil purifiers are most commonly used in lubrication systems supporting steam turbine machinery. Various technologies are employed to manage water contamination:

- Pleated cartridge filters: Efficiently remove water up to the oil's saturation threshold
- Coalescers and separators: Target free and emulsified water, also up to saturation level. These systems strike a balance between moderate capital investment and low operational costs
- Baseplate-mounted coalescers: Shown in Fig. 7.32, these are fixed installations
- Portable purification units: Widely used across industries, typically equipped with particulate and coalescing filters.

Advanced separation methods, including vacuum dehydrators, effectively remove free, emulsified, and dissolved water—bringing moisture levels below the oil's saturation threshold.

Oil conditioners are designed to process a small fraction of the total circulating oil volume—typically around 0.5%. For instance, a unit with a flow rate of 10 gallons per minute (38 L per minute) would be appropriate for an oil reservoir holding 2000 gallons (7571 L). The system's standard performance characteristics are depicted in Fig. 7.33.

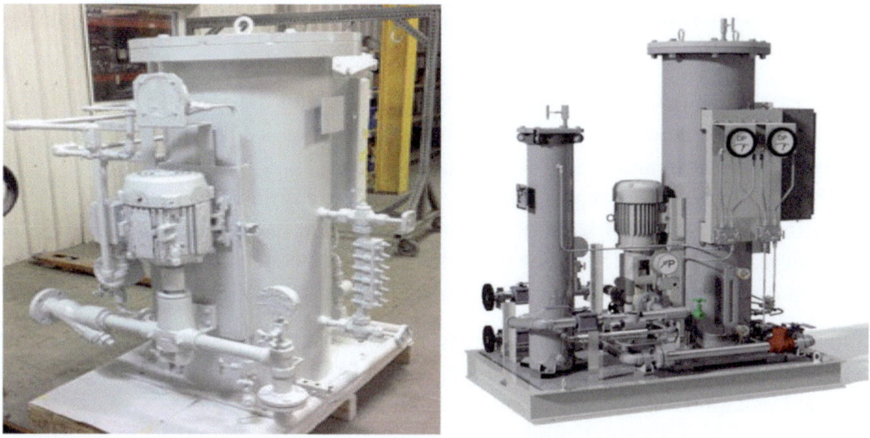

Fig. 7.32 Coalescing separators (Image courtesy of Hilliard Corporation, Elmira, NY)

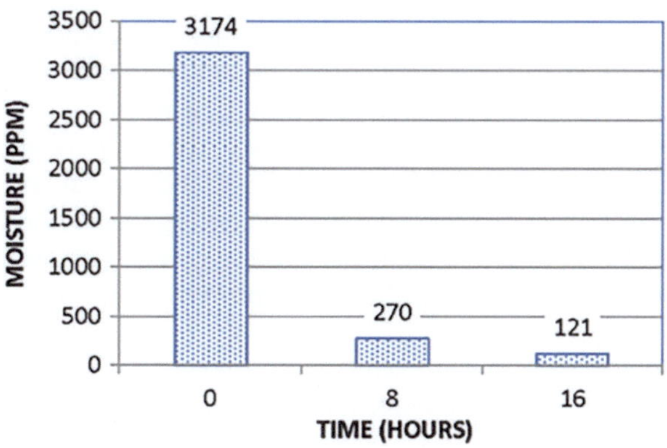

Fig. 7.33 Moisture removal performance of a 10 gpm (38 L/min) oil conditioner (Image courtesy of Hilliard Corporation, Elmira, NY)

7.17 Hazard and Operability Study for Lube Oil Systems

The lube oil system is a critical component examined during the hazard and operability (HAZOP) study of centrifugal compressor trains. During the HAZOP process, engineering and design flaws that may pose safety risks or hinder operability are identified and assessed. Each potential hazard is evaluated to determine whether the existing operational controls and procedures are sufficient.

A standard HAZOP worksheet (refer to Table 7.3) outlines key elements such as:

- Identified deviations
- Root causes of those deviations
- Consequences assuming no safeguards are in place
- Severity ratings
- Existing barriers and safeguards

- Recommended corrective actions, along with the responsible party for implementation.

Severity is typically rated on a scale (e.g., 1–5), and the HAZOP team assigns impact rankings across categories such as personnel safety, property damage, environmental harm, and other relevant factors.

Table 7.3 presents a representative HAZOP example for a lube oil system supporting the gear-motor train of a centrifugal compressor. Rankings in the table reflect increasing severity based on the potential consequences. The actual deviations reviewed during the HAZOP for this system included the following scenarios:

- Insufficient or no lube oil flow: May result from a clogged pump strainer, oil pump failure, blocked oil filter, accidental closure of pump discharge block valves, or closed isolation valves near pressure or temperature regulators.
- Excessive lube oil flow: Could be caused by a faulty low oil pressure transmitter or improper valve handling.
- Reverse oil flow: Typically due to backflow within the system.
- Incorrect flow direction of lube oil: Possible causes include open drain lines, joint leaks, or the rundown tank's manual fill valve being unintentionally left open.
- Abnormal pressure levels (high or low): Often associated with conditions similar to those causing high or low oil flow rates.
- Temperature deviations in the oil circuit or cooling loop: Influenced by extreme ambient temperatures, heater malfunction, failure of the oil temperature control valve, or restricted flow of the cooling medium to the oil cooler.
- Low oil level: Frequently linked to issues similar to incorrect oil flow direction.
- Startup anomalies: May arise from incorrect valve operations during manual unit startup.
- Emergency shutdown scenarios: Triggered by process disturbances, fire hazards, or gas-related emergencies.

Table 7.3 Sample HAZOP worksheet

Deviation	Cause	Consequence	Severity				Barriers
Low/no flow	Partial blockage of lube oil pump A suction strainer	Potential cavitation of the lube oil pump leads to its damage (this could be after a few hours)	0	1	0	0	No safeguards
		In the worst case, insufficient lube oil flows to the bearings of the electric motor, gear box, and compressor, leading to mechanical damage of the bearings	0	2	0	0	Pressure transmitter on the pump discharge will start the standby lube oil pump
							Pressure transmitter on the oil supply header will start the standby lube oil pump
							Pressure transmitters A/B/C (for 2-out-of-3 voting logic) on the lube oil header will trip the compressor motor (which breaks the power supply) and close the inlet and outlet valves on the compressor, so that oil will automatically flow from the rundown tank to provide coastdown lubrication and allow safe shutdown of the train

(continued)

Table 7.3 (continued)

Deviation	Cause	Consequence	Severity				Barriers
		Bearing damage could lead to other item damage and flying fragments of material (couplings), which could result in permanent injury to any operators in the area	4	3	0	3	Radial vibration trip on the compressor will trip the compressor motor (which breaks the power supply) and close the inlet and outlet valves on the compressor, so that oil will automatically flow from the rundown tank to provide coastdown lubrication and allow safe shutdown of the train
							Radial vibration trip on the gearbox will trip the compressor motor (which breaks the power supply) and close the inlet and outlet valves on the compressor, so that oil will automatically flow from the rundown tank to provide coastdown lubrication and allow safe shutdown of the train
		With reference to the preceding consequence, if bearing damage occurred, it would lead to high vibrations of the motor, gear box, and compressor, leading to their physical damage. In addition, there could be a loss of containment in the compressor, resulting in the release of process gas. Potential fire and/or explosion, potential asset damage, potential environmental impact, and potential reputation impact could result	4	4	2	3	Pressure transmitter on the pump discharge will start the standby lube oil pump
							Pressure transmitter on the oil supply header will start the standby lube oil pump

(continued)

Table 7.3 (continued)

Deviation	Cause	Consequence	Severity				Barriers
							Pressure transmitters A/B/C (for 2-out-of-3 voting logic) on the lube oil header will trip the compressor motor (which breaks the power supply) and close the inlet and outlet valves on the compressor, so that oil will automatically flow from the rundown tank to provide coastdown lubrication and allow safe shutdown of the train
							Radial vibration trip on the compressor will trip the compressor motor (which breaks the power supply) and close the inlet and outlet valves on the compressor, so that oil will automatically flow from the rundown tank to provide coastdown lubrication and allow safe shutdown of the train

Acknowledgements The images/graphics/figures in this chapter were provided by G.J. Oliver, Inc. (Phillipsburg, NJ), Standard-Xchange (Buffalo, NY), API Basco (Buffalo, NY), Yoshimine (Etajima, Japan), and Hilco/Star Products (Hilliard Corporation) (Elmira, NY). The section on accumulators, including the images in this section, were contributed by J. Dau from HYDAC Canada (Welland, Ontario, Canada), and in part by Charles (Chuck) Parker from G.J. Oliver, Inc. (Phillipsburg, NJ). The HAZOP example in this chapter was reproduced from the author's contributions to a project. This material was reprinted with permission of the publisher from Ghaisas N, Parker CA, Long M (2022) Modern Lubricating Oil Systems. Asia Turbomachinery and Pump Symposium (ATPS), Kuala Lumpur, Malaysia (May 2022).

References

1. API Standard 614 (2022) Lubrication, Shaft-sealing, and Oil-control Systems and Auxiliaries. 6th edn. American Petroleum Institute, Washington, DC
2. API Standard 611 (2022) General-purpose Steam Turbines for Petroleum, Chemical, and Gas Industry Services. 6th edn. American Petroleum Institute, Washington, DC
3. API Standard 676 (2022) Positive Displacement Pumps—Rotary. 4th edn. American Petroleum Institute, Washington, DC
4. API Standard 520 (2020) Sizing, Selection, and Installation of Pressure-relieving Devices Part I—Sizing and Selection. 10th edn. American Petroleum Institute, Washington, DC
5. API Standard 521 (2020) Pressure-relieving and Depressuring Systems. 7th edn. American Petroleum Institute, Washington, DC
6. API Standard 526 (2023) Flanged Steel Pressure-relief Valves. 8th edn. American Petroleum Institute, Washington, DC

Chapter 8
Flexible-Element Couplings

Abstract Couplings are essential components in machinery trains, yet they are often overlooked during the specification process by engineering contractors and owner companies. In many cases, users base their selection solely on basic parameters such as operating speed, torque, safety factor, and shaft dimensions. However, a comprehensive evaluation should also include additional critical factors—such as combined steady-state and cyclic stresses, angular misalignment tolerance, torsional rigidity, axial displacement capacity, axial natural frequencies, anticipated peak and transient torques, and both lateral and torsional critical speeds. To ensure a well-informed selection, this evaluation should be a collaborative effort involving the primary equipment vendor, subvendors for the driver and gear units (if applicable), the owner's machinery engineers, and the engineering contractor. Choosing the appropriate coupling is a key contributor to the long-term reliability and operational availability of machinery trains in process plants. This chapter highlights two of the most commonly used high-performance, non-lubricated coupling types in modern centrifugal compressor trains: metal diaphragm couplings and disk-pack couplings. It also covers standard selection methodologies, typical construction materials, balancing practices, torque monitoring strategies, and coupling guard design considerations.

8.1 Disk Couplings

A disk coupling is a mechanical device designed to transmit torque tangentially between a driving and a driven bolt positioned on a shared bolt circle. This torque transfer occurs through a series of thin, flexible disks. These disks also accommodate misalignment by deforming under load. Disk couplings are especially suited for high-speed turbomachinery systems, where minimizing weight and reducing the moment is critical for rotordynamic performance.

N. Ghaisas, *Centrifugal Compressors in Hydrocarbon Processing Industries*,
https://doi.org/10.1007/978-3-032-03687-2_8

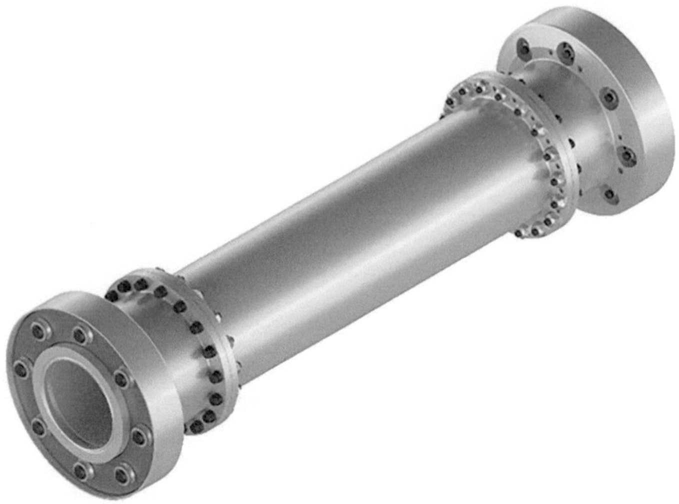

Fig. 8.1 Reduced-moment disk coupling (Image courtesy of Regal Rexnord, Milwaukee, WI)

Typically, a disk coupling features two flexible disk packs—one mounted at each end of a spacer tube or central section. Each pack comprises multiple thin metal membranes, with thicknesses ranging from 0.005 to 0.025 in. (0.127–0.635 mm), tailored to specific application needs. Common membrane materials include 300-series stainless steels, precipitation-hardened stainless steels, Inconel™, and Monel™ (refer to Sect. 8.11). To guard against corrosion and fretting, these membranes are often coated, with many coatings designed to be sacrificial.

Membranes come in various shapes, including scalloped, circular, and polygonal forms, with scalloped and polygonal designs being the most prevalent. Engineers optimize membrane profiles to ensure even stress distribution and minimal reaction forces on the bearings of connected equipment. Key design parameters for disk packs include the number and size of bolts, the bolt circle diameter, and the number of membranes required.

Figures 8.1 and 8.2 illustrate a reduced-moment disk-pack coupling and several membrane types used in these couplings. Each membrane typically features six, eight, or ten holes for torque transmission. Fewer holes result in lower torque capacity but increased flexibility. Figure 8.3 presents the fasteners used in disk-pack assemblies.

For a given coupling diameter, both torque capacity and misalignment tolerance are influenced by the number of bolts in the disk-pack design. As illustrated in Fig. 8.2, a three-bolt configuration features longer segments between bolt holes, which increases the effective bending length. This results in reduced stiffness and enhanced flexibility, allowing for greater misalignment accommodation. Conversely, a five-bolt arrangement distributes torque more evenly across the bolts, boosting torque capacity. However, the shorter link lengths increase angular stiffness, thereby limiting the coupling's ability to handle misalignment.

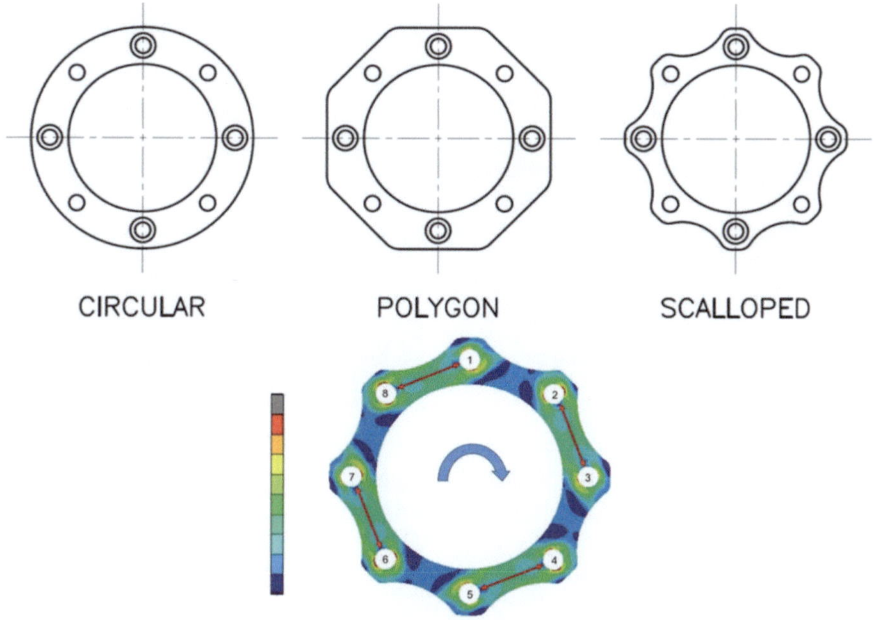

Fig. 8.2 Common disk-pack membrane profiles (Image courtesy of Regal Rexnord, Milwaukee, WI)

In essence:

- Three-bolt designs offer superior misalignment capability but lower torque transmission.
- Five-bolt designs deliver higher torque capacity but reduced flexibility.
- Four-bolt designs strike a practical balance between torque and misalignment performance, making them the most widely adopted configuration across diverse applications—from pumps to centrifugal compressors [1].

The bolt hole circle diameter determines the torque transmission capacity of a disk coupling. Its ability to accommodate misalignment depends on the chord length between bolts and the thickness of the disk or disk pack. Once manufactured, disk packs should remain intact and not be disassembled. Using piloted or factory-assembled disk packs enhances balance and minimizes the likelihood of fatigue-related failures.

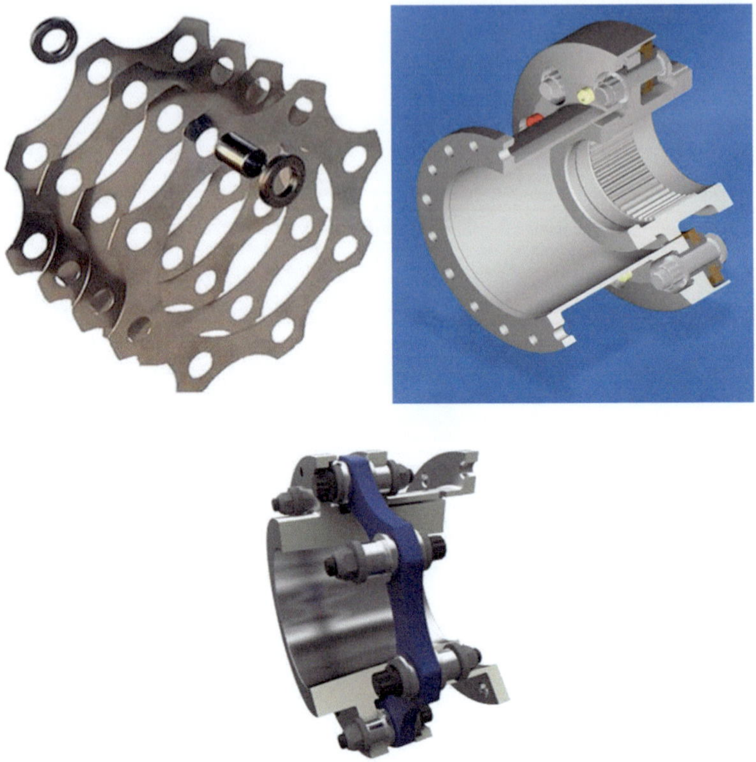

Fig. 8.3 Disk-pack fasteners (Image courtesy of Regal Rexnord, Milwaukee, WI)

High-performance flexible-element couplings offer several advantages over traditional lubricated couplings:

- Minimal to no maintenance required
- Virtually unlimited lifespan due to the absence of wear-prone components
- Reduced reaction forces on connected machinery—20–30% in bending, 10% in axial direction
- Consistently low and repeatable unbalance (no radial clearances)
- Stable unbalance levels over time
- No risk of lockup or spool cracking
- Lower chance of hub-to-shaft fretting
- Components can be interchanged in the field.

8.2 Diaphragm Couplings

High-performance diaphragm couplings typically come in two main configurations: single-element contoured designs (such as plate and flanged types) and multi-convoluted diaphragm styles. An additional variant features multiple straight diaphragms. The choice of material for these diaphragms—either vacuum-melted alloy steel or vacuum-melted stainless steel solid disks—depends on the required ultimate tensile strength.

To enhance durability and performance, diaphragms may undergo specialized treatments like shot peening. This process helps relieve residual stresses from machining and minimizes the risk of surface crack initiation.

In a standard single-element contoured diaphragm, the thickness tapers from the inner diameter (thicker) to the outer diameter (thinner). This gradual contouring ensures a uniform distribution of torsional shear stress across the diaphragm, resulting in lower stress levels during torque transmission under conditions of misalignment and axial deflection. These diaphragms typically have a minimum thickness of about 0.020 in. (0.508 mm).

All diaphragm couplings transmit torque from the outer diameter toward the inner diameter. For a given size, a thicker diaphragm can handle greater torque but offers reduced flexibility for angular and axial misalignments. Conversely, a thinner diaphragm supports higher misalignment but transmits less torque. Therefore, achieving an optimal diaphragm coupling design involves balancing key parameters such as continuous torque capacity, angular and axial deflection limits, and rated operating speed.

Figures 8.4 and 8.5 illustrate the configuration and profile of a single-element contoured diaphragm coupling.

A single-element flanged diaphragm features a bolted joint between the flexible component and the spacer, offering enhanced design versatility and simplified replacement when needed. These diaphragms, machined from a solid piece of steel, can be electron-beam welded both radially and axially at each end of the spacer tube, forming a unified, flexible assembly. When paired with two rigid or adapter flanges, the result is a streamlined coupling system with minimal connection points. This configuration is easily customizable to meet specific application requirements and enables rapid delivery of optimized solutions.

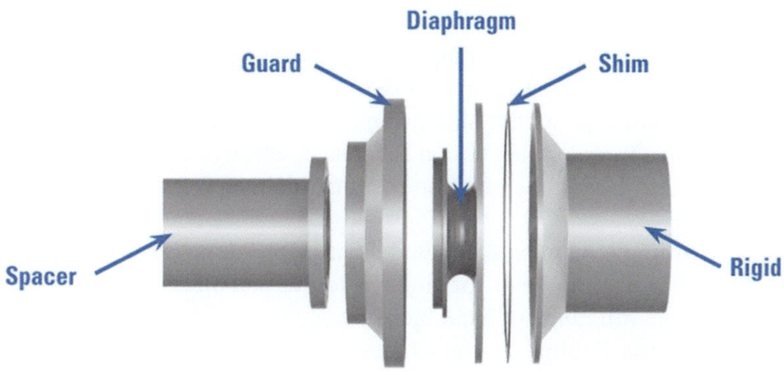

Fig. 8.4 Single-element contoured diaphragm coupling (Image courtesy of Regal Rexnord, Milwaukee, WI)

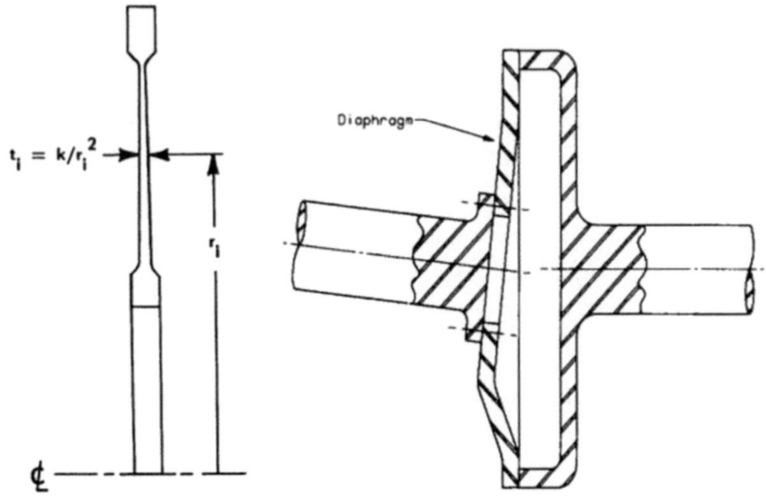

Fig. 8.5 Profile of a single-element contoured diaphragm (Image courtesy of Regal Rexnord, Milwaukee, WI)

An alternative design incorporates a stack of thin stainless steel plates. These are secured to an adapter using an interference-fit spine and a threaded clamp nut. Each diaphragm in the stack is uniformly thick but precision-machined to be thinner in the flexing zone, effectively preventing fretting.

This multi-diaphragm coupling configuration introduces a progressive failure safety mechanism. Due to offset stress, each diaphragm experiences different stress levels. In the event of misalignment-induced fatigue, the outer diaphragms are the first to fail. Subsequent diaphragms, subjected to lower alternating stress, continue functioning—similar to disk-pack couplings. These assemblies can remain operational until roughly half of the diaphragms have failed. Modern turbomachinery typically includes vibration monitoring systems capable of detecting unbalance caused by diaphragm failure.

This configuration places steel diaphragm packs at both ends of the spacer tube. Each pack, approximately 0.150 in. (3.8 mm) thick, consists of several thin flexing diaphragms or plates separated by filler rings on both the inner and outer diameters to prevent fretting. Torque transmission is achieved through end covers on the outer diameter and a clamping nut or bolt on the inner diameter, functioning as a single integrated unit.

Compared to single-element contoured diaphragms, thin convoluted diaphragms generate lower bending moments for the same torque and offer more consistent axial stiffness across the full range of axial movement.

In the event of misalignment, fatigue failure begins with the outermost diaphragm. If the system continues to operate, the next outermost diaphragm will fail, followed by gradual failure of the remaining diaphragms. This staged or progressive failure process—enabled by decreasing alternating stress—serves as a built-in safety feature.

8.3 Coupling Selection

Coupling manufacturers determine the optimal size and type of coupling for a specific application based on its duty requirements. Providing accurate and comprehensive information is essential to ensure the correct product is selected. As previously noted, the ideal approach involves collaboration among the unit vendor, subvendors supplying the driver and gears (if applicable), the owner's machinery engineers, and the engineering contractor. Together, they should define the service conditions and specify the appropriate coupling interface.

The key service conditions that influence coupling selection and design are outlined below:

1. Operating data:
   - Normal absorbed shaft power: Measured in horsepower (hp) or kilowatts (kW).
   - Maximum continuous shaft power: Also expressed in hp or kW.
   - Rotational speeds: Includes minimum, normal, rated, maximum continuous, and trip speeds, specified in revolutions per minute (rpm).
   - Momentary torque (if applicable): Specified in lb-in. (N·m).
   - Peak or maximum transient torque: Measured in lb-in. (N·m).
   - Cyclic torque (if applicable): Also given in lb-in. (N·m).
   - Service factor: As per API Standard 671 [2], the minimum required service factor is 1.5. A reduced service factor, based on normal absorbed shaft power, must not fall below 1.2. API 671 [2] defines the service factor as the "factor applied to the steady-state torque to allow for off-design conditions, cyclic and other variations, as well as equipment variations resulting in higher torque than that at the equipment normal operating point".
   - Factor of safety: Also defined in API 671 [2], this accounts for uncertainties inherent in coupling design or the "factor that is used to cover uncertainties in a coupling design."
   - Axial displacement: Includes both steady-state and transient axial movement of the driver and driven equipment, in. (mm).
   - Parallel offset: Steady-state and transient lateral misalignment, in. (mm).
   - Angular misalignment: Steady-state angular deviation, measured in degrees. API 671 [2] mandates a minimum of 0.2° under steady-state conditions.
2. Interface data for connecting shafts (driver and driven):
   - Shaft End Separation: Specify the distance between shaft ends, in. (mm).
   - Prestretch or Preload Requirements: Indicate the necessary preload, accounting for thermal expansion and mechanical growth of equipment and couplings, in. (mm).
   - Bore Dimensions and Configuration: Provide bore diameter and type—such as keyless tapered hydraulic fit, splined, or keyed cylindrical.

- Flanged Shaft End Geometry: Detail the diameter and thickness of the integral flanged shaft end, considering all expected load conditions.
- Tapered Shaft End Profile: Include measurements for both the large and small diameters, as well as the taper length.
- Diametral Interference Range: Define the minimum and maximum allowable diametral interference, in. (mm).
- Keying Details: State the number of keys and the dimensions of keyways for keyed shaft ends.
- Locknut Specifications: Specify the width and diameter of the shaft end locknut.

3. Optional information:

   - Service environment, for example, marine, presence of chlorine or hydrogen sulfide. If specified, a suitable protective coating can be selected based on the service environment.
   - Electrical insulation, if required.
   - Torque measurement, if specified.
   - Hardware such as a hydraulic installation tooling and moment simulator, if specified.

Based on the provided data, the selection of a suitable coupling typically follows these steps—though the sequence may vary depending on the manufacturer:

- Determine the normal continuous torque (steady-state torque) for the application.
- Calculate the minimum required coupling rating (steady-state selection torque) by applying the appropriate service factor (SF). Consider peak and momentary torque demands. Coupling ratings are commonly expressed in terms such as horsepower per 100 rpm (or kilowatts per 100 rpm), maximum continuous torque, and peak overload torque.
- Apply a reduced service factor if a significantly oversized driver is present, to avoid selecting an unnecessarily large coupling. Also, assess the torque capacity of the driven shaft and the interference fit of the hydraulically mounted coupling hub.
- Confirm the maximum hub bore capacity of the chosen coupling exceeds the specified shaft bore dimensions.
- Ensure the specified maximum operating speed is within the coupling's allowable speed range.
- Verify alignment capabilities—the selected coupling must accommodate the specified angular and offset misalignments, as well as axial movement between the driving and driven shafts.
- Check the required shaft-end spacing to ensure it meets or exceeds the minimum distance supported by the coupling.
- Review dynamic balancing requirements to maintain system stability and performance.

During the design phase of coupling components, the initially selected membrane type and size must be reassessed. Engineering factors specific to the application—such as weight, stress levels, lateral critical speed, torsional stiffness, maximum allowable diameter, material selection, and cost—can influence the sizing of the spacer tube, flanges, and fasteners. Any resulting dimensional adjustments must be reflected in the final membrane design. Figure 8.6 illustrates typical mass-elastic properties supplied by coupling manufacturers to the team responsible for conducting lateral, steady-state, and transient torsional analyses of the train system.

Table 8.1 presents a sample sizing calculation for a hydraulically fitted coupling hub, highlighting essential design parameters such as total interference, maximum transmissible torque via the interference fit, shaft contraction, and hub bore expansion. The ANSI/AGMA 9003-B08 standard (Flexible Couplings–Keyless Fits), published by the American National Standards Institute (ANSI) and the American Gear Manufacturers Association (AGMA), provides illustrative calculations to guide the design of hydraulically fitted hubs. The chosen interference fit must ensure reliable torque transmission without slippage or fretting at the interface between the shaft end and the inner bore of the coupling hub.

Fig. 8.6 Engineering data for a coupling (Image courtesy of Regal Rexnord, Milwaukee, WI)

Table 8.1 Hydraulically fitted coupling hub sizing calculations

Parameter	Value	Unit
Inputs		
Coupling hub length (L)	5	in.
Coupling hub diameter (D)	6	in.
Coupling hub average bore diameter (d)	4	in.
Coefficient of friction (μ)	0.300	
Operating speed	1800	rpm
Horsepower	19,000	hp
Yield strength of coupling hub material	95,000	psi
Yield strength of shaft	95,000	psi
Coupling interference rate	0.0015	in./in. of shaft diameter
Modulus of elasticity (€)	3.E+07	psi
Poisson ratio for steel	0.3	
Outputs		
Ratio of diameters	0.667	
Ratio of acceptable stress to yield stress (K)	0.870	
Maximum allowable combined stress at hub bore	82,650	psi
Maximum allowable hub contact pressure	25,678	psi
Contact area	62.832	$in.^2$
Required effective interference	0.003	in./inch
Loss of interference rate due to centrifugal force	0.000	in./inch
Required total interference rate	0.003	in./inch
Required diametral interference	0.012	In.
Maximum allowable transmitted torque	968,042	in.-lb
Ratio of maximum allowable torque to transmitted torque	1.456	
Coupling hub bore growth	0.010	in.
Shrinkage of shaft under coupling hub	0.002	in.
Growth of coupling hub outer diameter	0.008	in.

Before finalizing the type of coupling, several key performance parameters must be carefully assessed. When selecting between various diaphragm and disk coupling sizes for a specific service application, factors such as torque capacity, misalignment tolerance, angular stiffness, permissible misalignments, and torque-to-weight ratio should be thoroughly evaluated.

To illustrate this, Table 8.2 presents a comparative analysis of couplings used between compressor gears. The data clearly shows that the single-element contoured diaphragm coupling offers superior normal and peak torque capacities, and angular stiffness compared to the multiple disk-pack coupling. However, under the same

Table 8.2 Comparison of selected coupling types

Parameter	Values		Unit
	Single contoured diaphragm coupling	Multiple disk pack coupling	
Service conditions			
Power	3151		hp
Speed	11,970		rpm
Torque	16,595		in.-lb
Peak torque	33,190		in.-lb
Cyclic torque	± 16,595		in.-lb
Required service factor	2		
Characteristics			
Maximum diameter	7	6	in.
Normal torque capacity	47,706	39,386	in.-lb
Peak torque capacity	63,460	59,079	in.-lb
Actual service factor	2.9	2.4	
Angular misalignment	0.33	0.38	deg.
Lateral offset	0.14	0.17	in.
Angular stiffness	443	389	in.-lb./deg
Axial misalignment	0.07	0.09	in.
Axial load or force at misalignment	280	183	lbf
Drive end side coupling weight	18	15	lb.
Drive end side coupling center of gravity	3	2	in.
Driven end side coupling weight	18	15	lb.
Driven end side coupling center of gravity	3	2	in.

service conditions, the multiple disk-pack coupling outperforms the diaphragm coupling in all other evaluated aspects.

Disk couplings are often preferred by designers if a reduced-moment configuration is required in multibody centrifugal compressor trains. Their circumferential disk pack arrangement enables more efficient packaging than the radial layout of diaphragm flexible elements (refer to Fig. 8.7). Additionally, disk couplings are lighter, more compact, and exert lower forces on connected equipment for a given shaft size due to their multi-element design.

Some turbine and compressor manufacturers frequently opt for diaphragm couplings over disk couplings in high-horsepower applications where the system must tolerate substantial angular and axial deflections caused by thermal expansion.

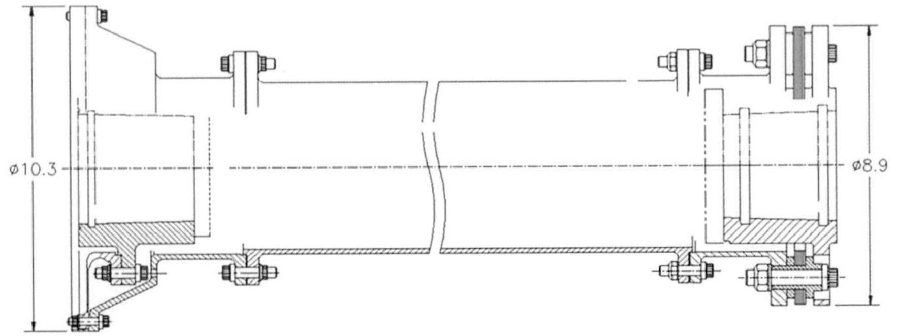

Fig. 8.7 Reduced-moment diaphragm versus disk coupling (Image courtesy of Regal Rexnord, Milwaukee, WI)

Diaphragm couplings are better suited to operate near their misalignment thresholds without the risk of fatigue failure from fretting or disk contact. They are also commonly chosen for synchronous motor setups, particularly during startup when cyclic torsional vibrations occur. Since motor torque profiles are not always clearly defined, the resulting peak cyclic torque can lead to material yielding or even failure during startup.

Neither diaphragm nor disk couplings inherently provide damping. When vibratory torque levels are considerable, a damping or hybrid coupling may be necessary to mitigate transient resonant torque effects. API TR 684 [3] outlines how coupling and component body modes influence torsional behavior. One method of torsional tuning involves using an oversized spacer tube to enhance both torsional and lateral stiffness. However, this improvement comes at the cost of increased mass and inertia. If a stiffness increase exceeding 10% is required, it typically necessitates upsizing the coupling or switching to a different type—thus demanding a higher service factor.

The design of flexible-element couplings can be affected by transient torque loads as well as the system's axial, lateral, and torsional natural frequencies. These factors are explored in detail in the sections that follow.

8.4 Peak Torque or Maximum Transient Torque

Peak torque—also referred to as maximum transient torque—is the highest torque a coupling is expected to endure across multiple transient events throughout its operational life (typically designed for 10^3 occurrences). Manufacturers usually publish peak torque ratings based on the coupling's maximum allowable misalignment. This rating plays a critical role in transient torsional analyses during startup and shutdown phases, particularly in evaluating the risk of low-cycle fatigue failure. Importantly, the peak torque must always remain below the coupling's rated torque.

According to API 671 [2], peak torque is defined as the torque level at which the most highly stressed component in the coupling reaches a safety factor of 1.15 relative to its yield strength. This definition accounts for the combined effects of rotational speed, angular misalignment, and axial displacement under all transient and momentary conditions.

As per API 671 [2], the factor of safety is determined by dividing the material's yield strength by the calculated stress. This stress is influenced by design-specific parameters, including torque, rotational speed, angular misalignment, and axial displacement.

8.5 Momentary Torque

As specified by API 671 [2], momentary torque refers to the torque level at which the most highly stressed component in the coupling reaches its material yield strength—corresponding to a safety factor of 1.0. This definition accounts for the combined effects of rotational speed, angular misalignment, and axial displacement. Momentary torque is typically linked to fault conditions in generators, such as short circuits or malsynchronization events, and is expected to occur only briefly.

8.6 Continuous Cyclic Torque

Continuous cyclic torque represents an alternating stress component—distinct from steady-state torque and typically associated with angular misalignment—that must be factored into coupling analysis. To ensure a sufficient Goodman safety factor, the analysis should consider the simultaneous occurrence of peak steady-state torque, maximum axial displacement, highest operating speed, and maximum angular misalignment.

This type of torque becomes a critical selection criterion when an adjustable-speed drive in the system imposes a continuous cyclic torque load exceeding 5%. It should not be confused with the cyclic torque generated during synchronous motor startup. Cyclic torque typically involves around 10^7 load cycles.

The continuous torque rating of a coupling is determined by its resistance to high-cycle fatigue and is essential for steady-state torsional response evaluations. Both steady-state and cyclic torque loads must remain within the coupling's rated or continuous torque capacity.

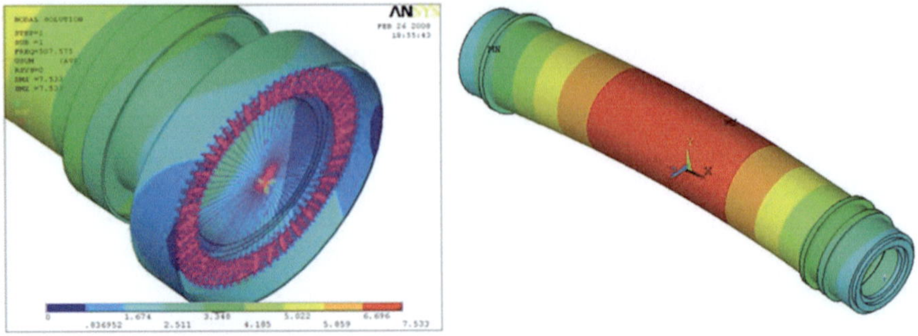

Fig. 8.8 Finite-element models of a coupling (Image courtesy of Regal Rexnord, Milwaukee, WI)

8.7 Axial Natural Frequency

The axial stiffness of most disk and diaphragm couplings exhibits nonlinear behavior, meaning the axial natural frequency changes with axial displacement. This frequency is proportional to the square root of the membrane pack's axial stiffness. When machines are connected, they can excite the coupling, causing it to resonate at its axial natural frequency.

According to API 671 [2], the axial natural frequency must not fall within ± 10% of the defined operating speed range. Notably, most dry couplings do not demonstrate axial natural resonance, making such occurrences uncommon in practical applications.

Equation 8.1 defines the axial natural frequency, which is typically evaluated at axial displacements of 0%, 50%, and 100%, as illustrated in Fig. 8.8. Manufacturer catalogs often provide axial deflection data based on the coupling's maximum continuous torque rating.

$$f = \frac{1}{2\pi}\left(\frac{k}{m}\right)^{0.5} \tag{8.1}$$

where f is the frequency (Hz [cycles/s]); m is the total mass of the center member, which is suspended between two flexible elements, plus 50% of the flexible elements and fasteners (lb [kg]); and k is the axial stiffness (lb/in. [N/m]).

8.8 Lateral Natural Frequency

When a coupling's operating speed aligns with its whirling speed, it can trigger resonance, resulting in significant lateral vibrations. To mitigate this, API 671 [2] recommends that the lateral natural frequency (N_c) of the coupling section—including the flex elements and assuming ideal (infinite) support conditions—should be at

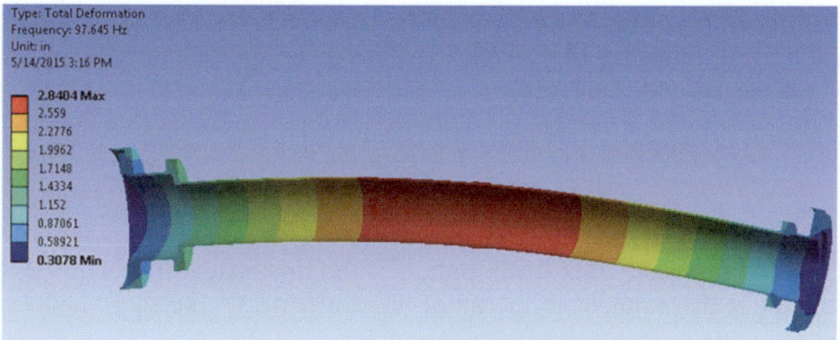

Fig. 8.9 Finite-element model of a complete coupling (Image courtesy of Regal Rexnord, Milwaukee, WI)

least twice the maximum specified operating speed when using the uniform-tube-equation approach outlined in ANSI/AGMA 9004/9104. Alternatively, if a more detailed finite-element analysis is performed using the actual geometry, N_c should be no less than 1.5 times the highest operating speed (refer to Figs. 8.6, 8.7, 8.8 and 8.9).

For spacer couplings analyzed via the simply supported tube method, the lateral natural frequency is determined by modeling the coupling as a uniformly thick tube supported at both ends. The first lateral natural frequency is calculated based on this simplified representation (Eq. 8.2).

$$f = \frac{\pi}{8L^2}\left[\frac{E(D^2 + d^2)}{\delta}\right]^{0.5} \tag{8.2}$$

where f is the frequency (Hz [cycles/s]), E is the modulus of elasticity of the spacer tube (psi] N/m^2]), D is the outer diameter of the spacer tube (in. [m]), d is the inner diameter of the spacer tube (in. [m]), L is the length between centers of disk packs or membranes (in. [m]), and δ is the density of the spacer tube material (lb/ft^3 [kg/m^3]).

8.9 Torsional Natural Frequency

Torsional analysis of a centrifugal compressor train accounts for the coupling(s) within the drive system. Since the torsional stiffness of the coupling is generally much lower than that of the other components in the train, it can significantly influence the torsional critical speeds. For this reason, the coupling is frequently used as a tuning element to modify the system's torsional natural frequency.

In configurations where the compressor train includes a single coupling, the torsional natural frequency of the entire system can be approximated by modeling it as a two-inertia system. This estimation is derived using Eq. 8.3.

$$f = \frac{1}{2\pi}\left[\frac{q(J_1 + J_2)}{J_1 J_2}\right]^{0.5} \tag{8.3}$$

where f is the frequency (Hz [cycles/s]), q is the torsional stiffness of the system (Nm/rad), J_1 is the moment of inertia of the driver (lb ft^2 [kg m^2]), and J_2 is the moment of inertia of the driven (lb ft^2 [kg m^2]).

8.10 Fatigue Diagram

The operational stresses experienced by the coupling must be evaluated using a fatigue diagram. While multiple valid approaches exist for this assessment, they all depend on decomposing the stresses into steady-state and alternating components. In standard applications, mean stresses typically include steady torque, axial displacement, and rotational speed, whereas alternating stresses encompass cyclic torque and angular misalignment.

These stress components are plotted on a modified Goodman diagram—or an equivalent fatigue diagram—based on the coupling material. To ensure safe operation, the resulting point must fall below the safe working stress line. The factor of safety (FS) for the coupling generally ranges from 1.25 to 2.0, depending on how conservatively the operating point is positioned within the acceptable region. An example of a modified Goodman diagram is illustrated in Fig. 8.10.

8.11 Materials of Construction

Coupling materials are selected based on factors such as load demands, weight limitations, cost considerations, and the operating environment. Additional criteria influencing material choice include torque capacity, rotational speed, and required torsional and axial stiffness characteristics.

Table 8.3 outlines commonly used materials for high-performance coupling components. Equivalent international materials may also be used, provided they meet the necessary performance and quality standards. If corrosion resistance is required, flexible elements are treated with protective coatings—such as SermeTel® applied to diaphragm profiles. Most coupling component materials exhibit a minimum ultimate tensile strength of approximately 130,000 psi (896 MPa).

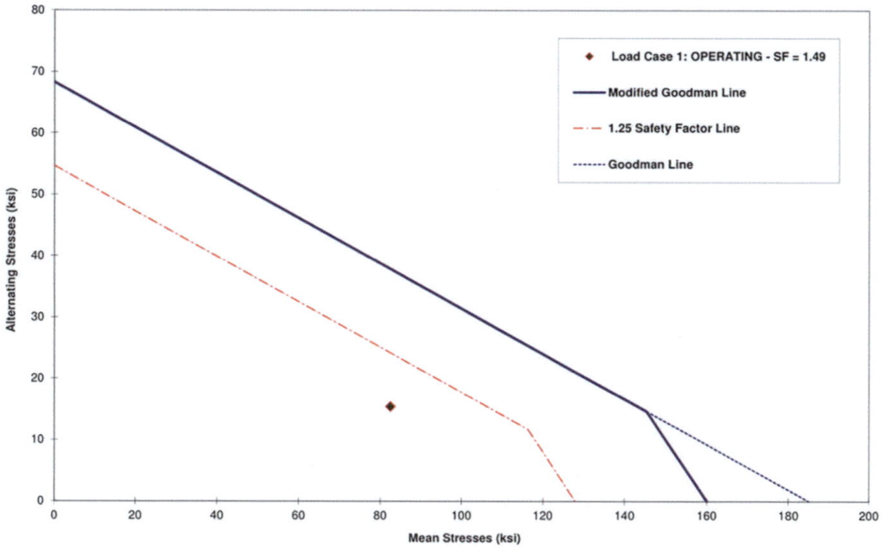

Fig. 8.10 Fatigue diagram (Image courtesy of Regal Rexnord, Milwaukee, WI)

Table 8.3 Materials for coupling components

Component	Material of construction
Disks or membranes	300 Series stainless steels, precipitation-hardened steels, high-strength nickel alloy steels
Single-element contoured and multiple-convoluted diaphragms	American Iron and Steel Institute (AISI) 4140 steels; AISI 4340 steels; 300 Series stainless steels; precipitation-hardened steels 15-5PH, 17-4PH, and Custom 455; high-strength nickel alloy steels
Spacer tube	AISI 4140 steels, AISI 4340 steels
Flanges	Forged AISI 4340 steels
Fasteners	Heat-treated A286 stainless steel, AISI 410 steels, AISI 4340 steels

8.12 Coupling Balance

API 671 [2] outlines three primary balancing methods for couplings:

- Method 1: Component balance
- Method 2: Component balance with assembly check balance
- Method 3: Component balance with full assembly balance.

The standard also addresses optional procedures such as repeatability checks and component interchangeability evaluations. For systems operating at speeds up to 3600 rpm, component balancing alone is typically sufficient. However, for service speeds exceeding 5000 rpm, many users require both component balance and assembly check balance to ensure reliability.

Coupling unbalance can arise from various sources, including balance tolerance limits, inherent unbalance, eccentricity in fixtures and mandrels, pilot clearance and eccentricity, keyway influences, shaft runout, and inaccuracies in the balancing equipment. To minimize residual unbalance in factory-assembled couplings, manufacturers often employ tighter tolerances, combine component and assembly balancing, use weight-matched fasteners, and apply matchmarking techniques. Matchmarking ensures that coupling components retain the same relative orientation used during the assembly balance process (refer to Fig. 8.11). A typical balancing setup is illustrated in Fig. 8.12.

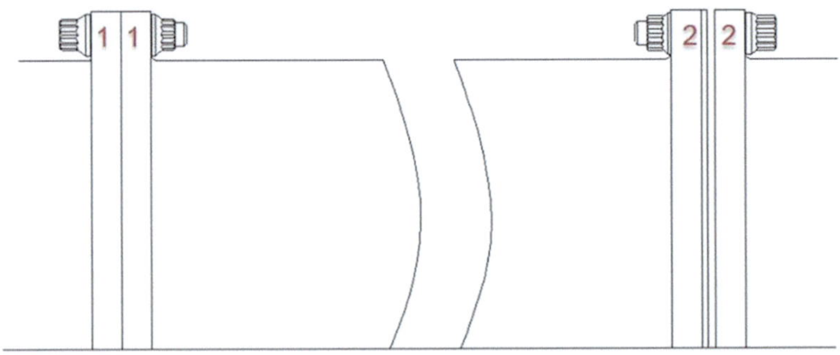

Fig. 8.11 Matchmarked coupling flanges (Image courtesy of Regal Rexnord, Milwaukee, WI)

Fig. 8.12 Assembled coupling on a balancing machine, API 671 method 3 [2] (Image courtesy of Regal Rexnord, Milwaukee, WI)

Potential unbalance refers to the theoretical unbalance that may be present within a coupling assembly. In contrast, residual unbalance denotes the actual unbalance measured in a component or assembly while it is undergoing evaluation in a balancing machine.

Field balancing, also known as on-site balancing, is occasionally performed during installation to fine-tune the balance of couplings equipped with trim-balance holes. The effectiveness of this process depends on several factors: the initial balance condition of the installed components, the precision of fits and clearances between mating parts, and the alignment of the coupling with the equipment's shafts and flanges.

8.13 Torque Measurement [4, 5]

In industrial applications, the two primary methods for measuring and monitoring torque are strain-gauge systems and phase-displacement systems. Both techniques assess either the strain or angular displacement of a torque-transmitting component—typically a spacer tube—to determine torque, rotational speed, and power output. Figure 8.13 illustrates a coupling based on the phase-displacement method.

Fig. 8.13 Phase-displacement torque-measuring coupling (Image courtesy of Regal Rexnord, Milwaukee, WI)

A strain-gauge torque measurement setup typically consists of:

- An instrumented spacer tube integrated into the coupling
- A rotary power transformer
- A signal conditioning and processing unit
- A readout monitor.

Strain gauges are strategically mounted on the spacer tube at specific angles relative to its rotational axis. Each gauge comprises numerous parallel insulated conductors. These gauges are firmly bonded to the tube to eliminate any relative movement and to ensure they receive power from the transformer.

After installing the electronic components within the spacer tube, final balancing of the strain-gauge coupling is essential. Proper alignment of the stator and rotor axes, along with precise radial and axial positioning of the stator coils relative to the rotor coils, is critical to prevent coil contact and potential damage.

When torque is applied to the spacer tube:

- The conductors in the strain gauges experience changes in electrical resistance.
- If the strain aligns with the conductors' orientation, compression occurs, reducing resistance.
- Conversely, tension increases the resistance.

Strain gauges under tension and compression are arranged in a way that produces a signal directly proportional to the applied torque.

The signal conditioning unit, located near the coupling, processes torque signals from the strain gauges via the power transformer and speed signals from stator-mounted sensors that detect pulses generated by rotating magnets on the spacer tube. These processed signals are then transmitted to the monitor, which displays real-time values for power, speed, and shaft torque.

There are two primary types of phase-displacement torque measurement systems: single-channel and dual-channel. These systems feature integrated signal-generating toothed flanges within the coupling design.

In a dual-channel configuration, two multipole stationary toothed rings are mounted around the rotating flanges inside the coupling guard. Each ring contains circumferentially embedded coils, powered by a 24 V DC supply. As the toothed flange rotates, it induces a sinusoidal voltage in each coil.

In contrast, a single-channel system performs this measurement in a single plane. The phase shift between the sinusoidal signals arises from the torsional deformation of the coupling element, which is directly proportional to the transmitted torque. A signal-processing unit analyzes the phase shift and frequency, conditioning the signals to deliver precise readings of speed, torque, and power.

Modern torque-measuring couplings—whether based on strain-gauge technology or phase-displacement methods—are known for their high reliability and accuracy. Choosing the appropriate coupling type depends on these factors:

- Maximum operating speed
- Required durability and service life
- Measurement precision
- Operating temperature conditions.

Typical applications include test benches, steam turbines, and gas turbine-driven centrifugal compressor trains operating in fouling environments.

8.14 Prestretching, Preloading, and Shimming

Thermal expansion of the connected shafts and the coupling itself, along with discrepancies between the designed and actual distances between shaft or flange ends, can generate axial loads on the coupling. These loads increase stress on the flexible elements and impose additional forces on the connected equipment. To mitigate this, couplings are prestretched during installation so that the flexible elements operate near their neutral positions under normal, steady-state conditions. Similarly, if one or both shafts are expected to move axially away from the coupling during operation, the coupling is preloaded (compressed) during installation to accommodate that movement.

The preload applied to coupling fasteners (bolts) is influenced by several factors: the coupling type, the resilience of the fastener material, the design of the flexible element, any applied coatings, and the friction at the flange face if it contributes to torque transmission. Typically, fasteners are preloaded to around 70% of their proof strength to ensure a secure bolted joint, promote uniform load distribution, and enhance resistance to loosening. Some manufacturers recommend replacing bolts and thread locknuts at the first sign of thread damage—whether on the bolt or the nut—to prevent progressive wear during repeated tightening and loosening cycles.

To address minor axial misalignments of tapered bore hubs that may arise during setup and installation, coupling manufacturers often supply two sets of shims. One set comes pre-installed with the coupling, while the second set is used to fine-tune the axial positioning of the hub flanges by adding or removing shims as needed. Figure 8.14 illustrates the hydraulic tooling used for installing coupling hubs, and Fig. 8.15 depicts the configuration of an axial stop device.

Fig. 8.14 Hydraulic tooling for the installation of a coupling hub (Image courtesy of Regal Rexnord, Milwaukee, WI)

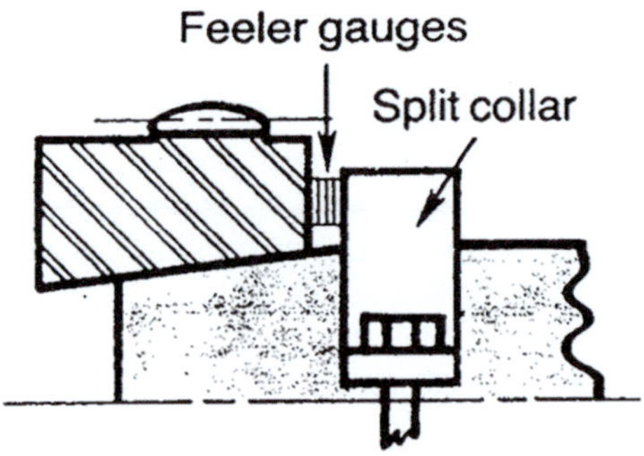

Fig. 8.15 Axial stop device (Image courtesy of Regal Rexnord, Milwaukee, WI)

8.15 Coupling Guards

Heat buildup within fully enclosed coupling guards is primarily caused by air shearing (as illustrated in Fig. 8.16) and turbulence. If peripheral velocities differ between the surfaces of large couplings and the sections of connected shafts enclosed by the guard, differential pressure can develop. This pressure unbalance may draw oil from adjacent bearing housings.

Such negative pressure conditions inside the guard are commonly observed in reduced-moment couplings. To mitigate windage effects, antiswirl baffles can be installed within the guard.

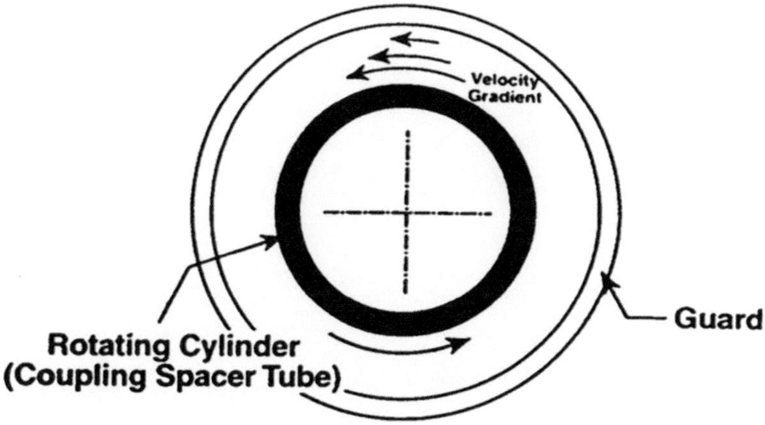

Fig. 8.16 Heat generated inside a coupling guard by the shearing of air (Image courtesy of Regal Rexnord, Milwaukee, WI)

The temperature rise inside coupling guards is influenced by several factors: the operating speed of the coupling, the dimensions of both the coupling and the guard, the ambient temperature, and the temperature of the shaft. Effective heat dissipation can be achieved through various design features, including strategically placed air vents, cooling oil nozzles, or external fins on the guard's outer surface. For optimal airflow, exhaust ports should be tangentially aligned to expel air efficiently, while intake vents should also be tangentially oriented and positioned to prevent the formation of stagnant air zones.

8.16 Coupling Failures

In typical cases of fatigue failure in single-element diaphragms caused by angular misalignment, cracks tend to initiate within the diaphragm and propagate circumferentially. As the crack spreads, system unbalance increases until the remaining material can no longer support the torque load—leading to complete rupture of the diaphragm.

Fig. 8.17 Diaphragm coupling failure caused by excessive axial misalignment (Image courtesy of Regal Rexnord, Milwaukee, WI)

To mitigate such failures, some couplings incorporate a backup mechanism, such as gear teeth with intentional clearance to prevent contact during normal operation. If the primary coupling fails, this backup drive engages to maintain torque transmission. However, gear teeth operating without lubrication degrade quickly under extended use. Failure of the flexible elements also results in misalignment and elevated vibration levels. Continued operation under these conditions, relying solely on the backup system, is strongly discouraged. Instead, the machinery should be safely shut down to prevent further damage.

Additional common failure modes in high-performance flexible-element couplings include torque overload, cyclic torsional stress, and fretting. Examples of these failure types are illustrated in Figs. 8.17, 8.18, 8.19, 8.20, 8.21, 8.22, 8.23, 8.24 and 8.25.

Fig. 8.18 Diaphragm coupling failure caused by excessive angular misalignment (Image courtesy of Regal Rexnord, Milwaukee, WI)

Fig. 8.19 Diaphragm and disk failures due to cyclic torque (Image courtesy of Regal Rexnord, Milwaukee, WI)

Fig. 8.20 Disk failure due to severe misalignment (Image courtesy of Regal Rexnord, Milwaukee, WI)

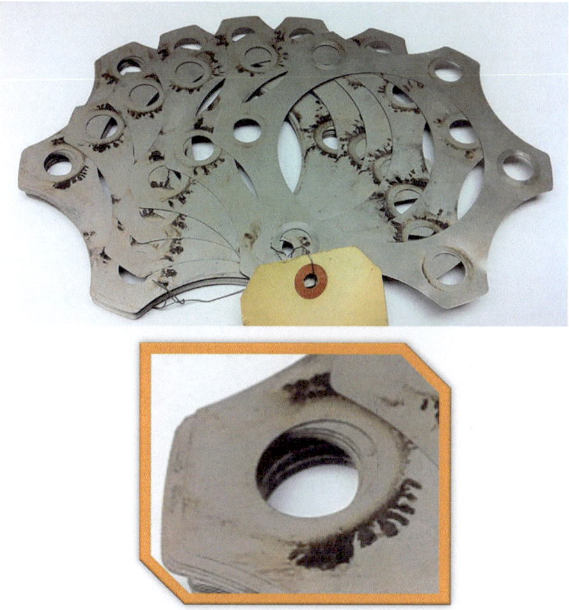

Fig. 8.21 Fretting damage on uncoated disks (Image courtesy of Regal Rexnord, Milwaukee, WI)

Fig. 8.22 Fracture at the center of a link and fretting in the link and at the bolt hole, indicative of torsional fatigue failure (Image courtesy of Regal Rexnord, Milwaukee, WI)

Fig. 8.23 Coupling bolt failure due to torque overload (Image courtesy of Regal Rexnord, Milwaukee, WI)

Fig. 8.24 Rippling on the diaphragm profile due to torque overload (Image courtesy of Regal Rexnord, Milwaukee, WI)

Fig. 8.25 Torque overload failure in a disk coupling (Image courtesy of Regal Rexnord, Milwaukee, WI)

Acknowledgements All of the images/graphics/figures in this chapter were provided by Regal Beloit America, Inc. (now Regal Rexnord), Milwaukee, WI.

This material was reprinted with permission of the publisher from Ghaisas N, Sakers C (2018) High-performance flexible element couplings for centrifugal compressor trains. CompressorTECH^2 magazine (July 2018), pp 36–42.

References

1. How disc couplings work. Form #10036E. Regal Rexnord Corporation, Milwaukee, WI
2. API Standard 671 (2020) Special-purpose couplings for petroleum, chemical, and gas industry services, 5th edn. American Petroleum Institute, Washington, DC
3. API TR 684-1 (2019) API standard paragraphs rotordynamic tutorial: lateral critical speeds, unbalance response, stability, train torsionals, and rotor balancing, 1st edn. American Petroleum Institute, Washington, DC
4. Riverhawk (Indikon) Torque Metering System Literature, provided by Regal Beloit America, Inc. (now Regal Rexnord), Milwaukee, WI
5. Torquemeters Limited Technical Catalogues (various), provided by Regal Beloit America, Inc. (now Regal Rexnord), Milwaukee, WI

Chapter 9
Shop Tests and Special Engineering Analyses

Abstract During the manufacturing of centrifugal compressors, numerous in-process examinations, inspections, and testing activities are conducted. Among these, the shop performance test and the mechanical running test hold particular importance for customers. The performance test serves as a key indicator of the compressor's actual thermodynamic efficiency and mechanical behavior, and helps validate the engineering design and analysis. The results from this test are instrumental in defining the head-capacity curve—its shape and limits—marked by surge at low flow rates and stonewall or stall at high flow rates. This chapter outlines the proper approach to scoping, planning, and executing both performance and mechanical running tests within the compressor manufacturer's facility. It includes a sample calculation demonstrating how to convert data from test at a certified point into predicted performance under specified operating conditions. Additionally, it covers specialized analyses required by certain users for large-scale compressors.

9.1 Test Facility Evaluation

Performance testing is typically included in the base scope of compressor purchase order by contractors and end users. However, during bid evaluation, the shop testing capabilities of prospective vendors often receive insufficient attention. This critical aspect is frequently overlooked until the manufacturer submits their test procedure. In some instances, performance testing cannot be conducted in full compliance with the American Society of Mechanical Engineers (ASME) Performance Test Code on Compressors and Exhausters (ASME PTC-10) [1] due to limitations within the test facility.

To avoid such issues, it is essential for both the user and manufacturer to proactively discuss and agree on the scope and constraints of the shop performance test. Ideally, this review should occur before issuing the purchase order and should address the following key considerations:

N. Ghaisas, *Centrifugal Compressors in Hydrocarbon Processing Industries*, https://doi.org/10.1007/978-3-032-03687-2_9

Test Setup and Conditions:

- Closed-loop testing with sub-atmospheric inlet pressure: May be necessary to reduce power demand, especially for high-power compressors with limited shop driver capacity.
- Torque measurement capabilities: Confirm availability of a properly rated device. If using the heat-balance method, request a detailed casing heat-loss calculation including convection and radiation losses, and specify sensor placement for ambient temperature measurement.

Temperature and Efficiency Calculations:

- Ambient temperature measurement: Should be taken near the compressor to accurately determine temperature differential, which affects polytropic efficiency (this temperature difference is the lost work input).
- Discharge temperature correction: Must account for radiation heat loss.

Flow and Leakage Considerations:

- Balance piston leakage: Should be measured or estimated and subtracted from inlet flow. If the balance line connects to the test loop, leakage must be included in performance data conversion.

Heat Load and Gas Handling:

- Heat exchanger suitability: Ensure shell-and-tube exchangers can handle the maximum heat load during testing.
- Test loop purging: Confirm inert gas purging capability before introducing test gas.
- Gas sampling: Must be performed prior to each measurement point.
- Gas mixing control: Verify procedures for volumetric blending of test gases.

Gas Composition and Safety:

- Online oxygen analysis: Required to monitor air content and adjust gas properties in real time. Variations in molecular weight affect calculated head. ASME PTC-10 [1] mandates continuous monitoring to prevent combustible mixtures.
- Maximum allowable working pressure: Ensure piping and equipment ratings meet test conditions.
- Overpressure protection: Identify installed safety devices.
- Abnormal condition detection: Confirm systems for alerting issues like coolant loss in heat exchangers. How is the lube oil cooling duty at full load established from the no-load, full-speed mechanical running test?

Lubrication System Integrity:

- Oil cleanliness procedures: Must meet oil system cleanliness criteria outlined in API 614 [2] and be verified before testing. Contaminants such as dirt, debris or foreign particles in the lubricant that exceed the bearing-to-rotor operating clearances can lead to abrasion of the babbitted bearing surfaces. Poor oil quality can result in localized heat buildup, disruption of the oil film, increased oxidation and accelerated component wear.

Integrated Unit Testing:

- Contract equipment compatibility: Verify shop can accommodate contract lube oil, dry gas seal systems, control panels, and other specified equipment.

Data Acquisition and Analysis:

- Vibration monitoring systems: Confirm availability of tools for Bodé plots, orbit diagrams, polar plots, frequency analysis, and waterfall spectra.
- Speed adjustment capability: Data system should correct test speed at guarantee point based on actual inlet temperature, i.e., adjust the calculated test speed at the guarantee point with respect to the actual gas inlet temperature prevailing on the day of the test. The inlet temperature used in the calculation of the test speed can be different from the actual compressor inlet temperature.
- Power input correction: Must account for actual inlet pressure. Specific volume ratios and stage deviations must comply with ASME PTC-10 tolerances [1]. Flow similarity requires volume flow coefficients, impeller-tip Mach numbers, and Reynolds numbers to remain within permissible limits.

9.2 Types of Performance Tests

The ASME PTC-10 test code [1] outlines two categories of performance tests. Type 1 tests are performed using the specified (or design) gas at the guarantee-point operating conditions. They are typically applied if the design gas is readily accessible—for example, in the case of air compressors.

Type 2 tests, on the other hand, permit the use of a substitute gas if testing with the original design gas is impractical. This approach is common if the design gas consists of hydrocarbon mixtures that are challenging to replicate under test conditions. Additionally, if the design gas is toxic, explosive, or highly flammable, a Type 2 test becomes necessary for safety and feasibility.

The substitute gas may be a pure component or a blend, selected to closely match the thermodynamic characteristics of the original design gas. Performance calculations can be based on either ideal or real gas laws, and the test code specifies acceptable deviations from ideal behavior for both the specified and substitute gases.

9.3 Test Gases

In closed-loop testing, commonly used test gases include carbon dioxide, helium-nitrogen mixtures, propane, R134a, and natural gas. Air is suitable only for open-loop testing and must be strictly avoided in closed-loop systems due to the risk of explosion if lubricating oil contaminates the air.

Test gases with low molecular weight are generally unsuitable for evaluating compressors intended for high-molecular-weight gas applications. This is because the required test speeds may exceed the compressor's maximum continuous operating speed. For example, refinery recycle gas has a molecular weight below 10, and centrifugal compressors used in this context are typically tested with helium-nitrogen mixtures. In contrast, compressors designed for high-molecular-weight gases—such as wet gas or chlorine—are usually tested with R134a or carbon dioxide.

When assessing the equivalency between test conditions and specified operating conditions, three key parameters must be carefully considered:

Volume Ratio:

- The volume ratio between the test gas and the specified gas must comply with the limits outlined in the ASME PTC-10 test code [1]. This ensures that the volume flow across all compressor stages remains consistent during performance testing. The volume ratio is directly influenced by molecular weight, rotational speed, and compressor head, and inversely affected by compressibility factor (Z), inlet temperature, and the specific heat ratio ($K = C_p/C_v$).

Machine Reynolds Number:

- This parameter reflects frictional losses within the compressor's gas passages. If the Reynolds numbers under test and specified conditions differ, the test efficiency must be adjusted accordingly, as mandated by the test code. The code also defines acceptable deviations for Reynolds number.

Machine Mach Number:

- Defined as the ratio of the gas velocity at the outlet of the first-stage impeller to the acoustic velocity at the compressor inlet, the machine Mach number indicates the compressor's peak capacity and relates to the stonewall effect. The test code specifies permissible deviations for this parameter. Ideally, the test gas should match or exceed the K value of the specified gas to closely replicate the design Mach number.

9.4 Sequence of Shop Acceptance Tests

The compressor performance test may be conducted either before or after the mechanical running test. However, if a compressor passes the mechanical running test but subsequently fails the performance test, the resources invested in the initial test are effectively wasted. To mitigate this risk, many users prefer to conduct the performance test first in the shop's witnessed test sequence, followed by the mechanical running test and any additional optional tests.

If a spare rotor (in an axially split compressor casing) or a spare rotor with a spare inner barrel assembly (in a radially split compressor casing) is purchased, the user or customer can specify whether the job rotor or the spare rotor is to be performance tested. It is common practice for compressor manufacturers to conduct internal factory testing before the customer- or user-witnessed tests take place.

The sequence of tests may vary depending on the scope of supply—whether it involves a single compressor or multiple units operating in parallel—and is often determined through mutual agreement between the user and the manufacturer. In cases where the compressor is steam turbine-driven, and both machines are manufactured by the same vendor at the same location, it is advisable to perform a full string test. This comprehensive test setup should include the compressor's performance test along with mechanical running tests for both the compressor and the steam turbine, using both the working and spare rotors (if supplied).

The following test sequence, outlined for a radially split casing compressor, is provided solely for illustrative purposes.

- Performance Test—Conducted using the spare rotor, inner barrel assembly, and coupling. The surge point (line slope) is validated by operating the compressor at 90 and 100% of the designated test speed.
- Noise-Level Survey—Performed during the performance test to record the sound power level (PWL, in dB) and the overall A-weighted sound pressure level (SPL, in dBA). Measurements include octave-band frequencies (in Hz) taken approximately 3 ft (1 m) from and near the casing centerline, along with background noise levels.
- Mechanical Running Test—Executed with the spare rotor, inner barrel assembly, and coupling. Includes verification of unbalance response.
- Disassembly Inspection—Visual examination of the spare rotor, inner barrel assembly, and coupling hub. Inspection covers bearings, seals, and internal components.
- Fit-Up Verification—Spare Coupling Hub—Contact check performed to confirm proper fit-up.
- Mechanical Running Test—Conducted using the job (working) rotor, inner barrel assembly, and coupling hub.
- Disassembly Inspection—Visual inspection of the thrust and radial bearings.
- Fit-Up Verification—Job Coupling Hub—Contact check performed to confirm proper fit-up.
- Static Gas Leakage Test—Performed on the fully assembled compressor using the job rotor and inner barrel assembly to verify sealing integrity.

9.5 Test Procedure Documents

Test procedures must be thoroughly documented to function as the sole reference during performance and mechanical running tests. They should contain the following details:

- Test Sequence and Schedule—A detailed timeline outlining the order of test activities. This shall be a visual representation of the test procedure, including a step-by-step graphical layout of the testing phases, the speeds involved and the time durations. Note that extended slow-roll operation—used to correct sag in large, heavy rotors before starting the mechanical running test—can adversely affect hydrodynamic oil film formation at low rotational speeds. This can result in polished contact zones or localized overheating and oxidation on the babbitted surfaces of radial bearings.
- Instrumentation Specifications—A comprehensive list specifying the type, location, quantity, and measurement accuracy of all instruments used during testing. Up-to-date calibration records for each instrument must be readily accessible.
- Test Setup Schematics— Clear diagrams illustrating the shop mechanical running test configuration, and performance test loop.
- Measurement Parameters and Sampling Frequency—A breakdown of key data points to be recorded during the mechanical running test, along with their respective sampling rates. These parameters may be tailored to suit specific applications and typically include:
  - Lubricating oil pressure at each supply point (psig/kPag)
  - Lubricating oil supply temperature (°F/°C)
  - Lubricating oil flow rate at each bearing (gpm/L/min)
  - Lubricating oil drain temperature at each return point (°F/°C)
  - Bearing metal temperature for each bearing, °F (°C)
  - Speed, rpm
  - Inlet pressure, psia (kPaa). Vacuum during the mechanical running test maintained by the shop ejector system
  - Inlet temperature, °F (°C)
  - Filtered and unfiltered shaft vibrations, mil peak-to-peak (μm peak-to-peak)
  - Axil displacement of rotor, mil (μm)
  - Buffer gas supply pressure to the dry gas seals, psig (kPag)
  - Sweep of frequencies at the maximum continuous speed and recording of vibration amplitude versus frequency range
  - Vibration Analysis During Coastdown Using Polar and Bode Plots
  - Unfiltered and filtered plots of vibration amplitude and phase angle versus rotor speed are examined during the coastdown phase. To capture the dynamic behavior across varying speeds—especially near critical speed transitions—both Polar and Bode plots are used. Polar plots, in particular, serve as essential diagnostic tools for identifying issues such as rotor unbalance, misalignment, and resonance.

  - Coastdown data offers enhanced diagnostic precision compared to ramp-up measurements, primarily because it is free from external influences like control system feedback, driver input, and load torque. This enables the compressor rotor to decelerate naturally across a wide speed range. Following a four-hour operation at maximum continuous speed, the rotor achieves thermal equilibrium, further improving the reliability of the coastdown data. For more detailed insights, refer to Sect. 14.5.1 in Chap. 14.
  - Determination of the first lateral critical speed
  - Slow roll runout.

- Test acceptance criteria.
- Schematic of the performance test arrangement.
- Tabulation of the test and specified (or design) conditions.
- Adjustment of the test conditions to contract or design conditions.

Figure 9.1 illustrates the string test configuration for a steam-turbine-driven compressor.

9.6 Sample Performance Test Calculations

The following example provides sample calculations that outline the basic procedure for converting test data to the specified or design point conditions, as defined by Eqs. 9.1–9.16. Table 9.1 lists the target conditions, while Table 9.2 presents the measured test data. All calculations adhere to the ASME PTC-10 code [1].

This is a five-point performance test, spanning from surge to overload. Test conditions are simulated to align with key parameters within the code's specified limits, including:

- Volume ratio (q_i/q_d)
- Capacity-to-speed ratio (q_i/N), which correlates with the flow coefficient
- Machine Mach number
- Machine Reynolds number.

This test case stipulates that the converted performance data shall include:

- A graphical comparison of actual versus test discharge flange pressures and power
- Overall performance curves based on flange-to-flange conditions
- A comparison between actual and test performance curves for the compressor.

Compressor efficiency must be calculated using total inlet and outlet conditions (flange-to-flange), incorporating mechanical losses to accurately reflect site operating conditions post-test.

The additional performance test tolerances for acceptance of this test are as follows:

- The flow at the surge point shall be $\pm$ 0% compared with the predicted surge flow.

- The head rise to surge shall be 5%.
- The flow at the end of the curve shall be − 2% compared with the predicted end-of-curve flow.
- The test operating speed shall be + 0%, − 2% compared with the calculated test speed.

Fig. 9.1 Setup for a centrifugal compressor string test (Image courtesy of Mitsubishi Heavy Industries Compressor, Hiroshima, Japan)

Table 9.1 Specified conditions

Parameter	Value
Compressor speed, N_{sp}	15,447 rpm
Inlet pressure, $P_{1,sp}$	1404 psia (96.80 bar A)
Inlet temperature, $T_{1,sp}$	127.22 °F (52.90 °C) 586.89°R (326.05 K)
Inlet density, ρ	0.9663 lbm/ft^3 (15.48 kg/m^3)
Inlet mass flow, wet, W_{sp}	2613 lb/min (19.75 kg/s)
Inlet volume flow, q	2705 ft^3/min (1.28 m^3/s)
Process gas	Refinery recycle gas
Gas molecular weight	4.50
Inlet Z_{sp}	1.037
Inlet $K_{sp} = C_p/C_v$	1.456
Discharge pressure, $P_{2,sp}$	1714 psia (118.18 bar A)
Pressure ratio, $P_{2,sp}/P_{1,sp}$	1.220
Polytropic head, H_{sp}	43,360 ft-lb/lbm (129.60 kJ/kg)
Polytropic efficiency	81.8%
Power at the driver shaft end	4326 bhp (3,326 kW)
Machine Reynolds number	6.4531×10^6
Machine Mach number	0.2542

9.6.1 Specified Conditions for the Compressor

See Table 9.1.

9.6.2 Test Data

See Table 9.2.

9.6.3 Calculation Procedure

Inlet Density

$$\rho_i = \frac{144 \times \text{MW} \times P_{1,\text{test}}}{Z_{\text{test}} \times 1545 \times T_{1,\text{test}}} \tag{9.1}$$

Table 9.2 Data recorded during test

Parameter	Value
Test gas	Helium
Gas molecular weight	4.002
Compressor speed, N_{test}	15,458 rpm
Inlet pressure, $P_{1,\text{test}}$	308.35 psia (21.26 bar A)
Inlet temperature, $T_{1,\text{test}}$	74.89 °F (23.83 °C) 534.56°R (296.98 K)
Inlet mass flow, wet, W_{test}	575.88 lb/min (4.35 kg/s)
Gas constant, R_{test}	386.30 ft•lbf/lbm•°R (2,078.6 J/kg•K)
Inlet Z_{test}	1.0112
Inlet h_{test}	219.80 Btu/lbm (510.77 kJ/kg)
Discharge pressure, $P_{2,\text{test}}$	376.87 psia (25.98 bar A)
Discharge temperature, $T_{2,\text{test}}$	132.02 °F (55.57 °C) 591.69°R (328.72 K)
Pressure ratio, $P_{2,\text{test}}/P_{1,\text{test}}$	1.222
Discharge Z_{test}	1.0124
Discharge h_{test}	291.45 Btu/lbm (677.35 kJ/kg)
Reynolds number correction in accordance with ASME PTC-10 test code [1]	1.0077

$$\rho_i = \frac{144 \times 4.002 \times 308.35}{1.0112 \times 1545 \times 534.56} = 0.213 \frac{\text{lb}}{\text{ft}^3} = 3.410 \frac{\text{kg}}{\text{m}^3}$$

Inlet Specific Volume

$$v_i = \frac{1}{\text{inlet density}} = \frac{1}{\rho_i} \tag{9.2}$$

$$v_i = \frac{1}{0.213} = 4.695 \frac{\text{ft}^3}{\text{lb}} = 0.293 \frac{\text{m}^3}{\text{kg}}$$

Inlet Volumetric Flow

$$q_i = \text{inlet mass flow} \times \text{inlet specific volume} = W_{\text{test}} v_i \tag{9.3}$$

$$q_i = 575.88 \times 4.695 = 2703.76 \frac{\text{ft}^3}{\text{min}} = 1.27 \frac{\text{m}^3}{\text{s}}$$

Discharge Specific Volume

$$v_{\mathrm{d}} = \frac{Z_{\mathrm{d,test}} \times 1545 \times T_{2,\mathrm{test}}}{144 \times \mathrm{MW} \times P_{2,\mathrm{test}}} \tag{9.4}$$

$$v_{\mathrm{d}} = \frac{1.0124 \times 1545 \times 591.69}{144 \times 4.002 \times 376.87} = 4.261 \frac{\mathrm{ft}^3}{\mathrm{lb}} = 0.266 \frac{\mathrm{m}^3}{\mathrm{kg}}$$

Head Developed by the Compressor

$$H_{\mathrm{test}} = \text{discharge enthalpy} - \text{inlet enthalpy} \tag{9.5}$$

$$H_{\mathrm{test}} = 291.45 - 219.80 = 71.65 \frac{\mathrm{Btu}}{\mathrm{lb\ mol}} = 166.65 \frac{\mathrm{kJ}}{\mathrm{kg\ mol}}$$

Polytropic Efficiency

$$n = \frac{\ln(\text{pressure ratio}, r_{\mathrm{p}})}{\ln\left(\frac{\text{inlet specific volume}}{\text{discharge specific volume}}\right)} = \frac{\ln(r_{\mathrm{p}})}{\ln\left(\frac{v_i}{v_{\mathrm{d}}}\right)} \tag{9.6}$$

$$n = \frac{\ln(1.222)}{\ln\left(\frac{4.695}{4.261}\right)} = 2.0496$$

Inverse of the Exponent

$$\frac{n-1}{n} = 0.5121 \tag{9.7}$$

Polytropic Head

$$H_{\mathrm{p,test}} = Z_{\mathrm{test}} \times \frac{1.545}{\mathrm{MW}} \times T_{1,\mathrm{test}} \times \frac{n}{n-1}\left[r_{\mathrm{p}}^{(n-1)/n} - 1\right] \tag{9.8}$$

$$H_{\mathrm{p,test}} = 1.012 \times \frac{1.545}{4.002} \times 534.56 \times \frac{1}{0.5121}\left[\left(1.222^{0.5121}\right) - 1\right]$$

$$= 44{,}097 \frac{\mathrm{ft\ lb}}{\mathrm{lbm}} = 56.667 \frac{\mathrm{Btu}}{\mathrm{lb\ mol}} = 131.81 \frac{\mathrm{kJ}}{\mathrm{kg\ mol}}$$

Polytropic Efficiency

$$\eta_{\mathrm{p,test}} = \frac{H_{\mathrm{p,test}}}{H_{\mathrm{test}}} \tag{9.9}$$

$$\eta_{\mathrm{p,test}} = \frac{56.667}{71.65} = 0.791$$

Gas Power

$$\text{Gas power}_{\text{test}} = \frac{\text{inlet mass flow} \times \text{polytropic head}_{\text{test}}}{33{,}000 \times \text{polytropic efficiency}_{\text{test}}} = \frac{W_{\text{test}} H_{\text{p,test}}}{33{,}000 \eta_{\text{p,test}}} \tag{9.10}$$

$$\text{Gas power}_{\text{test}} = \frac{575.88 \times 44{,}097}{33{,}000 \times 0.791} = 972.86\,\text{hp} = 725.75\,\text{kW}$$

9.6.4 Conversion of Test Data to Specified Conditions

Inlet Density

$$\rho = \frac{144 \times \text{MW} \times P_{1,\text{sp}}}{Z_{\text{sp}} \times 1545 \times T_{1,\text{sp}}} = 0.860 \frac{\text{lb}}{\text{ft}^3} = 13.78 \frac{\text{kg}}{\text{m}^3} \tag{9.11}$$

Speed Ratio

$$r = \frac{N_{\text{sp}}}{N_{\text{test}}} = \frac{15{,}447}{15{,}485} = 0.9993 \tag{9.12}$$

Inlet Volumetric Flow

$$q_{\text{conv}} = q_{\text{test}} r \tag{9.13}$$

$$q_{\text{conv}} = 2703.76 \times 0.9993 = 2701.6 \frac{\text{ft}^3}{\text{min}} = 1.275 \frac{\text{m}^3}{\text{s}}$$

Corrected Polytropic Efficiency

$$\text{Corrected polytropic efficiency} = \text{polytropic}_{\text{ test}} \times \text{Reynolds number correction} \tag{9.14}$$

$$\text{Corrected polytropic efficiency} = 0.791 \times 1.0077 = 0.797$$

Converted Polytropic Head

$$\text{Converted polytropic efficiency} = \text{polytropic head}_{\text{test}} \times (\text{speed ratio})^2 \times \text{Reynolds number correction} \tag{9.15}$$

$$\text{Converted polytropic head} = 56.67 \times 0.9993^2 \times 1.0077$$

$$= 57.03 \frac{\text{Btu}}{\text{lb mol}} = 44{,}379 \frac{\text{ft lb}}{\text{lbm}} = 132.65 \frac{\text{kJ}}{\text{kg mol}}$$

Converted Gas Power

$$\text{Converted gas power} = \frac{\text{inlet mass flow}_{\text{sp}} \times \text{ converted polytropic head}}{33{,}000 \times \text{converted polytropic efficiency}} \tag{9.16}$$

$$\text{Converted gas power} = \frac{2613 \times 44{,}379}{33{,}000 \times 0.797} = 4409 \text{ bhp} = 3288 \text{ kW}$$

where $K = C_p/C_v$ is the ratio of the gas specific heats, Z is the compressibility factor, h is the enthalpy, v is the specific volume, H is the compressor head, r_p is the pressure ratio, η_p is the polytropic efficiency, and MW is the molecular weight of the gas.

9.7 Full-Load, Full-Pressure, Full-Speed Test

Some users specify a full-load, full-pressure, full-speed (FLFPFS) test for compressors used in offshore modular installations and large-scale facilities such as liquefied natural gas (LNG) plants, reinjection systems, and ethylene production units. Key factors to consider when specifying an FLFPFS test for compressors include the following:

- Frame Sizes at Operational Limits—Equipment such as compressors, drivers, and intermediate components (e.g., gears in the drive string) operating near the upper thresholds of pressure, power, or flow on manufacturer experience plots should be carefully reviewed. Users are encouraged to verify whether similar frames have been successfully operating under comparable conditions elsewhere.
- Compressors in Demanding Applications—Units operating in high-speed, high-power, or high-pressure environments may require additional scrutiny due to elevated mechanical and thermal stresses.
- Compressors processing of High-Molecular-Weight Gases—Applications involving gases with molecular weights exceeding 30.
- Variable Operating Conditions—Compressors subject to multiple operating scenarios that directly influence plant output.
- Accelerated Commissioning Timelines—Projects with compressed commissioning schedules leave limited time for in-field performance adjustments. In such cases, investing in a thorough load test during the factory phase may prevent costly on-site corrections later.

- Rotordynamic Concerns—Known issues or potential instabilities identified through rotordynamic or stability analyses necessitate deeper investigation and possibly extended testing protocols.
- Multiple Sideload Configurations—Compressors featuring several sideloads may exhibit complex mechanical interactions that warrant comprehensive performance validation.
- Special Internal Features—Units equipped with honeycomb seals, shunt holes, or swirl brakes often require more rigorous testing than standard ASME PTC-10 Type 2 performance tests [1] or API 617 no-load mechanical running tests [3].
- Test Facility Capabilities—The ability of the test shop to perform a full unit test should be confirmed to ensure accurate simulation of field conditions.
- Project Schedule Flexibility—Availability of extra time within the project timeline can enable more extensive testing, reducing risk and enhancing reliability.

To enhance compressor stability, components such as damper seals, honeycomb seals, swirl brakes, and/or shunt holes may be incorporated. The anticipated performance of these features can be validated through testing under conditions that closely replicate actual field operation. Stability verification can be achieved via either a full-load, full-pressure, full-speed (FLFPFS) test or a full-load, full-pressure, full-power test, such as the ASME PTC-10 Type 1 test [1]. These tests assess the compressor's stability by comparing the measured log decrement with predicted values derived from rotordynamic analysis, which accounts for both test bed and foundation stiffness scenarios.

To prevent any ambiguity regarding the scope of FLFPFS testing, the compressor manufacturer and the user or purchaser must establish a clear, mutual agreement on the testing procedure beforehand. This agreement should explicitly outline the following elements:

- Test Duration—Typically conducted over a 4-h period, starting once the test loop parameters such as flow rates, pressures, and temperatures have stabilized.
- Driver Selection—Determine whether a contract driver or in-house (shop) driver will be used. Evaluate site conditions (environmental, utility access, and geographical factors) and shop capabilities to ensure suitability, especially for steam or gas turbine drivers. The agreement should also specify whether auxiliary systems like the contract lubricating oil system and seal gas control/support systems will be supplied.
- Equipment Scope—Define the range of shop and contract equipment to be used, including all necessary instrumentation.
- Lubrication Details—Specify lubricating oil characteristics, including flow rates and pressure levels at each supply point.
- Test Medium—Identify whether the test will use a single gas or a gas mixture. Hydrocarbon gases are preferred over inert gases (e.g., nitrogen) due to potential discharge temperature constraints.
- Dynamic Coefficients—Provide a table of expected cross-coupling and force coefficients at the impeller labyrinth for each stage, both during testing and at specified operating points.

- Log Decrement Analysis—Measure log decrement under all operating conditions and at the test point, which should be conducted at the maximum continuous speed.
- Vibration Data—Include generated plots such as orbits, shaft centerline, frequency spectrum, cascade plots, and polar diagrams.
- Data Acquisition—Define data collection intervals and the frequency range, typically spanning from 0.05 to 8 times the specified operating speed.
- Test Loop Diagram—Supply a schematic of the test loop setup.
- Test Sequence—Outline the sequence of operations, including variations in oil pressure and temperature.
- Critical Speed Verification—Confirm the system's critical speed during testing.
- Bearing Inspection—Conduct a post-test disassembly and inspection of thrust and radial bearings.
- Establish clear acceptance benchmarks for test success, including:
  - Final log decrement (must exceed 0.1)
  - Peak-to-peak shaft vibration amplitude
  - Peak-to-peak nonsynchronous vibration amplitude
  - bearing metal temperatures.

9.8 Special Engineering Analyses

Finite-element analysis has become a standard approach for verifying the design integrity of key components in large compressors—including casings, impellers, and inlet piping—used in applications such as ethylene and LNG production, as well as high-pressure reinjection systems. This section outlines the typical scope of such analyses to assist users in understanding and applying these methods effectively.

9.8.1 Finite-Element Analysis for Compressor Casings

Anticipated deformation—including elongation and contraction—at the casing midsection, joints, and pedestals is evaluated under the influence of extreme or maximum operating conditions, hydrostatic test pressures, and full vacuum scenarios (if applicable to the service). For horizontally (axially) split casings, dynamic stress analysis is performed on the casing cylinder, split line, end closures, shaft-end seal housings, and primary process nozzles. In contrast, radially split casings are assessed using finite-element analysis (FEA) techniques to evaluate stresses on the casing cylinder, shear ring(s), stop ring end closures, shaft-end seal housings, and main process nozzles.

It is important to note that a finite-element analysis report for compressor casings is not merely a formal deliverable—it serves as a critical technical record. The report should be detailed to support future troubleshooting efforts related to performance issues or mechanical failures. Typically, such reports encompass the following parameters, tailored to the specific casing configuration (horizontal or radial split):

- Design Parameters—Includes casing design pressure, tolerance for full vacuum conditions, and specified minimum and maximum design temperatures.
- Dimensional Specifications—Covers main process nozzle dimensions and the internal diameter of the compressor casing.
- Material Selection—Details the materials used for the casing, end closures, parting flange (in axially split designs), nozzles, and shaft-end seal housing.
- Stress Analysis—Evaluates von Mises stress levels in relation to the material yield strength.
- Internal Pressure Conditions—Specifies pressure values during hydrostatic testing, under vacuum, and at maximum or worst-case operating scenarios.
- Seal Integrity—Defines the maximum anticipated and allowable separation at the heads within the O-ring sealing area.
- Division Wall Behavior (if furnished)—Assesses expected deflection of the division wall seal in a dual-section single-casing compressor and its influence on system stability (predicted log decrement).
- Split Line Performance—Analyzes pressure at the split line of a horizontally split casing under peak operating conditions, minimum required contact pressure for effective sealing, and the percentage of yield stress experienced by split line bolts under maximum allowable working and hydrostatic test pressures.
- Radial Split Casing Dynamics—Reviews expected deflection at shear rings, stress levels on shear rings and seal housings, deformation of the outer casing during inner bundle installation (accounting for full bundle weight), and minimum permissible radial clearance between the inner bundle and casing bore.
- Nozzle Edge Stress Evaluation—Investigates stress concentrations at the edges of nozzle openings in the outer casing caused by roller traversal in radially split compressor configurations.

9.8.2 Finite-Element Analysis of Impellers

Finite-element analysis (FEA) should be stipulated for compressor impellers that are either prototypes or lack operational benchmarks. In such cases, the scope of analysis encompasses the entire impeller stage, including the impeller itself, the diffuser, the return bend, and the volute.

The static analysis of the impeller evaluates gas power and thrust at rated flow conditions, as well as centrifugal forces at the impeller tip corresponding to three operational speeds: 100% speed, maximum continuous speed, and trip speed.

If mechanical keys are used in addition to an interference fit at the impeller hub-to-shaft interface, they should be excluded from torque transmission calculations and thermal gradient assessments between the impeller and shaft. This is due to the differing material properties of the components involved.

The FEA report for the impeller should include the following elements:

- Specifications of the impeller, including bore diameter, tip diameter, material composition, and construction methodology.
- Details of the interference fit between the impeller and the shaft at the mounting location.
- Static structural analysis as previously outlined.
- Assessment of the reduction in design interference fit at rated, maximum continuous, and trip rotational speeds, along with the corresponding decrease in stage torque at each speed.
- Analysis of mechanical stresses including tensile, compressive, centrifugal, tangential, and those induced by thermal gradients. Evaluation of the design safety margin relative to the impeller material's minimum yield strength.
- Identification of forcing frequencies (up to the fourth harmonic and blade passing frequency), their separation margins, and the impeller's natural frequency. If a vaned diffuser is present, the influence of vane count is incorporated into this evaluation.

9.8.3 Dynamic Simulation

The scope and requirements for dynamic simulation are outlined in API 617 [3]. This section addresses various operational scenarios—normal startup, normal shutdown, emergency shutdown, and antisurge valve fail-open conditions—to aid reader comprehension. The objectives of dynamic simulation include:

- Verify system behavior under steady-state, normal, and upset conditions
- Assess the controllability of the antisurge valve
- Determine whether a hot-gas bypass valve is necessary to prevent compressor surge during coastdown in emergency shutdown situations, particularly when operating along the H–Q curve or surge control line
- Validate the sizing of the recycle (antisurge) valve
- Confirm that a minimum 10% margin exists between the load torque and the driver's speed-torque characteristics during startup.

Simulation models incorporate equipment specifications and time-based parameters to reflect transient conditions, as well as normal start and stop sequences. Data extracted from equipment modules within the simulation boundaries typically includes the following:

- Compressor performance across varying operating speeds, including head, inlet volumetric flow rate, polytropic efficiency, and shaft power consumption.
- Valve dynamics, covering flow-pressure relationships, flow coefficients (C_V values), valve behavior (linear or equal percentage), and actuation timing for opening and closing.
- Pressure profiles and holdup volumes within process vessels.

- Header system characteristics, such as pressure levels, holdup volumes, flow rates, and pressure losses throughout the piping network.
- Thermal and hydraulic behavior of heat exchangers, focusing on pressure drops and outlet temperatures of process fluids.
- Instrumentation and control systems for managing levels, flows, and pressures, along with their associated safety interlocks.

Dynamic simulation models integrate the compressor manufacturer's predefined time intervals, including those for standard startup, steady-state operation, emergency shutdown from normal conditions, and coastdown surge behavior during an emergency shutdown event.

Acknowledgements The image (Fig. 9.1) in this chapter was provided by Mitsubishi Heavy Industries Compressor, Hiroshima, Japan.

Some of the material in this chapter was reprinted with permission of the publisher from Ghaisas N (2011) Testing Centrifugal Compressor. CompressorTECH^2 magazine (Aug/Sept 2011).

References

1. ASME PTC 10-1997 (R2014) (2014) Performance test code on compressors and exhausters. American Society of Mechanical Engineers, New York
2. API Standard 614 (2022) Lubrication, shaft-sealing, and oil-control systems and auxiliaries. 6th edn. American Petroleum Institute, Washington, DC
3. API Standard 617 (2022) Axial and centrifugal compressors and expander-compressors. 9th edn. American Petroleum Institute, Washington, DC

Chapter 10
Grouting

Abstract The reliability of plant equipment plays a pivotal role in determining the profitability of industrial and manufacturing operations. Modern petroleum, petrochemical, and gas processing facilities rely on a wide array of rotating machinery used in both general-purpose and specialized process applications. Leading companies that operate dynamic equipment consistently aim to enhance efficiency and reliability, thereby reducing operational costs and boosting overall productivity—both of which directly benefit the bottom line. Long-term equipment reliability is influenced not only by the rotating machines themselves but also by associated piping, auxiliary systems, mounting methods, grouting, anchoring, and foundation design. Achieving optimal performance requires comprehensive engineering, stringent specifications, careful equipment selection, and adherence to best practices throughout design, manufacturing, installation, operation, and maintenance. Grouting, particularly epoxy grout, is a fundamental element of the equipment system. It establishes a monolithic bond between machinery and its foundation, ensuring effective load transfer, energy dissipation, and vibration damping. The importance of proper grouting in enhancing equipment performance, reliability, and minimizing life-cycle costs cannot be overemphasized.

10.1 Purpose of Grout

Grout functions in coordination with the machinery anchoring system—comprising mounting plates and anchor bolts—to preserve the precise alignment of equipment. It counteracts both vertical and lateral forces exerted on the concrete foundation. While cementitious grouts have long been the industry standard for machinery installation, epoxy grouts are increasingly favored due to their enhanced performance attributes, including:

- Superior resistance to shear, tensile, and compressive forces, ensuring structural integrity under demanding conditions.
- Reliable maintenance of equipment alignment, resisting displacement from vertical and horizontal loads.

N. Ghaisas, *Centrifugal Compressors in Hydrocarbon Processing Industries*,
https://doi.org/10.1007/978-3-032-03687-2_10

- Stronger adhesive bonding to concrete and steel than cementitious alternatives, enabling efficient transfer of loads and vibrations to the foundation for energy absorption and dissipation.
- Reduced shrinkage and fretting, offering better protection for porous concrete surfaces.
- Outstanding vibration dampening, aiding in the mitigation of dynamic forces (refer to Fig. 10.1).
- Robust resistance to chemicals, moisture, and chlorides, enhancing durability in harsh environments.
- Improved thermal resilience, withstanding heat and thermal cycling more effectively than cementitious grout.
- Rapid curing, allowing quicker attainment of service strength.
- Lower long-term cost, thanks to extended service life and superior mechanical performance.

In terms of vibration absorption, concrete and cementitious grouts offer 3–5 times the capacity of steel. However, low-modulus epoxy grouts outperform both, delivering approximately 30 times the vibration absorption of steel and 6–10 times that of cement-based materials.

Fig. 10.1 Prepared top surface of a concrete foundation for machinery (Image courtesy of Chinook Industrial Ltd, Calgary, Alberta, Canada, and ITW Performance Polymers, Montgomeryville, PA)

10.2 Machinery Foundations and Anchor Bolts

The American Petroleum Institute (API) publication API RP 686 [1] outlines recommended practices for machinery installation and design. Similarly, the American Concrete Institute (ACI) standard ACI 351.3R-18, titled Report on Foundations for Dynamic Equipment, is a widely recognized reference across the industry. The best practices presented in this chapter are designed to supplement these established standards.

Dynamic equipment foundations serve to transfer the machinery's dynamic forces into the underlying soil, making soil property analysis a vital aspect of foundation design. Key geotechnical parameters—such as soil unit weight, dynamic shear modulus, shear wave velocity, dynamic subgrade modulus, allowable bearing pressure or pile capacity, and Poisson's ratio—are gathered to assess soil or pile stiffness and damping characteristics. These values inform both static and dynamic foundation design.

A block foundation typically comprises a concrete mass block, piers, and a mat. Its total mass should be at least three times that of the rotating machinery it supports, although dynamic analysis may necessitate different mass ratios depending on the installation. The foundation's width should be no less than 1.5 times the vertical distance from its base to the centerline of the shaft or rotor.

To ensure structural integrity, the center of mass of the combined machine and foundation system should align with the centroid of the soil or pile group resistance. Horizontal eccentricity must be kept within 5% of the relevant foundation dimension. The concrete mat should have a minimum thickness equal to one-fifth of the smallest foundation dimension or one-tenth of the largest, whichever is greater.

Concrete must be fully cured before epoxy grout is applied. Low moisture content and minimal shrinkage are essential, with a standard curing period of 28 days typically yielding a design strength of at least 4000 psi (28 MPa). ASTM C 157 is the conventional test method used to measure moisture and shrinkage. Innovative technologies, such as embedded wireless sensors, allow for real-time monitoring of temperature and strength during curing. These sensors are placed approximately 2 in. (50 mm) below the surface prior to pouring and transmit data wirelessly to optimize curing conditions.

Prior to grouting, the foundation's top surface is prepared using handheld pneumatic chipping hammers to remove cement-rich laitance. All exterior edges are chamfered at a 45° angle. A well-prepared surface exposes the coarse aggregate, which provides the strongest bonding interface and promotes a monolithic connection with the grout. Figure 10.1 illustrates a properly prepared concrete foundation surface for machinery installation.

Anchor bolts counteract the upward and lateral forces generated by dynamic loads from mounted machinery, while epoxy grout resists the downward force caused by the equipment's deadweight. Working together, these components ensure efficient transmission of dynamic loads to the foundation, enabling energy absorption and dissipation, and preserving the precise alignment of the machinery.

Anchor bolts must be positioned accurately and within the tolerances specified in the foundation or construction drawings. For most applications, the typical embedment depth is ten times the bolt diameter, with an isolation length of eight bolt diameters to allow free stretch—particularly important in centrifugal compressor setups. Standard anchor bolts are manufactured to ASTM A193 Grade B7 specifications, paired with ASTM A194 Grade 2H nuts. A suggested layout for anchor bolt installation is illustrated in Fig. 10.2.

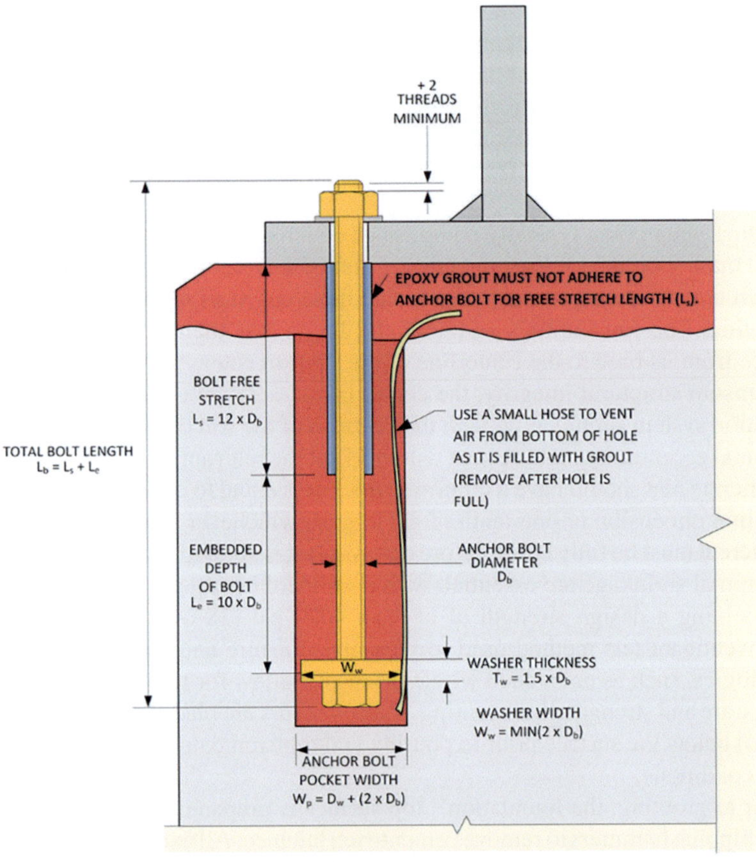

Fig. 10.2 Typical anchor bolt plan (Image courtesy of Chinook Industrial Ltd, Calgary, Alberta, Canada, and ITW Performance Polymers, Montgomeryville, PA)

10.3 Epoxy Grout

Temperature plays a critical role during the grout pouring process. When epoxy resin is combined with its hardener, a chemical reaction initiates that causes the epoxy molecules to cross-link, forming a solid, interlocking lattice structure—illustrated in Fig. 10.3.

The heat generated during the exothermic reaction between epoxy resin and hardener is influenced by several variables, including the quantity of epoxy used, ambient conditions, the specific chemical formulation, the initial temperatures of the components at mixing, and the thermal absorption capacity of surrounding materials.

Once polymerization begins, the epoxy gradually transitions from a liquid to a solid state—reaching what is known as the gel point. This stage marks the peak temperature of the reaction and represents a state of thermal equilibrium within the grout. Following this transformation of state, the temperature begins to decline. Full curing of the epoxy grout is achieved through ambient heat; however, if ambient conditions are insufficient, supplemental external heating may be required. The

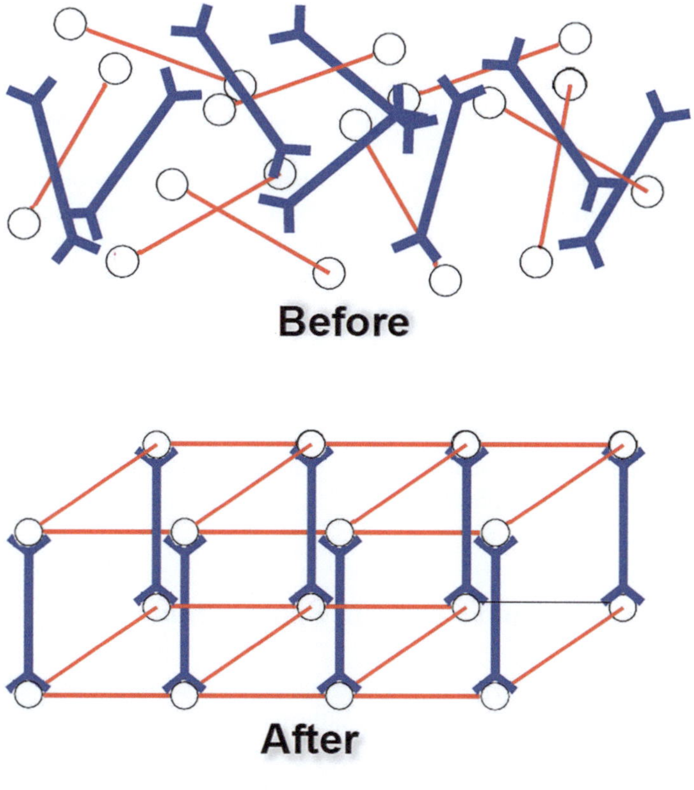

Fig. 10.3 Structural lattice of epoxy grout (Image courtesy of Chinook Industrial Ltd, Calgary, Alberta, Canada, and ITW Performance Polymers, Montgomeryville, PA)

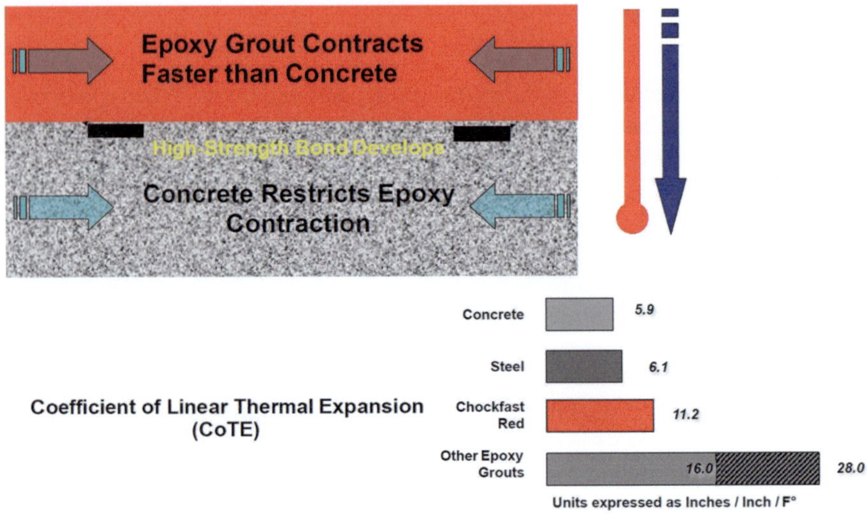

Fig. 10.4 Coefficient of linear thermal expansion for ITW Chockfast® Red epoxy grout (Image courtesy of Chinook Industrial Ltd, Calgary, Alberta, Canada, and ITW Performance Polymers, Montgomeryville, PA)

impact of linear thermal expansion after the temperature drops below the peak exothermic level is illustrated in Fig. 10.4.

A high exothermic cure temperature can be both beneficial and problematic during epoxy grout installation. On the one hand, elevated temperatures accelerate the curing process. On the other, they increase the risk of shrinkage and contraction as the grout cools down to ambient conditions. Figure 10.5 illustrates the exothermic profiles of two different grout formulations.

In conditions of extreme heat or cold, it is essential to shelter and precondition all grouting materials and equipment to maintain consistent product temperatures throughout the installation and curing process. Equally critical is safeguarding these materials from adverse environmental factors such as rain, snow, and wind. The designated preconditioning temperature must be reached and sustained for a minimum of 24 h prior to mixing and application.

10.4 Edge Lifting and Cracks in Epoxy Grout

Edge lifting can occur after grout placement when a drop in ambient temperature causes the epoxy grout to contract more rapidly than the bonded concrete. This differential movement generates tensile stresses at the foundation's edges and corners, just beneath the bond line between the grout and concrete. If these stresses surpass the

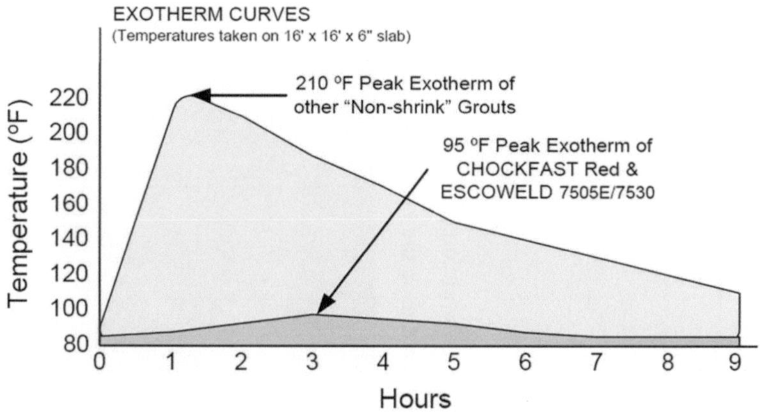

Fig. 10.5 Exotherm curves, ITW Chockfast® Red and Escoweld® epoxy grouts (Image courtesy of Chinook Industrial Ltd, Calgary, Alberta, Canada, and ITW Performance Polymers, Montgomeryville, PA)

concrete's tensile strength, perimeter cracking can result—typically appearing ½–8 in. (1.3–20 cm) below the bond interface. To mitigate or prevent edge lifting, a 45° chamfer can be applied around the foundation perimeter, or steel dowel pins can be installed by drilling along the edges. Figure 10.6 illustrates this phenomenon.

The difference between the peak exothermic temperature reached during grout placement and the lower ambient temperatures that may follow plays a significant role in the likelihood and timing of crack formation. Cracks can emerge immediately as the epoxy grout begins to cool after its initial thermal reaction. However, they may also develop long after installation—weeks, months, or even years later—if surrounding temperatures drop sufficiently to induce stress levels that exceed the cured grout's structural limits. This same mechanism applies to edge lifting.

While cracks are often viewed as a sign of grout failure, they are a recognized and common occurrence in materials like concrete, which undergo chemical reactions and heat-induced curing (refer to Fig. 10.7).

Thermal cracking in epoxy grout can be effectively anticipated and minimized by strategically placing prefabricated control joints—commonly known as expansion joints—throughout the installation. To determine optimal joint placement, the following factors should be carefully evaluated:

- Expected magnitude and frequency of thermal fluctuations in the operating environment.
- Thermal expansion coefficient of the epoxy grout.
- Conditions during the initial curing phase of the grout.
- Type and layout of the mounted machinery.
- Operating temperatures of the machinery train components.

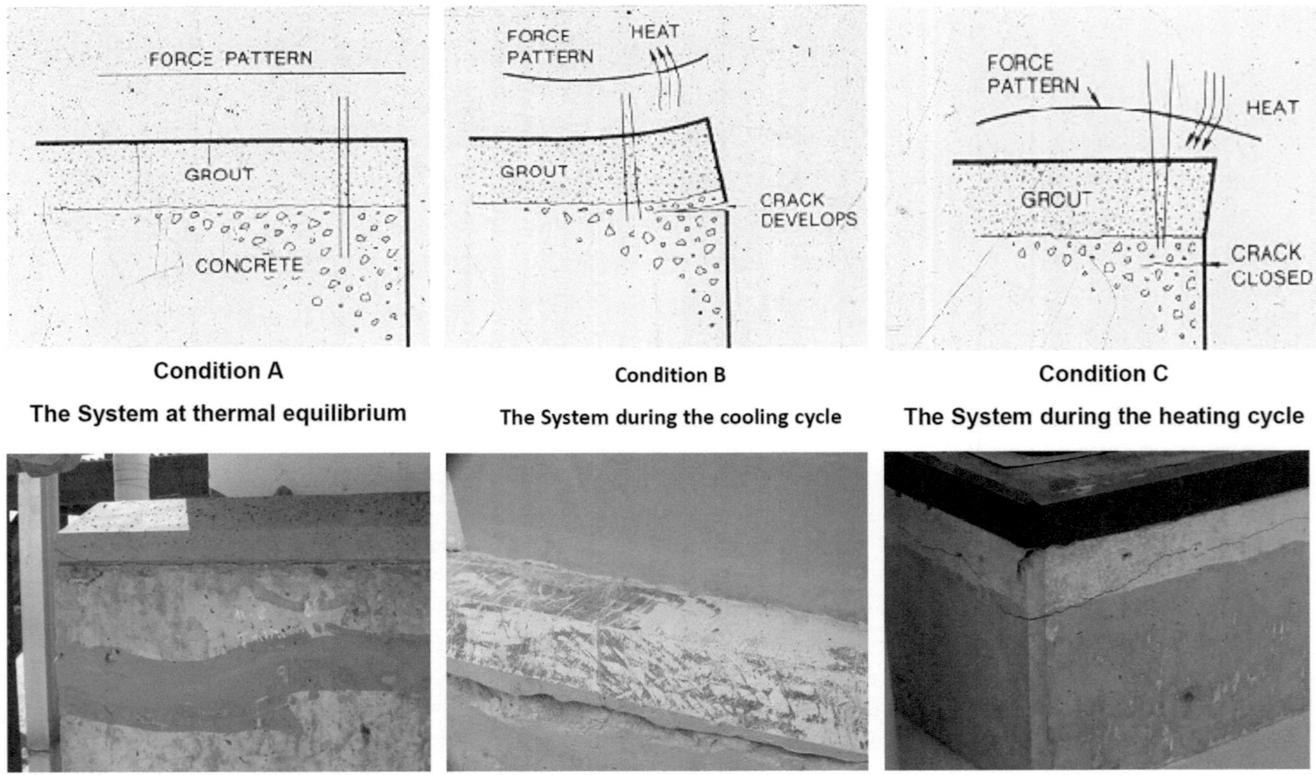

Fig. 10.6 Edge lifting in the foundation (Image courtesy of Chinook Industrial Ltd, Calgary, Alberta, Canada, and ITW Performance Polymers, Montgomeryville, PA)

Fig. 10.7 Temperature-induced cracks in concrete (Images courtesy of Chinook Industrial Ltd, Calgary, Alberta, Canada, and ITW Performance Polymers, Montgomeryville, PA)

When installed, control joints act as bulkheads, dividing large grout pours into smaller, more manageable sections. This segmentation enhances control over the placement process and improves overall efficiency. Additionally, control joints guide grout flow in preferred directions and provide flexibility to accommodate unexpected interruptions—such as material shortages or equipment failures—during installation.

Various commonly used control joint designs are illustrated in Fig. 10.8.

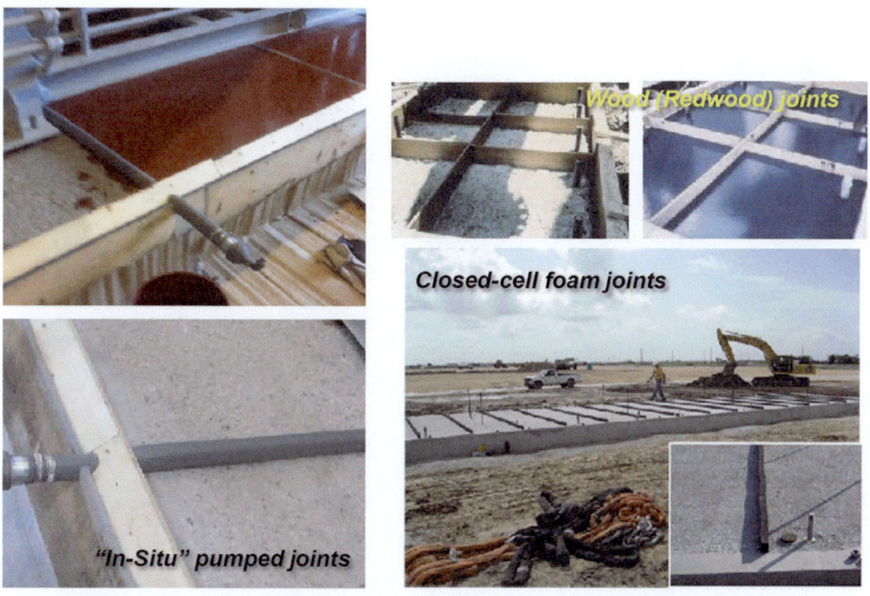

Fig. 10.8 Types of control joints in epoxy grout (Images courtesy of Chinook Industrial Ltd, Calgary, Alberta, Canada, and ITW Performance Polymers, Montgomeryville, PA)

10.5 Grout Forms and Pouring Techniques

Grout forms are positioned to hold the epoxy grout in place until it fully cures. These forms must be leak-proof and are commonly built using well-supported forming materials like ¾ in. (19 mm) plywood (refer to Fig. 10.9). To prevent adhesion, any surface of the form that comes into contact with the epoxy grout is treated with at least three layers of paste wax. Additionally, 45° chamfers are applied to both the internal and external corners to eliminate sharp edges.

Two-stage epoxy grout applications are frequently employed. The initial stage involves pouring grout between the concrete surface and the underside of the machinery baseplate to create a foundational mat. Once this layer has fully cured, a second epoxy grout pour is used to fill the remaining void between the baseplate and the top of the first pour (refer to Fig. 10.10).

In the single-stage grout pour technique, horizontal form covers are placed across the gap between the top of the machinery baseplate flange and the upper edge of the surrounding formwork. The grout is then poured in a single, continuous operation to complete the process (refer to Fig. 10.11).

To ensure consistent hydraulic head pressure during epoxy grout placement, equipment such as head boxes, standpipes, and troughs are utilized. This steady pressure aids in the controlled flow and precise positioning of the grout. Figure 10.12 illustrates wooden head boxes positioned on the foundation of a compressor package.

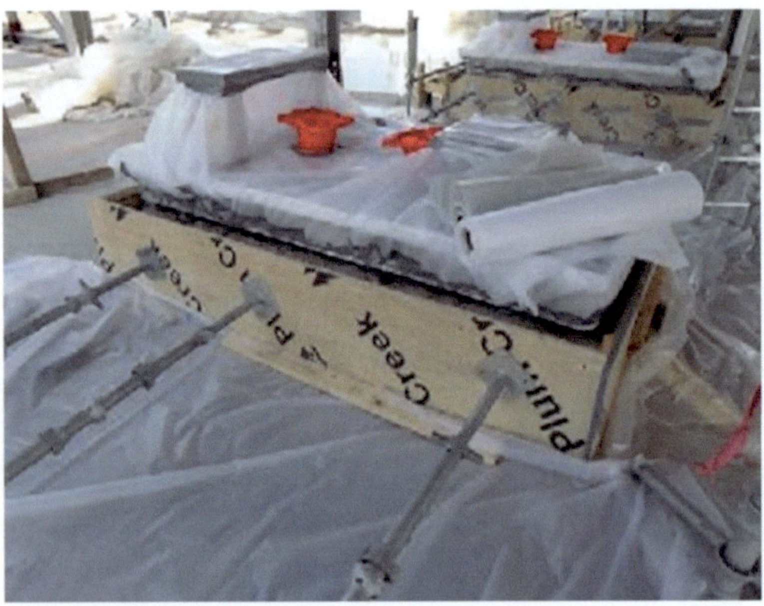

Fig. 10.9 Braced grout form (Image courtesy of Chinook Industrial Ltd, Calgary, Alberta, Canada, and ITW Performance Polymers, Montgomeryville, PA)

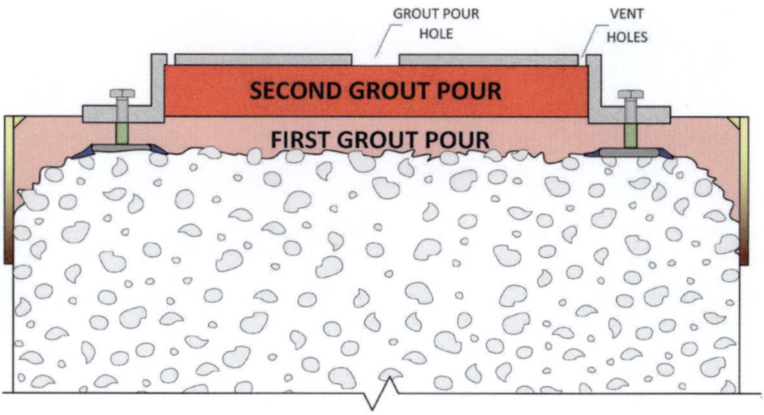

Fig. 10.10 Two-stage epoxy grout pour (Image courtesy of Chinook Industrial Ltd, Calgary, Alberta, Canada, and ITW Performance Polymers, Montgomeryville, PA)

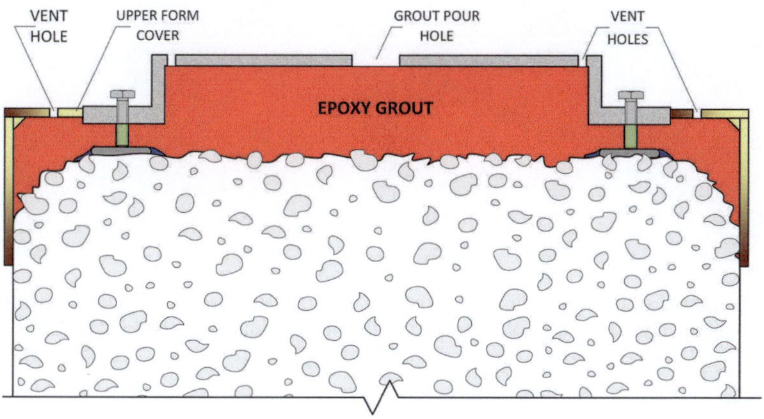

Fig. 10.11 Single-stage epoxy grout pour (Image courtesy of Chinook Industrial Ltd, Calgary, Alberta, Canada, and ITW Performance Polymers, Montgomeryville, PA)

To fill voids within the cured epoxy grout, liquid epoxy resin is injected into the affected areas. This process involves drilling two holes at opposite ends of the void—one designated for resin injection and the other functioning as a vent to allow air to escape and ensure thorough filling. Figure 10.13 displays the completed grout installation.

Fig. 10.12 Head boxes for an epoxy grout pour (Images courtesy of Chinook Industrial Ltd, Calgary, Alberta, Canada, and ITW Performance Polymers, Montgomeryville, PA)

Fig. 10.13 Epoxy grout under a compressor baseplate (Images courtesy of Chinook Industrial Ltd, Calgary, Alberta, Canada, and ITW Performance Polymers, Montgomeryville, PA)

10.6 Pumpable Epoxy Grouts

High-flow (HF), pumpable epoxy grouts have been engineered specifically for large-scale applications, offering several advantages over traditional pour methods. These advanced grouts streamline the placement process and enhance performance in demanding environments. Key benefits include:

- Pump-assisted placement: Grout can be delivered using pumping systems, eliminating the need for manual pouring.
- Optimized for tight spaces: Ideal for installations with restricted access or shallow clearance, and well-suited for extensive pour areas.
- Minimized material waste: Requires less volume to achieve desired results compared to conventional grouts.
- Superior performance at low depths: Maintains excellent workability and cured properties even in thin layers.

- Efficient batching: Supports larger batch sizes and simplifies the mixing and placement process.
- Accelerated production: Enables faster preparation and availability of grout for immediate use.

High-performance pumpable grouts also exhibit enhanced mechanical properties, including elevated compressive, tensile, and flexural strengths. Flexural strength—also known as transverse rupture strength—refers to the stress level the grout can withstand before deformation occurs. Figure 10.14 presents the specifications of a representative industrial-grade high-performance grout.

Compressive Strength:	13,900 psi (95.8 MPa)	ASTM C-579 MOD
Compressive Modulus:	1.6 x 10_6 psi (11.0 GPa)	ASTM C-579 MOD
Tensile Strength:	2,900 psi (20.0 MPa)	ASTM D-638
Flexural Strength:	5,200 psi (35.9 MPa)	ASTM C-580
Flexural Modulus:	1.55 x 10_6 psi (10.7 GPa)	ASTM C-580
Effective Bearing Area (EBA):	> 95 %	
Coefficient of Linear Thermal Expansion (CoTE):	19.1 x 10_{-6} in/in/°F (34.4 x 10_{-6} mm/mm/°C)	ASTM C-1339 ASTM C-531
Linear Shrinkage:	Negligible	ASTM D-2256
Bond – Steel:	2,050 psi (14.1 MPa)	ASTM C-1002
Bond - Concrete:	1,550 psi (10.7 MPa)	ASTM C-882
Fire Resistance:	Self Extinguishing	ASTM D-635
Density:	127.9 lb/ft³ (2048.8 kg/m³)	ASTM C-905

Fig. 10.14 Sourced data for ITW Chockfast® Red HF grout (Image courtesy of Chinook Industrial Ltd, Calgary, Alberta, Canada, and ITW Performance Polymers, Montgomeryville, PA)

Acknowledgements All of the images/graphics/figures in this chapter were provided by Chinook Industrial Ltd, Calgary, Alberta, Canada, and ITW Performance Polymers, Danvers, MA, and Montgomeryville, PA.

This material was reprinted with permission of the publisher from Ghaisas N, Matthews-Ewald C, Helfmann F (2020) Grout systems for rotating and reciprocating machinery. CompressorTECH^2 magazine (June 2020), pp 34–37.

References

1. API RP 686 (2009) (R2024) (E2025) Recommended practice for machinery installation and installation design. American Petroleum Institute, Washington, DC
2. Shade N (2012) Foundation and skid design considerations for large, high-speed reciprocating compressors. CompressorTECH^2 Magazine, Oct 2012. Reprint available at https://vdn.woodplc.com/assets/pdfs/Technical_Articles/CT2-2012_Foundations-and-Skids_large-high-speed-recips.pdf
3. European Forum for Reciprocating Compressors (2017) Guidelines for vibrations in reciprocating compressor systems. Available at https://www.recip.org/wp-content/uploads/2020/04/EFRC-Guidelines-for-Vibrations-in-Recipr.-Compr.-Systems_4th-edition.pdf
4. European Forum for Reciprocating Compressors (2017) Summary of international guidelines, standards, and best practices of foundations, anchor bolts, and grouting of reciprocating compressor systems. Available at https://www.recip.org/wp-content/uploads/2020/04/EFRC-RD-Report-Foundation-Anchor-Bolts-Grouting-October-19th-2017.pdf
5. Smalley AJ, Pantermuehl PJ (1997) Foundation guidelines. GMRC technical report No. TR-97-2. Gas Machinery Research Council, Dallas, TX
6. Kuly J, Adkins J, Krich R (2014) GMRC short course: best practices in compressor and gas-engine mounting. Presented at the gas machinery conference, Nashville, TN, 6–8 Oct 2014
7. Leary J, O'Malley R (2004) Grouting large skid mounted compressor packages. Presented at the 2004 gas machinery conference, Albuquerque, NM, 4–7 Oct 2004
8. Chockfast® Foundation Systems (1991) Electronic machinery grouting manual. ITW Philadelphia Resins, Montgomeryville, PA
9. Renfro EM (1985) Machinery foundations & grouting. Adhesive Services, Houston, TX
10. Kuly J (2010) Best practices in compressor mounting. Presented at the 7th conference of the EFRC, Florence, Italy, 20–22 Oct 2010
11. Harrison DM (2013) The grouting handbook: a step-by-step guide for foundation design and machinery installation, 2nd edn. Elsevier, Waltham, MA, p 6
12. Gigatec SmartRock™ literature

Chapter 11
Modularization

Abstract A module refers to a transportable unit composed of preassembled components from a process plant. The modularization approach aims to streamline facility layouts and shift a substantial portion of construction activities from the field to dedicated module yards, thereby minimizing on-site labor. In highly modularized setups, both small and large static equipment, as well as rotating machinery packages, are typically mounted on steel frameworks. A key element of modularization is the integration of equipment into these modules. Installing rotating or reciprocating machines on steel structures—rather than traditional concrete foundations—introduces distinct engineering challenges. Fabricated steel modules possess lower stiffness and damping characteristics compared to concrete, making it essential to assess how these structural properties affect the performance of mounted machinery. This requires specialized analytical techniques and tailored design strategies. This chapter explores the principles of modularization, various module types, structural configurations and loading conditions, and the methodologies used for structural analysis of modules.

11.1 Benefits of Modularization

The modular construction approach offers multiple benefits compared to traditional stick-built design methods.

11.1.1 Cost-Related Benefits

Modularization enables a significant transfer of labor hours from expensive on-site locations to more cost-effective fabrication shops or module yards. This strategic shift contributes to cost savings in several key areas:

- Minimizes the use of materials—particularly concrete, piping, and cabling—leading to a reduced total installed cost

N. Ghaisas, *Centrifugal Compressors in Hydrocarbon Processing Industries*,
https://doi.org/10.1007/978-3-032-03687-2_11

- Decreases the need for equipment grouting
- Relocates underground process drainage piping into above-ground installations completed in shop environments.

11.1.2 Schedule-Related Benefits

Modularization can significantly enhance project schedules through the following advantages:

- Enables concurrent fabrication and assembly processes, optimizing labor costs by leveraging the lower hourly rates of shop-based work compared to field labor
- Shifts a substantial portion of work to controlled shop environments, minimizing weather-related disruptions that typically affect site productivity
- Supports a faster construction timeline and facilitates early commissioning and startup of plant systems

11.1.3 Safety- and Quality-Related Benefits

Further benefits of modularization in enhancing safety and quality include:

- Enhanced workmanship quality due to a more controlled and consistent production environment.
- Reduced safety hazards related to working at heights and labor stacking.
- Less material clutter and movement on-site, improving overall site organization.
- Facilitates the replication of proven, safe engineering designs in future modular projects, while also allowing flexibility in space and equipment layout to mitigate issues like vibration transfer between machines sharing the same structural framework.

11.2 Equipment Modularization Concept

In every modularized project, the concept of machinery modularization is initiated during the scope definition phase through a feasibility study. This early evaluation determines whether modularization is practical and sets the foundation for further development and eventual implementation. Key motivators for conducting a module concept study include health, safety, and environmental (HSE) priorities; the need for long-term, reliable equipment operation; and specific performance criteria.

The study typically encompasses a broad range of considerations, such as:

- Compliance with local, national, and federal environmental regulations and standards

- Site-specific environmental conditions, including temperature extremes and the impact of wind, ice, snow, and rain on walkways
- Risks associated with hydrocarbon releases and emissions from both process and on-module equipment
- Noise levels generated by machinery and auxiliary systems, along with noise mitigation strategies
- Accessibility for operations, maintenance, inspections, and testing of equipment within the module
- Preliminary static and dynamic structural analyses, including transportation via sea or land
- Additional factors influencing the feasibility and scope of machinery modularization include:
  - Module dimensions and weight
  - Transportation constraints specific to the site
  - Construction and testing capabilities of the module fabrication yard or shop
  - Proximity of the module yard to the shipping port
  - Clearance requirements driven by operational, maintenance, and human factors
  - Cost–benefit analysis to justify modularization.

Projects where machinery modularization is a key driver should undergo this comprehensive assessment. A successful modular design must balance schedule and cost efficiency with uncompromised safety, operability, and maintainability.

The following section outlines module-specific requirements and considerations.

- Emergency Egress—Ensure clearly marked and unobstructed evacuation routes for safe exit during emergencies.
- Inspection Access—Provide adequate clearance for inspection ports, pull-out components, and equipment removal.
- Vertical Space—Maintain sufficient overhead clearance for operational and maintenance activities.
- Access Restrictions—Identify and mitigate obstructions such as panel doors, valves, cable trays, and conduits that may hinder accessibility.
- Maintenance Access—Incorporate walkways, ladders, platforms, and guardrails to facilitate routine inspections, adjustments, data logging, and servicing.
- Lighting and Utilities—Integrate on-module lighting and utility stations to support operational needs.
- Material Movement—Plan for efficient handling and transport of materials both within and outside the module.
- Operability Reviews—Conduct early-stage evaluations to ensure ease of operation and maintenance.
- Plot Optimization—Design module layout to accommodate laydown areas and allow unrestricted movement of bridge cranes across drop zones.
- Instrumentation Visibility—Position gauges, monitors, instruments, and isolation points to be easily accessible, visible, and clearly labeled.

- Thermal Protection—Apply appropriate hot/cold insulation and fireproofing measures.
- Electrical Integration—Strategically route cables and define entry points for junction boxes and motors.
- Component Placement—Determine optimal mounting heights and orientations for control panels and valves.
- Structural Design—Engineer flooring and grating to support the weight of equipment used during routine maintenance.

A comprehensive modularization strategy outlines the division of scope, responsibilities, tasks, and functions. This includes the supply of mounted equipment and associated auxiliary systems, electrical installations, instrumentation and control systems, heat tracing, grounding, mechanical and piping work—along with insulation and/or winterization measures. It also encompasses construction completion verification, commissioning, and on-module testing activities such as continuity checks, hydrotesting, flushing, and instrument calibration. Additionally, it details the logistics for transporting modules from the fabrication yard or shop to their final destination, along with a schedule for each phase.

11.3 Types of Modules

Modules are categorized by size—small, large, or very large—based on their design and transport requirements.

- Small modules are fully integrated plant components that can be moved via road or rail. These are typically self-contained units, ready to be installed on pre-prepared foundations and linked to adjacent modules. They generally require minimal precommissioning work.
- Large modules range in weight from approximately 100–600 tons (91–544 metric tons). Their maximum weight is determined by the lifting capacity available at the site using standard rigging techniques.
- Very large modules (VLMs) typically exceed 600 tons (544 metric tons) and can reach dry weights of up to 10,000 tons (9100 metric tons) in modern, large-scale ethylene and hydrocarbon processing plants. Due to their size, VLMs often necessitate specialized lifting, jacking, and transport solutions. They are typically moved by barge or heavy-lift vessels and positioned using self-propelled modular transporters (SPMTs).

11.4 Machinery Modules

Some machinery modules feature a single deck, while others incorporate both a main deck and a mezzanine level to accommodate auxiliary equipment. Depending on the layout, a machinery string may be installed on its own dedicated module, or two strings serving the same function can be positioned either longitudinally or transversely on a shared module floor. For roofless modules, an air-cooled lube oil heat exchanger can be integrated and mounted on top.

Elevated modules enable the use of vertically downward process nozzles on compressor casings, which allows for unobstructed movement of overhead bridge cranes—typically the preferred configuration.

In multibody centrifugal compressor strings, the high-pressure casing (radially split) is placed outside the low- and medium-pressure casings (axially split) to facilitate removal of the outer casing from the inner barrel assembly. As a result, the laydown area is positioned adjacent to the module to facilitate streamlined handling and maintenance activities.

Process and packaged equipment modules often include structural steel–mounted pumps and other machinery such as positive displacement blowers, centrifugal fans, and rotary screw compressors. Dedicated pump modules generally house pumps serving different processes, and their placement is carefully determined based on dynamic performance, accessibility for operation and maintenance, and ergonomic considerations.

Modules designed for large or complex rotating equipment—such as gas-turbine-driven compressor strings—typically feature elevated platforms, partially or fully enclosed side walls, removable roofs, and integrated auxiliary systems. Maintenance and handling are supported by bridge and jib cranes.

These integrated machinery modules require a detailed breakdown of tasks across various systems and components. At a higher level, this breakdown clarifies responsibilities and locations—whether in the module yard or at the job site—for all major work scopes and supply items, including shipped-loose parts or assemblies. The goal is to eliminate assumptions that could lead to scope gaps or delays. This planning should be finalized in parallel with the approval of the modularization concept to ensure seamless integration with the rest of the plant.

11.5 Module Structure

Structural steel plays a critical role in module design, often accounting for up to 50% of the module's total weight. Since weight directly influences transportation costs, optimizing structural steel usage is essential. The permanent steel framework of a module typically comprises machinery support beams, primary trusses, and deck plate girders—all of which enhance the module's overall stability and structural integrity. Welded connections are generally favored to ensure strength and rigidity.

In addition to the permanent structure, several key components are involved during ocean transport. These include:

- Grillage—A transitional framework between the permanent steel and the sea fastening system
- Sea fastening steel—Secures the module to the vessel's deck
- Internal sea fastening—Protects internal equipment during transit
- Ship-loose components—Installed on-site, such as external staircases, lightning protection rods, and mounting pads.

The following structural considerations should be carefully evaluated and finalized during the early stages of module design [1]:

- Connection Types: Welded vs. Bolted—The decision between welded and bolted connections is closely tied to the type of protective coating used. Welding galvanized steel poses health risks to workers, making bolted connections a safer alternative. However, bolted joints tend to be bulkier and may be prone to slippage. For module structures, weldable, high-strength, hot-rolled steel is typically used. The selection of steel profiles—such as T-sections, I-sections, or circular sections—and their dimensions depends on the material grade and load requirements. Key structural components like machinery support beams, main trusses, and deck plate girders are generally welded to ensure stability and structural integrity.
- Coating Options: Painted vs. Galvanized Steel—Painting is often favored by fabrication yards, especially those integrated with shipyards, as it allows for streamlined fabrication, coating, and assembly within the same facility. Galvanized steel, on the other hand, is preferred for sites with high corrosion risk, such as coastal environments. Projects along the US Gulf Coast frequently opt for galvanization, which typically necessitates bolted connections.
- Base Frame Configuration: Buried vs. Elevated—The height of base frames plays a critical role in module layout. Elevated frames are more commonly used due to their ease of installation. Buried frames, while less typical, offer advantages such as improved access and reduced need for operators to climb to reach equipment at the first level.
- Structural Flexibility: Rigid vs. Flexible Framing—The choice between rigid and flexible framing depends on the amount of bracing required. Rigid frames are often necessary for crane lifting operations or to address transportation constraints.
- Material Sourcing: Warehouse vs. Mill Order—In certain regions, sourcing steel directly from mills rather than warehouses ensures higher material quality and consistency.

Ladders, platforms, gratings, handrails, and cable tray supports are engineered to facilitate safe and convenient access for personnel working on the module. To ensure secure assembly, these elements are fastened using high-strength bolts and nuts.

Crane runway beams, floor stiffeners, pipe racks, and external platforms are dimensioned to maintain structural integrity under both static and dynamic loads. These components are typically joined through welding.

11.6 Loads on Machinery Module Structure

Structural loading for modular systems involves several additional factors beyond conventional design. These loads are generally classified into two categories: transportation loads and site loads.

- Transportation Loads include all forces acting on the module during transit and can vary significantly from site conditions. Key components are:
  - Transport empty weight, which may differ from the module's site empty weight
  - Stowage items, such as equipment or materials carried within or on the module
  - Wind loads, calculated specifically for transport scenarios
  - Marine forces, including wave-induced accelerations and inertia from vessel motion
  - Hogging and sagging, referring to the bending of the vessel deck due to wave action and uneven weight distribution
  - Land transport factors, such as vehicle acceleration and road gradient
- Site Loads—Once installed, modules are subject to loads similar to those in traditional stick-built structures:
  - Dead loads, primarily the weight of the steel framework, attachments, and concrete foundation
  - Live loads, including personnel, equipment, and temporary materials
  - Operational loads, from machinery, piping, cabling (with trays or conduits), and auxiliary systems
  - Environmental loads, such as wind, snow, and seismic forces
- Design Loads are tailored to the specific equipment and site conditions and include:
  - Live loads, like concentrated forces on floors, stairs, and roofs
  - Operating loads, driven by fluid weight, thermal expansion, and torque effects on equipment and piping (Eqs. 11.1 and 11.2), axial thrust loads in centrifugal compressors, and piping nozzle loads.

$$\text{Rated torque:}\quad T_{\mathrm{r}} = 5250 \times \frac{P}{f} \tag{11.1}$$

$$\text{Torque load:}\quad T_{\mathrm{f}} = \frac{T_{\mathrm{r}}}{d} \tag{11.2}$$

where T_{r} is the rated torque (lbf-ft), P is the transmitted power (hp), f is the rotational speed (rpm), T_{f} is the torque load or force (lbf), and d is the distance between the centerlines of the anchor bolts in the longitudinal direction (in.).

The analogous equations in Système international d'unités (SI) units are as follows:

$$\text{Rated torque:} \quad T_r = 9550 \times \frac{P}{f} \tag{11.3}$$

$$\text{Torque load:} \quad T_f = \frac{T_r}{d} \tag{11.4}$$

where T_r is the rated torque (N·m), P is the transmitted power (kW), f is the rotational speed (rpm), T_f is the torque load or force (N), and d is the distance between the centerlines of the anchor bolts in the longitudinal direction (m).

Loads associated with the movement of the on-module overhead crane are classified as operating loads. Variable loads refer to the dynamic forces produced by the machinery in operation. Environmental loads arise from external conditions such as wind speed, ice buildup, snowfall, seismic activity, and temperature fluctuations between the site's minimum and maximum ambient levels.

Variable loads impacting a module encompass a range of scenarios, including:

- Shock or impact forces during land transport, particularly during lifting and lowering operations
- Wind-induced acceleration loads encountered during ocean transit
- Vessel deck sagging, which causes vertical deflection in the module's structure
- Rigging-related loads during shop or yard handling, such as positioning the module onto the vessel, offloading, and final installation on its foundation at the jobsite.

The following factors are especially critical in the design of the module:

- Thermal Forces from Climatic Conditions—These arise from expansion and contraction caused by temperature fluctuations. They are particularly impactful in modules with large base frames situated close to ground level. Arctic modules present additional challenges, as their structures may be fully exposed to outdoor temperatures, entirely climate-controlled, or partially heated, each scenario introducing unique thermal stresses.
- Forces from Material Handling—Densely packed module layouts complicate material movement and logistics. Elements such as monorails, padeyes, wheeled carts, and designated drop zones influence both the spatial arrangement and structural design considerations.
- Vibration Forces from Machinery—Vibrations can be more pronounced in steel-framed structures due to their lower mass compared to traditional concrete foundations. This reduced mass makes them more susceptible to dynamic forces generated by operating equipment.

Machinery module structures are engineered to withstand a full range of conditions, including static loads, variable forces, and diverse operational scenarios—ranging from standard and peak operations to transient events such as the failure of an interference fit in a rotating component, which could lead to rotor disintegration. Both routine and emergency shutdown conditions are also accounted for. Comprehensive analysis of load combinations—including dead and variable loads—is conducted to inform the final design and sizing of steel structural elements. These load combinations follow the guidelines outlined in the American Concrete Institute (ACI) report ACI 351 [2].

11.7 Module Structure Analyses

For site-specific conditions, module design and structural analysis must comply with the applicable building codes established by law. However, these codes generally do not extend to module transportation. In such cases, industry standards—including those from the American Petroleum Institute (API), Det Norske Veritas (DNV, formerly DNV GL), and the International Organization for Standardization (ISO)—are typically referenced.

Support or boundary conditions vary significantly depending on the transportation method. Whether it's marine transport, SPMT (Self-Propelled Modular Transporter) movement, crane lifting, or on-site installation, each scenario presents distinct structural support requirements. Notably, transportation fatigue becomes a critical factor during extended marine routes, often impacting the integrity of structural connections.

Weight and center of gravity envelopes—commonly used in offshore module design—are also applicable to onshore modules. It is essential that design weights align with the module weight report to ensure accuracy and safety.

Machinery-Induced Structural Considerations—Modules equipped with machinery require specialized static and dynamic analyses of the steel framework. These assessments are vital to:

- Ensure equipment operates reliably within acceptable vibration thresholds across all operating conditions
- Safeguard personnel
- Maintain structural integrity, including stability and permissible deflections of frame members under applied loads.

Design and analysis [1] must also adhere to national, federal, and local codes and standards, all of which prioritize facility and personnel safety.

Advanced Structural Analysis—Comprehensive structural evaluations are conducted to:

- Determine deflections under load
- Measure relative displacements at the shaft ends of coupled machinery
- Calculate natural frequencies (including first and second harmonics)
- Identify mode shapes associated with these frequencies.

11.7.1 Static Analysis

Static analysis is applied across all structural types, including modular systems. While the focus is on static loads, certain dynamic loads are also treated as static for analysis purposes. Transportation-related loading encompasses marine transit, land-based movement, and crane lifting operations. The specific combination and requirements of these loads vary significantly depending on the application. For large-scale modules, marine loading is typically assessed by a naval architect.

Site-specific loads are evaluated similarly to those in traditional stick-built structures. Seismic forces are calculated using either the equivalent lateral force method or response spectrum analysis, and then applied statically within the structural model.

Acceptance criteria for static analysis are based on permissible inertia forces and the deflection limits of beams and columns under live and variable loads. During both transport and operation, machinery supports and the connections between primary structural elements (columns and beams) and the module's deck and roof experience a range of stresses. These components are evaluated against fatigue criteria to ensure long-term durability. Dynamic loads such as wind and seismic forces are represented through static equivalent analysis to simplify their impact assessment.

11.7.2 Dynamic Analysis

Key Factors in Dynamic Structural Analysis are:

- Structural Mass and Center of Gravity: These are derived from the analysis model. For machinery, this data should be obtained directly from the equipment manufacturer.
- Structural Stiffness and Damping: Stiffness is evaluated using the analysis model, while damping is incorporated using a conservative (low-end) estimate of the damping ratio.
- Foundation Characteristics: Both stiffness and damping values are established based on findings from the geotechnical report and the specifications of the proposed foundation system.
- Machine Support Rigidity: While always relevant, support rigidity becomes especially critical if the machine is mounted on steel framing. Adequate rigidity ensures proper shaft alignment and facilitates the neutralization of internal forces generated by the machine.

Dynamic analysis involves evaluating deflections or vibration amplitudes (X) by applying the principles of mass (m), stiffness (k), damping (C), and dynamic loading (F). This is considered a service-level assessment, where realistic—rather than overly conservative—values are preferred, except in the case of damping where caution is warranted.

The most widely used method is modal time-history analysis. This technique first identifies the system's natural frequencies through modal analysis, then applies time-history numerical simulation to calculate deflections. Accurate parameter selection—such as the number of modes, time step size, and total duration—is critical, as errors can significantly affect results. For large-scale models or high-speed machinery, this process can be computationally intensive.

Key dynamic design considerations for modules include:

- Mass and center of gravity
- Structural and foundation stiffness and damping
- Applied forces
- Resonant frequencies
- Acceptable vibration amplitudes
- Rigidity of the machine's support structure.

Dynamic analysis accounts for the interplay between mass, stiffness, damping, applied forces, and resulting deflections. Common excitation sources include rotor unbalance, misaligned couplings, and internal rubbing, all of which generate dynamic forces (refer to Eqs. 11.5–11.7 for unbalance force calculations [2]).

While time-history analysis reveals mode shapes and frequency responses over time, steady-state dynamic analysis focuses on identifying resonant natural frequencies.

$$\text{Unbalance force:} \quad F = m_{\mathrm{e}} e \omega^2 \tag{11.5}$$

$$\text{Horizontal component of force } F\text{:} \quad F_{\mathrm{h}} = m_{\mathrm{e}} e \omega^2 \cos \omega t \tag{11.6}$$

$$\text{Vertical component of force } F\text{:} \quad F_{\mathrm{v}} = m_{\mathrm{e}} e \omega^2 \sin \omega t \tag{11.7}$$

where F is the magnitude of the unbalance force (lbf [N]), m_{e} is the mass of the rotating part (lb [kg]), e is the eccentricity of the rotating part (in. [m]), and ω is the circular frequency of rotation (rad/s).

One of the key results derived from dynamic response analysis is the amplitude-versus-frequency plot, generated from the applied excitations. For designs where this is specified, the vibration amplitudes at the baseplate anchor bolt locations are evaluated to ensure they remain below the maximum allowable foundation vibration levels across the full operating speed range of the machine.

Steady-state dynamic analysis calculates deflections directly in response to harmonic loading. Resonance behavior is assessed by analyzing the system over a spectrum of machine speeds. To avoid interference, natural frequencies must be sufficiently separated—by at least 20%—from the operating speed range of the mounted machinery train. The structural integrity and stiffness (resistance to deformation) of module components are verified through both static and dynamic analyses.

ISO Standard 2631-1 [3] outlines fundamental criteria for assessing human exposure to vibration. This consideration plays a critical role in module design, particularly in zones routinely accessed by plant personnel during operation and maintenance. The standard also offers guidance on how vibrations may impact human health, sensory perception, and the potential for motion sickness.

Acknowledgements Some of the material in this chapter was reprinted with permission of the publisher from Ghaisas N (2021) "Module mounted machinery packages" CompressorTECH^2 magazine (21 June 2021).

References

1. Process Industry Practices (2017) PIP STC01015, Structural design criteria. Process Industry Practices, Austin, TX
2. ACI Committee 351 (2018) ACI 351 3R-18: report on foundations for dynamic equipment. American Concrete Institute, Farmington Hills, MI
3. ISO 2631-1:1997 (1997, confirmed 2021) Mechanical vibration and shock—evaluation of human exposure to whole-body vibration. Part 1: General requirements, 2nd edn. International Organization for Standardization, Geneva, Switzerland

Chapter 12
Lateral Rotordynamics

Abstract Rotating equipment plays a vital role across a wide range of industries, including oil and gas, refining, manufacturing, and chemical processing. Commonly referred to as turbomachinery, these systems encompass steam turbines, gas turbines, electric motors, and compressors. Many of these machines are essential to core operations and often run continuously for extended periods—sometimes years—before scheduled maintenance or downtime occurs. Any interruption in their operation can have a substantial impact on both productivity and financial performance. As industries demand greater power output, improved efficiency, reduced costs, and more compact designs, turbomachinery engineering has had to evolve rapidly. This evolution has driven the development of advanced analytical tools and more sophisticated design methodologies. Consequently, ensuring high reliability in these machines is not just desirable—it's imperative. This chapter aims to present a broad overview of rotordynamics, focusing on the physical principles behind the subject while avoiding deep theoretical detail.

Nomenclature

AF	Amplification factor
c, C	Damping coefficient
C_b	Bearing damping coefficient
c_c	Critical damping coefficient
C_eff	Effective damping
C_{xx}, C_{yy}	Direct bearing damping coefficient
C_{xy}, C_{yx}	Cross-coupling bearing damping coefficient
d, D	Diameter
E	Modulus of elasticity
F	Force
I	Area moment of inertia
k, K, K_i	Stiffness (lbf/in. [N/m])
K_b	Bearing stiffness (lbf/in. [N/m])

N. Ghaisas, *Centrifugal Compressors in Hydrocarbon Processing Industries*,
https://doi.org/10.1007/978-3-032-03687-2_12

K_{dyn_xx}, K_{dyn_yy}	Dynamic stiffness (lbf/in. [N/m]) in the x and y directions, respectively
K_{eff}	Effective stiffness (lbf/in. [N/m])
K_{xx}, K_{yy}	Direct bearing stiffness coefficients (lbf/in. [N/m])
K_{xy}, K_{yx}	Cross-coupling bearing stiffness coefficients (lbf/in. [N/m])
l	Length
m	Mass
MCOS	Maximum continuous speed (rpm)
N	Speed (rpm)
N_1, N_2	Speeds (rpm) at points 1 and 2, respectively
N_{c1}	First critical speed (rpm)
N_{c2}	Second critical speed (rpm)
N_{c1R}	First critical speed (rpm) on rigid supports
N_{c2R}	Second critical speed (rpm) on rigid supports
N_{mc}	Maximum continuous speed (rpm)
N_{min}	Minimum speed (rpm)
N_{peak}	Speed (rpm) at maximum amplitude
Q_0	Threshold cross-coupling
q_a	Anticipated cross-coupling of a given stage
Q_a	Total anticipated cross-coupling
SM	Separation margin
t	Time
x	Displacement
$\dot{x}$	Velocity
$\ddot{x}$	Acceleration
ξ	Damping ratio
ω	Angular frequency (rad/s)
ω_n	Natural frequency (rad/s)

12.1 Fundamentals of Rotordynamics

In general, a piece of rotating equipment consists of a shaft along with other mounted components, such as impellers or disks, that spins (rotates about the axis) and is supported by two or more bearings in the lateral direction. Because the rotor (shaft and mounted components) rotates in the bearings, the dynamics of the rotor-bearing system needs to be understood to successfully design the machine. One such engineering discipline for successfully designing and operating turbomachinery is rotordynamics. The subject of rotordynamics is the study of the dynamic characteristics of rotating machinery.

Three types of vibrations are associated with a rotating shaft: (a) lateral vibrations, involving the dynamics of the shaft perpendicular to the axis of rotation; (b) axial vibrations, involving the dynamics of the shaft along the axis of rotation; and (c)

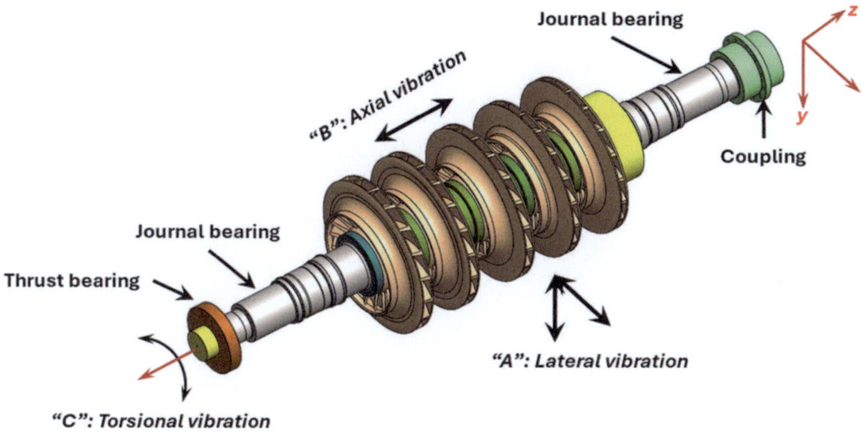

Fig. 12.1 Example of a compressor rotor with components, indicating types of vibrations

torsional vibrations, involving the dynamics of the shaft in the torsional direction of the axis (associated with the "twist" of the shaft). This section discusses the subject of lateral vibrations of a single rotor in the horizontal or vertical direction (to the axis of rotation). The effects of the coupled rotordynamics of multiple machines or gear dynamics are beyond the scope of this chapter.

An example of a turbomachine is a centrifugal compressor, in which a gas is compressed from a low pressure to a higher pressure. An example of a compressor is shown in Fig. 12.1. The three types of vibrations are illustrated in Fig. 12.1 and indicated by A for lateral vibration, B for axial vibration, and C for torsional vibration. This example shows a centrifugal compressor with five impellers supported by two radial bearings and one axial (thrust) bearing. The bearings that support the weight of the rotor are called journal bearings.

The journal bearings are usually rigidly mounted on pedestals. Alternatively, the journal bearings could also be mounted on flexible supports. In the following discussion, the term "bearing" refers to journal bearings, and the term refers to the assembled shaft and other rotating components that are mounted on the shaft.

Axial vibrations are generally absorbed by the thrust bearing. Typically, the more challenging type of vibrations are the lateral vibrations of the rotor. To successfully design and operate a turbomachine, understanding the basics of the lateral vibrations is a very important step. The design of the journal bearings and the influence of the bearings on the dynamics of the systems is the core of rotordynamic analyses. In this chapter, the influence of each component (bearings, dampers, supports, etc.) is explained, in order to shed light on the complex subject of rotordynamics.

Lateral rotordynamics is the study of rotor vibrations in the directions perpendicular to the rotational axis of the rotor. Before focusing in depth on the rotordynamics of turbomachinery, this chapter first discusses the basics of the subject of rotordynamics.

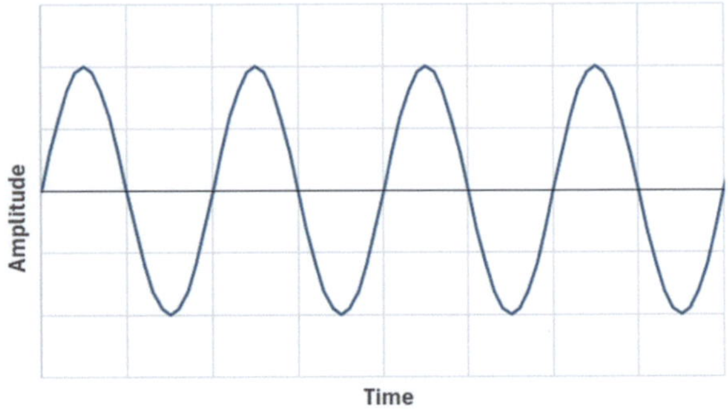

Fig. 12.2 Effect of stiffness

12.2 Key Terms

12.2.1 Rotor

A rotor consists of a shaft, impellers, sleeves, rotating labyrinths, a balance drum, and so on.

12.2.2 Stiffness (Denoted by K or k)

Stiffness is the property of a spring. This chapter considers only compression springs, that is, springs for which the load is generated by compressing the spring. Stiffness is the force required to compress the spring through unit distance. The unit of stiffness is the pound-force per inch (lbf/in.) in conventional units, or the newton per meter (N/m) in Système international d'unités (SI) units. If a spring is compressed and released, the stiffness of the spring causes a sinusoidal motion, as shown in Fig. 12.2. Theoretically, without any external resisting forces (such as friction), this motion is purely sinusoidal in nature.

12.2.3 Damping (Denoted by C or c)

Damping is the property of a damper (also known as a dashpot). Damping is the force required to compress the spring per unit velocity. The unit of damping is the pound-force second per inch (lbf s/in.) in conventional units, or the newton second per meter (N s/m) in SI units. If a pure damper is compressed and released, the motion is an exponentially decaying curve. However, in real systems, both stiffness and damping exist, causing an exponentially decaying sinusoidal wave, as shown in Fig. 12.3. Note that, for stable systems, the exponential curve would be decreasing, whereas for unstable systems, it would be exponentially increasing.

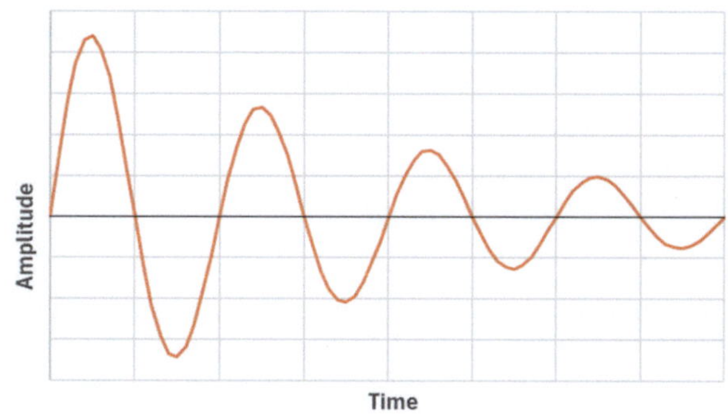

Fig. 12.3 Effect of damping

12.2.4 Amplification Factor

The amplification factor (AF) is calculated at the peak(s) of a response plot. It provides an indication of the "sharpness" of the peak and is defined using the half-power method (refer to Eq. 12.1). A typical unbalance response plot is shown in Fig. 12.4.

$$AF = \frac{N_{peak}}{N_2 - N_1}. \tag{12.1}$$

In Eq. 12.1, N_{peak} is the speed corresponding to the peak amplitude. If A_{peak} is defined as the amplitude of vibration at the peak rotor speed, then the speeds N_1 and N_2 can be obtained as the intercepts between the response curve and a horizontal line

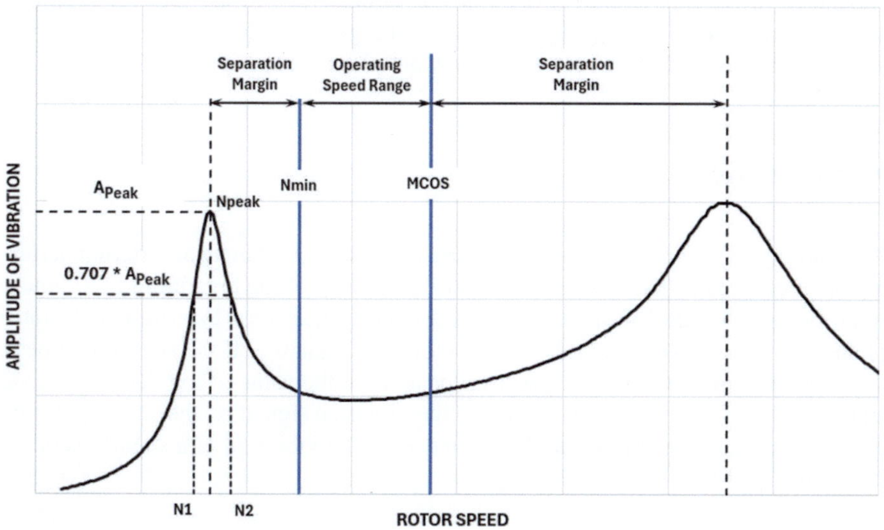

Fig. 12.4 Rotor response plot showing various parameters

at 0.707 × A_{peak}. The amplification factor is a dimensionless number. The higher the value of AF, the sharper the peak.

12.2.5 Critical Speed

The critical speed is also known as resonance and occurs when the running speed coincides with the natural frequency of the system. However, the word "critical" has a specific meaning. For example, for compressors and expanders, the term "critical" is defined by American Petroleum Institute (API) Standard 617 [1] and is based on the amplification factor.

12.2.6 Damped Natural Frequency

The damped natural frequency of a system is its natural frequency of system in the presence of damping, without any external excitation or forces.

12.2.7 Unbalance Response

The unbalance response is a steady-state behavior of a rotor-bearing system to an unbalance force. The output of the analysis is presented in the form of a plot of amplitude (at a given shaft location) versus running speed. This is often also referred to as a Bodé plot including both amplitude and phase versus speed.

12.2.8 Stability

The stability of a system is defined as the reaction of the system to an external perturbation or excitation. The stability is measured in terms of the logarithmic decrement (referred to as "log dec"), where log dec is defined as the natural logarithm of the ratio of any two successive amplitudes of the vibration signature. If the subsequent peak has a lower amplitude than the previous peak, then the value of log dec is positive, and hence, the system is stable. Based on the vibration signature, the stability can be classified as stable, neutrally stable, or unstable. The vibration of a stable system is shown in Fig. 12.5.

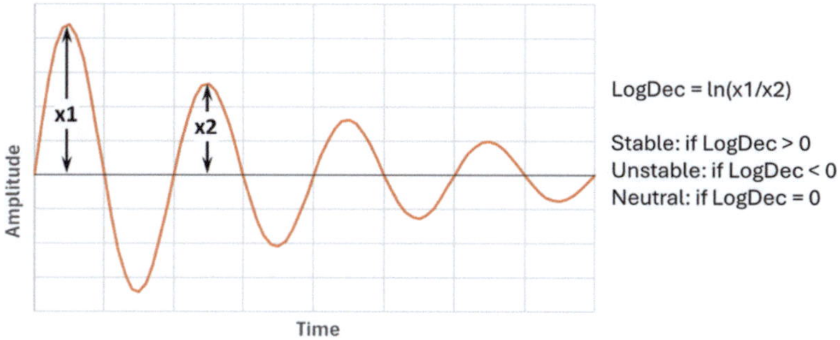

Fig. 12.5 The vibration signature of a stable system after external perturbation

12.3 Basic Vibrations (Free Vibrations)

To understand the fundamentals of rotordynamics, it is essential to understand the principles of basic vibration theory. The simplest model to represent a vibrating system, shown in Fig. 12.6, consists of a mass *m*, supported by a spring of stiffness *k*. If the mass is set in motion along the axis of the spring, the result is an oscillatory motion, as shown in Fig. 12.7. No external forces act on this system, and hence, the system is considered to be under free vibration.

The frequency of oscillation is given by solving the differential equation $m\ddot{x}+kx=0$. The circular natural frequency is given by Eq. 12.2

$$\omega_{\mathrm{n}} = \sqrt{\frac{k}{m}}. \tag{12.2}$$

Fig. 12.6 Simple spring–mass system

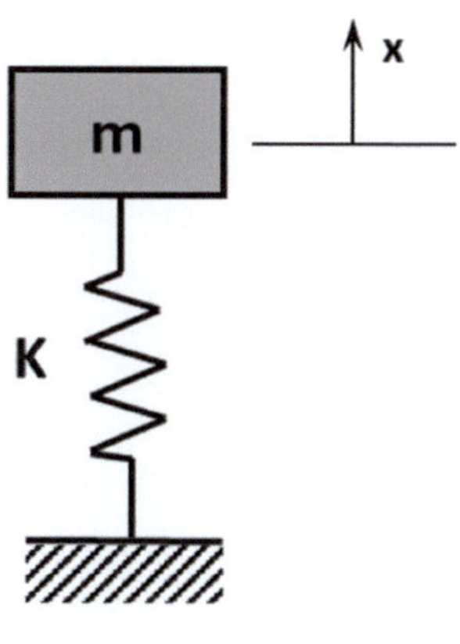

Fig. 12.7 Oscillation of the mass for the system depicted in Fig. 12.6

This equation gives the fundamental frequency of the simple, one-degree-of-freedom, spring–mass system. Although the rotordynamic model of an actual machine is far from this simple representation, this equation provides a very intuitive understanding of the basic principles.

A more realistic system would include some form of damping as generated by friction, material damping, or an external source such as an oil film. Figure 12.8 shows a spring–mass–damper system, for a mass m, a spring with stiffness k, and a dashpot with damping constant c.

When the mass is set in motion, the oscillation consists of two components: a sinusoidal wave (because of stiffness) and an exponentially decreasing part (assuming positive damping).

Figure 12.9 shows the combined effect of the spring and damper, which results in an exponentially decreasing sinusoidal wave. The frequency of oscillation for free vibration (without external forces or forcing functions) is given by solving the second-order differential Eq. 12.3

$$m\ddot{x} + c\dot{x} + kx = 0. \tag{12.3}$$

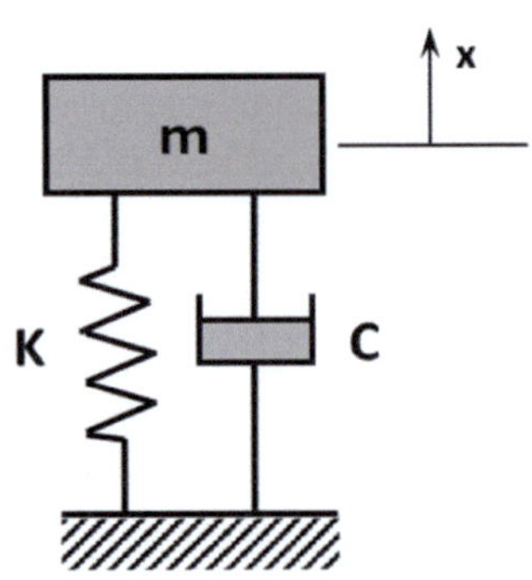

Fig. 12.8 Simple spring–mass–damper system

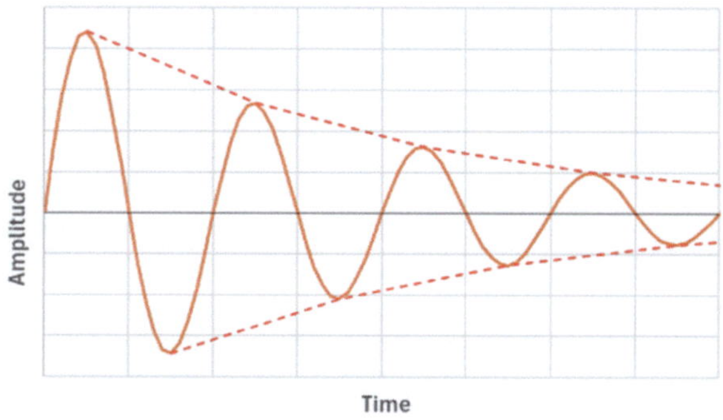

Fig. 12.9 Oscillation of the mass for the system depicted in Fig. 12.8

The resonant frequency is termed the damped critical speed and is given by Eq. 12.4

$$\omega_{\mathrm{d}} = \omega_n\sqrt{1-\zeta^2}. \tag{12.4}$$

where $\zeta = c/c_{\mathrm{c}}$ is the damping ratio and $c_{\mathrm{c}} = 2\sqrt{km}$ is the critical damping constant.

12.3.1 Application of Basic Vibration Principle to Rotordynamics

One of the first models used to discuss rotordynamic behavior in more detail is the Jeffcott rotor, shown in Fig. 12.10, which is named after Henry Homan Jeffcott, who formulated the model in 1919 [2]. It consists of a slender shaft (of negligible mass) with a heavy disk at the midspan that is supported on rigid mounts at its ends. Because the shaft is slender and the supports are rigid, the deflected shaft in motion is as shown by the dotted line in Fig. 12.10.

From beam theory, for a simply supported shaft with a circular cross section, the bending stiffness is given by Eq. 12.5

$$k = \frac{48EI}{l^3}, \tag{12.5}$$

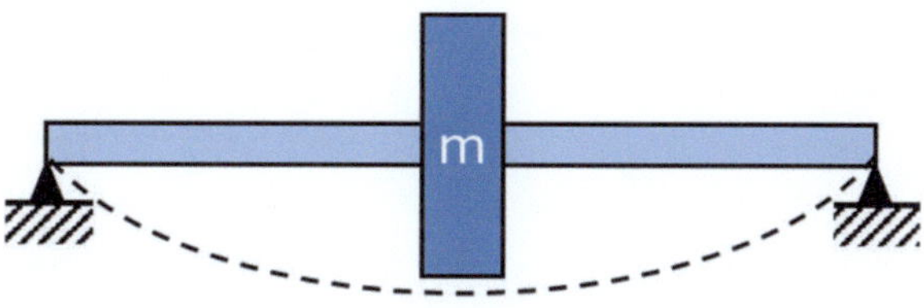

Fig. 12.10 Jeffcott rotor

where E is the elastic modulus, I is the area moment of inertia, and l is the length between supports.

Replacing the value of k in Eq. 12.2 gives the first natural frequency of the rotor as

$$\omega_n = \sqrt{\frac{k}{m}} = \sqrt{\frac{48EI}{ml^3}}.$$

Using $I = \pi d^4/64$, the frequency can be written as (Eq. 12.6)

$$\omega_n = \sqrt{\frac{3\pi E d^4}{4ml^3}}. \quad (12.6)$$

Even though Eq. 12.6 is based on a very simple rotor model with a single degree-of-freedom, it provides a very intuitive explanation of what happens to the natural frequency when the diameter or length of the shaft (the most common parameters used in the design of turbomachinery) is varied.

12.4 Basic Vibrations (Forced Vibrations)

In Sect. 12.3, the system that was described did not have any forcing functions (hence the 0 on the right-hand side of the equation). In the real world, it is impossible to eliminate the unbalance of the rotor. The remaining unbalance (also known as residual unbalance) could be due to various factors such as imperfections in the material, porosity and so on. All machines operate with some residual unbalance. This residual unbalance acts as a forcing function on the rotor system.

The Jeffcott rotor helps explain the behavior of the rotor with an unbalance. Figure 12.11 shows a disk with an unbalance, in which the center of mass m is not located at the geometric center of the disk C. The offset r between the center of mass and the geometric center results in a centrifugal force when the shaft begins to spin. This force tends to bend the shaft by pulling it away from the centerline, thereby causing the rotor to bend. In effect, the shaft centerline moves in an orbit called a whirl orbit.

The equation of motion, including the stiffness and damping terms, is given by Eqs. 12.7 and 12.8. In each of these equations, the term on the right-hand side is the forcing function due to the centrifugal forces.

$$m\ddot{x} + c\dot{x} + kx = m\omega^2 r \cos \omega t \quad (12.7)$$

$$m\ddot{y} + c\dot{y} + ky = m\omega^2 r \sin \omega t. \quad (12.8)$$

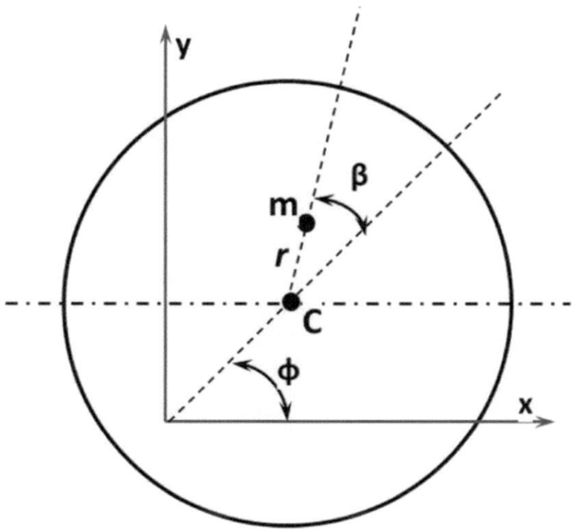

Fig. 12.11 Disk with unbalance

The solution to this system of equations is given by

$$x = \frac{\omega^2 r}{\sqrt{\left(\frac{k}{m} - \omega^2\right)^2 + \left(\frac{c\omega}{m}\right)^2}} \cos(\omega t - \beta_1)$$

$$y = \frac{\omega^2 r}{\sqrt{\left(\frac{k}{m} - \omega^2\right)^2 + \left(\frac{c\omega}{m}\right)^2}} \sin(\omega t - \beta_1)$$

$$\beta_1 = \tan^{-1}\left[\frac{c\omega}{m\left(\frac{k}{m} - \omega^2\right)}\right] \tag{12.9}$$

A plot of the synchronous whirl as a function of shaft speed is shown in Fig. 12.12. The amplitude of vibration (peak-to-peak) is the displacement of the shaft from its initial reference position.

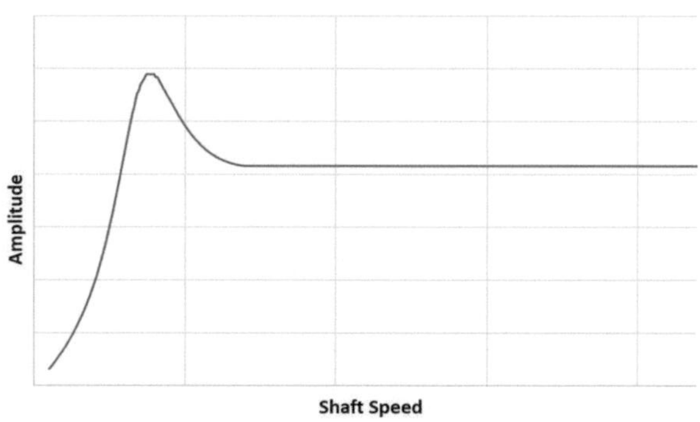

Fig. 12.12 Plot of amplitude versus shaft speed

12.5 Support Flexibility

Contrary to the simplified Jeffcott approach, the supports in machines are not completely rigid and do exhibit some pliability. For example, the most common flexibility is introduced in the rotors by the oil film in the journal bearings that support the rotor [3, 4]. The oil film inherently provides both stiffness and damping properties. In addition to the stiffness and damping of the oil film, there could be other sources of stiffness (and damping) that are in series with the oil film, such as (1) the bearing pad or sleeve (an example is the stiffness of the journal bearing pad of a tilt-pad journal bearing), (2) the damper bearing O-rings, and (3) the pedestals that support the rotor.

To develop an intuitive understanding of the effect of support stiffness, consider the simple system of a rotor supported on bearings as shown in Fig. 12.13. Figure 12.14 shows two springs in series—for example, this could be the rotor supported by journal bearing. This example considers only the stiffness of the shaft and the stiffness and damping of the bearing oil film to explain the concept.

When two springs are connected in series, the effective spring rate is given by Eq. 12.10

$$\frac{1}{K_{\text{eff}}} = \frac{1}{K_1} + \frac{1}{K_2} \tag{12.10}$$

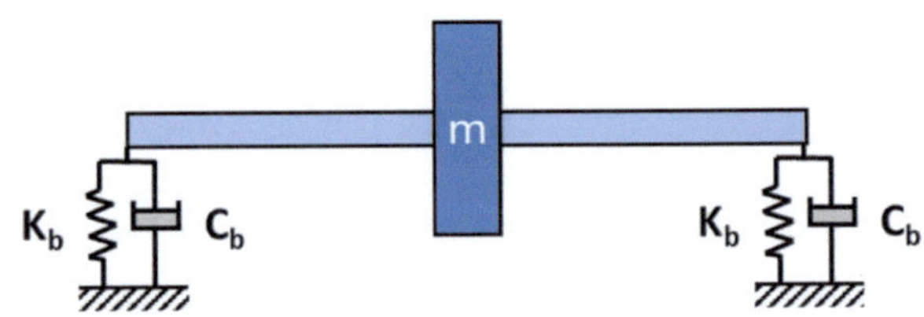

Fig. 12.13 Rotor supported on bearings

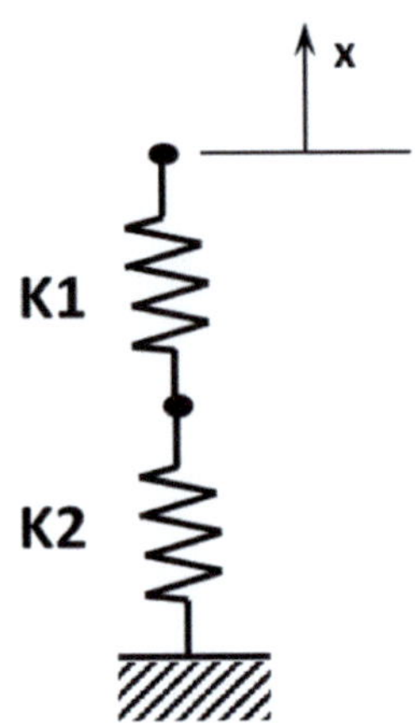

Fig. 12.14 Springs in series

from which one can readily obtain

$$K_{\text{eff}} = \frac{K_1 K_2}{K_1 + K_2}$$

From this equation, it can be inferred that, when two springs are in series, the effective spring rate is less than the lowest spring rate. Using the basic definition of critical speed, it can be intuitively concluded that when the support is not rigid and has some flexibility, the critical speed is reduced (compared to the critical speed on rigid supports).

A flexible support also results in lower forces being transmitted to the base structure, which minimizes the vibration of the rotor with respect to the supports.

12.6 Journal Bearings

The function of a journal bearing is to support the rotating shaft with minimal wear between the bearing inner surface and the rotating shaft. The two most common classifications of bearings are rolling-element bearings and fluid-film bearings. In this section, fluid-film bearings will be covered in detail. An understanding of how a fluid-film bearing functions forms the basis of understanding the dynamics of turbomachinery. The journal bearings, because of the properties of the oil, have a significant influence on the rotordynamic behavior of the machine.

12.6.1 Basic Concept of a Journal Bearing

The simplest journal bearing is a plain sleeve that is fitted in a bearing housing (Fig. 12.15). The shaft that is supported by the bearing is slightly smaller than the inner diameter of the sleeve, and the clearance volume is filled with oil. The most common types of oil classes used in journal bearings are International Organization for Standardization viscosity grade (ISO VG) 32 and ISO VG 46. When the shaft is not spinning, the shaft is resting at the bottom dead center of the clearance space. As the shaft starts to spin, the metal-to-metal friction is quickly overcome, and the hydrodynamic forces generated "lift" the shaft.

Because of the clearance profile, the rotating shaft forms a convergent wedge in the direction of rotation. This generates a pressure in the oil film that supports the shaft. The rotation of the shaft acts as a "self-actuating pump," and oil is circulated under the shaft continuously as the shaft spins. Hence, the term "hydrodynamic bearing" as another name for such journal bearings. Because of its viscosity, the oil inherently has properties of stiffness and damping associated with it. This stiffness and damping play a critical role in the behavior of the rotating machinery.

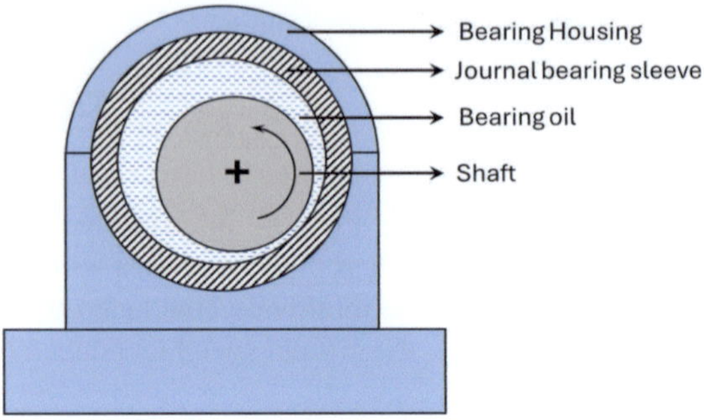

Fig. 12.15 Example of a plain cylindrical journal bearing

12.6.2 *A Brief History of Journal Bearing Designs*

Based on the application, field experience, and technology, the bearing design process has evolved to meet the higher demands of industry [5]. Figure 12.16 shows the development of the commonly used designs over time, beyond the simple plain cylindrical bearing.

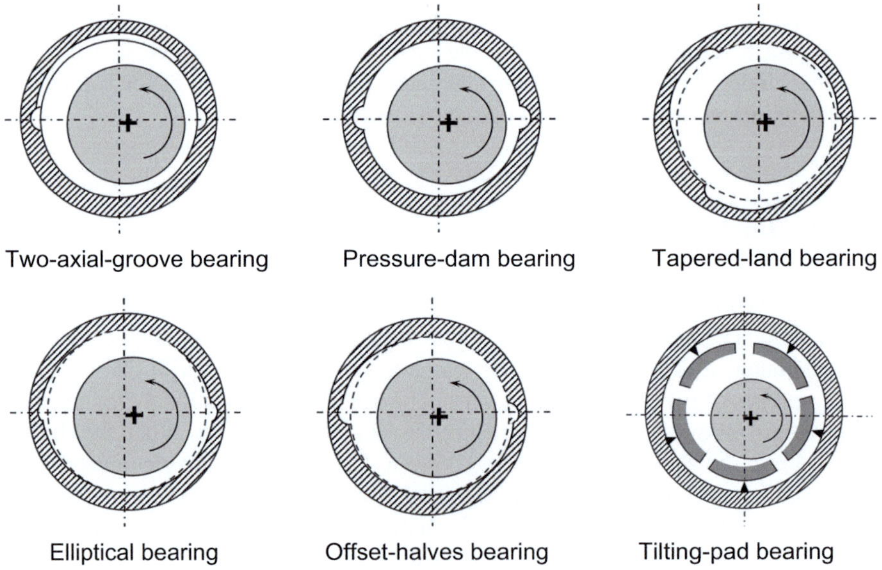

Fig. 12.16 Examples of different types of journal bearings

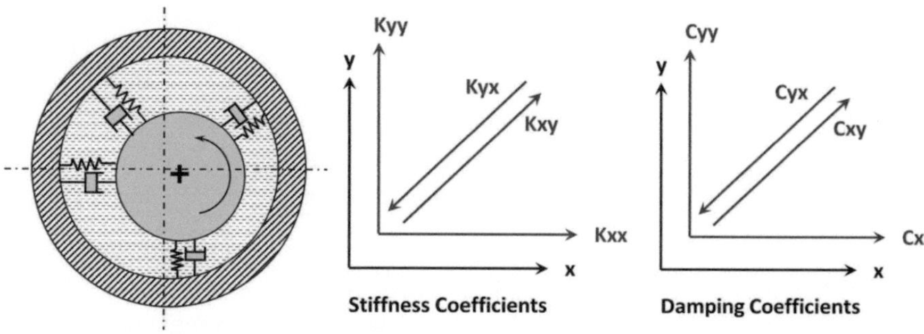

Fig. 12.17 Dynamic coefficients of a journal bearing

12.6.3 Journal Bearing Properties and Their Significance

As the needs of machines are pushed toward higher speeds and higher power, the dynamic stability of the machine becomes a critical parameter. A journal bearing spinning in the clearance space filled with oil inherently generates stiffness and damping coefficients. To understand the importance and effects of these coefficients, consider a plain sleeve bearing.

Figure 12.17 shows a plain sleeve bearing, with oil stiffness and damping properties. In general, there are four stiffness coefficients (K_{xx}, K_{xy}, K_{yx}, K_{yy}) and four damping coefficients (C_{xx}, C_{xy}, C_{yx}, C_{yy}). The coefficients with subscripts of *xx* and *yy* are called direct coefficients, whereas the coefficients with subscripts of *xy* and *yx* are called cross-coupling coefficients.

K_{xx} represents the stiffness of the oil film in the *x* direction with a force applied in the *x* direction. K_{xy} and K_{yx} are the coefficients for the stiffness of the oil film that is generated by motion in a direction perpendicular to the direction of the applied force. Hence, the cross-coupling coefficients K_{xy} and K_{yx} act in the tangential directions. These cross-coupling coefficients play a significant role in the dynamic stability of the journal bearing and, hence, of the rotor system. If the magnitudes of K_{xy} and K_{yx} are different, the rotor whirls in an elliptical orbit.

12.6.4 Significance of Cross-Coupling Coefficient (K_{xy})

Under lightly loaded conditions, when the rotor weight is trying to move the rotor in the clearance space, in the *y* direction, the stiffness K_{yy} is generated. However, because of the inherent properties of the oil film, the K_{yx} stiffness also acts on the shaft in the tangential direction, trying to move the shaft in the *x* direction. In turn, because of the motion in *x* direction, the shaft moves in the *y* direction. This coupled movement continues until the effective damping brings the system to equilibrium (Eq. 12.11).

Table 12.1 Relative stabilities and load capacities of four types of journal bearings

Stability	Design	Load capacity
Least stable	Plain cylindrical	High
More stable	Elliptical, pocket, pressure dam	↓
More stable	Offset halves, multilobe	
More stable	Tilting-pad design	Moderate

If the effective damping is not sufficient to overcome the cross-coupled stiffness, the system vibration grows in amplitude and becomes unstable, eventually reaching a limit cycle restricted by the bearing inner diameter. This unstable vibration behavior of the plain sleeve bearing under lightly loaded conditions limits the operation of the compressor. This fact also triggered the evolution of the bearing to achieve better performance and improve its stability.

$$C_{\text{eff}} = C_{xx} - \frac{K_{xy}}{\omega}. \tag{12.11}$$

The tilting-pad bearing design, which has been used successfully since 1912 [6], is the preferred bearing of choice in newer turbomachinery. Tilting-pad bearings have almost zero cross-coupled stiffness and, hence, are relatively the most stable bearings. Because each pad can tilt independently based on the direction of the load, the cross-coupled forces are low. This design has helped the turbomachinery industry to address the instability caused by journal bearings and push the limits of turbomachines. Table 12.1 shows the relative stabilities of the common types of journal bearings.

12.6.5 Journal Bearing Pad or Pivot Orientation

In a tilt-pad bearing, depending on the direction of the load with respect to the pivot, the configuration is classified as load between pivots (LBP) or load on pivot (LOP). Note that the terms "pad" and "pivot" are used interchangeably in the literature.

The two configurations are shown in Fig. 12.18. The LBP configuration provides more symmetric stiffness and damping coefficients. Symmetric support properties provide circular orbits, whereas asymmetric coefficients could result in elliptical orbits. In general, circular orbits are preferred. This is because circular orbits result in smaller vibration amplitudes as the rotor is going through a resonance (where the running speed coincides with the natural frequency of the rotor system), compared to the larger major axis of the ellipse for asymmetrical bearing properties. In general, the LBP configuration is good for forced response, and the LOP configuration is good for stability. In the design process, one needs to select the pad-load orientation based on the stability and response results.

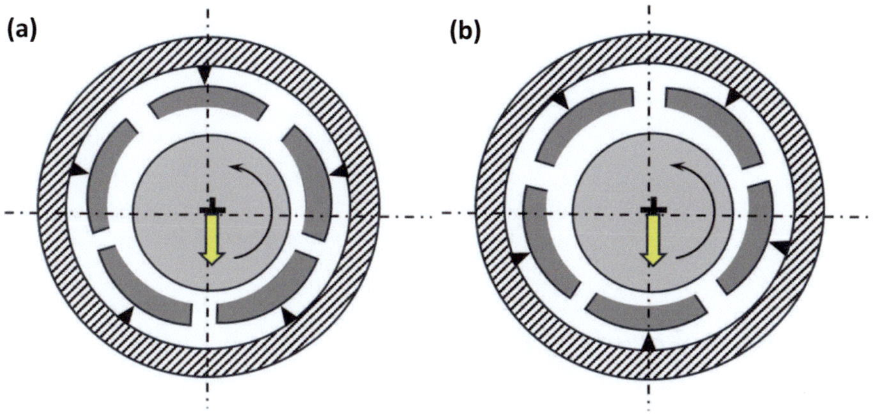

Fig. 12.18 **a** Load between pivots and **b** load on pivot configurations

12.6.6 Journal Bearing Pivot Location

In tilt-pad bearings, the location of the pivot behind the pad affects the rotordynamic behavior of the system. The two types of pivot locations are shown in Fig. 12.19. The pivot or pad offset is the ratio between the distance of the leading edge of the pad to the pivot and the total arc length of the pad. When the pivot is at the center, it is referred to as a center pivot.

Pivot offsets generally are between 0.5 ($\theta_1/\theta = 0.5$) and 0.65 ($\theta_2/\theta = 0.65$). An offset pad has a larger converging taper for the oil film to travel, and hence, this larger film thickness results in a decreased operating temperature. Also, offset pivots have higher load-bearing capacities. However, in general, the stability of the rotor system could decrease with an offset pivot.

The amplification factor of the first critical speed increases with pivot offset. A pivot offset of less than 0.5 is not desirable because it results in a diverging film and almost no load-carrying capacity.

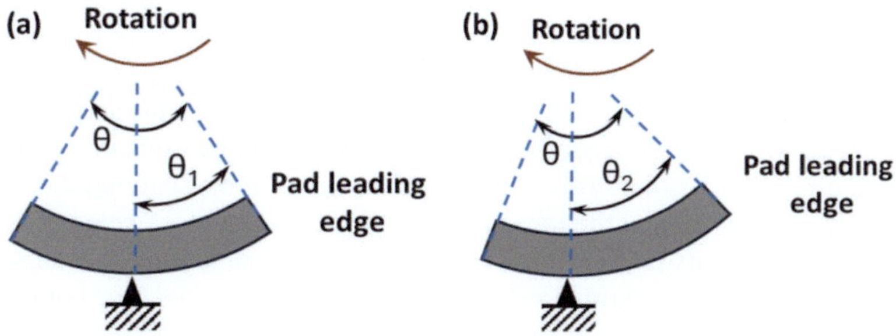

Fig. 12.19 **a** Center pivot and **b** offset pivot configurations

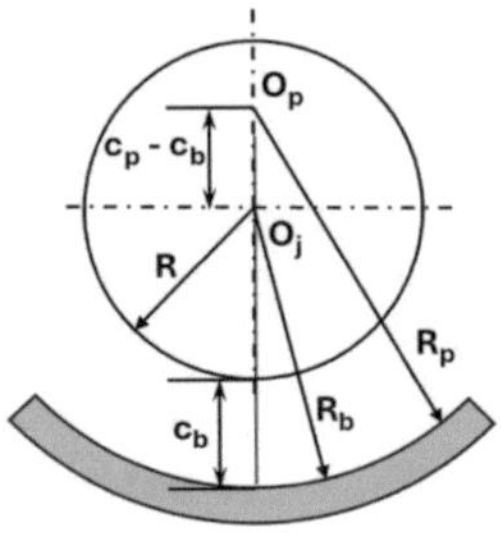

Fig. 12.20 Definition of bearing preload

12.6.7 *Journal Bearing Preload*

Another design parameter that can be varied for a tilt-pad journal bearing is the preload. Figure 12.20 shows the calculation of the preload for a tilt-pad journal bearing [7]. Preload is a geometric factor defined as follows: preload $= 1 - (C_b/C_p)$, where C_b is the assembled bearing clearance and C_p is the pad machined clearance. If a bearing has a preload, then the center of curvature of the pad (O_p) and the geometric center of the journal bearing (O_j) are not coincident. In other words, preload can be expressed as a percent reduction in the clearance.

A preload of zero means that, if the shaft is centered, then the clearance is the same around the circumference of the shaft. When a pad has a preload greater than zero (typical values are between 0 and 50%), a converging oil film forms under the journal. This results in hydrodynamic forces that support the shaft. However, if the preload is less than zero (negative), this would result in a diverging film that results in poor or low load-carrying capacity of the bearing. In tilt-pad bearings, under conditions of a light or negative preload, the top pads could become unloaded and start to flutter, which causes vibration of the rotor and, in addition, could result in pivot wear.

12.6.8 *Oil Whirl and Oil Whip*

Oil whirl is a fluid-induced instability and is classified as a self-excited rotor vibration. The oil whirl phenomenon stems from the interaction between the fluid and the rotor in the bearing clearance space. When the shaft is spinning in the clearance space, the oil is dragged along as a result of viscous forces. The hydrodynamic action supports the weight of the rotor. The reaction forces are in the vertical direction and in the tangential direction. The fluid in contact with the surface of the shaft has the velocity of the shaft, and the fluid at the inner surface of the bearing has a velocity of zero. This results in a velocity profile across the thickness of the fluid film, with an average velocity close to 50% of the rotational speed [8]. As the rotational speed of the shaft increases, the tangential component of the force increases to the point where the resulting cross-coupling stiffness exceeds the effective damping of the bearing. This results in a circular orbit in forward precession at a frequency of almost 50% of the

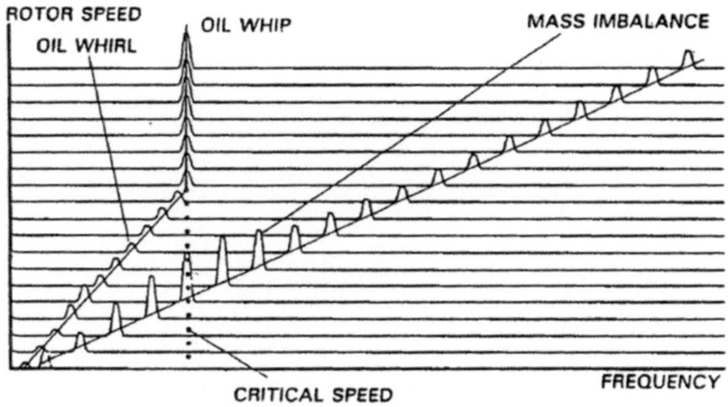

Fig. 12.21 Classic waterfall diagram showing the oil whirl and oil whip phenomena

rotational speed. This subsynchronous whirl is called oil whirl. The oil whirl tracks the rotational speed at close to 50% of its value (average velocity of the oil film).

When the rotor speed reaches almost twice the first bending mode of the rotor, the oil whirl changes to oil whip. This transition occurs because the frequency of the oil whirl reaches the first natural frequency of the rotor system. Oil whip is an unstable condition where the amplitude of vibration increases and eventually reaches a limit cycle. Because the oil whip excites the first bending mode, it can be damaging [9]. For beam-style machines (where most of the mass is between the bearings), the amplitude at midspan increases significantly, and this could potentially be detrimental to the bearing and, hence, to the machine. The oil whirl and oil whip phenomena are shown in the classic waterfall plot in Fig. 12.21 [10].

12.7 Shaft Motion

When a turbomachine is in operation, the shaft is spinning in the clearance space in the journal bearing. Figure 12.22 shows a diagram of a shaft in motion in the clearance space in the journal bearing. For simplicity, a plain sleeve bearing is depicted in the figure, and the bearing housing is not shown. As shown in the figure, when the shaft is rotating, the marker on the shaft moves about the center of the shaft. This motion is called rotation and corresponds to the running speed (usually measured in rotations per minute [rpm]) of the machine.

However, because of the gyroscopic effects of the rotating (or spinning) shaft and the nonsymmetric bearing properties, the center of the shaft (indicated by +) is not at a fixed position for a given speed. This center of the shaft, while spinning, moves in an orbit in the clearance space that is filled with oil. This motion is called whirl. Depending on the direction of the center of the shaft with respect to a fixed frame of reference, the whirl is further classified as forward whirl or backward whirl.

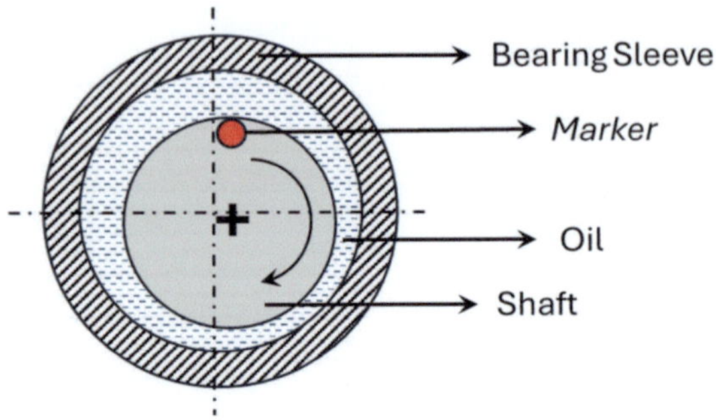

Fig. 12.22 Rotation of a journal bearing shaft

Figure 12.23 shows a freeze-frame representation of forward whirl, with t_1 through t_8 representing arbitrary time increments. For simplicity, only the shaft with a marker is shown. Using a fixed frame of reference, one can see from the figure that the marker moves in a clockwise direction (whirl) with respect to the center of the spinning shaft (rotation), which is also moving in a clockwise direction. Because the rotation and whirl are in the same direction, this whirl is known as forward whirl.

Figure 12.24 shows a freeze-frame representation of backward whirl, with t_1 through t_8 again representing arbitrary time increments. Using a fixed frame of reference, one can see from the figure that the marker moves in a counterclockwise direction (whirl) with respect to the center of the spinning shaft (rotation), which is moving in a clockwise direction. Because the rotation and whirl are in opposite directions, this whirl is known as backward whirl.

The orbit of the center of the shaft is called the whirl orbit and is shown by a dotted line in Figs. 12.23 and 12.24. This whirl orbit is the "amplitude" as measured by the probes in a turbomachine. The shape of the orbit depends on a large number of parameters such as the weight of the rotor, the bearing coefficients, the running speed, and so on. In general, forward whirl is more common, as it is stimulated

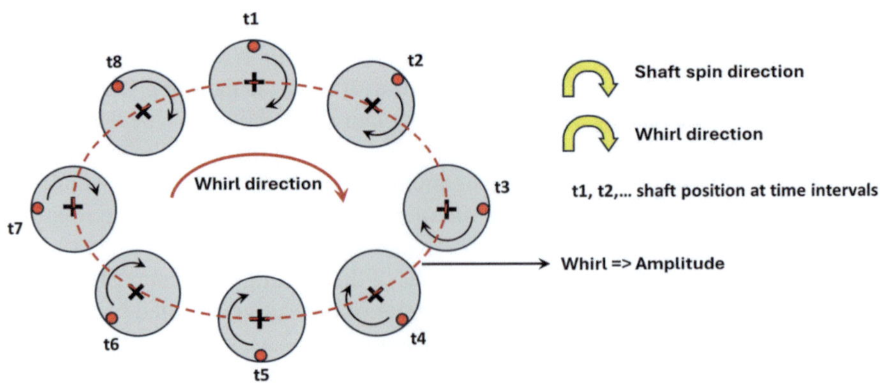

Fig. 12.23 Journal bearing shaft positions at arbitrary time increments t_1–t_8, indicating forward whirl

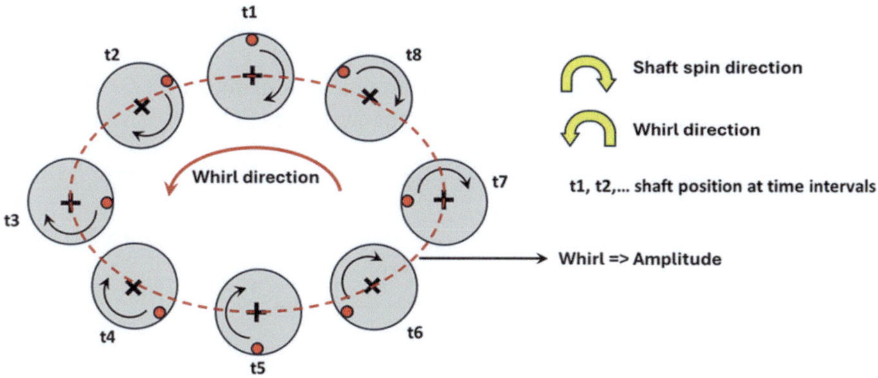

Fig. 12.24 Journal bearing shaft positions at arbitrary time increments t_1–t_8, indicating backward whirl

by unbalance forces. Backward whirl is seen when a mechanical rub occurs in the machine.

The frequency associated with forward whirl is called the forward whirl frequency, and the frequency associated with the backward whirl is called the backward whirl frequency. Because the split is primarily governed by gyroscopic effects (which are a function of speed), at low speeds, the forward and backward frequencies are almost the same. As the speed increases, the forward and backward frequencies start to diverge. This can be seen in the damped eigenvalue analysis (refer to Sect. 12.9.5).

12.8 Sommerfeld Number

The Sommerfeld number (S) [11] is a dimensionless number related to bearing static performance characteristics and is written as (Eq. 12.12)

$$S = \frac{\mu NLD}{W}\left(\frac{R}{C}\right)^2, \tag{12.12}$$

where μ is the fluid viscosity, N is the shaft speed (rev/s), L is the pad length, D is the bearing diameter, W is the external load, R is the bearing radius, and C is the pad clearance.

Equation 12.12 is for long journal bearings (generally for those for which the bearing pad axial length-to-diameter ratio is greater than 1.0). This simple equation for the Sommerfeld number is used to quickly design a journal bearing (calculate the journal eccentricity and attitude angle). For example, a high load, a low journal speed, or a low viscosity produces the same (high) operating journal eccentricity. The journal eccentricity and attitude angle defining the static performance of the journal bearing are shown as functions of operating speed, lubricant viscosity, and the applied load.

However, for short bearings (for which bearing length/diameter < 1.0), Eq. 12.13 for the modified Sommerfeld number (σ) can be written as [12]

$$\sigma = \pi S\left(\frac{L}{D}\right)^2 = \frac{\mu\omega LR}{4W}\left(\frac{L}{C}\right)^2$$

$$\sigma = \frac{\left(1-\varepsilon^2\right)^2}{\varepsilon\sqrt{\left[16\varepsilon^2 + \pi^2\left(1-\varepsilon^2\right)\right]}}, \tag{12.13}$$

where ω is the rotational speed (rad/s), e is the eccentricity, and ε is the eccentricity ratio ($= e/c$).

Under given conditions, the bearing geometry parameters (L, R, c) are known, the viscosity is known, and the external load is known. Hence, Eq. 12.13 can be iteratively solved for the eccentricity ratio that is required to support the external load.

The Sommerfeld number is applicable for laminar flow and does not account for bearing preload or changes in viscosity due to temperature.

The ideal range of the Sommerfeld number for journal bearings is from 1 to 10. The probability that the oil film pressure is not able to support the load is higher if the Sommerfeld number is less than 1 or greater than 10. In these cases, journal-to-bearing rub, localized overheating, and bearing damage can occur.

According to the charts shown in Fig. 12.25 [13–16], for a given bearing geometry, load, and oil type, the following variables can be determined:

- Minimum film thickness.
- Lubricant flow.
- Oil film pressure.

Note that, in Fig. 12.25, the Sommerfeld number is expressed more simply as $\mu N/P$, where μ and N have the same meanings as in Eq. 12.12 and P is the load per unit of bearing area.

12.9 Rotordynamic Analyses

In this section, a compressor is used as an example of a turbomachine for explaining the principles of rotordynamics. Rotordynamic analyses are performed in different phases of building the machine by the original equipment manufacturer, by end users, or by different engineering groups. Indeed, rotordynamic analyses of the compressor typically start during the quotation phase, well before the machine is designed in detail. This prevents the need for any major changes later during the manufacturing process. Rotordynamic calculations can be grouped into the following analyses:

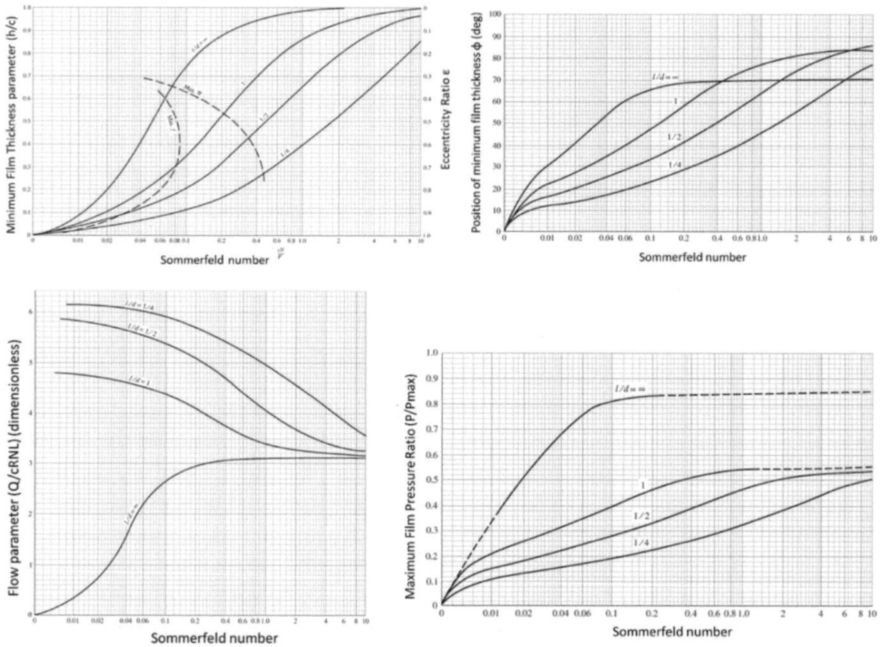

Fig. 12.25 Various design parameters plotted as a function of the Sommerfeld number

- Undamped critical speeds.
- Unbalance response analysis.
- Damped critical speeds.
- Stability analysis.
- Steady-state torsional analysis.
- Transient torsional analysis.
- Lateral rotordynamics of the train.

In this chapter, the first four types of analyses, which pertain to lateral rotordynamics, are discussed. Torsional analysis is presented in Chap. 13.

12.9.1 Rotordynamic Analysis Procedure

In rotordynamics, all of the components of a piece of rotating machinery are primarily modeled as axisymmetric elements. A typical rotordynamic analysis of a piece of rotating machinery begins with the definition of the rotating components such as the motor spindle, turbine stages, impeller stages, or balance drum (also known as a balance piston) and the shrunk-on components such as sleeves, bearings, labyrinth seals, and couplings. For example, in the case of a centrifugal compressor, the

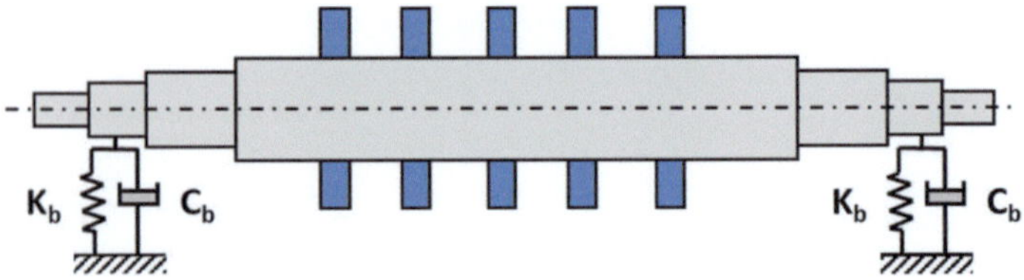

Fig. 12.26 Typical rotor model used to illustrate the rotordynamic analysis procedure

components would be the shaft, impellers or stages, bearings, seals, and other components.

The rotordynamic solver is a software program that contains a mathematical representation of the rotor-bearing system and is used to (1) determine the eigenvalues (natural frequencies) and the corresponding mode shapes, (2) calculate the response of the rotor to a given unbalance, and (3) calculate the stability of the system to external perturbations [17]. Depending on how the rotordynamic solver is built, the rotor is divided into segments or sections, each with a length and diameter. (Note that the mass diameter can be different than the stiffness diameter in the case of shrink fits and line-to-line fits.) Optionally, point masses can be added for some components such as couplings and the thrust disk. A model of a typical system used for rotordynamic analysis is shown in Fig. 12.26.

12.9.2 Steps in a Rotordynamic Analysis

1. The rotor model is discretized (with each segment of the shaft represented by length, diameter, and mass properties).
2. The geometries of the journal bearings (pad geometry, clearance, preload, etc.) are obtained.
3. The geometries of the labyrinth teeth and damper seals (if any) (number of teeth, pitch, clearance, etc.) are obtained.
4. Aerodynamic conditions for the compressor are obtained.
5. The data obtained in steps 1–4 are typically used as inputs to the rotordynamic solver/program to run the following (steady-state lateral) analysis:

    a. Undamped critical speed map/analysis.
    b. Damped eigenvalue analysis.
    c. Unbalance response analysis.
    d. Stability analysis.

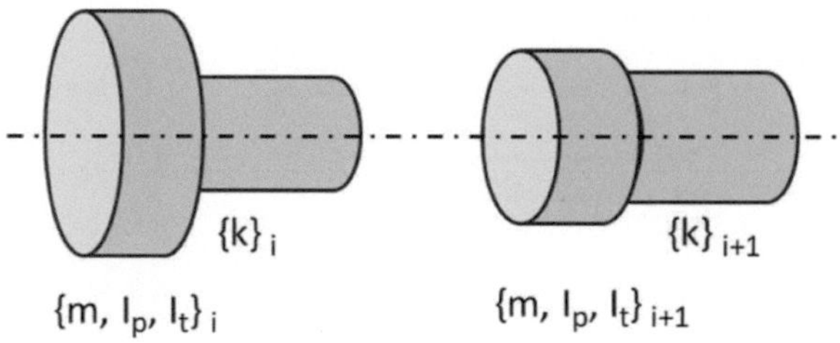

Fig. 12.27 Model of the transfer matrix method

12.9.3 *Methods Used in Rotordynamics*

Two common methods are used in most industrial programs to formulate the rotordynamic equations, namely, the transfer matrix method and the finite-element method.

12.9.3.1 Transfer Matrix Method

The transfer matrix method is the classic method used in rotordynamic solvers. Lund [18–20] was one of the pioneers in the field of rotordynamics. In the transfer matrix method, the rotor is divided into stations or elements. Each element has a uniform diameter and uniform material properties. A state vector is built for each element, consisting of lumped mass and inertia properties at the left side of the element, followed by a massless shaft element that has the stiffness properties. Typically, the matrix for each element is 8×8. Starting with the boundary conditions on one side of the rotor system, the element matrix is transferred from one element to the next until it reaches the other end of the rotor using transfer matrices. The solution is obtained by satisfying all of the boundary conditions at the relevant points. Because the matrix size is a constant, this method is computationally very fast. Figure 12.27 shows a high-level schematic of the transfer matrix formulation.

12.9.3.2 Finite-Element Method

With the advent of computers that can perform computations at a very high rate, the finite-element method has gained popularity in solving static and dynamic engineering problems [21–24]. In this method, the rotor is divided into smaller sections or stations. The element is modeled using distributed properties. Each element is modeled as a line element with an associated mass and stiffness matrix at the nodes (or ends) of the element.

The successive elements are interconnected at the nodes. A second-order differential equation is written for each element or section of the rotor, and the differential equations are assembled in the form of a matrix. The size of the final matrix is dependent on the number of elements or nodes in the rotor model. Hence, the size of the

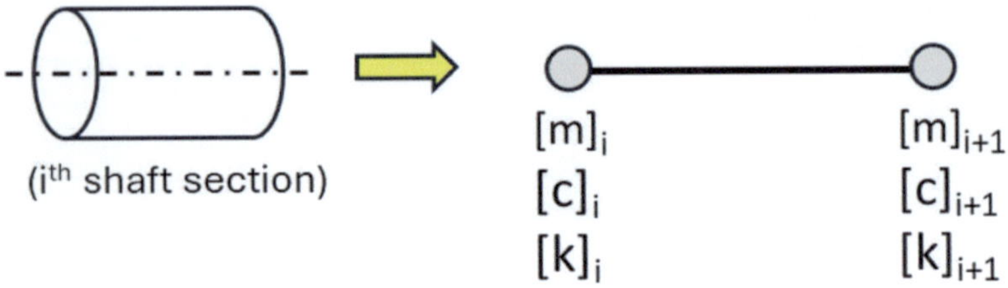

Fig. 12.28 Model of the finite-element method

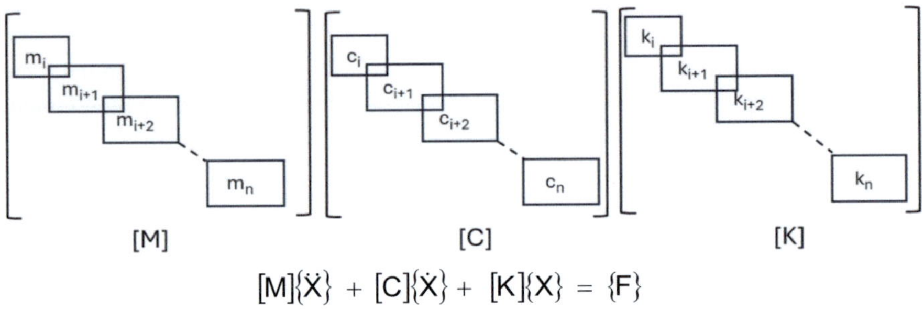

Fig. 12.29 Assembled matrices of the system of differential equations for the finite-element model

matrix can become large, thereby increasing the computational time significantly. However, the finite-element method makes it easier to model parallel shafts, concentric rotors, and support structures than the transfer matrix method. Figures 12.28 and 12.29 show high-level schematics of the concepts of the finite-element method.

The equation in Fig. 12.29 is solved for the eigenvalues by setting $\{F\} = \{0\}$. To solve the unbalance response, the vector $\{F\}$ is set equal to the forcing function.

12.9.4 Undamped Critical Speed Analysis

As the name indicates, the purpose of the undamped critical speed analysis is to find the undamped critical speeds of the rotor-bearing system. Although damping is present in various forms such as fluid-film damping, damper bearings, air friction, and viscous forces, this basic analysis for finding the undamped critical speeds of the system provides insight into the potential locations of the actual (damped) critical speeds and helps in identifying potential problems [25].

Because of the mathematical complexity and duration of solving the equations for rotordynamic response and stability analyses, before the advent of high-speed computers, the undamped critical speed was the only analysis that was done to determine the quality of a machine. Moreover, many manufacturers use the results of the undamped critical speed analysis to compare a new design to their experience.

Hence, even though this analysis does not include damping, it still plays an important role in accessing the rotordynamics of the machine.

12.9.4.1 Undamped Critical Speed Map

The results of an undamped critical speed analysis are shown as a plot of the critical speed of the machine versus the support stiffness on a log–log scale diagram. The basic governing equation for the undamped analysis is (Eq. 12.14)

$$m\ddot{x} + kx = 0. \tag{12.14}$$

Because it is an undamped analysis, there is no damping (c) term in the equation. The right-hand side of Eq. 12.14 is zero because there are no forcing functions.

An example of an undamped critical speed map is shown in Fig. 12.30 [26]. The dynamic stiffness of the journal bearing is given by Eq. 12.15, where K_{xx} is the direct stiffness in the x direction, C_{xx} is the direct damping in the x direction, and ω is the rotational speed in radians per second (rad/s).

$$K_{\text{dyn\_}xx} = \sqrt{K_{xx}^2 + (\omega C_{xx})^2}. \tag{12.15}$$

This bearing parameter (because it contains both the stiffness and damping properties) is used in the undamped critical speed plot. The bearing dynamic stiffness is superimposed on the map to provide an idea of the potential critical speeds of the machine.

12.9.4.2 Generating the Undamped Critical Speed Map

Now consider how the undamped critical speed map is generated and what each curve means.

Step 1

Start with a blank plot (log–log) of stiffness along the x-axis and critical speed along the y-axis, as shown in Fig. 12.31.

Step 2

Replace the journal bearings with a relatively soft spring with a stiffness denoted by K_1. Excite the rotor (e.g., hit the rotor with a hammer). This will excite the rotor, which will vibrate at its natural frequencies. Because the rotor-bearing system is not a single degree-of-freedom system, there are many natural frequencies. Calculate the first four natural frequencies, and plot them on the graph at the x-axis value of K_1 (Fig. 12.32). The vertical line represents the support stiffness K_1, and the four points on this line represent the first four critical speeds.

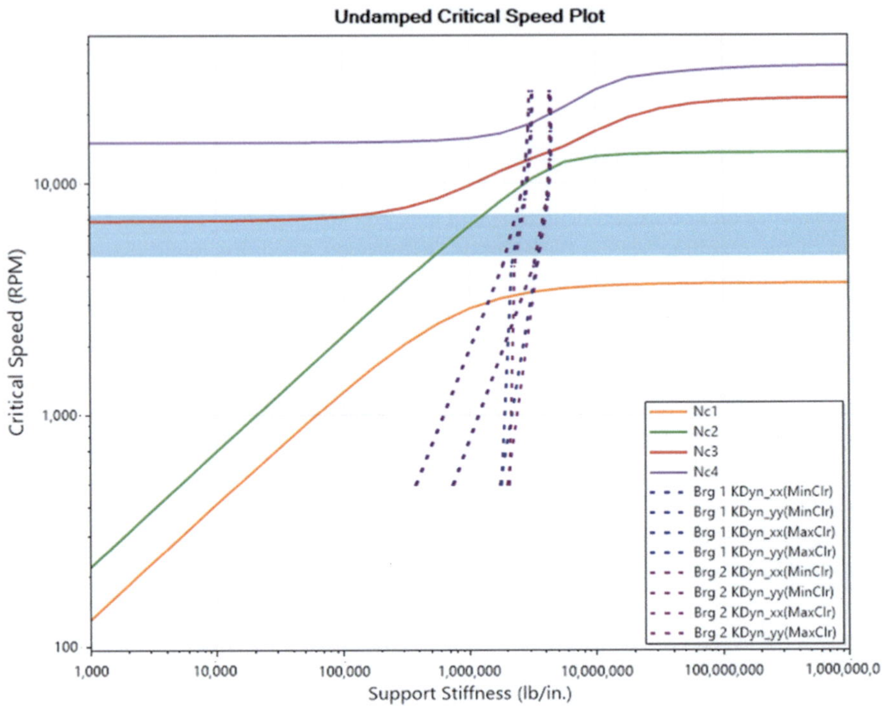

Fig. 12.30 Undamped critical speed map (*Note* Brg = bearing, MaxClr = maximum bearing clearance, MinClr = minimum bearing clearance)

Step 3

Change the support stiffness from K_1 to a stiffer spring with stiffness K_2, and excite the rotor. Calculate the first four natural frequencies, and plot them on the graph at the x-axis value of K_2 (Fig. 12.33).

Step 4

Change the support stiffness from K_2 to a stiffer spring with stiffness K_3, and excite the rotor. Calculate the first four natural frequencies, and plot them on the graph at the x-axis value of K_3 (Fig. 12.34).

Step 5

Change the support stiffness from K_3 to a stiffer spring with stiffness K_4, and excite the rotor. Calculate the first four natural frequencies, and plot them on the graph at the x-axis value of K_4 (Fig. 12.35).

Step 6

Join the lowest four points, then the next set of points, and so on until four curves are obtained. Each curve corresponds to a natural frequency. The lowest curve is

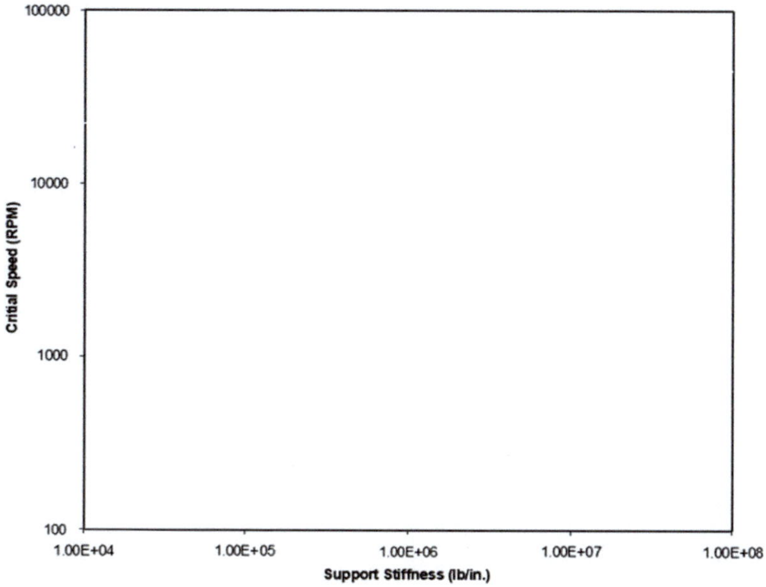

Fig. 12.31 Blank undamped critical speed map

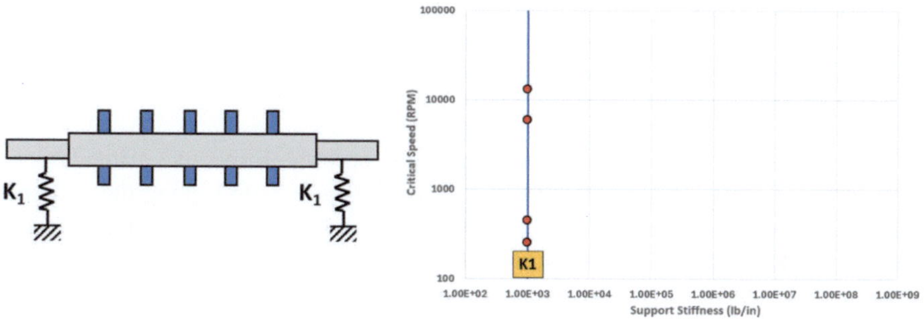

Fig. 12.32 First four critical speeds at stiffness K_1

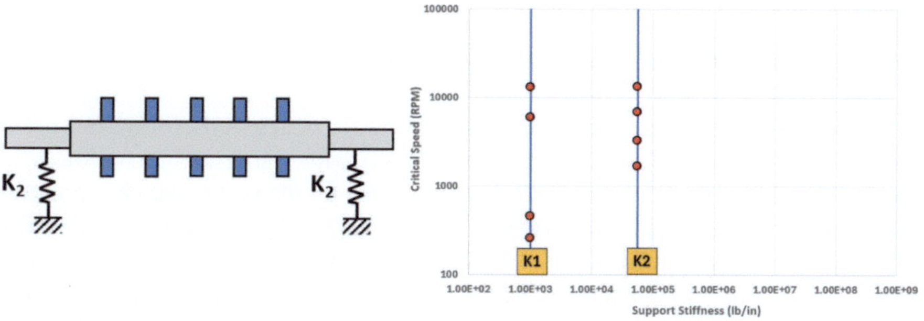

Fig. 12.33 First four critical speeds at stiffness K_2

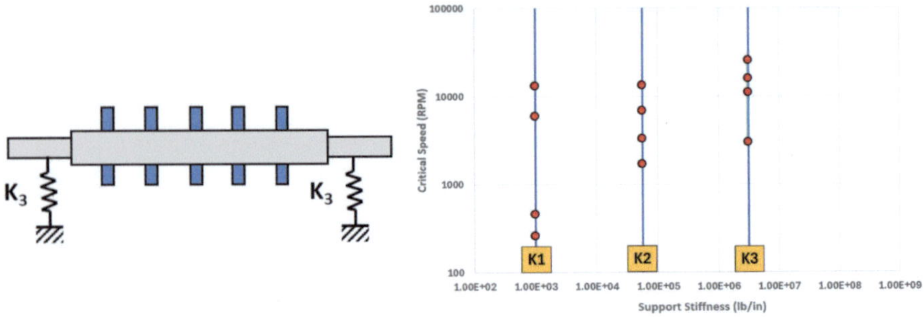

Fig. 12.34 First four critical speeds at stiffness K_3

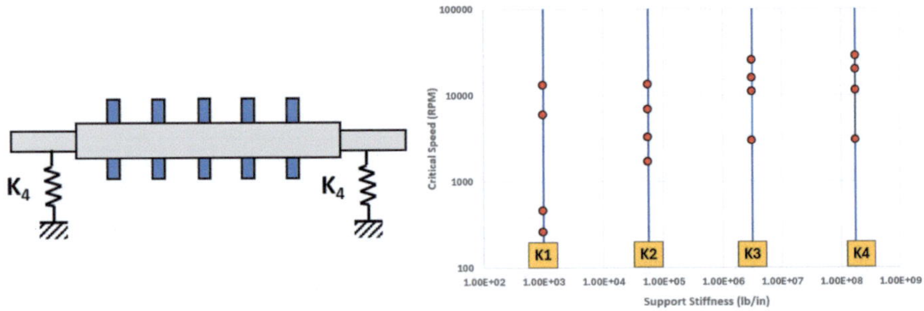

Fig. 12.35 First four critical speeds at stiffness K_4

for the first critical speed. Each curve indicates how the respective critical speed varies with support stiffness. On this plot, draw the curve representing the dynamic stiffness as a function of speed. For the sake of simplicity, only one bearing dynamic stiffness is plotted on the graph. Add horizontal lines for the operating speed range of the machine (Fig. 12.36). The intersections of the bearing stiffness line with the critical speed lines are the potential critical speeds of the machine. This completes the construction of the undamped critical speed map.

12.9.4.3 Understanding the Undamped Critical Speed Map

To understand what is happening to the rotor and the critical speeds, follow one of the curves in Fig. 12.36 from the left end (low stiffness) to the right end (high stiffness) of the map. For example, consider the bottommost line on the graph, which represents the first critical speed and is reproduced in Fig. 12.37. At the left end of the map (point A), the support stiffness is relatively low with respect to the shaft stiffness. Hence, the rotor does not bend and traces the shape of a cylinder as it rotates about the undeformed centerline. This rigid-body motion is called the cylindrical mode.

As the support stiffness increases, the rotor is pulled downward at the bearing locations, as shown by points B and C in Fig. 12.37. As the support stiffness increases

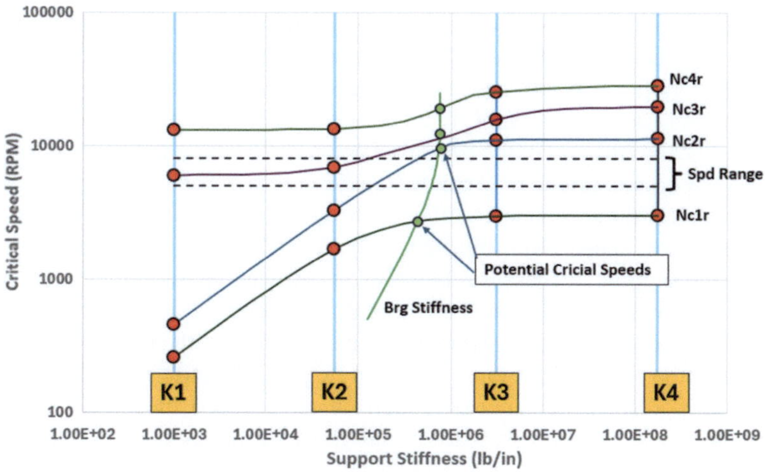

Fig. 12.36 Complete undamped critical speed map

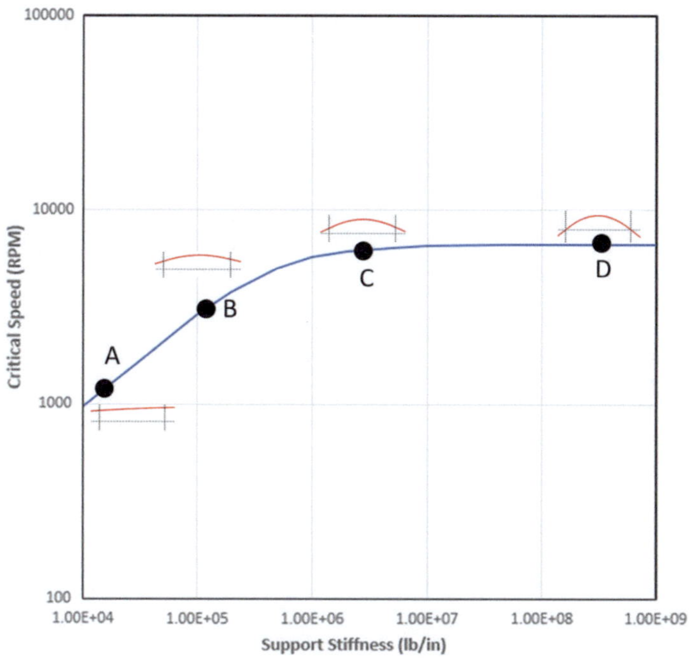

Fig. 12.37 Variation in mode shapes of the first critical speed

to high values (point D), the rotor shows very minimum displacement at the support locations. Between points C and D, the displacement at the bearing locations does not change significantly, and hence, the frequency remains the same. The critical speed at point D is called the first rigid bearing critical speed of the rotor, N_{c1R}.

At point D, the stiffness of the bearings is greater than the shaft bending stiffness. This is an important parameter in the basic design of the machine.

Next, consider the second lowest line on the graph in Fig. 12.36, which represents the second critical speed and is reproduced in Fig. 12.38. At the left end of the map (point A), the support stiffness is relatively low. Hence, the rotor does not bend and traces the shape of inverted cones connected at the ends. This rigid-body mode is called the conical mode. As the support stiffness increases, the rotor is again pulled downward at the bearing locations, as shown by points B and C in Fig. 12.38.

As the support stiffness increases to high values (point D), the rotor has no significant displacement at the support locations. Between points C and D, there is not much displacement at the bearings, and hence, the frequency remains the same. The critical speed at point D is called the second rigid critical speed of the rotor, N_{c2R}.

The third and fourth critical speeds are generally well above the operating range of most beam-style compressors (in which the impellers or stages are between the journal bearings).

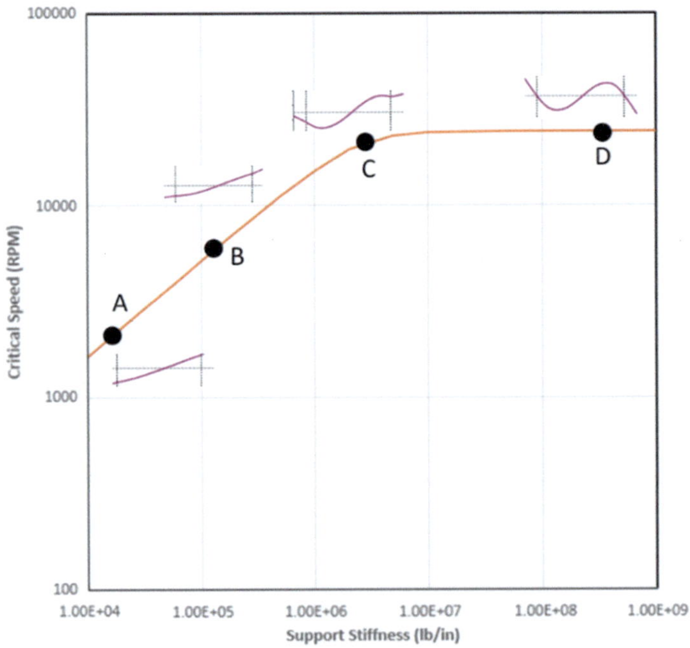

Fig. 12.38 Variation in mode shapes of the second critical speed

12.9.4.4 Significance of the Bearing Characteristic Line

Next, consider the bearing coefficient line that was superimposed on the undamped critical speed map (refer to Fig. 12.36). Figure 12.39 shows a close-up of the bearing dynamic stiffness line on the plot. The dynamic stiffness is given by Eq. 12.15 (repeated here), where K_{xx} is the direct stiffness in the x direction, C_{xx} is the direct damping in the x direction, and ω is the rotational speed in radians per second.

$$K_{\text{dyn\_}xx} = \sqrt{K_{xx}^2 + (\omega C_{xx})^2}.$$

In general, the dynamic stiffness in both directions, $K_{\text{dyn\_}xx}$ and $K_{\text{dyn\_}yy}$, can be different and will depend on the bearing geometry (minimum and maximum values of the bearing clearance). For simplification, the trend in only one dynamic bearing stiffness is shown in the figure. The bearing properties depend on the thickness of the oil film (in addition to the oil viscosity, temperature, etc.).

The thickness of the oil film depends on the rotational speed. Hence, the bearing stiffness and damping depend on the rotational speed. Therefore, the dynamic stiffness of the bearing can be plotted as a function of speed on the undamped critical speed map. The intersections of the line for the dynamic bearing stiffness with the lines for N_{c1} and N_{c2} (cf. Figure 12.36) are the potential critical speeds of the rotor. The values obtained should be close to the actual (damped) critical speed that is shown in an unbalance response plot (cf. Figure 12.4). However, because the plot of the undamped critical speed does not include any damping, these intersection points should be considered only as potential critical speeds.

If these intersection points fall within the operating speed range (horizontal lines for the minimum speed [N_{min}] and maximum continuous operating speed [MCOS] on the plot), as shown in Fig. 12.40, then there could be a potential critical speed in the operating range. However, note that, because damping is not considered here, these modes in the operating range could be partially or completely damped out.

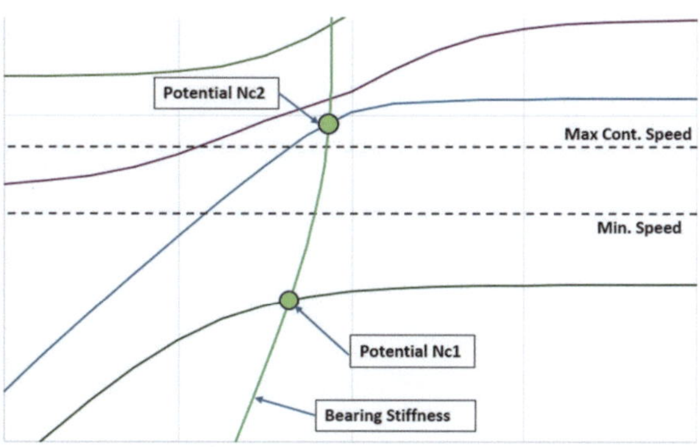

Fig. 12.39 Bearing dynamic stiffness line

Fig. 12.40 Potential critical speed in the operating range

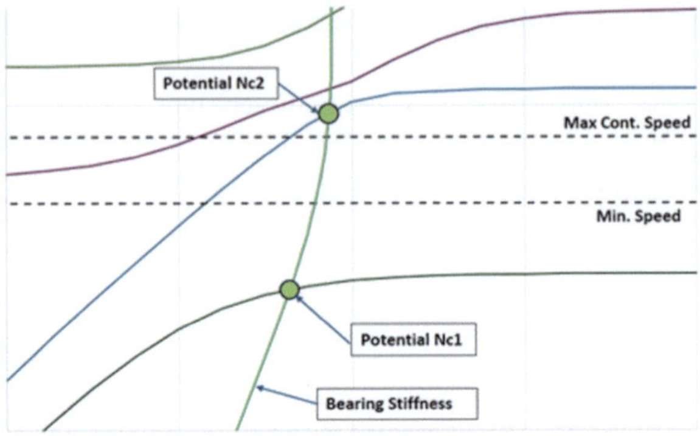

12.9.4.5 Significance of N_{c1R} and N_{c2R}

The first rigid critical speed (N_{c1R}) is an important design and screening parameter. API Standard 617 [1] outlines the level of stability analysis that must be completed based on the average gas density and the flexibility ratio (MCOS/N_{c1R}), as shown in Fig. 12.41. In addition to the API requirement, many manufacturers have their own experience plots based on the flexibility ratio and N_{c1R} that are used to screen the machine before any detailed design work is undertaken.

Generally, most compressors operate between the first and second critical speeds. Hence, it is considered to be good practice to place the bearing characteristics in the sloping portion of a plot of the shaft orbit (refer to Fig. 12.45 later in this chapter). In

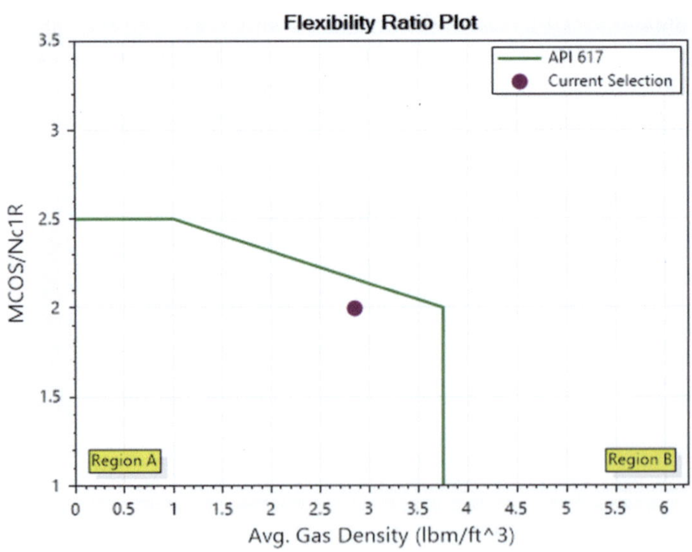

Fig. 12.41 Flexibility ratio plot in accordance with API 617

this way, the bearing design parameters can the used to fine-tune the rotordynamics. If the bearing characteristics are far to the right on the map, that means that the bearings are very rigid and do not control the rotordynamics, which is generally not desirable. Therefore, the undamped critical speed map provides a broad spectrum of insights into the behavior of the rotor and the selection and design of the journal bearings.

12.9.5 Damped Eigenvalue Analysis

The previous section discussed the simplified approach and the process for finding the undamped natural frequencies. This section discusses damped eigenvalue analysis. In all real-world systems, some amount of damping will be present, originating from viscous damping, frictional contact, internal material forces, fluid interaction, windage, and so on. Hence, it is important to determine the natural frequencies of a machine with actual stiffness and damping values. The purpose of this damped eigenvalue analysis is to determine the damped natural frequencies and the corresponding damping (again described in terms of the logarithmic decrement, or log dec) of each eigenvalue [20, 27, 28]. This is a free-vibration problem, meaning the right-hand side (forcing function) is zero. The governing differential equation is (Eq. 12.16)

$$m\ddot{x} + c\dot{x} + kx = 0. \tag{12.16}$$

To understand how this analysis is done, consider a typical centrifugal compressor rotor. The rotor is supported on fluid-film bearings. The speed is varied from zero to a high value (typically 2–3 times the running speed of the machine), in discrete steps. At each speed increment, the natural frequencies of the rotor and the corresponding log dec values are calculated.

Typically, the first six to eight modes are considered. Because of gyroscopic and other effects, each mode has a forward and backward mode (F indicates forward whirl, and B indicates backward whirl) [27]. The output of a typical damped eigenvalue analysis is shown in Fig. 12.42, for the first six modes. The x-axis displays the rotor speed, and the y-axis represents the damped natural frequencies (in rpm).

A synchronous line, for which the x- and y-axis values are the same, is drawn on the plot. The lines representing the minimum operating speed and maximum continuous operating speed (MCOS) are plotted as horizontal lines. The significance of the synchronous line is that the intersections of the synchronous line with the frequency lines indicate the resonance conditions.

From the example in Fig. 12.42, the rotor crosses the first natural frequency at the point labeled A. As the speed of the rotor increases, it eventually reaches the operating speed (the 100% speed of the rotor), as shown in Fig. 12.43. This figure includes only the first two modes to maintain simplicity and to explain the importance of the damped eigenvalue plot.

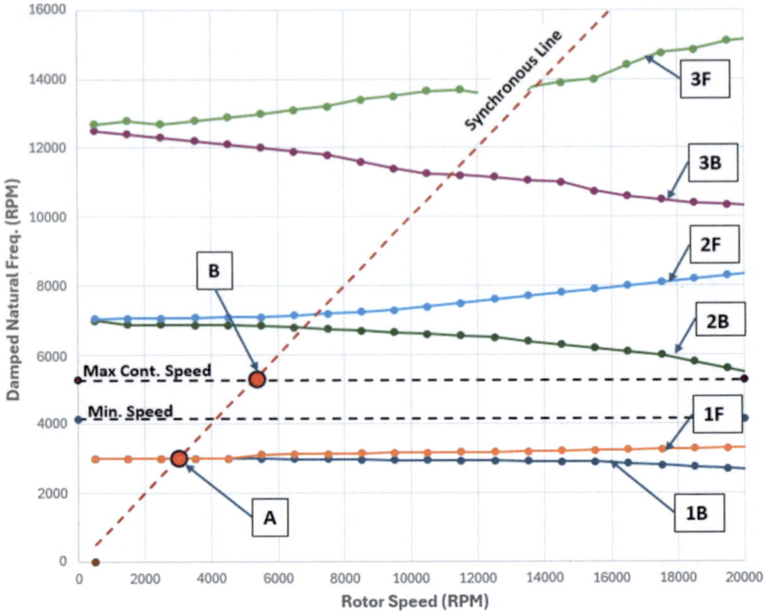

Fig. 12.42 Damped eigenvalue plot

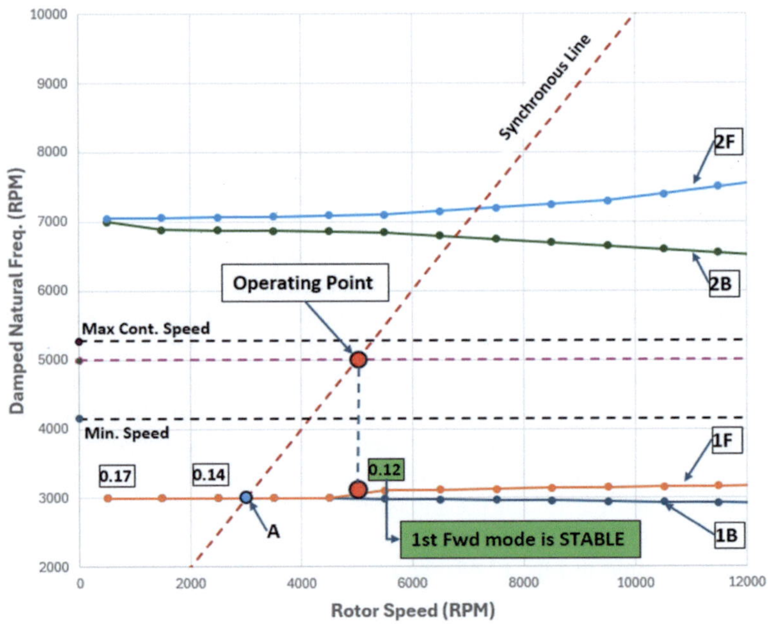

Fig. 12.43 Example of a stable system

When the rotor is operating at the 100% speed (indicated by "operating point" in Fig. 12.43), the first natural frequency has a log dec value of 0.12. This subsynchronous frequency is well damped because the log dec value is greater than zero. This is an example of a stable system.

Now consider a system that has low stability. Figure 12.44 shows a system that could potentially be unstable. As the rotor begins to rotate from zero speed, it crosses the first natural frequency indicated by the point A. In this example, at A, the rotor system has a positive log dec value of, say, 0.05, though the system is barely stable. However, when the rotor reaches the operating point, the first natural frequency has a log dec value of − 0.11.

This indicates a potentially unstable system. This means that, when the rotor is running at the 100% speed, there may be a subsynchronous frequency that is equal to the first natural frequency and has a negative log dec value. The conclusion is that the rotor supported on bearings could be inherently unstable without any forcing functions. Hence, this plot would provide an indication of the potential instability of the rotor, which might prompt design changes at an early stage.

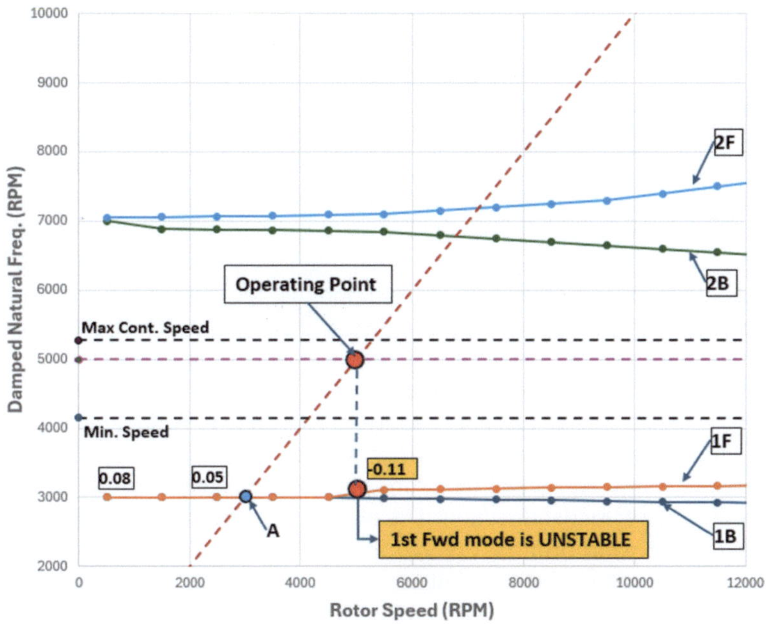

Fig. 12.44 Example of a system with negative log dec value at the first natural frequency

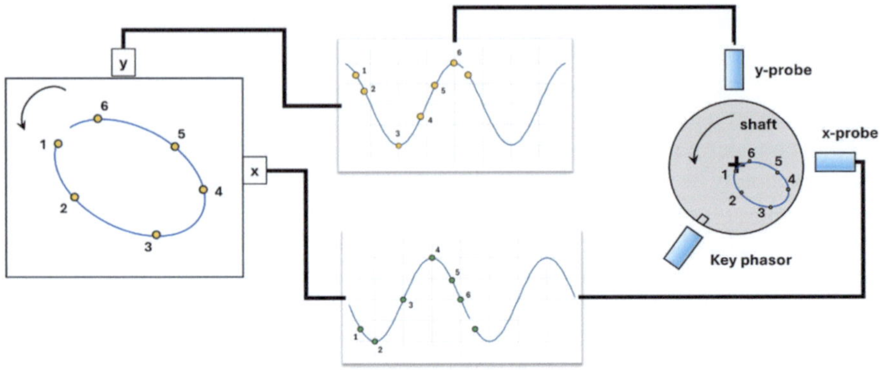

Fig. 12.45 Probe and associated instrumentation showing the shaft orbit

12.9.6 Unbalance Response Analysis

During the design of turbomachinery, the critical speeds of the machine need to be evaluated. In accordance with API requirements, the critical speeds closest to the operating range must be calculated along with other parameters. Most industrial centrifugal compressors operate between the first and second critical speeds, and some centrifugal compressors operate well above the first critical speed of the rotor system.

In cases where the machine is operating between the first and second critical speeds, it is necessary to calculate the location of the (first) critical speed, through which the rotor will pass to reach the operating speed, and to calculate the peak amplitude at the critical speed [29]. In addition, the location of the second critical speed must be calculated. According to API 617 for compressors, there are three attributes that need to be calculated: the critical speed, the amplification factor, and the separation margin. Figure 12.4 (refer to Sect. 12.2.4) shows the parameters that need to be calculated from a response analysis according to API 617 [1].

Typically, a centrifugal compressor has two sets of probes located close to the two journal bearings. At each of the probe planes, there are two probes located 90° apart in the circumferential direction. The two probes are required to ensure that the magnitude of the whirl orbit is captured in two orthogonal directions. The probes with the instrumentation and the orbit are shown in Fig. 12.45.

12.9.6.1 Definitions of Terms on a Response Plot

The response diagram is a plot of rotor speed versus vibration amplitude (at steady state), as measured at the probe location on the machine. This diagram is created for a given unbalance configuration. In accordance with API 617 [1], the definitions of the various parameters that need to be evaluated are provided in the following sections.

Amplification Factor

The amplification factor is calculated at the peak(s) of the response plot. It provides an indication of the sharpness of the peak. The half-power bandwidth method is used to determine the amplification factor (AF) as (Eq. 12.17)

$$\mathrm{AF} = \frac{N_{\mathrm{peak}}}{N_2 - N_1}, \tag{12.17}$$

where N_{peak} is the speed corresponding to the peak amplitude. In accordance with the definition of the half-power bandwidth (or half-power point) method, the half-power point is when the peak drops by 3 dB (on a logarithmic scale). This translates to a point that is equal to $N_{\mathrm{peak}}/\sqrt{2}$ on a linear scale. The two half-power points N_1 and N_2 can be obtained by drawing a horizontal line on the curve at $0.707 \times N_{\mathrm{peak}}$ (where 0.707 is equal to $1/\sqrt{2}$); the intercepts are the speeds N_1 and N_2. The amplification factor is dimensionless number. The higher the number, the sharper the peak.

Critical Speed

According to API 617, all peaks on the response plot are not necessarily critical speeds. The term critical is applied if the amplification factor at the peak is greater than or equal to 2.5. If the amplification factor exceeds 2.5, then the peaks must have a minimum separation margin from the nearest operating speed.

Separation Margin (for Critical Speeds Below the Operating Range)

If the amplification factor is greater than or equal to 2.5, the critical speed is below the minimum operating speed, and the minimum required separation margin (SM) is given by Eq. 12.18

$$\mathrm{SM} = 17\left(1 - \frac{1}{\mathrm{AF} - 1.5}\right). \tag{12.18}$$

Required Separation Margin (for Critical Speeds Above the Operating Range)

If the amplification factor is greater than or equal to 2.5, the critical speed is above the maximum continuous operating speed, and the minimum required separation margin (SM) is given by Eq. 12.19

$$\mathrm{SM} = 10 + 17\left(1 - \frac{1}{\mathrm{AF} - 1.5}\right). \tag{12.19}$$

The following sections explain the method used to find the first and second critical speeds for a machine whose operating speed range is between the first and second critical speeds. Typically, to plot the response diagram (at the probe locations), the rotor is supported on fluid-film bearings. An unbalance is added at the appropriate location, and the speed of the rotor is varied from 0 rpm to a high value (usually 2–3 times the MCOS). At each speed, the amplitudes at the probe locations and any other points of interest (e.g., the midspan) are calculated. The amplitude is plotted as a function of speed at the designated locations.

12.9.7 Finding the First Critical Speed

According to the preceding discussion of critical speeds and mode shapes, the first critical speed tends to bend the rotor in the form of a bow. Hence, to simulate the first bending mode, an unbalance is added at the midspan of the rotor in the analytical model. The magnitude of the unbalance is specified in API 617 [1]. This unbalance creates a centrifugal force of $F = mr\omega^2$, where m is the mass of the rotor, r is the offset between its center of mass and the geometric center, and ω is its angular frequency. This force pulls the rotor in the direction of the force, hence simulating the first mode. The speed of the rotor is varied, and at each speed point, the amplitude is calculated.

Figure 12.46 shows a schematic of a rotor model with a midspan unbalance. Figure 12.47 [26] shows the response plot of the vibration amplitude at the probe location (e.g., the intake-end probe). Note that the bearing properties (stiffness and damping) are functions of speed and are calculated at every speed point. In this example, the bearing properties in the horizontal and vertical directions are almost identical, and hence, the two probes (x and y) have similar amplitudes. Figure 12.48 [26] shows the phase diagram at the same location. A phase change of 180° can be seen as the rotor passes through the critical speed. Figure 12.49 [26] shows the three-dimensional mode shape at the critical speed of 3300 rpm.

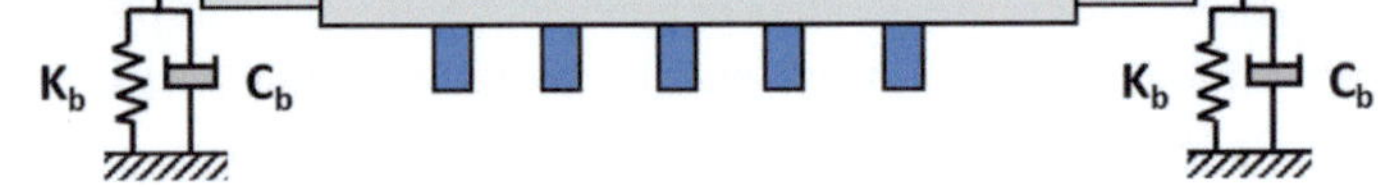

Fig. 12.46 Simulation of the first critical speed: midspan unbalance

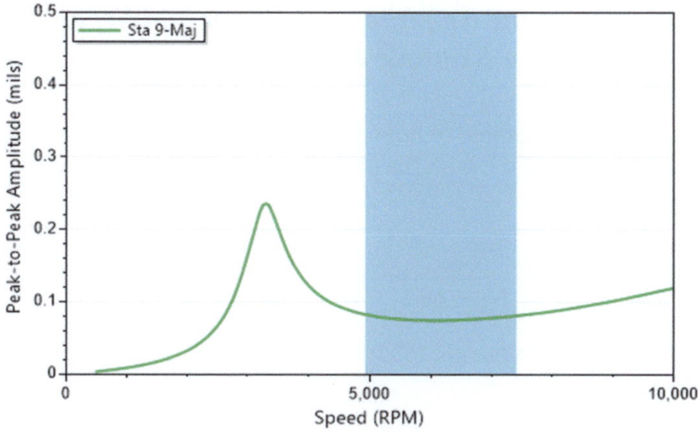

Fig. 12.47 Rotor response plot at the intake-end probe for a midspan unbalance

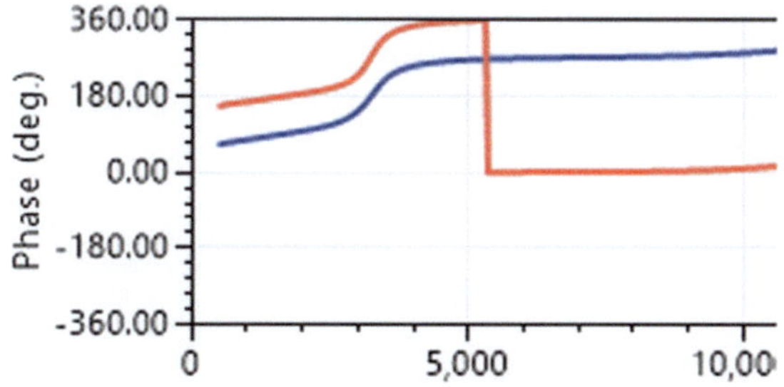

Fig. 12.48 Phase plot at the intake-end probe for a midspan unbalance

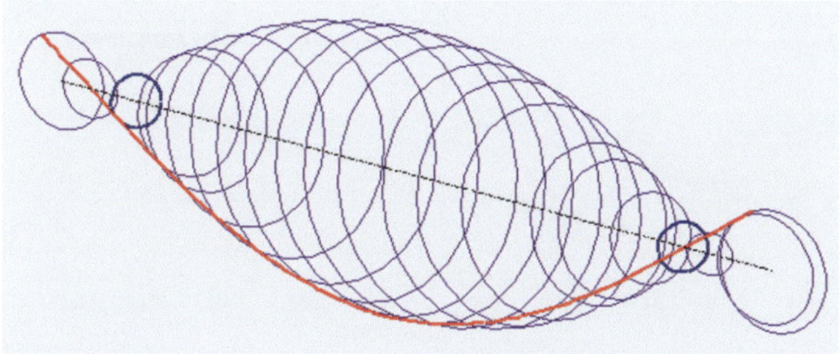

Fig. 12.49 Three-dimensional deflected shape of a rotor at its first critical speed

12.9.8 Calculating the Amplification Factor and Separation Margin for an Example Rotor

Figure 12.50 shows an example of how to calculate the amplification factor for the response shown in Fig. 12.47, with the peak shown in a close-up view.

From the figure

$$\mathrm{AF} = \frac{N_{\text{peak}}}{N_2 - N_1} = \frac{3300}{3700 - 3300} = 4.714.$$

The required separation margin (SM) for the first critical speed can be calculated using Eq. 12.18 as

$$\mathrm{SM} = 17\left(1 - \frac{1}{\mathrm{AF} - 1.5}\right) = 17\left(1 - \frac{1}{4.714 - 1.5}\right) = 11.7\%.$$

Hence, the required (minimum) separation margin of the first critical speed is 11.7% according to API. From Fig. 12.47, the actual separation margin is 32%. Therefore, the separation margin meets the API requirements.

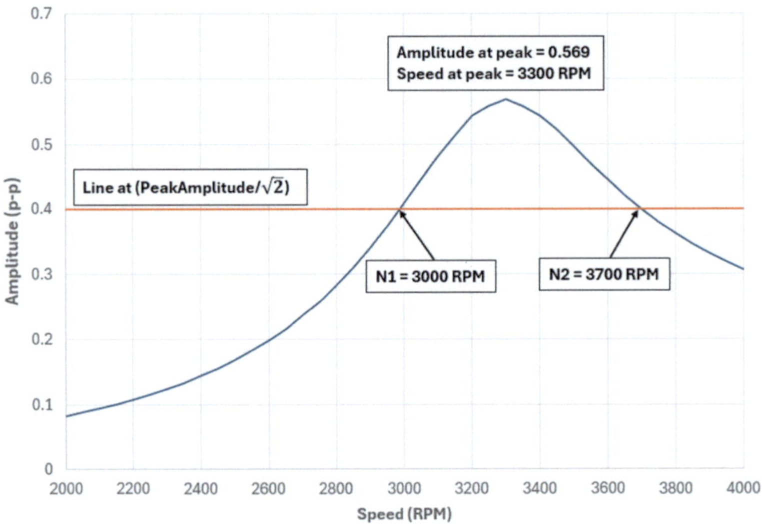

Fig. 12.50 Calculation of the amplification factor

12.9.9 *Finding the Second Critical Speed*

According to the discussion of critical speeds and mode shapes, the second critical speed tends to bend the rotor in the form of a full sine wave. Hence, to simulate the second bending mode, unbalances are added, out of phase, at one-quarter span and three-quarters span of the rotor in the analytical model. The magnitude of the unbalance is specified in API 617. This unbalance creates a centrifugal force of $F = mr\omega^2$ that pulls the rotor in the direction of the force, hence simulating the second mode. The speed of the rotor is varied, and at each speed point, the amplitude is calculated. Figure 12.51 shows the rotor with the unbalance locations. Figure 12.52 [26] shows the response of the rotor at the probe location, and Fig. 12.53 [26] shows the three-dimensional mode shape at the second critical speed.

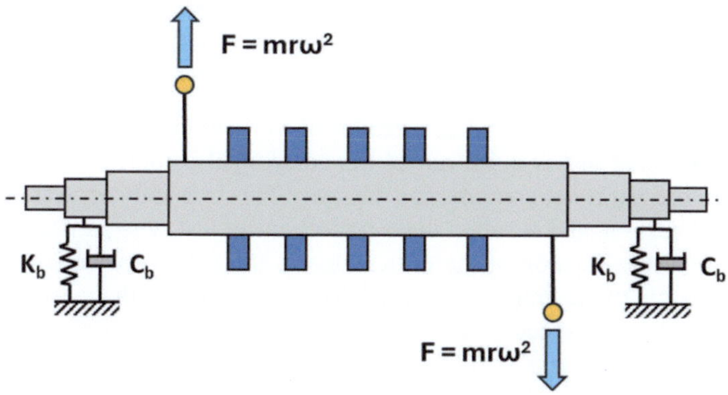

Fig. 12.51 Simulation for of the second critical speed: quarter-span unbalances

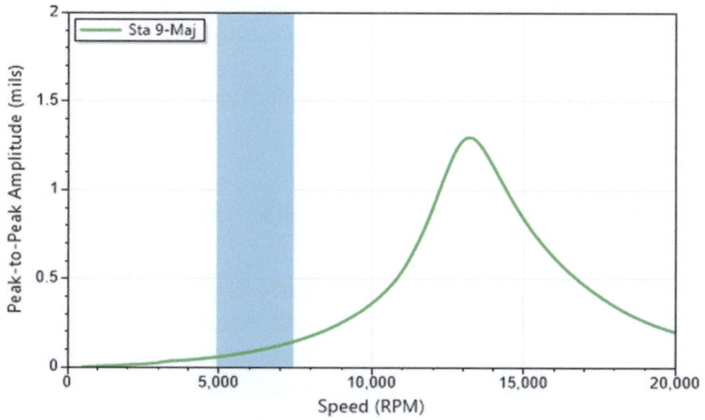

Fig. 12.52 Rotor response plot at intake-end probe for quarter-span unbalances

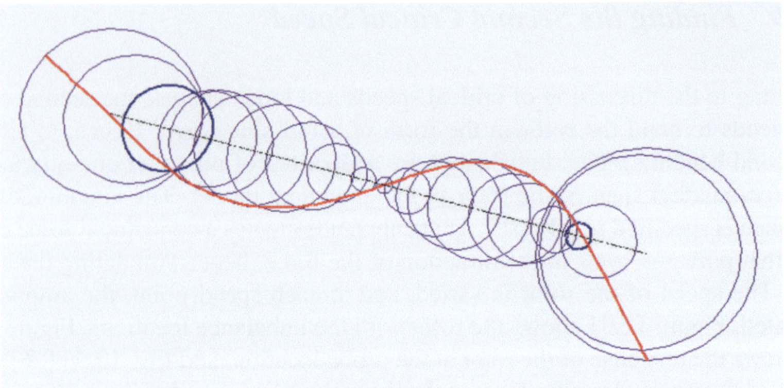

Fig. 12.53 Three-dimensional deflected shape of a rotor at its second critical speed

12.9.9.1 Test Stand Verification

API Standard 617 requires that a mechanical verification test be performed. A mechanical running test is a 4-h run of the centrifugal compressor under no load (i.e., the compressor is not compressing any gas). This is usually done at the running speed of the machine and is called the 4-h mechanical test. The purpose of this test is to determine the location of the critical speed, the amplification factor, and the separation margin of the first critical speed. The description of the mechanical test is given in API 617. The speed is increased from 0 rpm to the running speed, and the speed is held along the run at different points for a finite amount of time to obtain the steady-state response.

For centrifugal compressors, the unbalances at the midspan and quarter-span were used on the model to analytically calculate the critical speed, amplification factor, and separation margin. On the test stand, unbalances cannot be located at the midspan or quarter-span. Hence, the most common place to locate an unbalance is in the coupling. The amount of unbalance to be used is given in API 617. Figure 12.54 shows an example of the test stand results of a mechanical test [30].

12.10 Effect of Temperature on Elastic Modulus

The modulus of elasticity is a mechanical property of solids that gives a material its stiffness properties. The modulus of elasticity of a material depends on the temperature. At a microscopic level, when a material's temperature increases, the vibrations of the atoms increase, which increases the distance between atoms in the crystal structure. Hence, the internal forces among atoms decrease, leading to a decrease in the modulus of elasticity. In turn, this decrease in elastic modulus reduces the critical speed of machinery comprised of this material. Depending on the temperature inside

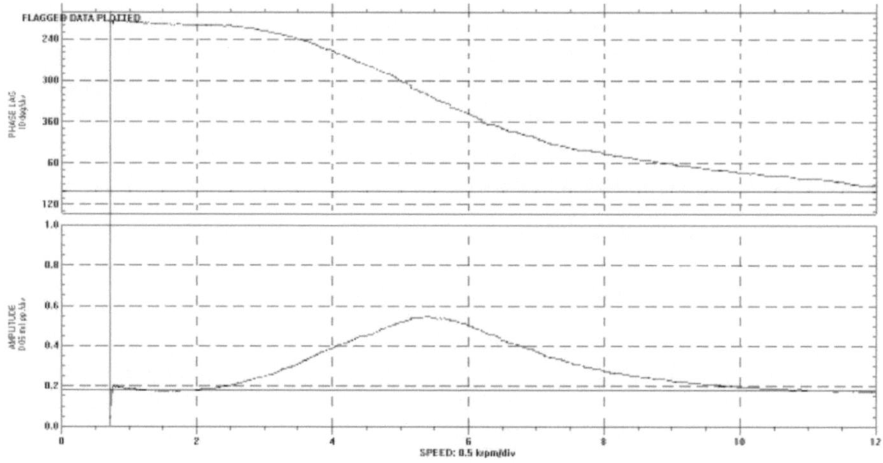

Fig. 12.54 Example of the results of a mechanical running test of a centrifugal compressor

the machine, the change in critical speed could be significant. For example, in the case of steam turbines, the temperature that the shaft encounters is extremely high (anywhere between 500 and 800 °F [between 260 and 427 °C]), and hence, the change in the modulus of elasticity due to the high temperature needs to be considered in the calculation of the machine's critical speeds.

12.11 Rotordynamic Stability

One of the most important parameters to consider in rotordynamic design of a rotating machine is the stability of the system. This is important because the machine (for example, a centrifugal compressor) is subjected to constant excitation forces coming from the gas being compressed. Rotordynamic stability is the study of the reaction of a rotor to external perturbations. With the push toward higher speeds, higher powers, and smaller machines, stability needs to be examined closely because it might limit a machine at these higher speeds and powers. Instability in rotating machinery usually manifests as a subsynchronous vibration that could initiate the shutdown of the machine. This could cause a great deal of downtime and turn out to be expensive because of potential redesigns required to solve the instability issue. Hence, it is extremely important to understand the underlying principles of rotordynamic stability.

In the field of rotordynamics, one of the more complicated analyses (to model) is stability. Extensive work has been done in this area [31–33]. Even though the modeling of the various parameters becomes complex, methods are being used to understand the nature of the problem and mitigate the risks.

In a compressor, potential sources of instability have been identified. However, this does not mean that, if a machine contains one or more of these sources, the machine is automatically unstable. Rather, the stability of the machine greatly depends on the magnitude of the destabilizing forces and the countermeasures that have been employed in the design. The following list notes the most common (but certainly not all) sources of instability in compressors:

- Aeroflow-path-induced instability (impeller–diffuser interactions).
- Mechanical rub.
- Seals (labyrinth seal, honeycomb seals, damper seals, etc.).
- Plain sleeve bearings.

As discussed in the section on journal bearings (Sect. 12.6), a plain sleeve bearing could potentially cause the rotor to become unstable as a result of the cross-coupling forces generated in the oil film. With the use of tilting-pad journal bearings, oil film-induced instability is currently no longer a major source of instability.

To quantify stability, the logarithmic decrement (shortened to log dec) is used. As mentioned in Sect. 12.2.8, log dec is defined as the natural logarithm of ratio of any two successive amplitudes of a machine's vibration signature. The definition of log dec (repeated here) is shown in Fig. 12.55. If the subsequent peak has a smaller amplitude, the log dec value is positive, and hence, the system is stable. Based on the vibration signature, the stability of a machine can be classified as stable, neutrally stable, or unstable, as shown in Fig. 12.56.

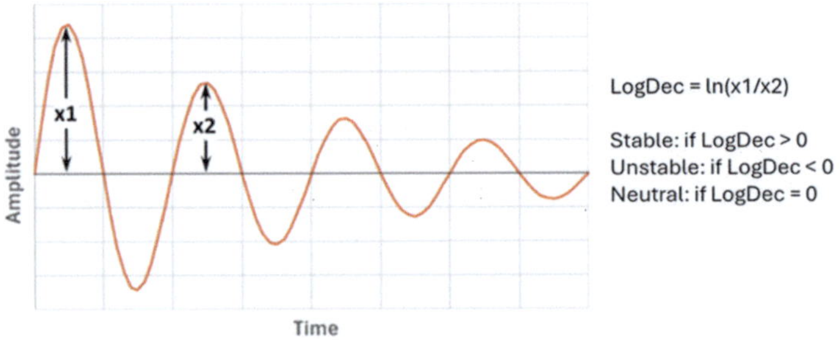

Fig. 12.55 Definition of the logarithmic decrement (log dec)

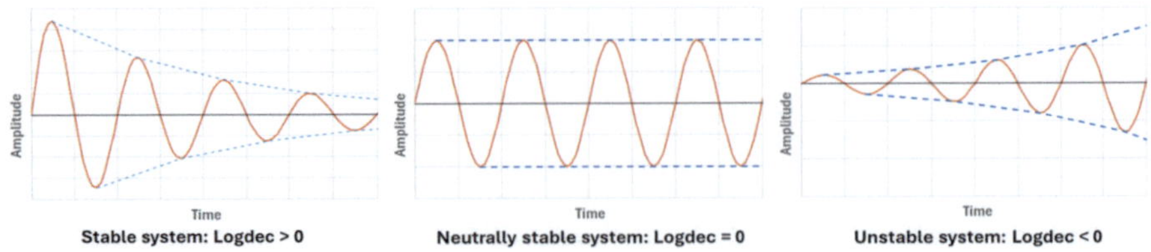

Fig. 12.56 Classification of the stability of a system

12.11.1 Aerodynamic Cross-Coupling: Self-Excited Vibration

As mentioned in the section on journal bearings (Sect. 12.5), cross-coupled forces lead to a decrease in the stability of a system. One of the major sources of cross-coupling in a compressor is the labyrinth seal. The gas flowing under the relatively tight clearance space of a toothed labyrinth seal results in cross-coupling forces on the rotor. As a result of the cross-coupling forces, the rotor exhibits nonsynchronous vibrations. These are self-excited vibrations because they result from the whirling motion of the rotor. Although journal bearings could be a source of cross-coupling, with the use of tilting-pad journal bearings, the cross-coupling forces of the bearings are almost negligible.

Modeling of the aeropath and the flow is highly complex. Hence, based on experimental data and experience, each OEM has developed its own method for determining stability and assessing the goodness of the machine. API 617 outlines an empirical formula (based on the modified Wachel number) [1, 34] for calculating the anticipated aerodynamic cross-coupling forces (denoted by q_{a}) for each impeller of a compressor, which is given by Eq. 12.20

$$q_{\mathrm{a}} = \frac{\text{power} \times B \times C \times \text{discharge density}}{\text{speed} \times \text{impeller tip diameter} \times \text{tip width} \times \text{suction density}}, \tag{12.20}$$

where $B = 3$ and $C = 63$ in US customary units ($C = 9.55$ in SI units).

This anticipated cross-coupling could be viewed to have the same effect. It is a dynamic phenomenon, caused by the gas moving in the compressor, and hence, it exists all the time when the machine is running. Given this cross-coupling (destabilizing force), a rotor-bearing system would need to be designed to overcome this existing cross-coupling force.

12.12 Stability Analysis

API 617 describes how to perform the stability analysis of a compressor. Given the total cross-coupling, Q_{a}, of the rotor (which is the sum of the q_{a} values for all of the stages), the aim of the stability analysis is to determine the amount of cross-coupling that the system can withstand before becoming unstable (denoted by Q_0), that is, before the log dec value of the system goes from positive to zero. Of course, if the log dec value of the system is less than or equal to zero to begin with, then the system is inherently unstable and will need to be redesigned accordingly.

To obtain the maximum cross-coupling the system can withstand, an "artificial" cross-coupling is applied at the midspan of the rotor (for analysis purposes only; no physical bearing is added at this location), as shown in Fig. 12.57. The rotor is supported on actual bearings, and the analysis assumes that the running speed is MCOS.

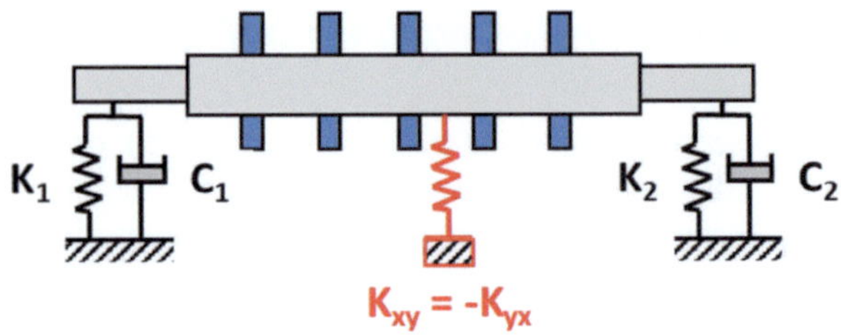

Fig. 12.57 Cross-coupling applied at the rotor midspan to determine rotor stability

The cross-coupling that is applied at the rotor midspan is varied from zero to a high value. At each cross-coupling value, the rotating machine is excited, and the log dec value of the first forward mode is recorded. Because the first mode is the lowest-energy mode, in most cases, the first forward mode is the one that becomes unstable. Figure 12.58 shows a plot of the log dec value of the first forward mode as a function of the cross-coupling applied at the rotor midspan.

As the cross-coupling increases, the log dec value decreases. The reason for this is that, as is known, direct damping (C_{xx}, C_{yy}) of a system helps control a system's vibrations. In other words, the exponential decay in Fig. 12.55 (for a stable system) is due to the damping in the system (Eq. 12.21).

$$C_{\text{eff}} = C_{xx} - \frac{K_{xy}}{\omega}. \tag{12.21}$$

Also, from Eq. 12.21, the cross-coupling forces (i.e., K_{xy}) reduce the effective damping. The effect of the cross-coupling forces in terms of the vibration signature is shown in Fig. 12.59.

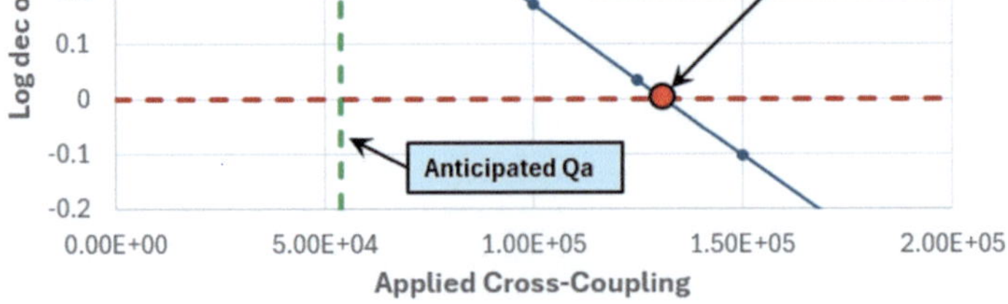

Fig. 12.58 Plot of cross-coupling versus log dec of the first mode

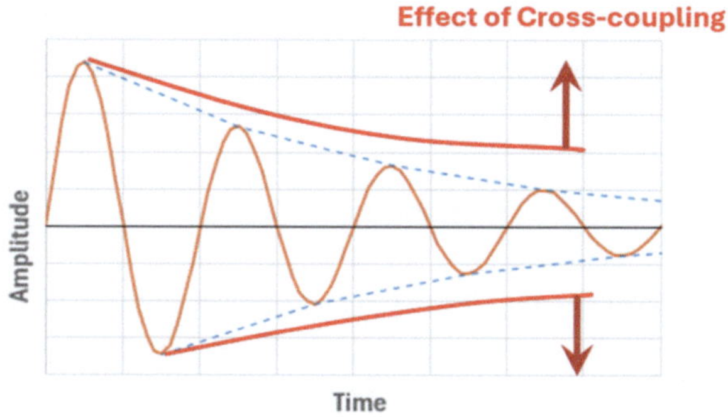

Fig. 12.59 Effect of increasing the cross-coupling forces

The aim of this stability analysis is to determine (a) the threshold cross-coupling at which the log dec value becomes zero and (b) the log dec value at Q_a (the anticipated cross-coupling). This value is denoted by Q_0. The ratio Q_0/Q_a gives the safety factor for the stability of the rotor system.

In general, any cross-coupling in a system acts as a destabilizing force and reduces the effective damping. API 617 requires that all forms of excitation need to be included in the stability analysis. In compressors, the most common form of excitation (cross-coupling) comes from toothed labyrinth seals. These seals are used to restrict leakage and recycle flow, specifically at the impeller cover locations, the interstage locations, and the balance drum locations. Because toothed labyrinths run at tight clearances, the gas flowing in the small space between a stationary surface and a rotating surface creates cross-coupling forces that could act to destabilize the system. The expected log dec value at Q_a can be found using the log dec plot in Fig. 12.58, which includes all of the sources of excitation. API 617 states that, for a machine to be API-compliant, this value should be greater than or equal to 0.1. Each OEM could also have their own minimum limits, which may be larger than the API minimum requirement.

12.13 Methods to Improve Stability

In cases in which the stability is lower than expected or does not meet API requirements or the design criteria of the customer, purchaser, or manufacturer, the design of the machine might have to be changed, for example, through an increase in the shaft diameter or in the frame size of the machine. Such design changes could add a significant cost to the project. However, there are methods that can be used to improve stability without increasing the machine size.

One of the main parameters that governs stability is the effective damping. There are two ways that stability can be improved: (1) decreasing the source of instability (i.e., reducing the cross-coupling forces) and (2) adding positive damping to the system (i.e., increasing the direct damping forces, C_{xx} and C_{yy}). This section outlines the most common methods and hardware used to improve stability.

12.13.1 Swirl Brakes on Seals

Labyrinth seals are used extensively in turbomachinery. In compressors, they are used in the aeroflow path (impeller cover, balance drum, etc.) to reduce recycle leakage that would add to internal losses (which manifests as an increase in power), because more work needs to be done to compress this additional flow. In addition to damping, labyrinth seals inherently produce cross-coupling forces that could act as destabilizing forces.

The longer (and more teeth on) the seal, the higher the cross-coupling forces. One of the significant reasons for the high cross-coupling forces of labyrinth seals is due to the inlet swirl of the gas entering the high-pressure side of the seal. Swirl or the swirl ratio is defined as the ratio of the tangential velocity of the gas to the circumferential velocity of the shaft. This inlet swirl can be reduced by using swirl brakes at the inlet of the labyrinth seal [35]. These swirl brakes act as "gates" to guide the gas in the axial direction and remove or at least reduce the tangential component of the velocity. Gas flows with high entrance swirl and low entrance swirl are shown in Fig. 12.60 [26]. Even though the swirl brake is in the aeroflow path, it has a negligible effect on the aeroflow performance. However, it has a significant benefit in improving the rotor stability [36].

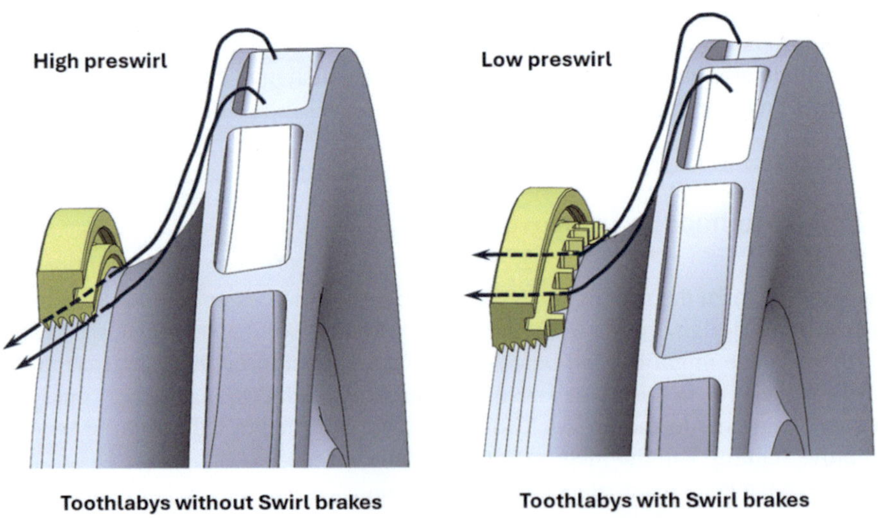

Fig. 12.60 High swirl and low swirl at the entrance of the impeller toothed labyrinth seal

Fig. 12.61 Damper seal with swirl brakes

The use of swirl brakes provides a benefit in improving the stability of the machine. It is highly cost-effective and does not require much axial space. Hence, in most cases, swirl brakes can be retrofitted into existing or new machines to improve stability. The swirl brakes act by cutting down the source of instability, namely the cross-coupling forces developed by the labyrinth seals.

12.13.2 Damper Seals

A damper seal (also known as a hole pattern seal) is a type of seal that can be used instead of a toothed labyrinth seal. In most cases, damper seals are used at the balance drum (or balance piston) location. The damper seal is made of solid metal (usually aluminum) with holes at the inner surface. A damper seal with swirl brakes is shown in Fig. 12.61 [37]. When operating under a pressure differential, the holes in damper seals trap gas and create a damping effect. Although damper seals do produce stiffness and cross-coupled stiffness, the significant amount of direct damping generated by damper seals outweighs the effects of the cross-coupling forces. Swirl brakes are typically applied at the front or high-pressure end of damper seals (leading edge where the flow enters the seal) to reduce the swirl of the incoming gas.

Although damper seals offer improved rotor stability, they may not be applicable in all cases. In general, a damper seal could have a higher gas leakage than a labyrinth seal, which might impact aeroflow performance. The challenge in such cases is to strike a balance between the aeroflow performance and the rotor stability. A damper seal increases the stability by adding damping to the system [35, 38].

12.13.3 Damper Bearings

Another component that improves stability by adding direct damping to the system is the damper bearing [39–43]. Because oil, on account of its viscosity, naturally has damping properties, the principle of a damper is to add a film of oil at a location on the rotor so that the oil film generates direct damping. The most common location for adding a damper bearing is behind (in series with) a journal bearing. Damper bearings are also known as squeeze film dampers. Although there are different designs of damper bearings, the underlying principle is the same, as described next.

Figure 12.62 shows a cross section of a damper bearing [44, 45]. The damper bearing consists of adding a spring or damper element in series with the journal bearing. The small axial gap between the cage and the housing allows the cage to move freely in the housing. The cage is prevented from rotating by antirotation pins. The cage is supported by O-rings or mechanical springs. This ensures that the oil film is trapped in the clearance space between the cage and the bearing housing. The O-rings provide stiffness to the damper bearing in the horizontal and vertical directions (perpendicular to the axis). As for a journal bearing, the parameters of a damper bearing (clearance, distance between O-rings) can be varied to tune the damper for each application. For functionality, centering of a squeeze film damper is important. Different designs of centering mechanisms are used in practice, including O-rings, additional metal springs, or integral designs of squeeze film dampers.

The lube oil (same as the oil supply to the journal bearings) is supplied at a high pressure into the annular gap of the damper through a feed groove usually at the midpoint between the O-rings. When movement occurs at the journal bearing as a result of vibration, the bearing cage moves, squeezing the oil film in the damper. This generates direct damping that reduces (or dampens) the vibration level. The damper bearing not only reduces the amplitude and amplification factor of the response as a machine going through a critical speed, but also adds a significant amount of direct damping that improves the stability of the machine. The rotor-support system is

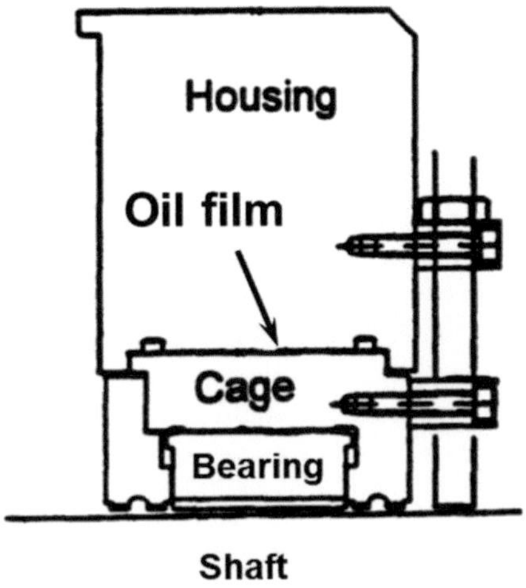

Fig. 12.62 Damper bearing in series with a journal bearing

softened, resulting in less shaft bending at the first critical speed. In many cases, this mode becomes a critically damped translatory or cylindrical mode.

Because of their advantages in improving machine stability and response, squeeze film dampers have been used widely in nearly all aircraft gas turbines (typically in combination with roller element bearings) and many other types of turbomachinery [44], such as centrifugal compressors. The use of damper bearings has allowed compressors to push the limits of operation, enabling operating at higher speeds and higher powers.

12.14 Thermal Instability: The Morton Effect

An instability need not occur at nonsynchronous frequencies. One instability that occurs at the synchronous speed of the rotor is thermal instability [46]. This phenomenon was first studied by Morton and Keogh [47, 48] and, hence, is also known as the Morton effect. This synchronous instability is a result of the temperature gradient at the journal surface, which is due to viscous shearing within lubricant film. To obtain an understanding of the Morton effect, assume that the rotor is acted upon only by unbalance forces (which is true in most cases because of the residual unbalance).

When the shaft is whirling in the journal bearing clearance space, as shown in Fig. 12.63, there is one point on the journal that is at the minimum film thickness and the opposite point (180° from the minimum film thickness) is at the maximum film thickness. Under specific load and speed conditions, as the shaft goes around in the orbit, these points always remain the same; that is, the point on the journal with minimum thickness remains the same, and the point on the journal with maximum thickness remains the same. Because the point of minimum thickness is associated with high viscous shear stress, this produces a higher local temperature.

Similarly, the point of maximum thickness is at a lower temperature. This sets up a thermal gradient across the journal cross section. When the shear stresses in the shaft become relatively larger, this causes localized bending of the shaft at the journal section. This thermal bending results in a slightly different load distribution and causes a decrease in the bearing clearance. This, in turn, elevates the temperate gradient, hence resulting in an increase in the synchronous instability amplitude. Once thermal equilibrium is attained, the load distribution changes, and the amplitude decreases. This triggers the thermal gradient, and the hot and cold spots re-occur in the journal, repeating the thermal bow and increasing synchronous vibration. The cyclic increases and decreases in synchronous vibrations are a characteristic of the Morton effect.

In most cases, the Morton effect is seen in heavily loaded overhung rotors [49, 50] (where the impeller is outside the bearing span). However, the Morton effect has also been observed in compressors and other turbomachinery where most of the mass is between the journal bearings (beam-style) [47]. Although the Morton effect occurs under specific load and speed conditions, the synchronous instability can

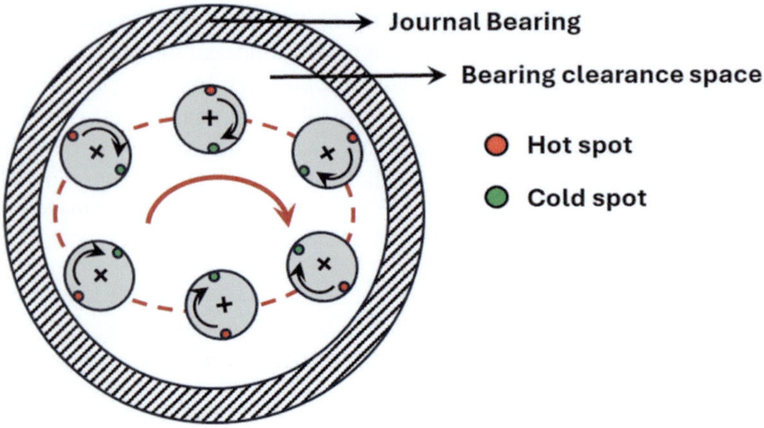

Fig. 12.63 Morton effect: hot and cold spots on the journal section

usually be reduced by reducing the overhung weight. For beam-style machines, the coupling overhung weight on the rotor is typically reduced by either downsizing the coupling or changing the coupling spacer from steel to titanium. The other method is to change the design of the journal bearing by modifying its clearance, pad width, or oil temperature.

12.15 Aeroflow-Induced Instability: Rotating Stall

In Sects. 12.6.1 and 12.6.2, the rotordynamic instability resulting from cross-coupling forces from the labyrinth seals and the gas flowing in the tight clearance spaces between the rotor and the stator. In this section, another form of instability caused by aerodynamic forces is discussed. The flow disturbances caused by impeller-to-diffuser misalignment, impeller stall, and diffuser stall manifest in the form of nonsynchronous forced (as opposed to self-excited) vibration of the rotor.

Before a discussion of stall, it is important that the differences between surge and stall be pointed out. Surge is a system-level phenomenon that includes not only the compressor, but also the connected piping, pressure vessels, valves, and so on as the control volume. It is defined as a condition where a full flow reversal occurs. Under surge conditions, the compressor experiences high vibration levels and fluctuations in the inlet and discharge pressures and the temperature. In contrast, stall is a localized phenomenon occurring in the compressor. It is characterized by a local region of reduced momentum, lower pressure, and so on. It can occur within a component of a stage and manifests as a subsynchronous vibration. Although stall can occur at any operating point on the compressor performance map, it is usually associated with very high or very low-flow rates.

In general, there are two types of stalls: stationary stalls and rotating stalls [51]. Stationary stalls cause nonuniform pressure fields that result in unbalanced forces on the rotor. These stall cells can form in return channels, guide vanes, vaned diffusers, and other locations. When a compressor is operating far from the design conditions, the angles of incidence of the gas approach and the vane setting (because of the design) can be very high. This causes flow separation, resulting in a nonuniform pressure field and, hence, stall cells. The stalled field results in vibrations at the blade-vane passing frequency. Because this stall is at a fixed location of the compressor, it is called a stationary stall.

A rotating stall, on the other hand, occurs in the impeller and diffuser portion of a compressor. A rotating stall in an impeller is described as a nonuniform pressure field (circumferential) that rotates, usually at a speed less than the running speed and in the direction of the shaft rotation. This rotating pressure field creates a subsynchronous vibration frequency. Although there are many theories about how these pressure fields are formed, the two most popular (initial) works on this topic [52, 53] suggested that a rotating stall in an impeller results from flow perturbations at the exit of the impeller that prevent the flow from following the blading curvature.

A diffuser rotating stall is due to low-momentum fluid coalescing along the diffuser walls, which occurs as a result of the flow lacking sufficient radial momentum to escape the diffuser. The stall can form in a single stall cell or multiple stall cells. The radial vibration frequency in this case is a function of the number of stall cells.

In general, an impeller rotating stall can manifest as a subsynchronous frequency ranging from 50 to 90% of the running speed. For a diffuser rotating stall that occurs in the vaneless portion of the diffuser, in most cases, the frequency is between 6 and 33% of the running speed. Unlike self-excited vibrations in which the frequency appears at the first natural frequency of the rotor, stall frequencies track the speed of the rotor. Stall phenomena are complex subjects, and this section should be treated as a basic introduction to the topic.

Acknowledgements All of the images/graphics/figures in this chapter are courtesy of Siemens Energy Inc., Houston, TX. They were provided by Krishnaswamy (Krish) Ramesh, who is the contributing author of this chapter. Krish thanks Mark Kuzdzal, David Ransom, Klaus Steff, and Glenn Grosso for all their advice and for taking the time to review this chapter.

References

1. API Standard 617 (2022) Axial and Centrifugal Compressors and Expander-compressors, 9th edn. American Petroleum Institute, Washington, DC
2. Jeffcott HH (1919) XXVII. The lateral vibration of loaded shafts in the neighbourhood of a whirling speed—the effect of want of balance. Philos Mag, Ser 6 37(219):304–314
3. Nicholas JC, Barrett LE (1984) The effect of bearing support flexibility on critical speed prediction. ASLE Trans 29(3):329–338
4. Kirk RG, Reedy SW (1988) Evaluation of pivot stiffness for typical tilting-pad journal bearing designs. ASME J Vib, Acoust, Stress, Reliab Des 110(2):165–171

5. Nicholas JC (1996) Hydrodynamic journal bearings—types, characteristics and applications. Mini course notes. 20th Annual Meeting, The Vibration Institute, Willowbrook, IL, p 76
6. Simmons JEL, Advani SD (1987) Michell and the development of tilting pad bearing. In: Dowson D, Taylor CM, Godet M, Berthe D (eds) Fluid film lubrication—Osborne Reynolds Centenary: proceedings of the 13th Leeds–Lyon symposium on tribology. Tribology series, vol 11. Elsevier Science, Amsterdam, pp 49–56
7. Nicholas JC (1994) Tilting pad bearing design. In: Proceedings of the twenty-third turbomachinery symposium, Turbomachinery Laboratory, Texas A&M University, College Station, TX, pp 179–194
8. Newkirk BL, Taylor HD (1924) Shaft whipping due to oil action in journal bearings. Gen Elec Rev 28(8):559–568
9. Verma AK, Sarangi S, Mahesh K, Banerjee S (2012) Oil whip detection using stator current monitoring. In: 2012 IEEE symposium on computers and informatics, ISCI 2012, Penang, Malaysia, 18–20 March 2012. IEEE Press, Piscataway, NJ, pp 119–124. https://doi.org/10.1109/ISCI.2012.6222678
10. Vance JM (1988) Rotordynamics of turbomachinery. Wiley, Hoboken, NJ
11. San Andrés L (2006) Hydrodynamic fluid film bearings and their effect on the stability of rotating machinery. Paper presented during the AVT-143 RTO AVT/VKI lecture series, Rhode St. Genèse, Belgium, 20–23 March 2006. Educational notes RTO-EN-AVT-143, Paper 10. RTO, Neuilly-sur-Seine, France
12. Childs D (1993) Rotordynamic introduction to hydrodynamic bearings and squeeze-film dampers. In: Turbomachinery rotordynamics: phenomena, modeling, and analysis. Wiley, New York, chap 3, pp 132–226
13. Raimondi AA, Boyd J (1958) A solution for the finite journal bearing and its application to analysis and design: I. Trans ASLE 1(1):159–174
14. Raimondi AA, Boyd J (1958) A solution for the finite journal bearing and its application to analysis and design: II. Trans ASLE 1(1):175–193
15. Raimondi AA, Boyd J (1958) A solution for the finite journal bearing and its application to analysis and design: III. Trans ASLE 1(1):194–209
16. Budynas RG, Nisbett JK (2010) Shigley's mechanical engineering design, 9th edn. McGraw-Hill Series in Mechanical Engineering. McGraw-Hill, New York
17. Ramesh K (2002) A state-of-the-art rotor dynamic analysis program. In: Proceedings of the 9th of international symposium on transport phenomena and dynamics of rotating machinery, Honolulu, HI, 10–14 Feb 2002
18. Lund JW (1980) Review of analytical methods in rotor-bearing dynamics. Tribol Int 13(5):233–236
19. Lund JW (1964) Spring and damping coefficients for the tilting-pad journal bearing. ASLE Trans 42(4):342–352
20. Lund JW (1974) Stability and damped critical speeds of a flexible rotor in fluid bearings. Trans ASME, Ser B 96(2):509–517
21. Nelson HD, McVaugh JM (1976) The dynamics of rotor-bearing systems using finite elements. ASME J Eng Ind 98(2):593–600
22. Nelson HD (1985) Rotordynamics equations in complex form. ASME J Vib, Acoust, Stress, Reliab Des 107(4):460–461
23. Ruhl RL, Booker JF (1972) A finite element method for distributed parameter turborotor system. ASME J Eng Ind 94(1):128–132
24. Nelson HD (1980) A finite rotating shaft element using Timoshenko beam theory. ASME J Mech Des 102(10):793–803
25. Ruddy AV, Summers-Smith D (1980) An introduction to the influence of the bearings on the dynamics of rotating machinery. Tribol Int 13(5):199–203
26. Images courtesy of Fox Innovation & Technologies, La Porte, TX
27. Kirk RG (1980) Stability and damped critical speeds: how to calculate and interpret the results. CAGI Tech Dig 12(2):1–14

28. Gunter EJ (1966) Dynamic stability of rotor-bearing systems. Tech. Rep. NASA SP-113. National Aeronautics and Space Administration, US Government Printing Office, Washington, DC
29. Lund JW, Orcutt FK (1967) Calculations and experiments on the unbalance response of a flexible rotor. ASME J Eng Ind 89(4):785–796
30. Soulas TA Colby G, Desai AR, Griffin T (2011) CO_2 compression for capture and injection in today's environmental world. In: Proceedings of the first Middle East turbomachinery symposium, Doha, Qatar, 13–16 Feb 2011
31. Bently DE, Muszynska A, Littrell NB (1996) Margin of stability. Orbit 15(2):4–6
32. Wachel JC (1983) Case histories of rotor dynamic instabilities in the field. Short course in Rotordynamics of Turbomachinery. Turbomachinery Laboratory, Texas A&M University, College Station, TX
33. Bansal PN, Kirk RG (1975) Stability and damped critical speeds of rotor-bearing systems. ASME J Eng Ind 97(4):1325–1332
34. Memmott EA (2000) Empirical estimation of a load related cross-coupled stiffness and the lateral stability of centrifugal compressors. In: Proceedings of the 18th machinery dynamics seminar, CMVA, Halifax, Nova Scotia, Canada, 26–28 April 2000, pp 9–20
35. Moore JJ, Soulas TA (2003) Damper seal comparison in a high-pressure re-injection centrifugal compressor during full-load, full-pressure factory testing using direct rotordynamic stability measurement. In: Proceedings of the ASME 2003 international design engineering technical conference and computers and information in engineering conference. Volume 5: 19th biennial conference on mechanical vibration and noise, parts A, B, and C. Chicago, IL, 2–6 Sept 2003, pp 1319–1326
36. Moore JJ, Hill DL (2000) Design of swirl brakes for high pressure centrifugal compressors using CFD techniques. In: Proceedings of the 8th international symposium on transport phenomena and dynamics of rotating machinery, ISROMAC-8, vol II, Honolulu, HI, 26–30 March 2000, pp 1124–1132
37. Eldridge TM, Soulas TA (2005) Mechanism and impact of damper seal clearance divergence on the rotordynamics of centrifugal compressors. In: Proceedings of ASME turbo expo 2005: power for land, sea, and air, vol 4: Turbo Expo 2005. Reno, NV, 6–9 June 2005, pp 907–915
38. Memmott EA (1994) Stability of a high-pressure centrifugal compressor through application of shunt holes and a honeycomb labyrinth. In: 13th machinery dynamics seminar, CMVA, Toronto, Canada, Sept 12–13 1994, pp 211–233
39. Ramesh K, Kirk RG (1995) Design procedure for evaluating stability of turbomachinery supported on squeeze film dampers. ASME J Tribol 117(4):742–744
40. Ramesh K, Kirk RG (1999) Nonlinear response of rotors supported on squeeze film dampers. In: Proceedings of the ASME 1999 design engineering technical conferences. Volume 7A: 17th biennial conference on mechanical vibration and noise, Las Vegas, NV, 12–16 Sept 1999, pp 1079–1085
41. Memmott EA (1987) Damper bearings and stability of a CO_2 compressor. Presented at the Seventh annual rotating machinery & controls industrial research conference, San Diego, CA, 21–25 June 1987
42. Memmott EA (1992) Stability of centrifugal compressors by applications of tilt pad seals, damper bearings, and shunt holes. In: Proceedings of the 5th international conference on vibrations in rotating machinery, Bath, England, 7–10 Sept 1992, pp 99–106
43. Marshall DF, Hustak JF, Memmott EA (1993) Elimination of subsynchronous vibration problems in a centrifugal compressor by the application of damper bearings, tilting pad seals, and shunt holes. Paper presented at the Rotating machinery conference and exposition, Somerset, NJ, 10–12 Nov 1993. Paper No. NJIT-ASME-HI-STLE
44. Kuzdzal MJ, Hustak JF, Sorokes JM (1994) Identification and resolution of aerodynamically induced subsynchronous vibration during hydrocarbon testing of a 34,000 HP centrifugal compressor. In: Proceedings of the 4th international conference on rotordynamics, IFToMM, Chicago, IL, 7–9 Sept 1994

45. Kuzdzal MJ, Hustak JF (1996) Squeeze film damper bearing experimental vs. analytical results for various damper configurations. In: Proceedings of the twenty-fifth turbomachinery symposium, Turbomachinery Laboratory, Department of Mechanical Engineering, Texas A&M University, College Station, TX, 17–19 Sept 1996
46. Hu Q, Zhu M, Yang J (2020) Study on thermal unstable vibration of rotor under journal whirl with large amplitude in journal bearing. Int J Rotating Machinery 2020:1980759. https://doi.org/10.1155/2020/1980759
47. de Jongh FM, Morton PG (1994) The synchronous instability of a compressor rotor due to bearing journal differential heating. In: Proceedings of the ASME 1994 international gas turbine and aeroengine congress and exposition. The Hague, The Netherlands, 13–16 June 1994. ASME Paper No. 94-GT-35
48. Keogh PS, Morton PG (1994) The dynamic nature of rotor thermal bending due to unsteady lubricant shearing within bearing. Proc R Soc, London A 445:273–290
49. Berot F, Dourlens H (1999) On instability of overhung centrifugal compressors. In: Proceedings of the ASME international gas turbine & aeroengine congress & exhibition, Indianapolis, IN, 7–10 June 1999. ASME Paper No. 99-GT-202
50. Kirk RG, Guo Z, Balbahadur AC (2003) Synchronous thermal instability prediction for overhung rotors. In: Proceedings of the thirty-second turbomachinery symposium, Turbomachinery Laboratory, Texas A&M University, College Station, TX, 8–11 Sept 2003, pp 121–135
51. Sorokes JM, Kuzdzal MJ, Marshall DF (2016) A review of aerodynamically induced forces acting on centrifugal compressors, and resulting vibration characteristics of rotors. Paper presented at the 45th turbomachinery and 32nd pump symposia, Houston, TX, 12–15 Sept 2016. https://doi.org/10.21423/R16K5W
52. Abdelhamid AN (1980) Analysis of rotating stall in vaneless diffusers of centrifugal compressors. In: Proceedings of the ASME 1980 international gas turbine conference and products show, vol 1B: general, New Orleans, LA. 10–13 March 1980. ASME Paper No. 80-GT-184
53. Frigne P, Van den Braembussche R (1984) Distinctions between different types of impeller and diffuser rotating stall in a centrifugal compressor with vaneless diffuser. ASME J Eng Gas Turbines Power 106(2):468–474

Chapter 13
Torsional Analysis

Abstract This chapter presents a practical approach to torsional analysis. It covers fundamental techniques for determining the stiffness and inertia of mechanical components. Examples range from basic two-mass and three-mass systems to more intricate configurations that require computer-based modeling. The chapter also explores the transient response of torsional systems under forced conditions, such as during synchronous motor startup and short-circuit scenarios. Fatigue assessment methods and life expectancy projections are examined in detail. Given the complexity of torsional issues, theoretical models often need validation through real-world measurements. This approach enhances the reliability of system modifications and ensures machinery is engineered for safe and effective operation.

13.1 Introduction

Even though a minimum of two inertias connected by a spring (e.g., a motor driving a fan through a flexible-element coupling) are required, torsional analysis of rotating equipment is less complicated than lateral rotordynamics analysis. Torsionally, it is easier to simplify the mechanical parts of a machine train into stiffnesses and inertias. Bearings and damping have little effect on the frequency calculations. For simple systems, closed-form equations can suffice to yield accurate torsional resonant frequency values. It can be difficult to obtain accurate inertia values for complex geometries such as centrifugal impellers, but these values can also be measured experimentally. The torsional stiffness values of flexible-element couplings are available from the manufacturers but should be double checked. Elastomeric couplings are sometimes applied and introduce nonlinear elements into the analysis. Moreover, gears may have backlash, which can introduce additional nonlinear effects.

Many different mechanisms can excite torsional resonances in rotating equipment trains. Some examples of prime movers that produce torque pulsations are synchronous motors, variable-frequency drives, and reciprocating engines. Cylinder "misfires" can induce large internal shaft stresses. Speed-changing gears also can produce torque pulsations. The driven equipment may also be a continuous torsional

N. Ghaisas, *Centrifugal Compressors in Hydrocarbon Processing Industries*,
https://doi.org/10.1007/978-3-032-03687-2_13

exciter, as with positive-displacement pumps and compressors and with blade-pass frequencies in centrifugal equipment. Off-design operation can induce torsional vibrations as well as process upsets such as compressor surge. This chapter uses English customary units, followed by International System of Units (Système international d'unités [SI]) equivalents in parentheses, and includes a unit conversion chart for common torsional analysis units.

13.2 Basic Torsional Information

To begin the torsional modeling of a rotating equipment train, one must know the system geometry. It is possible to do a lumped-parameter analysis in which the total inertia of the main components is connected by single-value stiffness springs. However, this approach is often questionable, as it relies on the calculations of others and should be done only as a last resort. It is impossible to determine the original system geometry from a reduced lumped-mass model. It is much better if the analyst has access to the complete system geometry and simplifies judiciously. This usually means having access to good rotor and coupling drawings and correctly calculated or measured weights and inertias for mounted components.

Obtaining rotor drawings is sometimes difficult, especially for older equipment. For new equipment, the purchaser must ask the manufacturer to provide the drawings before construction begins so that an independent analysis can be performed. Improper modeling is not unusual and is very difficult to detect without access to raw data and the assumptions of the analyst.

With transient- or forced-response analysis, which is critical in predicting component life, the applied steady and alternating loads must be accurately defined. Often, the equipment manufacturer will supply the corresponding data, but it may be necessary to measure the actual torque pulsations under normal operation, including different loading requirements. Fatigue analysis requires knowledge of the material properties, which may take some research to obtain. For this reason, fatigue analysis is usually done with a very conservative approach.

13.3 Determining the Polar Moment of Inertia

The most basic data value required for torsional analysis is the polar mass moment of inertia, J. Figure13.1 shows a simple circular disk that would be mounted to a shaft, and Eq. 13.1 provides the required value in terms of known or measurable quantities

$$J = \frac{\rho \pi L\left(D_{\mathrm{o}}{}^{4} - D_{\mathrm{i}}{}^{4}\right)}{32}, \tag{13.1}$$

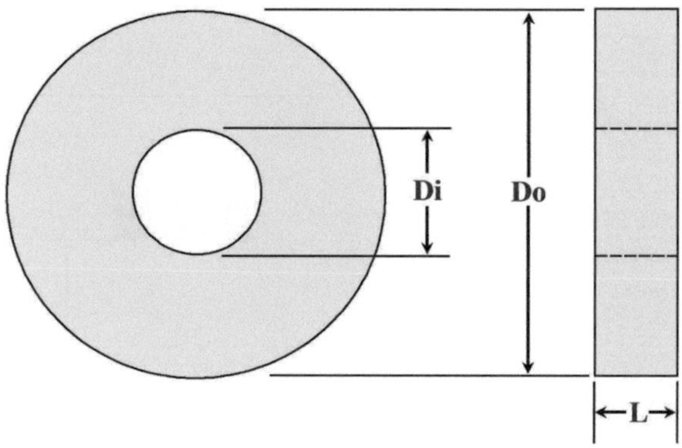

Fig. 13.1 Geometry of a simple uniform circular disk

where D_i and D_o are the inner and outer diameters, respectively, of the disk; L is the disk length; ρ is the density of the disk material; and J is given in units of lb in.^{2} (kg m^2).

If the weight, W, of the part is known, the polar moment of inertia can be expressed as

$$J = \frac{W\left(D_o{}^2 + D_i{}^2\right)}{8}, \tag{13.2}$$

where, again, J is obtained in units of lb in.^{2} (kg m^2).

If the disk is not a solid object, but rather a more complicated object such as an impeller, calculating the polar moment of inertia can be difficult. There are several ways to determine the correct values. Sometimes, the manufacturer will provide the data. Alternatively, a previous lateral or torsional analysis may be available, or it can be determined experimentally. A good estimate for modifying the density, ρ in Eq. 13.1, for closed centrifugal impellers is as 21–25% of the actual material density. For open impellers, the estimated density can range from 15 to 25% depending on the construction and complexity of the vanes. It helps to accumulate a file of known impeller properties. If an impeller is available to be measured, computer-aided design (CAD) programs are available that can calculate the inertia. For reciprocating parts, the polar inertia is one-half of the reciprocating mass times the square of the radial throw distance.

Often, the total polar inertia of a rotor is available from the manufacturer. Occasionally, the published inertia values of parts or rotors are suspect, and a more thorough investigation is warranted. In this chapter, the radius of gyration (wr^2) values are used. If diametral values are supplied as GD^2 numbers, they are 4 times the wr^2 values. This will lead to substantial calculation error if not fully understood. Another area of confusion is the units associated with inertias. The English mass-equivalent inertia unit is lb in. s^2 (kg m s^2), and the weight equivalent unit is lb in.^{2} (kg m^2). It is critical to keep the units straight. Providing inertia values without any units at all

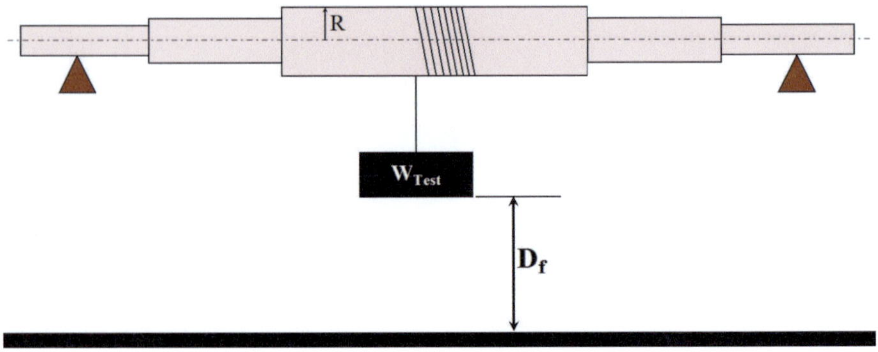

Fig. 13.2 Timed rundown test for determining rotor inertia

will leave the analyst guessing; a guess is not good when one is doing meticulous engineering analysis.

If a good rotor inertia value is not available and if a complete rotor is available, the total rotor inertia can be determined using an experimental method. This method involves mounting the rotor in rollers such as a balance machine. The test setup is diagrammed in Fig. 13.2. Then, a thin wire (of radius r_{Wire}) is wrapped several times around the rotor (of radius R_{Rotor}). On the free end of the wire, just enough weight is added to start the rotor turning in the rollers. This weight is the amount required to overcome friction. Then, an additional weight (WTest) is added and released. The time, t, for the weight to drop a measured distance, Df (about 20–40 in. [50–100 cm] is a good drop distance to use), is then measured. This procedure should be repeated several times to ensure repeatability. The times are then averaged, and Eq. 13.3 is used to calculate the total inertia of the rotor.

$$J_{\mathrm{Total}} = W_{\mathrm{Test}}(R_{\mathrm{Rotor}} + r_{\mathrm{Wire}})\left(\frac{t^2}{2D_{\mathrm{f}}} - \frac{1}{386.088}\right), \tag{13.3}$$

where J_{Total} is obtained in units of lb in. s^2 (kg m s^2).

This method takes a little practice, and it is recommended that some expertise be developed on a rotor with a known inertia or a bare shaft for which the inertia can be calculated exactly. Once the total rotor inertia has been determined, the inertia of impellers or other attachments such as motor windings can be calculated by subtracting the values for those parts whose inertia can be calculated.

An experimental method also exists for determining the polar inertia of components such as impellers and other irregular parts. This method involves hanging a platform from the ceiling with three thin wires at a precise radius R from the exact center of the platform, as shown in Fig. 13.3. The weight of the table and the weight of the test piece must be known. Their combined weight is WT. It is easy to calibrate this test with a disk of known inertia.

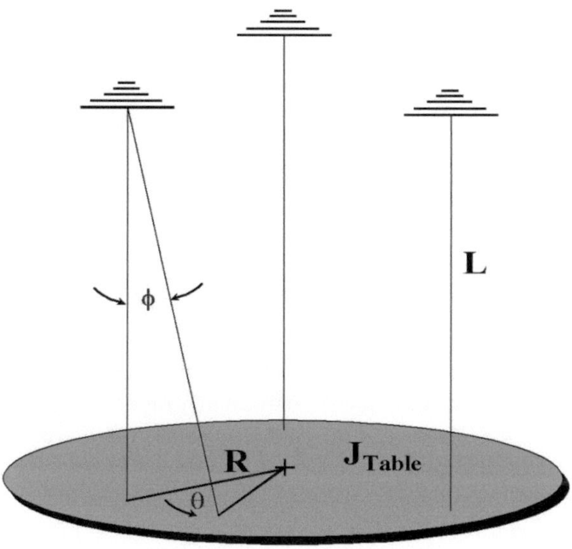

Fig. 13.3 Trifilar pendulum

The wires should be as long as possible to minimize error. The table is usually made of sheet metal capable of supporting the parts to be measured. The inertia of the table by itself can be determined from Eq. 13.4

$$J_{\text{Total}} = \frac{R^2 W_{\text{T}}}{4\pi^2 L f^2}, \tag{13.4}$$

where, again, J_{Total} is obtained in units of lb in. s^2 (kg m s^2).

A part is very carefully centered on the table and given an initial angular displacement, θ, of less than 5°. Once released, the assembly will oscillate, and the frequency of the oscillation, f, in cycles per second, is measured over 10 or more cycles. Using Eq. 13.4, the total inertia of the table plus the test piece can then be determined.

Then, the inertia of the table (JTable) is subtracted to obtain the inertia of the test piece. As mentioned previously, the units of the polar inertia resulting from this test are lb in. s^2 (and the corresponding SI units are kg m s^2).

13.4 Determining Torsional Stiffness

The other primary information required for torsional analysis is the stiffness of the connections between inertias. If these sections are circular shafting as shown in Fig. 13.4, then Eq. 13.5 will provide the required values. In this equation, Φ is the twist angle, in radians, which results from an applied torque, T. I_{P} is not the same as J. It is the cross-sectional area polar moment of inertia [$\pi(D_{\text{o}}{}^4 - D_{\text{i}}{}^4)/32$], and G is the shear modulus of the material. G is approximately 11.6×10^6 psi (7998 MPa)

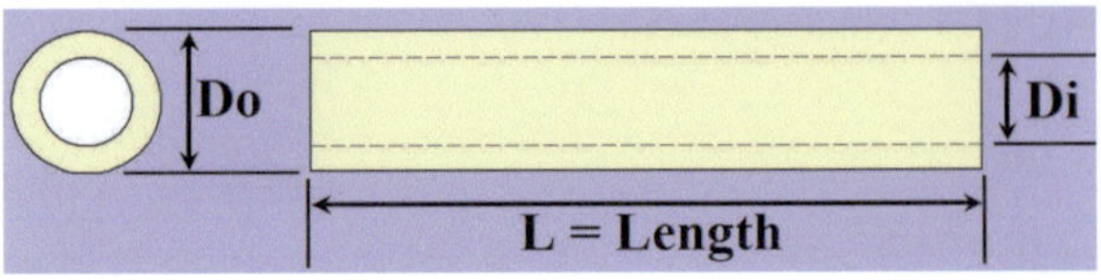

Fig. 13.4 Typical circular shaft cross section

for steel.

$$\Phi = \frac{TL}{I_P G} \therefore K_T = \frac{T}{\Phi} = \frac{I_P G}{L}, \tag{13.5}$$

where K_T is the stiffness and is obtained in units of in. lb/rad (N m/rad).

Mounted components, particularly those with an interference fit, alter the effective torsional stiffness of that section. This is often seen with mounted coupling hubs, as illustrated in Fig. 13.5. This figure shows a shrink length, *L*, and it is assumed, based on testing, that one-third of the shaft penetration under the hub is free to turn and two-thirds is not. This technique is for typical mechanical mounting practices in the rotating equipment industry.

More precise methods of determining the exact penetration effect are available (see, for example, [1]), but they are rarely necessary. For keyed parts with little interference, the penetration effect is approximately one-half of the total length. The softest torsional members in a rotating equipment train are usually the flexible-element couplings. The torsional stiffness values supplied by the coupling manufacturers always include the penetration effect.

It must be noted that penetration effects also exist where changes in the shaft diameter take place in the torque path. As torque "flows" from one diameter section to the next, there is a slight decrease in the effective diameter at the change in

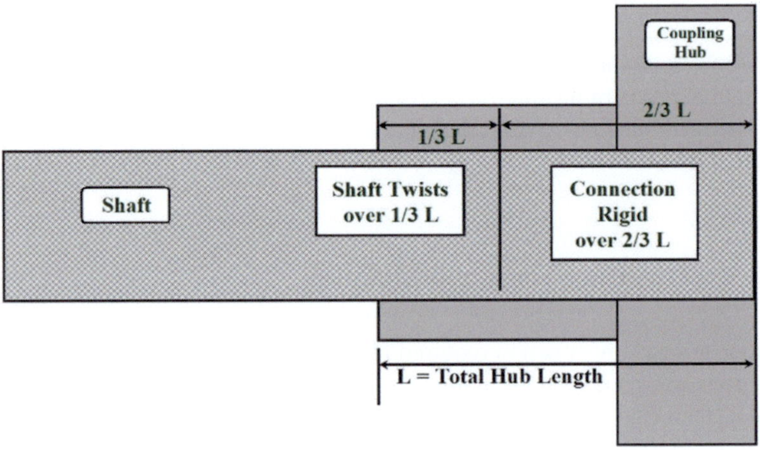

Fig. 13.5 Shaft penetration effect

diameters because the torque does not radiate to the corners of the larger diameter. This effect is usually not considered unless the diameter change is greater than 2 to 1, as the frequency calculation results are altered by less than 1%.

13.5 Modeling of Equipment Trains

Once each rotor in a train has been defined, a calculation method is selected. The easiest method is the lumped inertia and spring method, as illustrated in Fig. 13.6.

Determining the stiffnesses between inertias is a process of summing the stiffnesses of each shaft section. In Fig. 13.6, the shafting between J_1 and J_2 consists of three segments. The stiffness, or K value, of each segment is determined by Eq. 13.5, and this process is repeated for each spring in the train. In segments with a flexible-element coupling, the total stiffness will be less than the stiffness of the coupling alone. Because the shaft diameters near couplings are the smallest in the machine, their torsional stiffness is fairly low and must be included to avoid large errors. The technique for adding shaft stiffness values is illustrated in Fig. 13.7. The equivalent stiffness, KE, is an inverse sum of the individual stiffnesses, as shown in Eq. 13.6. This procedure works for any number of segments.

$$\frac{1}{K_{\mathrm{E}}} = \frac{1}{K_{1\mathrm{A}}} = \frac{1}{K_{1\mathrm{B}}} = \frac{1}{K_{1\mathrm{C}}}, \tag{13.6}$$

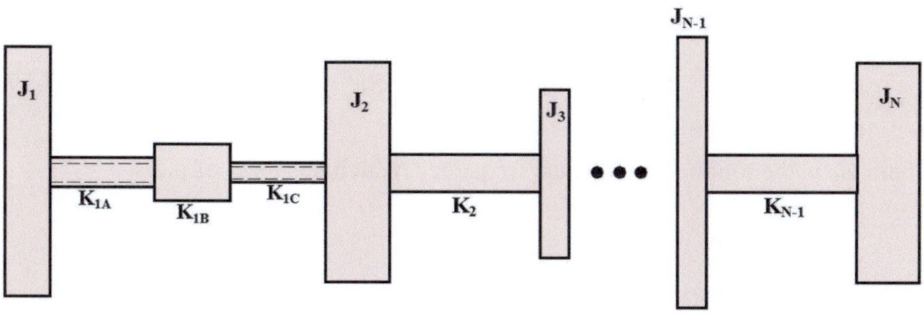

Fig. 13.6 Lumped stiffness and inertia modeling

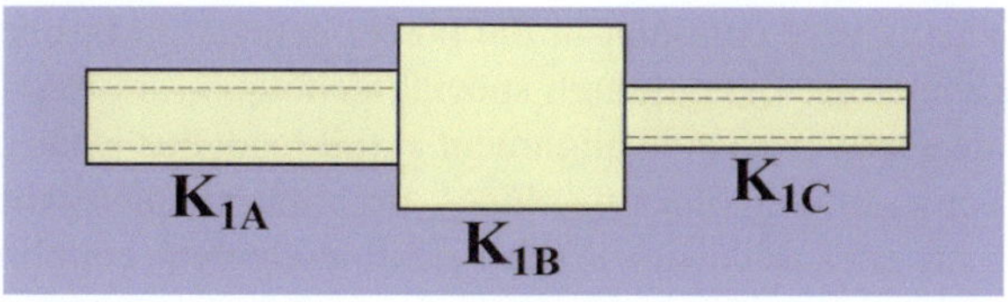

Fig. 13.7 Multisegment shafting

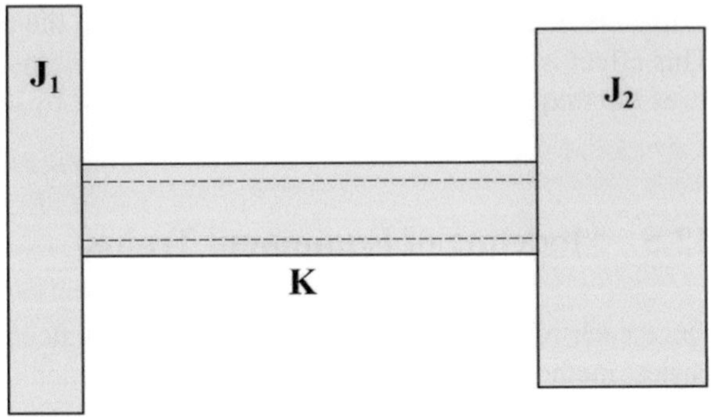

Fig. 13.8 Simple two-mass system

where K_{1A}, K_{1B}, and K_{1C} are the stiffness values of segments A, B, and C, respectively, and $1/K_E$ has units of in. lb/rad (N m/rad).

13.5.1 Calculating Torsional Frequencies

There are always $J_N - 1$ important torsional resonances. A two-mass system has only one torsional resonance. A three-mass system has two torsional resonances, and so forth. Because calculation of the necessary inertia and stiffness values can be done according to the preceding equations, the actual torsional resonance frequencies can be calculated. For the simple two-mass system shown in Fig. 13.8, the resulting equation (Eq. 13.7) is not complex.

$$\omega_n = \sqrt{K\frac{J_1 + J_2}{J_1 J_2}}, \tag{13.7}$$

where ω_n is the torsional resonance frequency, which has units of rad/s.

13.5.2 Couplings

Two or more machines can be joined together with couplings of various types. The simplest type of coupling is direct flange-to-flange mounting, called a rigid coupling. This is very common in the power generation business where large turbines drive generators. Even though special techniques are required to ensure that the flanges are nearly perfect in alignment at full load, this is the most reliable type of coupling. Some, mostly older machines are close-coupled with only a narrow gap between shaft ends and with a short flexible-element coupling attached to each shaft end with mounted hubs. However, in most modern applications, the shaft ends are well

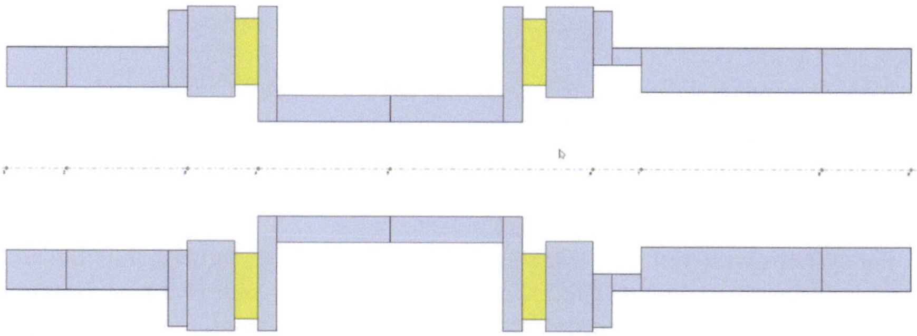

Fig. 13.9 Typical high-performance American Petroleum Institute (API) Standard 671 flexible-element coupling

separated, allowing more lateral flexibility to the coupling and easier access to piping, bearings, and seals.

Flexible-element couplings are used to connect a multitude of types of machinery, from small pump drives to very large process machines in refineries and chemical plants. Figure 13.9 shows a typical cross section of a finite-element model of a high-performance flexible-element coupling. American Petroleum Institute (API) Standard 671 provides a thorough guide to the selection and application of couplings. There are many different types of flexible-element couplings including quill shafts, hollow spacer tubes (illustrated), flexible links, grid-type couplings, and elastomeric couplings. One of the primary benefits of flexible-element couplings is that they allow for successful operation under different operating conditions and are able to accommodate some misalignment.

Many machines are purposely misaligned when cold to adjust to thermal growth during startup and different load and/or speed conditions. Regarding torsional analysis, couplings are, by far, the components that are most often manipulated to tune torsional resonances away from system operating frequencies. In general, it is usually better to design couplings to be as torsionally compliant as possible. This will place the primary torsional resonance below the operating speed. Sometimes, couplings with elastomer elements are used to obtain soft torsional stiffness. These couplings can add significant damping as well. The damping converts torsional vibration to heat, and this must be considered. They also tend to be large and heavy, which can affect lateral critical speeds.

13.5.3 Geared Systems

Many rotating equipment trains include speed-reducing or speed-increasing gear sets. To calculate the torsional resonance frequencies of a geared train, the stiffnesses and inertias at one of the train speeds must be converted or reflected to the other train speed by the square of the speed ratio. Figure 13.10 shows a typical geared system.

In this figure, $N1$ is the rotor speed at $J1$ and $J2$, and $N2$ is the rotor speed at $J3$ and $J4$.

It does not matter which speed is used as the reference speed. Conceptually, turning the pinion side of a speed reducer gives one a mechanical advantage and reduces both the effective stiffness and inertia on the low-speed side. When the low-speed bull gear is turned, it is much more difficult to accelerate the rotor on the high-speed side, so the apparent stiffness and inertia are greater. Figure 13.11 illustrates this process.

For systems with multiple gear sets, this process cascades through the train with a factor of the speed ratio squared applied sequentially for each gear pair. Torque is altered by the speed ratio directly, rather than the speed squared.

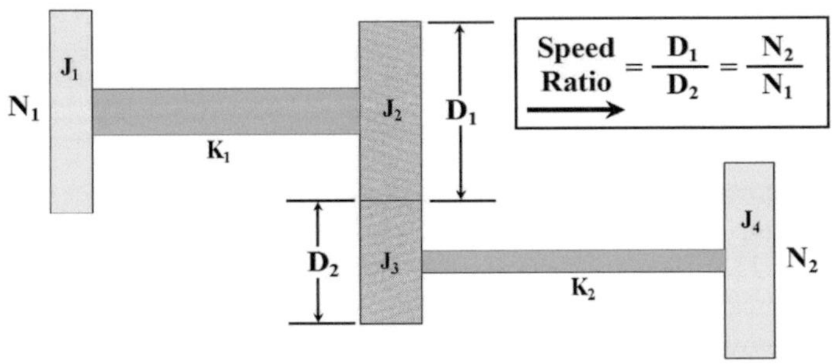

Fig. 13.10 Typical geared torsional system

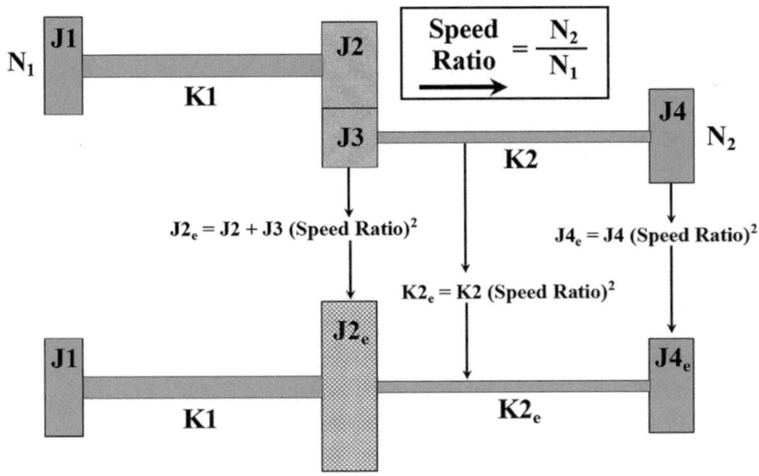

Fig. 13.11 Reducing a geared system to an equivalent system

13.6 Torsional Mode Shapes

For each torsional resonance frequency (called an *eigenvalue*) in a machine train, there is an associated torsional resonance mode shape (called an *eigenvector*). Torsional mode shapes are different than lateral mode shapes and require a different perspective.

The shapes of torsional modes are presented as the dimensionless relative twist amplitude, with a maximum of + 1 and a minimum of − 1. Actual twist amplitudes and shaft stresses can be calculated only from a forced-response analysis with a defined forcing function. For the simple two-mass system, the resonant torsional twist amplitude at the inertia J_1 will be greater by the ratio J_2/J_1, with a single node point between the masses. Figure 13.12 presents the calculated *animated* mode shape for the example from Fig. 13.8.

An animated mode shape shows several cycles of the relative twist amplitude across the equipment train superimposed over the physical outline of the train. In this case, the maximum twist amplitude is on the left-hand side with the smaller inertia. The smallest inertia has less resistance to oscillatory forces and is more easily rotationally displaced. There is a node point near the center of the weakest connecting link between the two inertias. The slope of the mode shape line indicates the degree of twist in the corresponding segment.

The shear stresses are greatest in the shaft section with the greatest mode shape slope. Here, the shear stress at the torsional resonance will be a maximum in the middle tube section. The sections next to the center tube have a slight slope with less twist. The sections on the end have no slope to the mode shape line. This means that these sections are acting as rigid bodies or pure lumped inertias. When the torsional resonance is measured, strain gauges will produce the maximum output when placed on the most highly stressed part. However, many of the other test methods that

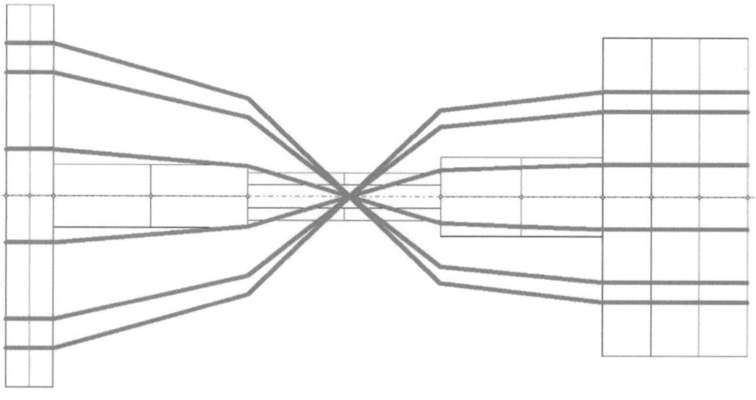

Fig. 13.12 Animated mode shape for the simple two-mass example

measure torsional amplitude will be most responsive at the places where the twist amplitude is a maximum.

13.7 Interference Diagrams

There are often multiple torsional resonances and multiple torsional excitation sources. To evaluate these potential interferences, a special interference diagram has been developed to show the whole picture at once. Often mislabeled as other types of charts, the interference diagram may be shown in linear format or logarithmic format.

The logarithmic format has several advantages, including increased clarity, and it is often more easily understood. Figure 13.13 presents a typical interference diagram with all system frequencies on the horizontal axis and all torsional resonances on the vertical axis.

Although the 2*X* line frequency is shown here as a potential excitation mechanism, the author has never seen a documented case of line frequencies causing significant torsional excitation except in exciting blading tangential modes in large turbine-generator sets.

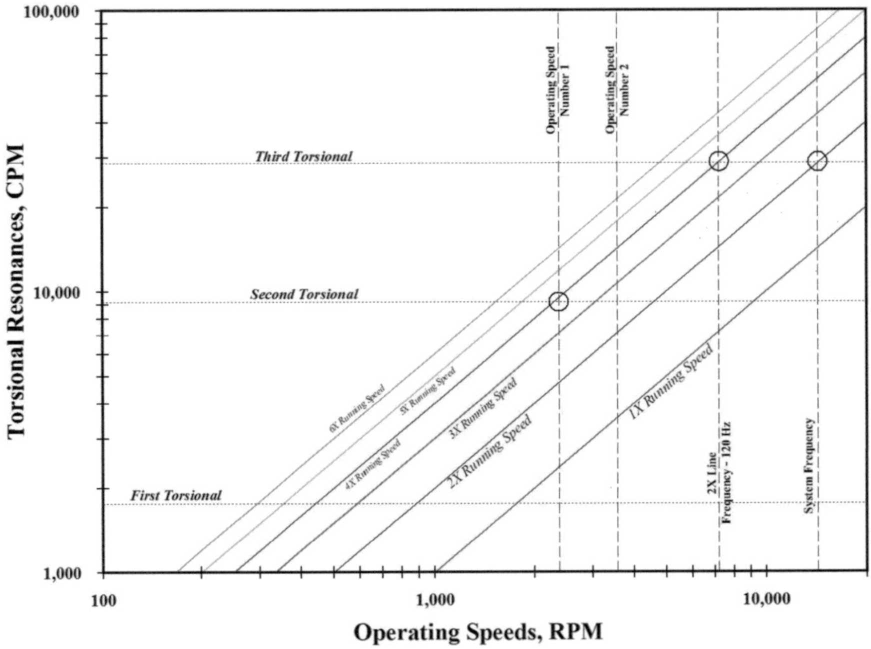

Fig. 13.13 Typical torsional interference diagram

13.8 Forced Response of Torsional Vibration

Forced-response analysis has two subcategories: steady-state excitation and transient excitation. Steady-state forced response usually means constant speed, with one or more components in the system producing constant torque pulsations. Transient torsional excitation is produced by changing conditions such as speed or load or electrical interruption. To analyze the transient torsional response of a train, the forces, as a function of speed or time, must be defined. Although the avoidance of torsional resonances is usually a design goal, there are systems in which torsional resonances will be encountered during normal operation. In these cases, a torsional analysis is performed that will predict the probable shaft stress levels. Then, the designer can allow for these stresses when selecting shaft sizes and materials. The remainder of this chapter details some of the most common torsional excitation mechanisms.

13.8.1 Synchronous Motors

A transient condition is encountered during across-the-line startup of synchronous motors. These motors produce torque pulsations between startup and synchronous speed. The pulsations start at twice the line frequency (120 or 100 Hz) and decay to 0 Hz when the motor synchronizes. Thus, all torsional resonances between 0 and 7200 cycles per minute (cpm) will be excited in reverse order every time the motor is energized.

These torque pulsations can be several times the rated full-load torque of the motor and have been known to cause very dangerous machinery failures. The only real source of information about the nature of motor torques is the motor manufacturer. Fortunately, the number of high-stress cycles is limited during each start, and it is usually possible to design the system components to be strong enough to withstand a reasonable number of start events. A detailed example of a synchronous motor train analysis is provided here. Figure 13.14 shows a compressor train that consists of a six-pole 5500 hp (4100 kW) synchronous motor driving an air compressor through a gear speed increaser.

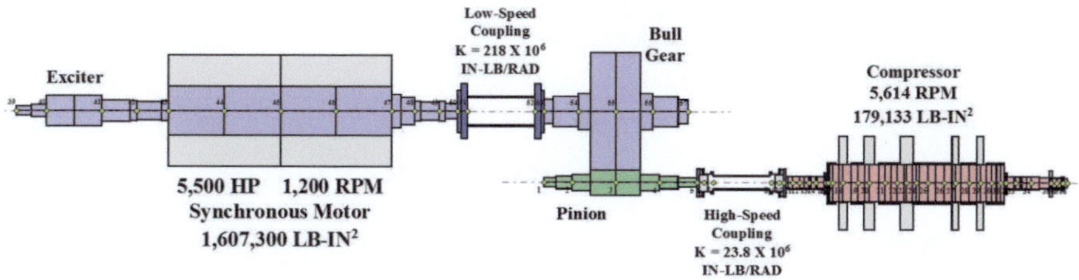

Fig. 13.14 Synchronous motor–gear–compressor train

Every synchronous motor generates alternating torque impulses during a startup. These pulses begin at twice the line frequency (7200 cpm in the USA) and decline inversely with speed to zero when the synchronous speed is attained. This means that every torsional resonance between 0 and 7200 cpm will be excited during each startup. These torque pulsations are not trivial and have been the cause of many machinery failures. Thus, any machine train with a synchronous motor must be evaluated with a transient torsional analysis.

A transient analysis consists of creating a spring–mass model and defining all of the torque producers and absorbers. The torque information is generally available only from the motor vendor, who must communicate with the manufacturer of the other machinery and produce a set of torque curves similar to those in Fig. 13.15. A forced-response procedure is applied to the model to simulate the startup. At time zero, the driving, oscillatory, and load torques are applied simultaneously. This causes the train to accelerate in speed. Then, at very small time steps, a new speed is calculated, along with the torsional oscillation amplitude. Ultimately, the alternating torque amplitudes are converted to the shear stresses in the shafts.

One important calculation coming from this analysis is the time required to reach full speed, as can be seen in Fig. 13.16. Anything longer than 20 s could be damaging to the motor. After synchronization, the motor torque is balanced by the load torque,

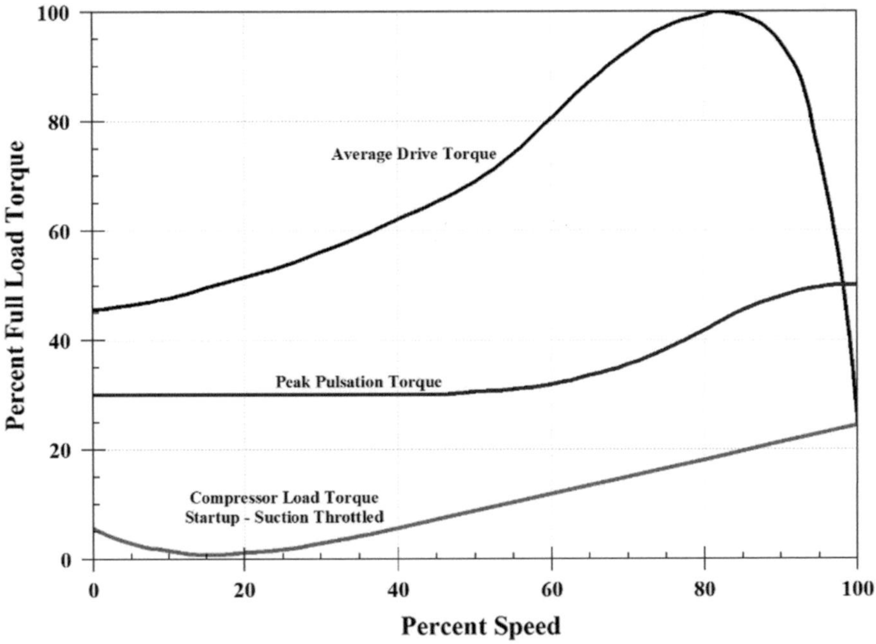

Fig. 13.15 Speed versus torque curves for a 5500 hp (4100 kW) synchronous motor

and there are no more alternating torque pulsations from the motor. The very-low-frequency oscillations often seen immediately after synchronization are due to the interaction of the mechanical system with the electrical grid.

The ripples in the time–speed curve of Fig. 13.16 are actual speed fluctuations caused by the torsional resonances. The resonance that is encountered at 11.8 s into the startup is the second torsional resonance at 3260 cpm. The first torsional resonance, at 1275 cpm, is encountered 14.5 s into the startup. Figure 13.17 presents the animated mode shape of the first torsional resonance.

This mode shape shows that all of the twist amplitude is between the motor and the bull gear. Where the slope of the mode shape line is greatest is where the maximum shaft shear stresses will occur. For this reason, on this compressor train, the coupling

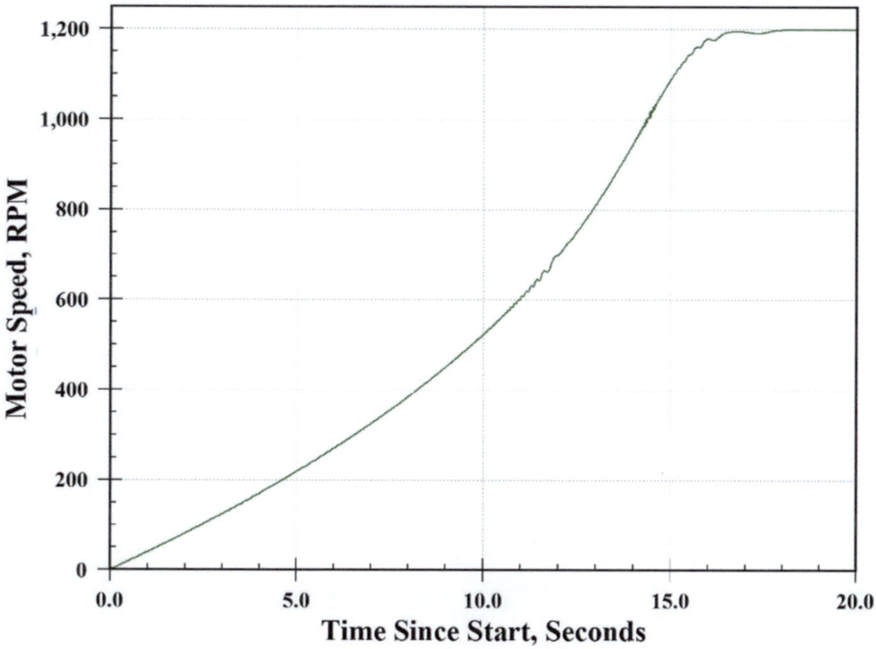

Fig. 13.16 Startup time calculation for a synchronous motor–gear–compressor train

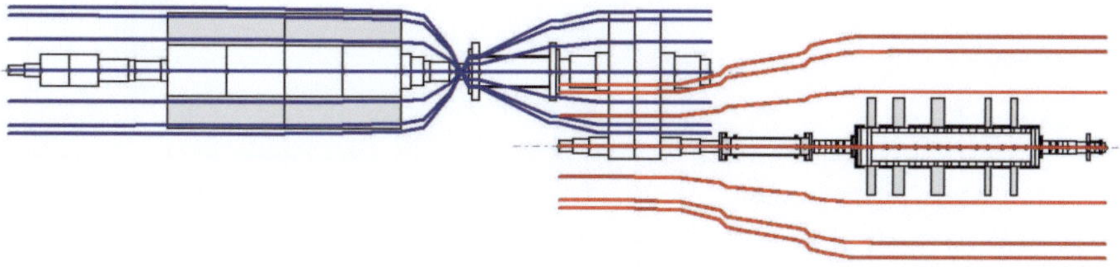

Fig. 13.17 First torsional resonance mode shape for example 3 (1275 cpm)

flanges on both the motor and the bull gear were made integral with the shafts. This avoids any stress concentrations associated with shrunk-on hubs or keyways. Keyways can have stress concentration factors as high as 4 and are a common place for shaft failure to initiate.

Figure 13.18 shows a plot of the alternating shaft shear stresses in the motor during a portion of the startup. Because of the presence of the driving torque, Fig. 13.18 shows the shear stresses offset below zero by the amount of the steady driving torque. The first torsional resonance with the 10,000 psi (69 MPa) peak occurs near 14.5 s. In this train, only the first torsional resonance was of any concern.

It is possible to use these plots to determine the number of shear stress cycles that will accumulate and potentially fatigue the shaft. A number of techniques are available for doing this. The ultimate goal would be an infinite life, but it is typical to settle for a long but finite life provided that it is economically feasible and the machine does not accumulate a large number of normal startups over its lifetime. This example is a case where good design practices resulted in a machine train with an estimated 2000 startups before a failure would occur. This train (as documented in API TR 684 [6]) ran extremely smoothly for 15 years, with no shaft stress being indicated before the train was retired for a larger-capacity unit. It is a credit to the designers that no problems occurred throughout the lifetime of the train even though the interval between the first torsional resonance at 1275 cpm and the 1200 rpm operating speed was less than 7%.

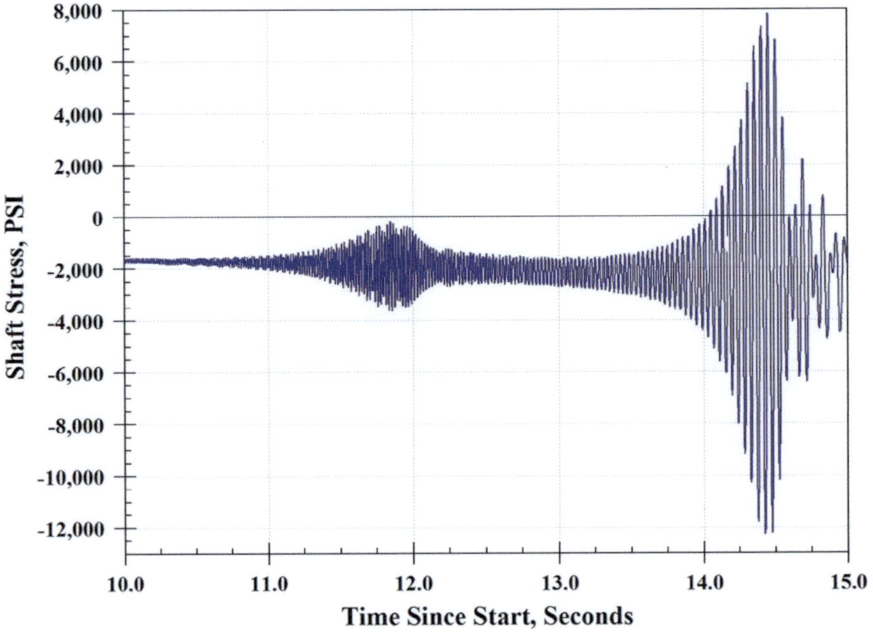

Fig. 13.18 Maximum low-speed shaft stresses during startup for example 3

13.9 Electrical System Interaction and Short-Circuit Analysis

Analysis of interactions with the electrical system is a complex subject because it involves coupling the mechanical system to the electrical grid. This is important in large turbine-generator sets and is discussed in detail by Walker [2]. Normally, a mechanical system equivalent to the electrical grid is constructed with analogs for stiffness, damping, and inertia. Electrical faults such as short circuits can induce extreme torsional excitations. Bad contactors and switches have also been known to excite torsional resonances. Line frequency and twice line frequency can function as a torsional exciter. Short-circuit evaluation is examined in this section. Consider the machine train in shown in Fig. 13.19. This is the transient torsional short-circuit analysis of an unusual machine train. It consists of a two-pole variable-frequency drive (VFD) motor coupled to a generator that is specifically designed to evaluate short circuits. Figure 13.19 provides a cross-sectional view of the train computer model.

The first torsional resonance occurs at 6263 cpm. A dimensionless plot of the stress field at the first resonance, Fig. 13.20 indicates that the maximum resultant shaft shear stress is in the 4.92 in. (12.5 cm) diameter section of the drive motor shaft. This is the area of concern in stress and fatigue evaluation.

The short-circuit static and alternating air-gap torques (in kiloNewton-meters, where 1 kN m = 8850.745 in. lbf) were provided for the generator as specified in Eqs. 13.8 and 13.9 (where $\omega = 377$ rad/s or line frequency). Note that these equations should not be used for any other analysis! Other forms of these equations are also sometimes supplied.

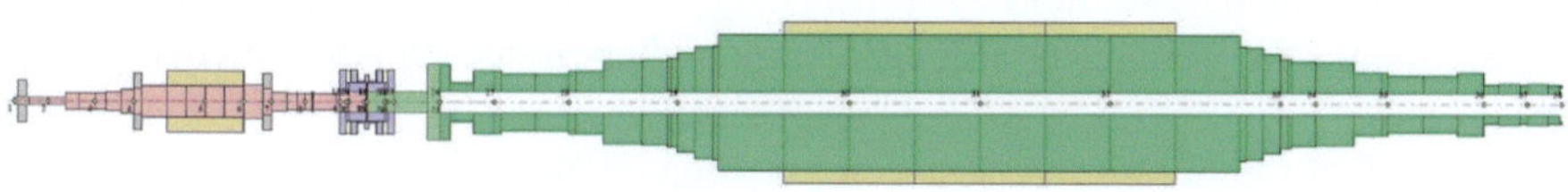

Fig. 13.19 Finite-element computer model: train cross section

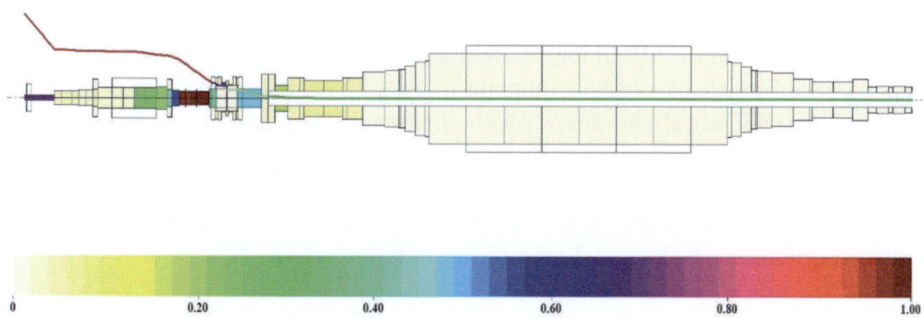

Fig. 13.20 Dimensionless stress field in the motor–generator train at the first resonance

13.9.1 Three-Phase Short Circuit

$$T_{\text{Air gap}} = 4300e^{(-t/0.105)} \sin \omega t + 1720e^{(-t/0.043)}. \tag{13.8}$$

13.9.2 Line-to-Line Short Circuit

$$T_{\text{Air gap}} = 4300e^{(-t/0.087)} \left[e^{(-t/0.091)} \sin \omega t - 0.5e^{(-t/0.087)} \sin 2\omega t\right] + 1720e^{(-t/0.043)}. \tag{13.9}$$

These equations were used to produce files of instantaneous torques at 0.00001 s intervals. These files were then imported as the transient torque inputs to the torsional analysis program and applied at the center of the generator. The primary focus of the transient torsional analysis is to determine the magnitude of the alternating stresses in the shafting as well as the number of high-stress cycles during each short-circuit event. With this information, one can estimate the number of safe events before component shear fatigue failure would occur. It is important to note that only a 60 Hz operating speed (3600 rpm) was considered during this analysis, as indicated by the short-circuit equations provided. Other speeds, if used, may yield different results.

Figure 13.21 presents the fast Fourier transform (FFT) spectrum of the three-phase air-gap torque derived from Eq. 13.8. Note that a three-phase short circuit has a DC component and a 60 Hz component only. Using Eq. 13.8 as the excitation input, a time transient response analysis is run with a time step of 1×10^{-6} s for 0.5 s. This creates a large output data file. Although this "sampling rate" is finer than absolutely required to define the output stress field waveform (Fig. 13.22), it produces a smoother curve and eliminates the possibility of missing any higher-frequency components without any aliasing.

The FFT of the line-to-line short circuit (Eq. 13.9) shows that this type of event contains a DC component, a 60 Hz component, and a 120 Hz component. Figure 13.23 shows the FFT spectrum of the line-to-line air-gap torque. Running the time transient analysis produces the time waveform (Fig. 13.24) of the shaft shear stresses produced by the line-to-line short-circuit event.

For this application, the three-phase short circuit does not produce any shear stress oscillations that would accumulate increasing shaft fatigue. The line-to-line short circuit, which is always more severe than the three-phase event, produces two cumulative fatigue cycles for each test with a stress concentration factor of 1.6. To better understand how cumulative shaft fatigue occurs as a result of torsional vibration and how it is evaluated, it is necessary to first discuss torsional damping, stress concentration factors, and identifying fatigue cycles.

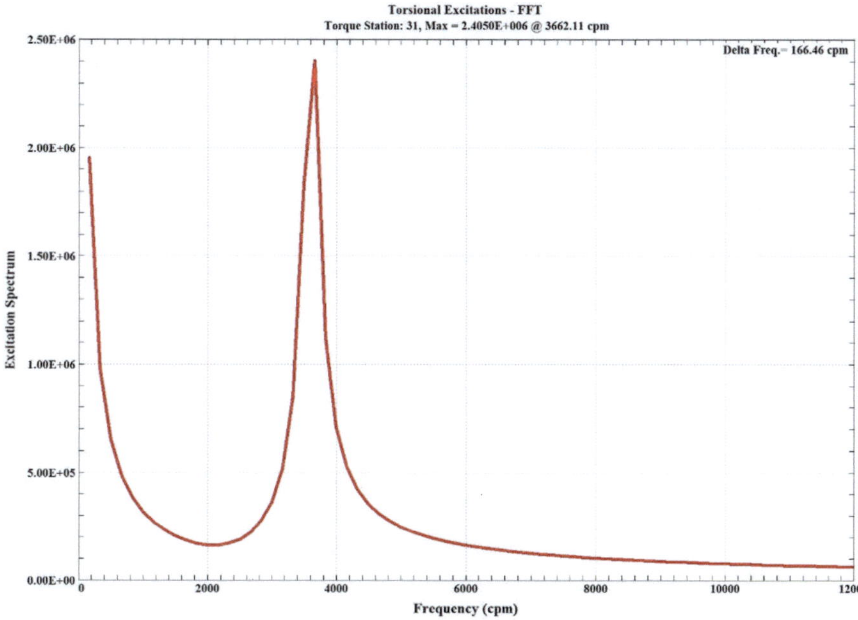

Fig. 13.21 Spectrum of the air-gap torque due to a three-phase short circuit

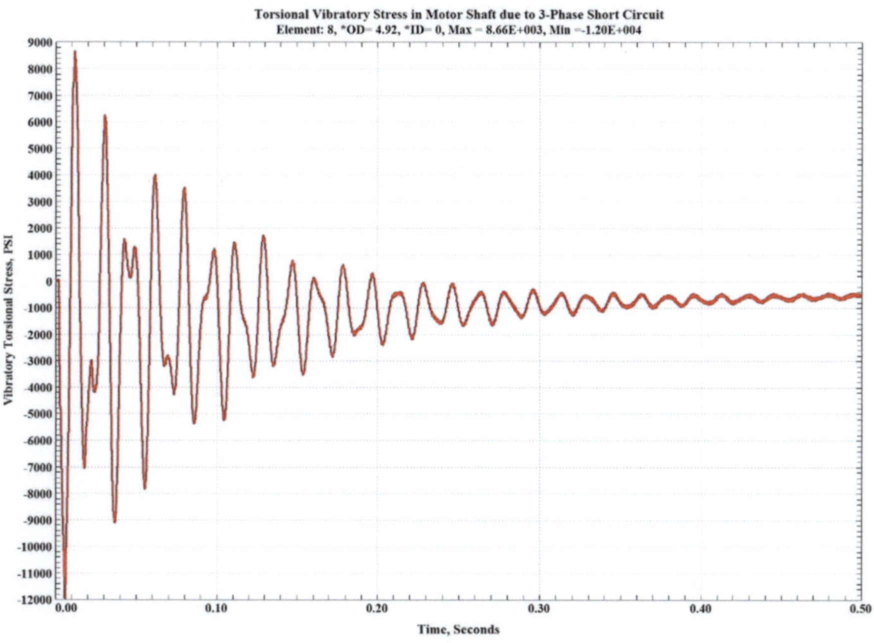

Fig. 13.22 Shear stress induced in the motor shaft during a three-phase short-circuit event

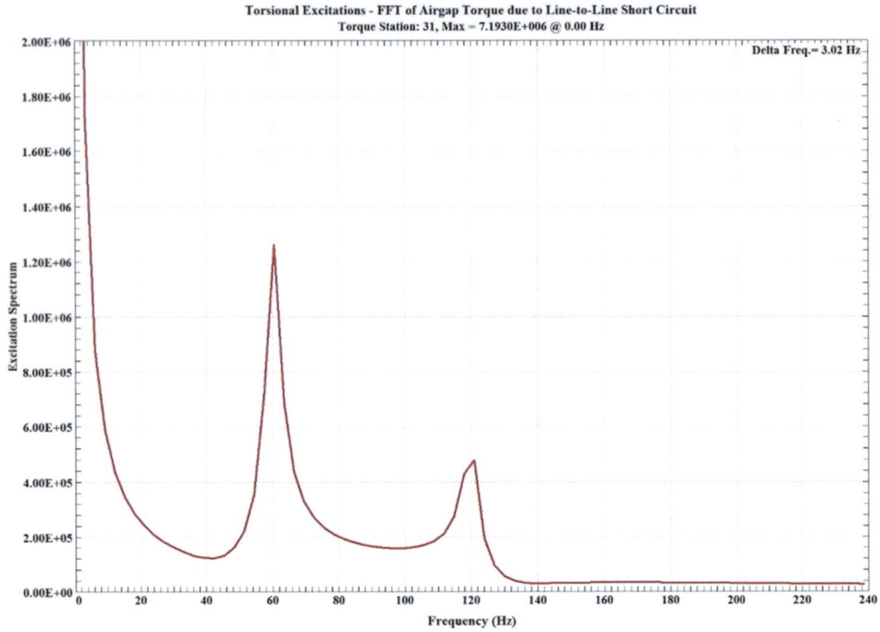

Fig. 13.23 Spectrum of the air-gap torque due to a line-to-line short circuit

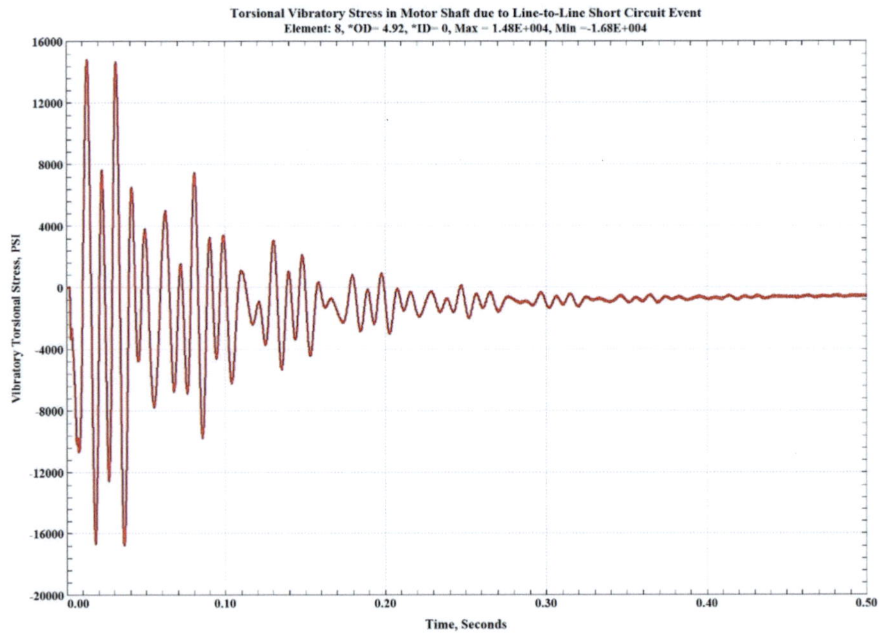

Fig. 13.24 Shaft shear stress induced in the motor shaft during a line-to-line short-circuit event

13.10 Fatigue Analysis

Fatigue analysis is an extremely complicated subject and is limited here to shafting in shear. Torsional damping is always very limited in most rotating equipment trains. There are two types of torsional damping. First, discrete damping occurs at individual points and is usually difficult to quantify. Second, modal damping is a system property and has different values for each torsional resonance.

Some examples of discrete damping are the attached electrical system, fluid film bearings (mostly in gear sets because of lateral–torsional coupling), frictional slippage, and so-called elastomeric couplings. These damper couplings have an elastomer as part of the torque path that can be either in compression, such as rubber blocks, or in shear. The compression types are highly nonlinear, whereas the shear type is more linear with load and deflection. If damper couplings must operate in or near torsional resonances, and many can do this, they must be sized to dissipate the heat generated from the oscillations. Figure 13.25 shows resilient damper coupling.

Gear type and grid-type couplings also add small amounts of torsional damping. The downside is that, when these couplings act as a damping mechanism, wear occurs, and these parts will need regular lubrication and maintenance. There are also some viscous or fluid coupling devices that add discrete damping, and they are very soft torsionally. Modal damping, usually associated with hysteresis within the material, is the technique often applied in the analysis of the torsional forced response of rotating equipment trains. It is a percentage of critical damping for each torsional mode. The actual modal damping is very difficult to measure and is assumed to be 1–3% in most cases; however, it has been reported to be as low as 0.2% or as high as 7% depending on the system. This is often a judgment call by the analyst. Some of the sources of modal damping are material hysteresis and working fluid interaction.

13.10.1 Shaft Shear Stress and Fatigue

When shafting fails as a result of fatigue from oscillatory torsional stresses, there is a characteristic 45° break, as illustrated in Fig. 13.26. This is the weakest plane

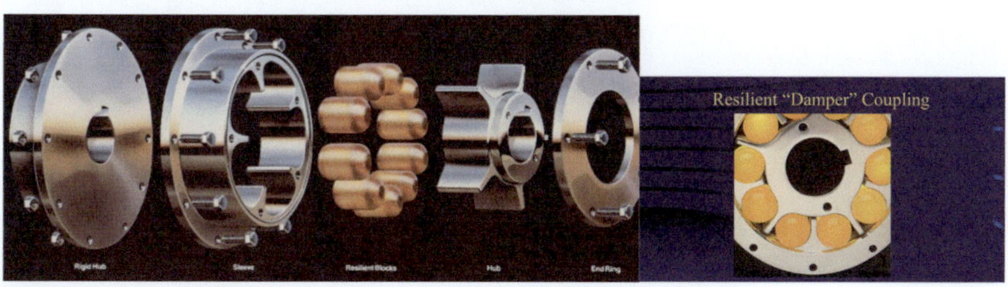

Fig. 13.25 Resilient damper coupling

in torque shear. Quite often, the fatigue failure starts in a keyway, sharp corner, or other stress riser point. Good machine design practices try to minimize stress risers and lower stress concentration factors. The forced-response analysis attempts to predict the torsional stresses throughout the train during transient events such as a synchronous motor start or at a steady speed.

Shear stress, S_s, in psi (kPa), in a solid circular shaft is given by Eq. 13.10

$$S_s = \frac{16T}{\pi D^3}, \tag{13.10}$$

where D is the outside diameter in in. (cm) and T is the torque in in.-pounds (Newton-meters). For a hollow shaft for which D_o is the outside diameter and D_i is the inside diameter, the shaft stress, S_s, in psi, is calculated as (Eq. 13.11)

$$S_s = \frac{16TD_o}{\pi \left(D_o{}^4 - D_i{}^4\right)}. \tag{13.11}$$

In the aircraft industry, where weight is a major consideration, the acceptable steady shear stress limit often exceeds 20,000 psi (138,000 kPa) for shafting. In the petrochemical industry, it is fairly rare to find steady torsional stresses exceeding 15,000 psi (103,000 kPa), and in severe-duty situations or in cases where corrosion is a factor, the steady stress is often limited to less than 10,000 psi (69,000 kPa).

Depending on the material and the service and environmental factors, machinery designers select shaft sizes based on horsepower and, when significant alternating torques are present, an assumed endurance limit. The endurance limit is the stress level at which (theoretically) an infinite number of cycles are possible without failure. In reality, there are often too many material variables to set an exact endurance limit. For this reason, guidelines given for particular situations may be hard to find, and engineering judgment must be used. There are hundreds of publications addressing stresses and fatigue in hundreds of situations.

Fig. 13.26 Shaft failure due to cyclical torsional vibration

Stress concentration factors (SCFs) must also be included when the acceptable stress level is being set for any system. Peterson's Stress Concentration Factors [5] is the industry standard for evaluating stress concentration factors in a very wide variety of geometric situations. SCF values in most references are for the situations where the part being evaluated experiences more than one million cycles. Machines with continuous torque pulsations such as reciprocating engines and compressors will easily exceed one million cycles, and the published stress concentration factors are valid. When other machines experience intermittent or occasional stress cycles that exceed the endurance limit such as a synchronous motor start, the stress concentration factors must be reduced by about 25–50% to achieve realistic results. Every situation is going to be different.

All materials will break if numerous applications of stresses of sufficient magnitude are applied. One way to evaluate stresses is the von Mises criterion, also known as the maximum distortion strain energy criterion. A physical interpretation of the von Mises criterion is that yielding begins when the elastic energy of distortion in a shaft or other part reaches a critical value. Elaborating further about the theory of elastic and plastic distortion is well beyond the scope of this chapter (refer to [18–21] for details). From a practical standpoint, if one assumes that the endurance limit of steel in tension is 30,000 psi (207 MPa), then the von Mises criterion applied to a torsional shear is 0.577 times this value, or 17,310 psi (119 MPa).

Experience has shown that applying this value as the limit yields very good correlations with experimental data and real in-service failures for cases involving alloy steels. All fatigue cycles exceeding the von Mises limit are cumulative and build up over the life of the component. Low cycle fatigue occurs when the stress is high enough to cause plastic deformation. High cycle fatigue involves the accumulation of elastic deformations.

Figure 13.27 displays an actual plot of the shear stress level versus the number of cycles to failure (*S–N*) for a typical shafting material in the high-cycle-fatigue range. In practice, the net shaft stresses the vector sum of the axial, torsional, and lateral bending stresses. Often a three-dimensional finite-element program, such as ANSYS or Nastran, is needed to evaluate a complex geometry situation. The validity of any such analysis is highly dependent on the assumptions made, the accuracy of the applied forcing functions, and the experience of the analyst.

The enlargement of a 1 s segment of time during a synchronous motor startup in Fig. 13.28 shows the calculated shaft shear stress versus time in the motor shaft, allowing the individual positive and negative peaks to be evaluated. Beginning with the maximum and minimum peaks, each pair of peaks before and after the primary pair is added together; divided by 2; and then multiplied by the stress concentration factor, any material factors, and a factor of safety. These factors must be chosen carefully because the use of overly aggressive factors will predict early failure whereas underestimation of the factors will predict too long of a component life. Figure 13.29 shows how one cycle that exceeds the von Mises limit is evaluated on the *S–N* curve shown earlier in Fig. 13.27. Thus, if this resolved alternating stress cycle is 32,206 psi (222 MPa), then this equates to 12,000 cycles to failure. The same procedure can be used for each cycle that exceeds the von Mises limit. With the resulting values, a

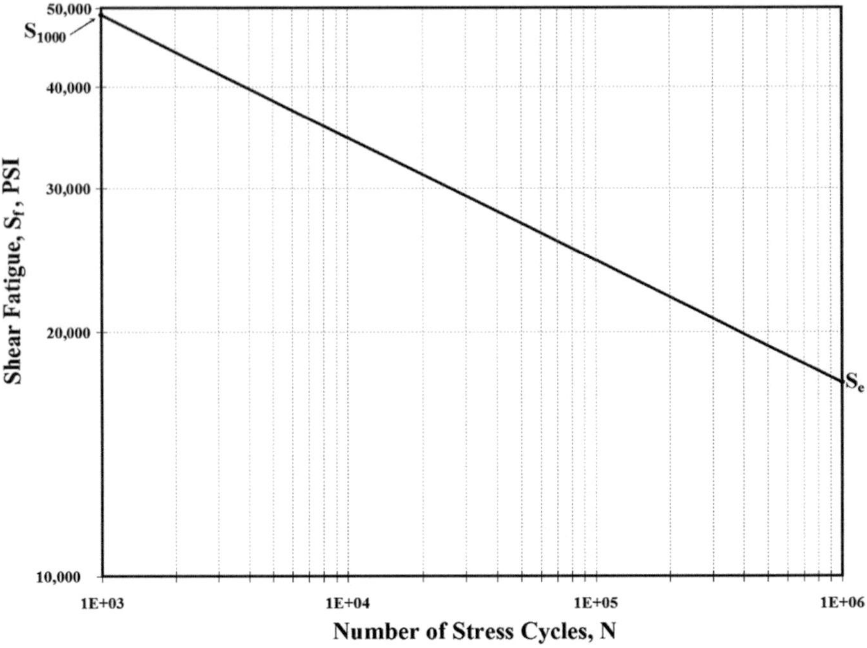

Fig. 13.27 Typical shear fatigue–stress cycle plot for steel

table can be developed (Table 13.1) that defines all of the damaging cycles and their corresponding cycles to failure. These are added inversely using Miner's law. The inverse of the resulting sum is the predicted number of cycles to failure for the shaft in question. In this case, the inverse of the inverse sum is 8983. As an additional conservative measure, this value is often divided by 2, giving 4491 safe starts in this case.

13.10.2 Variable- (or Adjustable-) Frequency Drives

In the early days of VFD motors (also called adjustable-frequency drive [AFD] motors), a fairly crude six-step synthesized power waveform was fed to the motor as a stepped square wave. This produced numerous torque harmonics and caused high torsional vibrations, stresses, and shaft failures. More modern VFD technologies, such as pulse-width modulation, have reduced this problem. If the user includes such a provision in the purchase document, the drive manufacturer will be obligated to supply a list of torque harmonics along with amplitude and phase information. Details on VFDs can be found in Refs. [13–16].

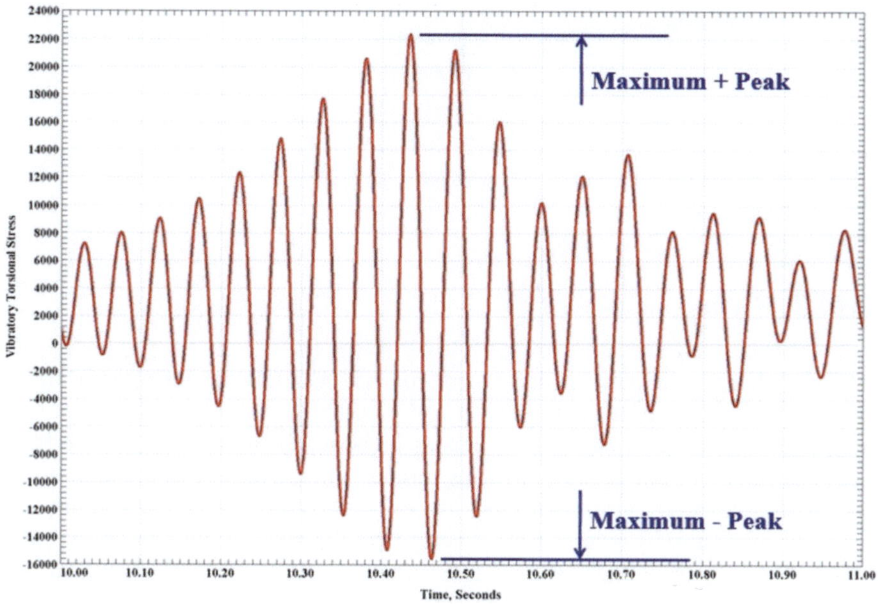

Fig. 13.28 Counting stress cycles from a synchronous motor startup

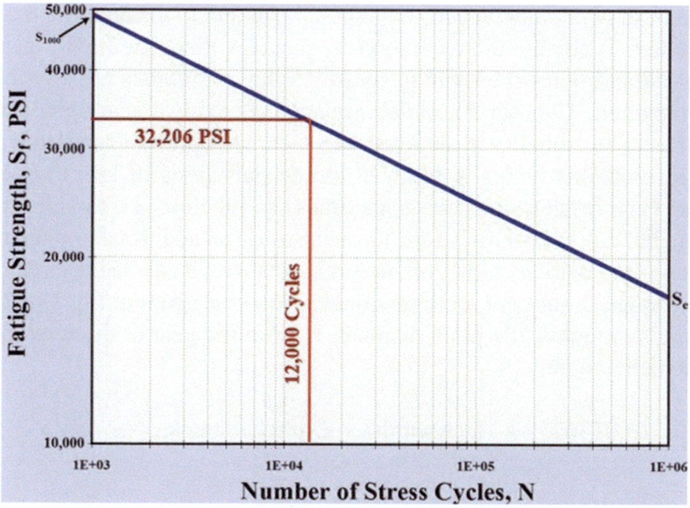

Fig. 13.29 Translation of an individual stress cycle to number of cycles to failure

Table 13.1 Relating stresses to number of cycles to failure

Cycle	Shear stress (psi [MPa])	Cycles to failure
1	19,563 (135)	3.9×10^5
2	26,914 (186)	8.8×10^4
3	32,206 (222)	1.2×10^4
4	26,684 (184)	8.8×10^4
5	20,088 (139)	3.7×10^5

13.10.3 Gears

Gear sets change the train speed (except for 1:1 gear ratios) and rotation direction. Gears also take pure rotary torque and produce lateral and tangential torque components. The lateral component creates high steady radial loads on the bearings supporting the gears. Any alternating torques will induce alternating radial load and measurable lateral vibration. Thus, the only place in most machine trains that torsional vibration will manifest itself is through the gears. Gears themselves can cause unsteady torque transmission due to accuracy errors in the gear manufacturing process. Some of the conditions that can cause a gear to function as a torsional exciter are pitch-line runout, tooth-spacing errors, excessive backlash, low-quality gears, and worn gears.

The backlash phenomenon adds a nonlinear element to the analysis that is sometimes included. Typically, the excitation frequencies associated with gears are the running speed and twice the running speed. Higher harmonics are possible but are rarely encountered. The user's best defense against having a gear torsional problem is to buy the highest-quality gears available. Gears bought in accordance with API 613 [8] specifications typically produce torque pulsations of less than 1% of the load torque. Gear mesh frequencies sometimes can function as a torsional excitation mechanism, but this is extremely rare. Gears also add an additional torsional stiffness spring that is often ignored. However, in some instances, it can be important. The API TR 684 [6] equation for gear tooth torsional stiffness is given in Eq. 13.12, where E is the Young's modulus, the pitch diameter is from the gear at the reference speed, and λ is the helix angle.

$$K_{\text{Mesh}} = (0.013625) \times (\text{face width}) \times E(\text{pitch diameter})^2 \times \cos^2 \lambda. \quad (13.12)$$

13.10.4 Sudden Load Changes

Every time an electric motor is energized, a torque shock is generated that will "ring" the torsional resonances of any machine train. Fortunately, the duration of the resulting oscillations is short and does not normally cause any harm. On complex

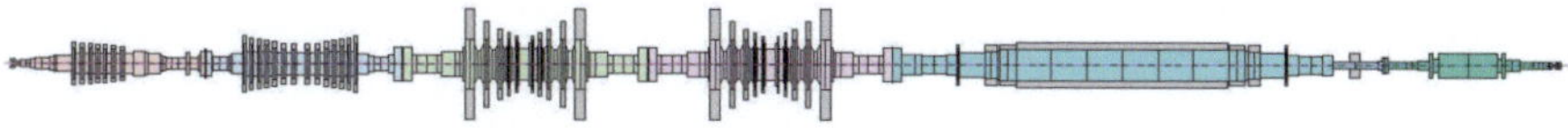

Fig. 13.30 800 MW steam turbine-generator train

Table 13.2 Conversions of common torsional analysis units

Multiply	By	To get
Length (in.)	25.4	Mm
Polar moment of inertia (lb in.^{2})	1/386.088	lb in. s^2
Polar moment of inertia (lb in.^{2})	2.929×10^{-4}	kg m^2
Torsional stiffness (in. lb/rad)	0.11298	N m/rad
Weight (lbm)	1/386.088	Mass, lb s^2/in
Weight (lbm)	0.45359	kg
Force (lbf)	4.4482	N
Power (hp)	745.7	W
Stress (psi)	6894.8	N/m^2 (Pa)

systems, very high-stress levels have been measured and caused the use of a "soft-start" system that applied initial torque gradually. Sudden load removal is also a torque pulse and occurs every time a motor/generator is turned off or a turbine trips.

These events do not cause excessive torsional oscillations. One of the worst types of situations is a total load shed on a large multimegawatt turbine-generator train similar to that depicted in Fig. 13.30. This train is nearly 100 ft (30 m) long and the rotating mass is more than 235 tons (213 metric tons [t]).

Because this train is so long and has so many torsional modes, the transient stresses can be sufficient to damage or break a shaft or the long, low-pressure turbine blades. Torsional resonances near 60 or 120 Hz can also excite tangential resonances of turbine blades. Historically, there have been very serious train wrecks due to these problems.

13.11 Common Unit Conversions

Unit conversions that are commonly used in torsional analysis are provided in Table 13.2.

Acknowledgements All of the images/graphics/figures/tables in this chapter were provided by Malcolm Leader, Applied Machinery Dynamics, Durango, CO, who is the contributing author of this chapter.

References

1. Nestorides EJ (1958) A handbook on torsional vibration. Cambridge University Press, Cambridge, UK
2. Walker DN (2003) Torsional vibration of turbomachinery. McGraw-Hill, New York
3. Corbo MA, Melanoski SB (1996) Practical design against torsional vibration, tutorial. In: Proceedings [5]of the Twenty-Fifth Turbomachinery Symposium. Texas A&M University, College Station, TX, pp 189–222
4. Ker Wilson W (1959) Practical solution of torsional vibration problems, vol 1–5. John Wiley & Sons, New Yorkilkey WD,
5. Pilkey WD, Pilkey DF (2008) Peterson's stress concentration factors, 3rd edn. John Wiley & Sons, New York
6. API TR 684-1(2019) API Standard Paragraphs Rotordynamic Tutorial: Lateral Critical Speeds, Unbalance Response, Stability, Train Torsionals, and Rotor Balancing. 1st edn. American Petroleum Institute, Washington, DC
7. Jackson C, Leader ME Design and commissioning of two synchronous motor–gear–air compressor trains. Presented at the 13th Texas A&M Turbomachinery Symposium, Houston, TX, 13–15 Nov 1983
8. API Standard 613 (2021) Special Purpose Gear Units for Petroleum, Chemical, and Gas Industry Services. 6th edn. American Petroleum Institute, Washington, DC
9. Holdrege J, Subler W, Frasier W (Jan 1983) AC induction motor torsional vibration consideration – a case study. In: IEEE Transactions on Industry Applications, vol IA-19, no. 1, pp 68–73.
10. Simmons H, Smalley A (1984) Lateral gear shaft dynamics control torsional stresses in turbine driven compressor train. ASME paper 84-GT-28 presented at the 29th International Gas Turbine Conference and Exhibit, Amsterdam, The Netherlands, 3–7 June 1984
11. Vance J, French R (1984) Measurement of torsional vibration in rotating machinery. ASME paper 84-DET-55 presented at the 7th Annual Conference on Design Engineering, Birmingham, UK, 25-27 Sept 1984
12. Simmons H, Smalley A (1989) Effective tools for diagnosing elusive turbomachinery dynamics problems in the field. ASME paper 89-GT-71 presented at the ASME 1989 International Gas Turbine and Aeroengine Congress and Exhibition, Toronto, Ontario, Canada, 4–8 June 1989
13. Dashefsky GJ (1930) The elimination of torsional vibration. Presented at the Third National Oil and Gas Power Meeting, The Pennsylvania State College, University Park, PA, 12–14 June 1930
14. Huetten V, Zurowski RM, Hilscher M (2008) Torsional interharmonic interaction study of 75 MW direct-driven VSDS motor compressor trains for LNG duty. In: Proceedings of the Thirty-Seventh Turbomachinery Symposium. Turbomachinery Laboratory, Texas A&M University, College Station, TX, pp 57–66.
15. Kaiser TF, Osman RH, Dickau RO (2008) Analysis guide for variable frequency drive operated centrifugal pumps. In: Proceedings of the Twenty-Fourth International Pump Users Symposium. Turbomachinery Laboratory, Texas A&M University, College Station, TX, pp 81–106.
16. Dickau R, Perera L (2000) Mechanical vibration problems with variable frequency drives. In: Proceedings of the International Pipeline Conference, Calgary, Alberta, Canada, October 1–5, 2000, vol 2. ASME, New York, pp 1415–1423
17. Tanaka K, Nemoto H, Takahashi N, Fukushima Y, Akita Y, Tobise M (2010) Measurement and simulation of forced torsional vibration with inter-harmonic frequencies in variable speed drive motor driven compressor. In: Proceedings of the Eighth IFToMM International Conference on Rotor Dynamics, Seoul, Korea, 12–15 Sept 2010. IFToMM, Fredericton, New Brunswick, Canada, pp 844–851.
18. Young WC, Budynas RG, Sadegh AM (2012) Roark's formulas for stress and strain, 8th edn. McGraw Hill, New York

19. Lee YI, Barkey ME, Kang HT (2012) Metal fatigue analysis handbook. Elsevier, Amsterdam, The Netherlands
20. Rice RC (ed) (1988) SAE fatigue design handbook, AE-10, 2nd edn. SAE International, Warrendale, PA
21. Rice RC (ed) (1997) SAE fatigue design handbook, AE-22, 3rd edn. SAE International, Warrendale, PA

Chapter 14
Condition Monitoring and Diagnostics

Abstract Vibration monitoring has long been recognized as a reliable method for safeguarding compressor trains. Modern monitoring systems not only offer protection but also incorporate advanced features for diagnostics and condition assessment. This chapter explores how condition monitoring technologies are applied to compressors, with particular emphasis on vibration monitoring as a key diagnostic and analytical tool. While other techniques are briefly discussed, they may hold equal significance within a comprehensive condition monitoring strategy. The chapter presents a high-level overview of this expansive subject, intentionally omitting detailed and specialized content for the sake of brevity.

14.1 Condition Monitoring

Condition monitoring is a broad term used to describe the practice of monitoring machinery operating parameters over time with the intention of identifying changes in these parameters. The changes are then evaluated to detect potential problems with the machinery. Condition monitoring incorporates vibration, temperatures, pressures, flow rates, and other process parameters, as well as machinery design factors.

Condition monitoring can be viewed to include predictive maintenance (PdM) practices and technologies beyond vibration. For example, infrared thermography is widely used to evaluate the condition of electrical equipment, boiler insulation, and rotating equipment. Some examples of PdM tools are as follows:

- Vibration monitoring (periodic and/or continuous).
- Infrared thermography.
- Lube oil condition monitoring.
- Electrical testing of motors.
- Acoustic monitoring.

Although the focus of this chapter is on vibration monitoring and diagnostics, vibration should not be evaluated in the absence of other operational parameters. Likewise, the other PdM technologies should be used with vibration when applicable.

N. Ghaisas, *Centrifugal Compressors in Hydrocarbon Processing Industries*,
https://doi.org/10.1007/978-3-032-03687-2_14

A singular focus on vibration is not desirable for the optimum outcome with respect to machinery reliability.

14.2 Protection and Condition Monitoring Systems

Monitoring systems can be divided into the following three broad categories:

- Continuous monitoring and protection systems.
- Scanning systems.
- Portable data collection systems.

The monitoring system may be selected based on a variety of factors, including machine criticality and cost. The monitored points and measurements may vary to some extent with the monitoring system. Table 14.1 provides a comparison of parameters typically monitored using systems in the three categories. This table is not intended to be all-inclusive. Depending on the specific monitoring system, the measurements and points may be different.

A continuous monitoring system is typically used when machinery protection is required. Often, American Petroleum Institute (API) Standard 670 [1] is used to define these systems. The system is designed to trip the machine when the vibration exceeds set points. Trip logic may be used in some systems to allow a combination of set points from multiple measurements to be used for the trip or to allow the use of redundant measurements. For example, redundant axial position measurements are often used, and trip logic of two out of three measurements might be required to cause a machine trip.

Table 14.1 Comparison of common parameters monitored with continuous, scanning, and portable systems

Parameter	Continuous	Scanning	Portable
Radial relative vibration: direct	√	P	X
Radial relative vibration: overall	X	P	√
Radial seismic bearing vibration: direct	P	P	X
Radial seismic bearing vibration: overall	X	P	√
Axial rotor position (thrust)	√	X	X
Axial rotor vibration from thrust probes	P	P	P
Axial bearing seismic vibration	X	P	√
Radial relative vibration: 1X amplitude and phase	√	P	X
Radial vibration relative probe DC gap voltage	√	P	P

√ the parameter is often monitored using the referenced system, *X* the parameter is likely not monitored using the referenced system, *P* the parameter may be monitored using the referenced system

Continuous monitoring systems may be connected to a condition monitoring database system. These systems work with the monitoring system to collect and store historical data. The continuous monitoring system may also provide data to other computer systems through built-in communication functionality. The systems may provide a hardwired output (4–20 mA or other scale) to use as an analog input to a recorder or to another system such as a distributed control system (DCS).

Scanning systems may be used when continuous monitoring systems are not required. These systems collect data periodically from the monitored instruments. For example, they may take a sample once per hour, once per day, once per two hours, and so on. The data are then output to a condition monitoring system on a server.

A third option for condition monitoring is periodic manual vibration data collection. This option is used to save on upfront costs, and it can be a cost-effective option when the initial investment of a continuous monitoring or scanning system is not justified. To implement this type of program, the user collects data on some frequency (monthly, quarterly, etc.) using a handheld data collector with temporarily mounted transducers. The normal program involves the collection of almost all vibration data from a temporary accelerometer. However, other vibration transducers may be used, and data from a continuous monitoring system may be sampled by connecting to designated, buffered signal outputs.

For the purposes of condition monitoring, data from the monitoring system are transmitted to a computer system. The condition monitoring software allows the user to produce plots to evaluate the data further and provides additional functionality. Some of the functions and features of typical condition monitoring programs are as follows:

- Storing data from a variety of vibration monitoring systems.
- Importing and exporting data from and to process control and monitoring systems such as DCSs and plant historians.
- Setting software alarms.
- Establishing baseline data to use as a reference on plots with current data.
- Creating logic to drive alarms using data from multiple measurements.

14.3 Vibration Measurement Locations and Transducers

Vibration data can be measured using a variety of transducers including the following:

- Proximity probe systems.
- Accelerometers.
- Piezoelectric (or "piezo") velocity sensors
- Moving-coil velocity transducers.

Because of the nature of mechanical systems, vibration responses vary across a range of frequencies. At lower frequencies, displacement is more significant than

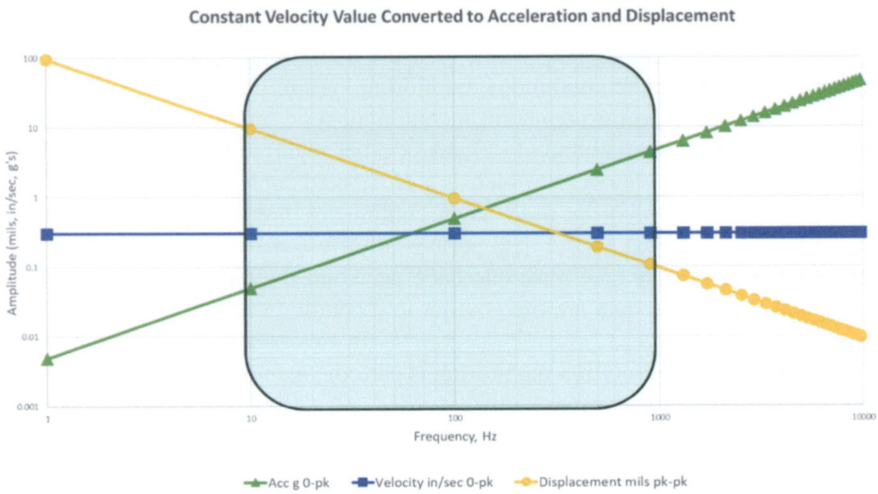

Fig. 14.1 Comparison of acceleration, velocity, and displacement measurements over a frequency range, with velocity held constant

acceleration. At higher frequencies, acceleration is more significant than displacement. Figure 14.1 illustrates this principle. For a given velocity level, the calculated displacement values and acceleration values are plotted.

General-purpose accelerometers, piezo velocity transducers, and proximity probes are often used to monitor vibrations in the range of 10–1000 Hz. Within most of this range (highlighted), Fig. 14.1 shows that all of the measurements could be used. When there is a need to measure low- or high-frequency acceleration, special-purpose transducers may be required.

Proximity probe systems use eddy currents to sense the distance between the probe tip and the target surface without contacting the target. For vibration measurement, the target is the moving shaft. Because the signal varies with the distance, the resulting measurement is the distance between the probe and the target in displacement units. The average distance is used to determine the average position of the rotor within the bearing clearance, and the dynamic part of the signal is used to determine the vibration.

Proximity probes are often installed in *X*–*Y* orthogonal pairs. This configuration allows for the plotting of shaft orbits that depict the dynamic path of the shaft centerline. Similarly, the average location of the shaft can be plotted to show the position of the rotor within the bearing.

Proximity probes differ from accelerometers and velocity transducers with respect to the following characteristics:

- Proximity probe measurements are relative to their mounting location. For example, if the probe is attached to the side of the bearing, the measurement is shaft motion relative to the bearing. Accelerometers and velocity transducers measure absolute motion relative to Earth.

- Proximity probe measurements are valid down to 0 Hz. They are often used for position measurements in addition to vibration measurements. Acceleration and velocity measurements roll off at frequencies above 0 Hz, although special low-frequency transducers are available.
- Proximity probes are sensitive to shaft imperfections, and because they can measure low frequencies, shaft bow and other sources of mechanical and electrical runout are present in the signal along with the vibration signal. For some analyses, it is desirable to subtract these low-frequency parts of the signal. This process is called compensation.

Slow-roll compensation of a proximity probe signal is performed by taking a measurement at a speed where there are no measurable contributions to the signal from dynamic motion of the shaft. Slow-roll data can be recognized on a Bodé plot. The slow-roll measurement is then subtracted from all other measurements, providing compensated data. See Sect. 14.4.4 for additional information on recognizing and using compensation on Bodé plots. After the data have been compensated, the results can be displayed on many plots, depending on the conventions used by the manufacturer of the data collection system and analysis software.

Glitch is a term that is sometimes used to refer to the combination of mechanical and electrical runout that affect proximity probe data. This terminology implies that all slow-roll data comprise error or noise, and this implication is incorrect. In fact, slow-roll data can be useful for diagnosing problems in some cases, and slow-roll data can be mostly noise in some cases. For example, slow-roll data resulting from residual magnetism represent error or noise, but slow-roll data resulting from a residual shaft bow or a coupling problem are the result of a real mechanical problem. Often, the analyst cannot discern between meaningful slow-roll data and slow-roll error. Therefore, slow-roll compensation should be used only as an analysis technique, not as a tool to remove "error" from the signal [2, 3].

Accelerometers use a piezoelectric crystal material that changes properties in response to changes in forces applied to it. The resulting measurement is the acceleration of the transducer relative to Earth, also known as the absolute acceleration. Because they are measuring absolute motion, accelerometers are a type of seismic transducer.

Piezo velocity transducers consist of an accelerometer with a built-in integration system to convert the output to velocity units. Like accelerometers, they are a type of seismic transducer because they measure velocity relative to Earth.

Another type of seismic transducer is a moving-coil velocity transducer. They produce a velocity output when a coil is moved relative to a magnet. Moving-coil transducers were popular before piezoelectric accelerometers and velocity transducers became popular. These transducers remain in use today for some applications such has high-temperature measurements. Because they are self-powering, they can be easily used with some temporary monitoring systems.

Table 14.2 provides a comparison of four common types of transducers. This table is intended to be a simplified, general comparison. A wide variety of specialized transducers are available for consideration when required.

Table 14.2 Comparison of vibration measurement transducer systems

Characteristics	Proximity probe	Accelerometer	Piezo velocity sensor	Moving-coil velocity transducer
Measures relative shaft vibration	√	X	X	X
High-frequency measurements	X	√	P	X
Low-frequency measurements	√	P	P	√
Good across a wide frequency range	P	√	√	P
Self-powering	X	X	X	√
Suitable for high-temperature environment	P	√	X	√

√ this system is the preferred or only choice, *X* measurement is not possible or not a good option for this system, *P* data may be recorded with this system, but it is probably not the best choice

Vibration is typically monitored at each bearing in the machine train. The specific location of the measurements varies according to the monitoring system.

A typical compressor machine train with continuous monitoring containing a steam turbine driving a compressor through a gearbox might include the following measurements:

- *X*–*Y* proximity probes at each bearing mounted at 45°L and 45°R from top dead center on each compressor and turbine radial bearing.
- *X*–*Y* proximity probes on each gearbox bearing, if possible. Depending on gearbox design, the probes might be mounted at angles other than 45°L and 45°R. For example, some probes may be mounted at 135°L and 135°R.
- Redundant thrust probes monitoring axial position and vibration at each thrust bearing.
- One or two accelerometers per gearbox to monitor high-frequency vibrations on the gearbox casing.

For the same machine train monitored by a portable vibration system using an accelerometer, measurements might include the following:

- Horizontal and vertical measurements on each radial bearing taken at true vertical and true horizontal directions.
- Axial vibration on at least one bearing per machine case.
- Multiple horizontal, vertical, and axial measurements on the gearbox.

14.4 Vibration Data Acquisition and Signal Processing Fundamentals

The acquisition of "vibration" data does not consist of a simple, singular measurement. Rather, it comprises a family of measurements, calculated parameters, and graphical representations of the vibration signal. Because the signal is inherently dynamic, special processing must be applied to characterize the signal using measurements. The signal is also analyzed graphically, and the digital capture of the signal presents challenges that must be addressed to produce plots that accurately represent the signal.

Because of the complex nature of measuring and capturing the vibration signal, comparisons from one instrument to another can be difficult. Even measurements made with a single manufacturer's family of devices can vary greatly depending on the configuration of the measurements. The discussion provided herein is not intended to be a complete explanation of signal processing. However, the user must be aware of the basics of signal processing when attempting to evaluate vibration measurements. References provided at the end of the chapter include additional information on the details of signal processing.

14.4.1 Measurements

Some common measurements made on vibration signals include the following:

- Direct amplitude.
- Overall vibration.
- *n*X amplitude and phase.
- Not 1X.
- Gap voltage.
- Bias voltage.

Direct vibration amplitude is a measurement of the amplitude of the vibration signal before most processing. For displacement signals, peak-to-peak (denoted pk-pk) units are used. For acceleration and velocity signals, 0-peak (0-pk) or root-mean-square (rms) units are used. The actual measured values depend on the rise and fall time of the peak detector used, the frequency range of the measurement, and other parameters. Another consideration is that all systems do not calculate 0-pk values in the same manner. Bently Nevada, for example, defines a 0-pk value as one-half of the pk-pk value, regardless of the symmetry of the signal.

Root-mean-square (rms) measurements are an indication of the energy in the signal. It is sometimes believed incorrectly that the relationship between the peak-to-peak and rms values is pk-pk $= 2\sqrt{2} \times$ rms. This is true only for a signal with a single sine wave frequency. However, this convention was adopted by instrument manufacturers when vibration measurements were becoming common. Therefore, some meters measure the rms value and multiplied it by a conversion factor to obtain a 0-pk or pk-pk measurement despite the possibility of large errors.

Overall vibration is a measurement that characterizes the amplitude of the vibration signal. However, the term "overall" should be used with context. For example, the overall vibration is often taken from the square root of the sum of the squares of the spectral peaks. In some cases, it is taken from the waveform.

The nX amplitude and phase measurement includes an amplitude of the signal filtered to a single frequency and a phase angle measured from a point in time to the first positive peak of the signal in the time waveform. The time is converted to degrees based on defining one vibration cycle as 360°.

The 1X amplitude and phase is the most frequently used nX amplitude and phase measurement. For a 1X amplitude and phase measurement, one vibration cycle is equal to one rotation of the shaft. As a result, the 1X phase angle is the location on the shaft of the high spot, where the high spot is the location on the shaft under the probe when the shaft makes its closest approach to the probe. In terms of the filtered vibration signal, the high spot corresponds to the first positive peak in the signal after the once-per-turn reference pulse occurs.

The not 1X vibration is the part of the signal including all content except the 1X vibration. This measurement is made by a filter or combination of filters that block the 1X content while allowing the rest of the signal to pass through the filter.

Gap voltage is the DC component of a proximity probe signal. In most proximity probe systems, the power supply is − 24 V DC. Older systems may have used a − 18 V DC power supply. When the proximity probe system is connected to the power supply, the signal output is dependent on the distance between the probe tip and the shaft or any metal part that is large enough to affect the probe signal. When the shaft is touching the probe, the signal is more positive (about − 2 V DC). When the shaft is too far away from the probe to affect the signal, the signal is more negative (about − 20 V DC). Gap voltage is used to provide position measurements and can be an indication of probe system health.

Bias voltage is the DC component of a seismic sensor such as an accelerometer or a piezo velocity sensor. This measurement can be trended to track the health of the sensor.

14.4.2 Dynamic Sampling and the Fast Fourier Transform

The shapes, patterns, and frequency content of vibration signals are useful for interpreting machine behavior. Prior to the widespread use of computers, analog devices such as oscilloscopes and swept filters were used to view and evaluate the signal. Modern condition monitoring and vibration analysis systems capture and record samples of the signals. These systems provide plotting tools to allow an analyst to conveniently evaluate the dynamic samples.

The frequency content of the vibration signal may not be obvious from observation of the signal. Fourier analysis is the main tool used to evaluate the frequencies present in the signal. In the early 1800s, Jean Baptiste Joseph Fourier claimed in a paper on heat propagation that any continuous periodic signal could be represented as the sum of sinusoidal waves. This claim formed the basis for Fourier analysis [4, p 141]. In 1965, James W. Cooley and John W. Tukey published a paper that provided a detailed process for performing a complex, discrete Fourier transform on a computer. The algorithm became known as the fast Fourier transform (FFT) [4, p 225; 5].

The FFT is the mathematical implementation of Fourier analysis that transforms the signal from the time domain to the frequency domain, as shown in Fig. 14.2. The FFT determines how much of the signal content is attributable to the frequencies within the span.

The process of converting an analog signal into a digital signal is known as sampling. A high-level block diagram of the process is shown in Fig. 14.3. Both low-pass filters provide alias protection. The initial analog filter is fixed at the maximum frequency of the instrument. The digital filter is configured based on the resample frequency. Some or all of the samples that result from the process are stored in a database for plotting and further analysis.

The resample step is often the collection of an asynchronous sample configured on the basis of a frequency span and the desired resolution of the spectrum that would be calculated by the FFT process. Because the asynchronous sample settings are based on the FFT process, it is logical that the samples are configured to provide optimal spectral plots.

Another resampling technique is synchronous sampling. This process involves collecting a fixed number of samples per shaft rotation. The sampling speed depends on the rotor speed. At full speed, synchronous sampling is normally performed at a higher rate than asynchronous sampling, resulting in better-quality timebase plots (orbits and waveforms). The high sampling rate results in low-resolution spectral plots.

14.4.3 Data Analysis

This section explains the various trend plots and their use in condition monitoring.

14.4.4 Plot Types Used to Display Vibration Data

Vibration data can be viewed in plots of several types including the following:

- Trend.
- Time waveform.
- Spectrum.
- Orbit.
- Average shaft centerline.

- Bodé.
- Polar.
- Waterfall.
- Cascade.

A trend plot (Fig. 14.4) is simply a plot of values over time. With vibration data, the typical measurements (direct, 1X amplitude and phase, etc.) discussed previously may be plotted. Often, vibration trend data are correlated with trend data from process parameters such as pressure, temperature, and flow to gain insight into machine behavior. Trend plot data are typically collected at a rate of about one sample per second. However, the actual collection rate can vary greatly between systems and is dependent on the configuration of the system. Some systems capture high-density trend data in response to alarms. For example, one well-known vibration monitoring system captures data at a rate of one sample every one-tenth of a second in response to an alarm.

Another plot of vibration amplitude versus time is a time waveform plot, as shown in Fig. 14.5. This plot is a graphical reproduction of the dynamic vibration signal. It displays data that have been sampled from the digitized raw vibration signal. The data are sampled at a much higher rate than a trend plot. For example, consider a compressor operating at 10,000 rpm. A typical synchronous sample rate configuration for this machine would be 128 samples per revolution. This equates to 1,280,000 samples per minute or 21,333.33 samples per second. A typical asynchronous sample would be configured for a frequency span of 500 or 1000 Hz. The sample rate for a 1000-Hz sample is 2560 samples per second.

For the analysis of vibration data, capturing a faithful reproduction of the signal is important. The shape of the signal can provide information that helps to characterize the nature of the vibration. For example, a clean sine wave may indicate that most of the vibration is caused by unbalance. Impacting can be seen in the time signal, and certain impact patterns may be related to rolling-element bearing faults or abnormal signals from gearboxes.

A spectrum plot (Fig. 14.6) shows the amplitudes and frequencies resulting from the FFT process. The analyst can use the frequency content to find potential issues with the machine. A subsynchronous peak may indicate a rotor stability problem. A peak at 5X could indicate a vane passing frequency. For many reasons, the spectrum plot has become one of the most important and widely used plots in vibration analysis.

Despite its almost universal appeal, the spectrum plot has some limitations. The nature of the FFT and the processing techniques do not allow impacting to be clearly indicated in a spectrum. Also, the amplitude of high frequencies in the spectrum is likely not accurate. The sampling must be configured with enough resolution to show closely spaced frequencies. Some of the limitations of the FFT arise from the assumption that the continuous signal is the same as the sample and that it repeats infinitely. Stated another way, the signal goes from negative infinity to positive infinity and matches the sample. In short, the analyst must understand the potential issues and limitations of the spectrum to be able to use it effectively.

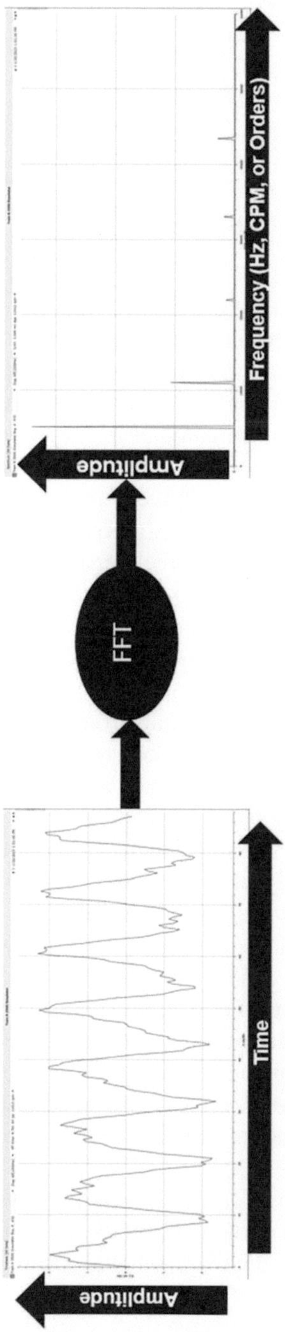

Fig. 14.2 FFT converts the signal from the time domain to the frequency domain

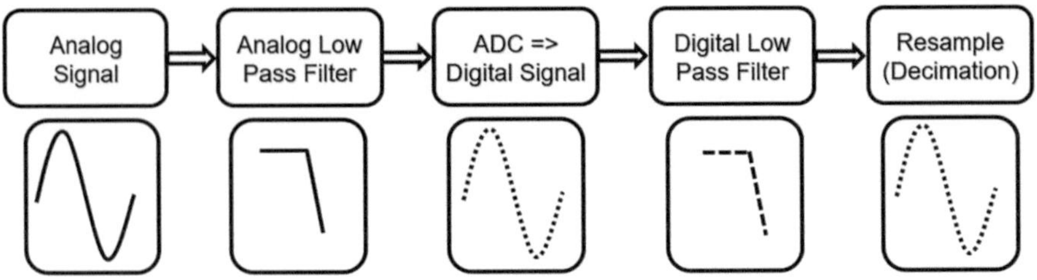

Fig. 14.3 Block diagram of the vibration data sampling process

An example of a variation of the typical spectrum plot is the full-spectrum plot shown in Fig. 14.7. The full-spectrum plot is constructed from two signals. The signals are typically from two proximity probes observing the same axial location on the shaft and mounted in orthogonal locations. Because the full spectrum is based on two signals, the plot provides information beyond the spectrum of a single signal, sometimes called the half spectrum. The additional information provided in the full spectrum is the precession direction and ellipticity of a filtered orbit at each frequency in the spectrum. Because the full-spectrum plot is derived from two probes and provides information pertinent to the orbit, the plot is sometimes defined as the spectrum of an orbit. In Chap. 8 and Appendix 5 of *Fundamentals of Rotating Machinery Diagnostics* [6], Bently and Hatch provide additional details on the full-spectrum plot, including discussions on the processing that is used to create the plot.

An orbit plot (Fig. 14.8) displays the motion or the path of the shaft centerline. The path is displayed by combining the signals from two probes observing the same axial location, sometimes identified as the probe target, on a shaft. The orbit is often displayed with the two signals that are combined to generate the orbit. In these cases, it may be called an orbit timebase plot.

Orbit plots are important in vibration analysis because they reflect the forces that control the rotor response. Typically, the unbalance force is the cause of most vibration, and the unbalance force is a symmetrical force. If the forces that control the rotor response are symmetrical (isotropic), then the rotor response should be circular. Other factors, however, cause orbits to be elliptical rather than circular. Most bearing characteristics are not isotropic, and these uneven bearing stiffnesses result in elliptical orbits. Other factors include the bearing support design, piping attachments, and foundation design. In addition to the design factors, preloads applied to the rotor system cause the orbit shape to vary. Preloads are steady-state unidirectional forces applied to the rotor, including process loads and misalignment forces. Preload forces can distort the normal orbit path.

A plot that is closely related to an orbit plot is the average shaft centerline position plot. This plot shows the position of the shaft centerline within the bearing clearance. The shaft position is measured by the DC gap voltage from the proximity probe. The shaft orbit is the dynamic position of the shaft centerline measured by the variable component of the probe signal. The average shaft position is the average of the orbit.

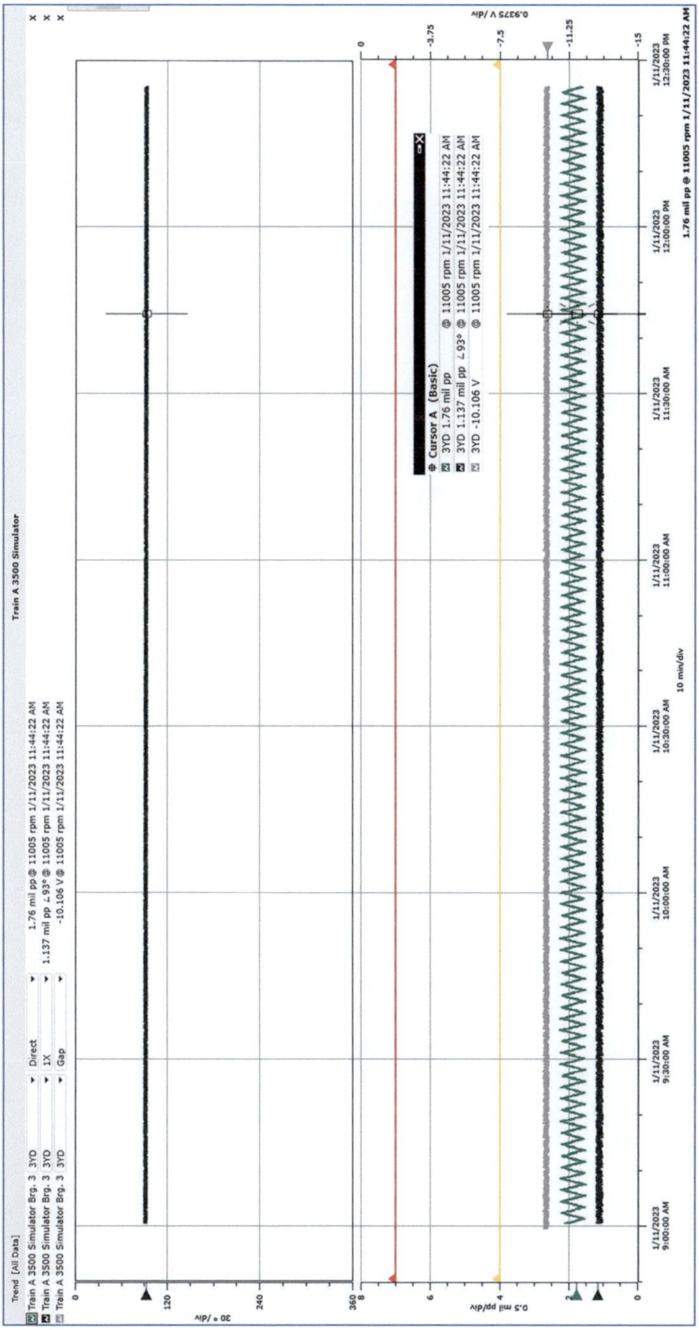

Fig. 14.4 Example trend plot with direct vibration (green), 1X amplitude and phase vibration (black), and gap voltage (gray), along with alarm (yellow) and danger (red) levels for the given application

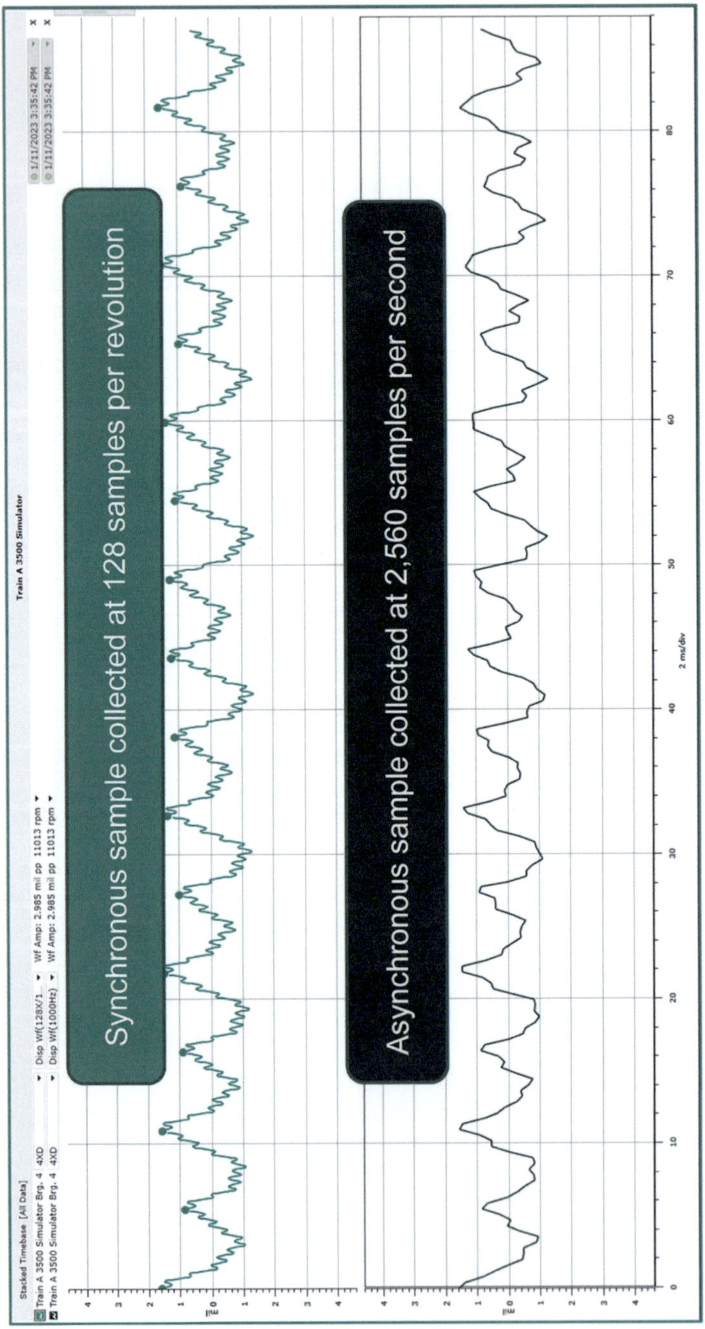

Fig. 14.5 Example time waveform plots

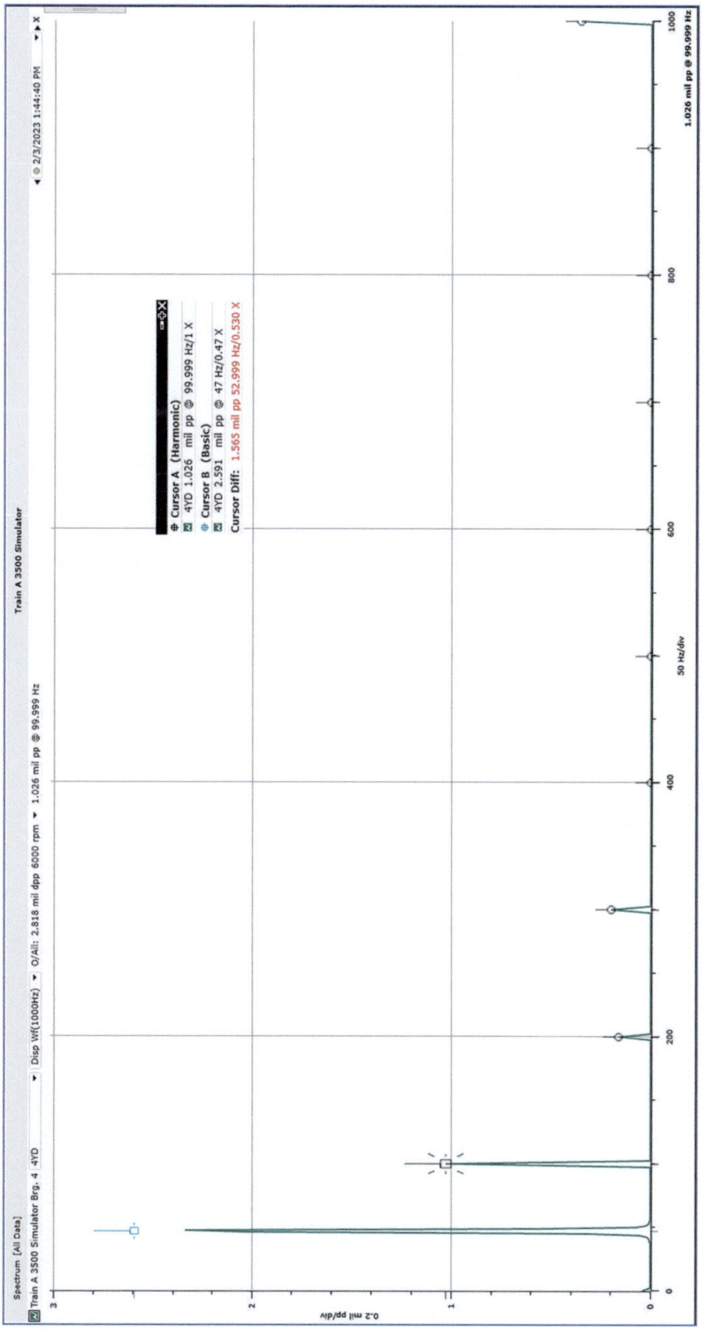

Fig. 14.6 Example spectrum plot

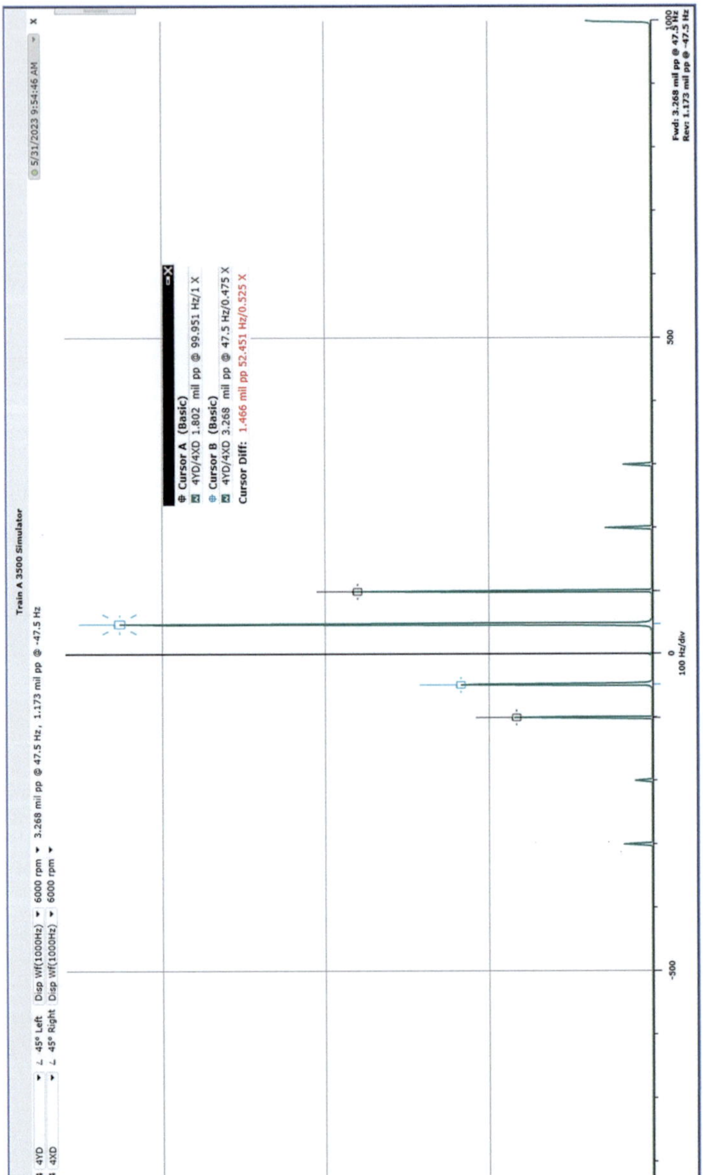

Fig. 14.7 Example full-spectrum plot

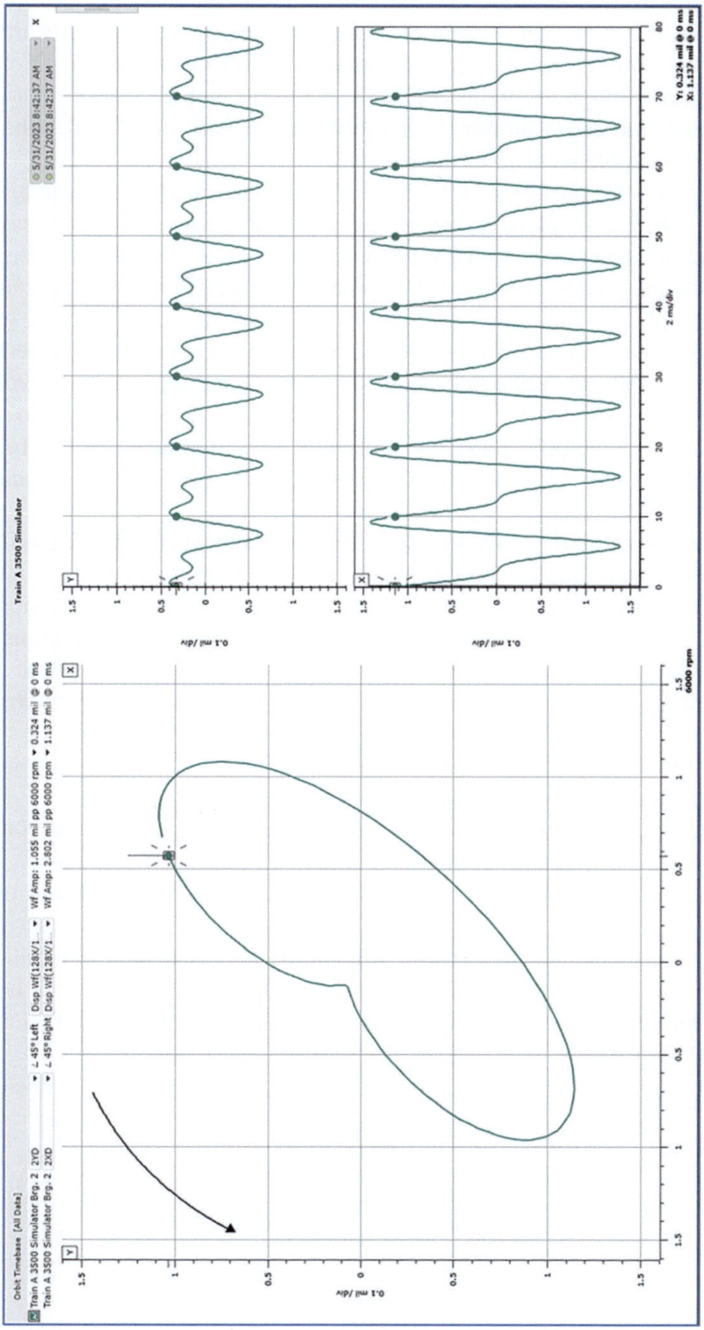

Fig. 14.8 Example orbit timebase plot

Figure 14.9 provides an example of an average shaft centerline plot. The plot also contains an orbit overlay from the full-speed operation of the machine.

A Bodé plot is a plot of the vibration amplitude versus the phase in Cartesian coordinates. The most common parameter plotted on a Bodé plot is the 1X vibration. However, any filtered vibration (2X, 3X, 4X, etc.) may be shown on a Bodé plot. Sometimes, the direct amplitude is plotted along with the filtered vibration component(s). Example Bodé plots are provided in Figs. 14.10 and 14.11.

The Bodé plot is the preferred plot for identifying slow-roll data that can be used for compensation. Figure 14.10 shows an uncompensated Bodé plot with the slow-roll region identified on the plot. Figure 14.11 shows the same data on a compensated Bodé plot. Note that, in the system used to generate the plots, the direct data are not compensated.

Figure 14.12 provides two example polar plots that were compensated. To generate the plots, the filtered amplitude and phase data are plotted on a polar coordinate system. Polar plots display the changes in phase and amplitude with respect to speed and/or time. Because the polar and Bodé plots contain similar information, they are often used together. The loops on polar plots correspond to peaks and phase changes on Bodé plots.

Unlike Bodé plots, polar plots are useful during steady-state conditions. Under most conditions, phase angles should remain relatively stable and constant during steady-state conditions. However, a malfunction such as a rub, a shaft crack, or a loss of mass could result in changes in the phase angle during steady-state conditions. The phase angle changes could then be evaluated on a polar plot. The same phase angle changes on a Bodé plot would be a vertical line that would be practically impossible to evaluate.

Changes in frequency content over time and speed can be observed on waterfall and cascade plots. Examples are provided in Figs. 14.13 and 14.14.

The cascade plot shows spectra plotted against speed. Because speed is constant under steady-state conditions, spectra plot over each other, rendering the plot useless for evaluating steady-state data. Transient data, however, are shown clearly on the cascade plot.

In contrast, transient data can become difficult to see clearly on a waterfall plot. The waterfall plot shows changes in spectra versus time. The waterfall plot is used to evaluate steady-state data, and it can be used to evaluate transient data if a suitable time range is selected.

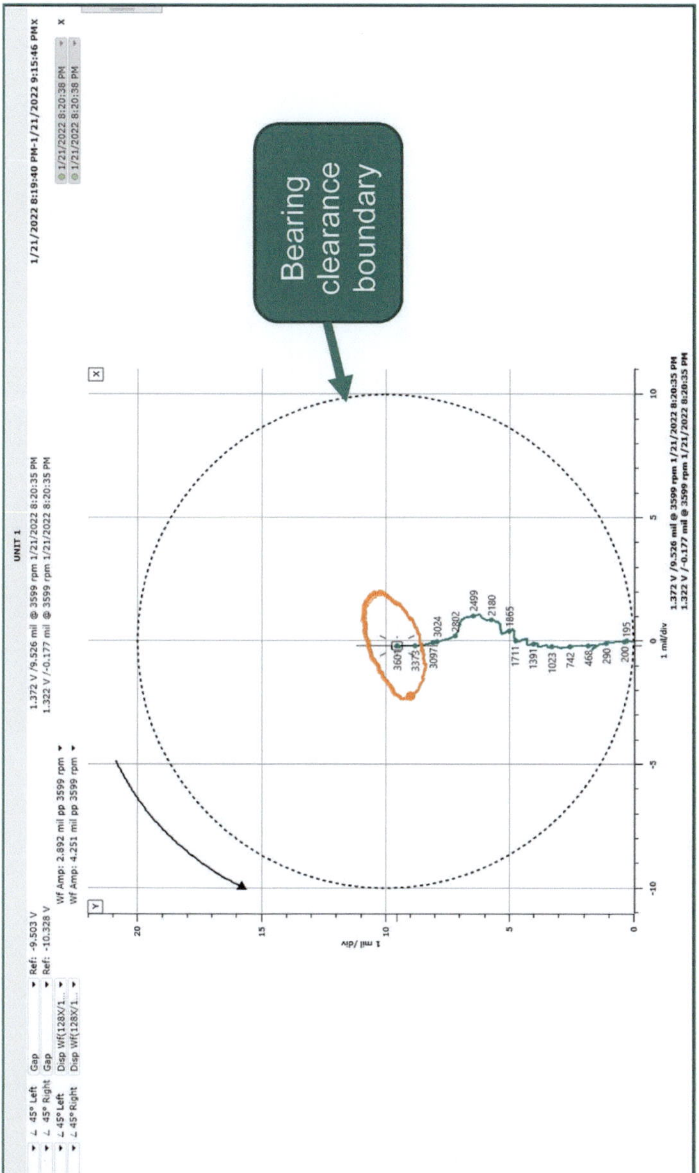

Fig. 14.9 Example shaft centerline position plot with orbit overlay

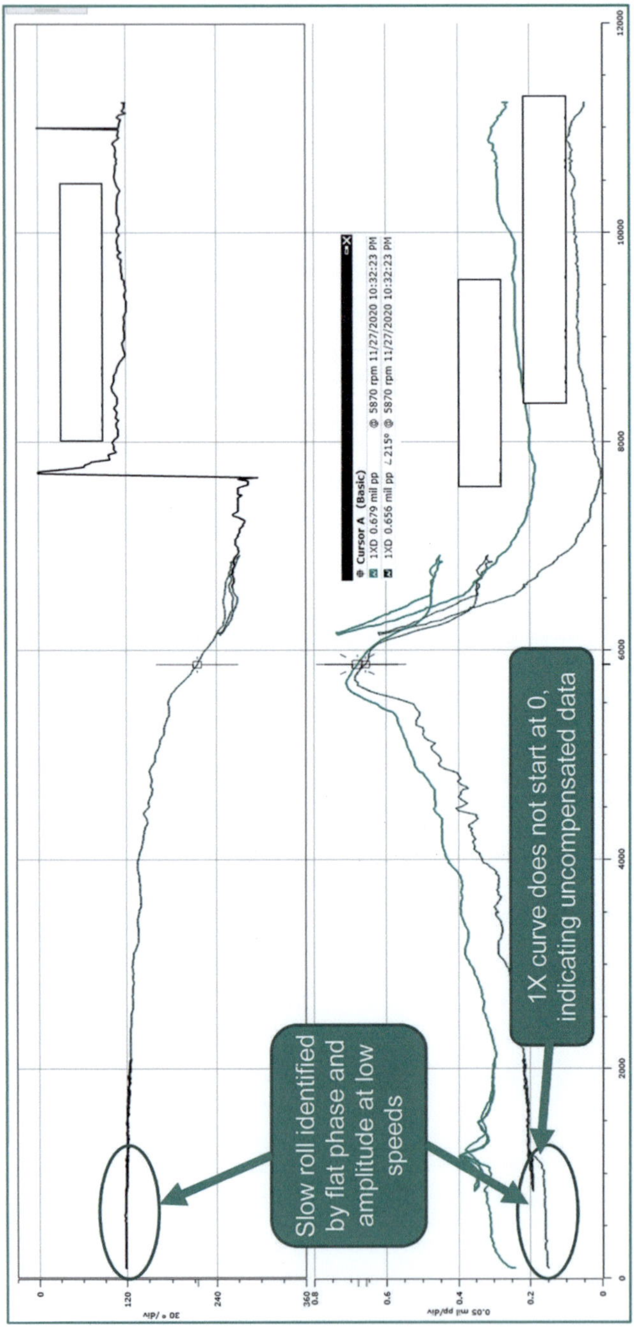

Fig. 14.10 Example of an uncompensated Bodé plot

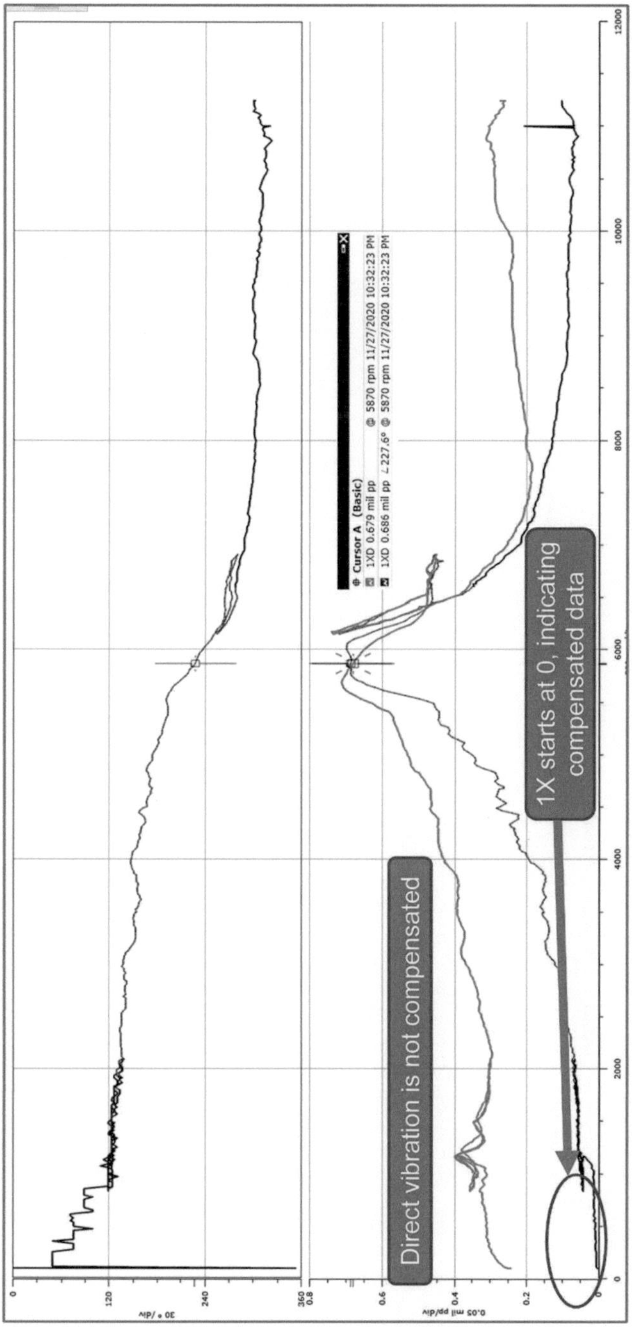

Fig. 14.11 Example of a compensated Bodé plot

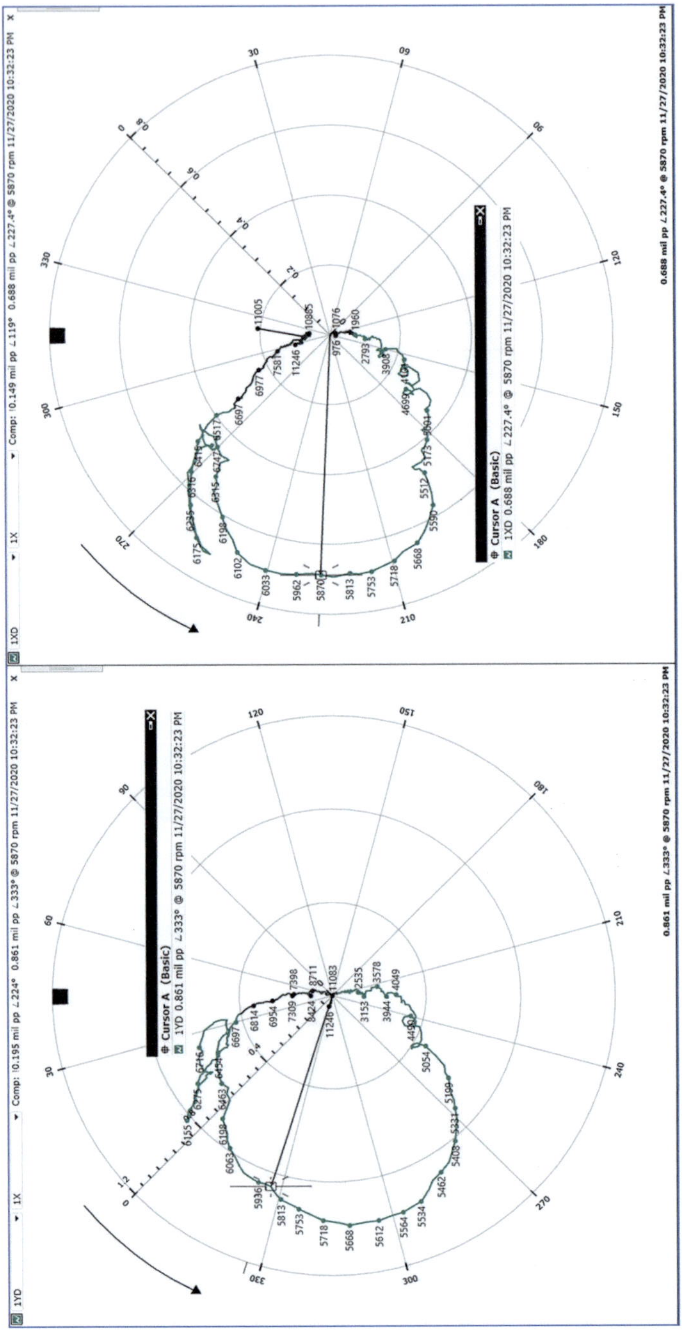

Fig. 14.12 Example polar plots (compensated)

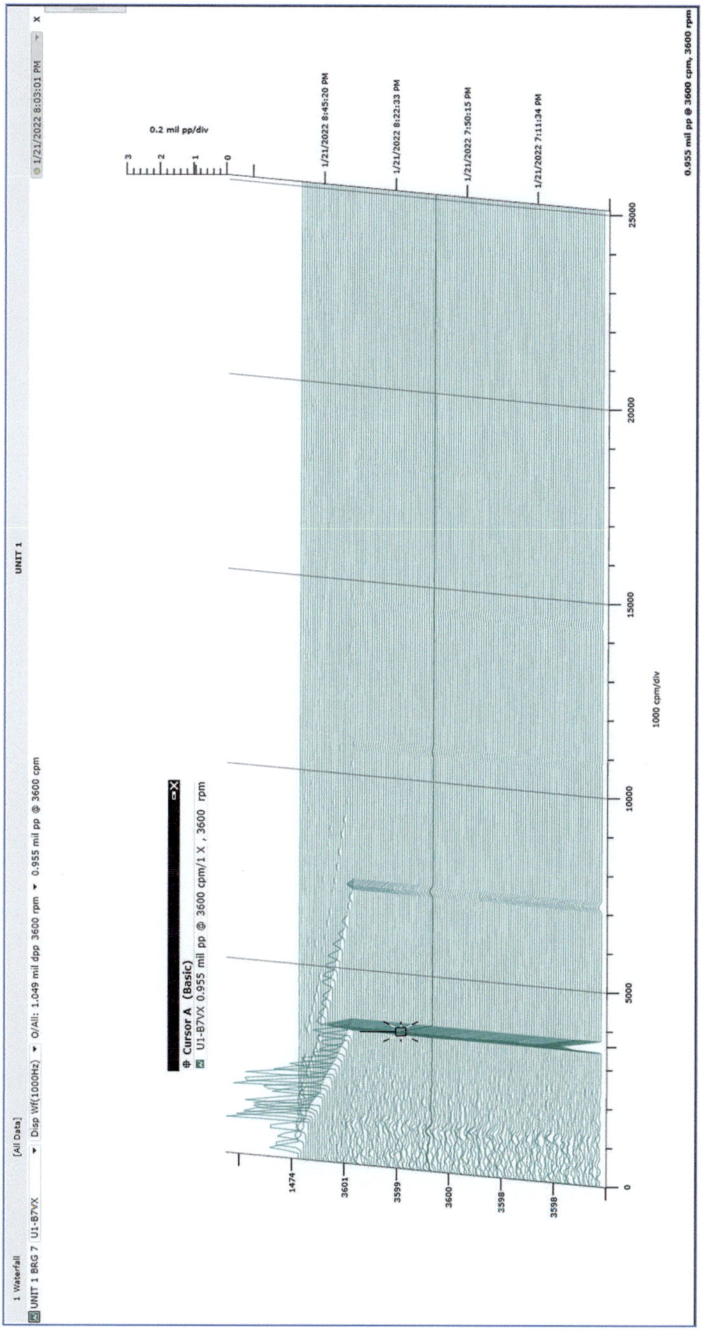

Fig. 14.13 Example waterfall plot

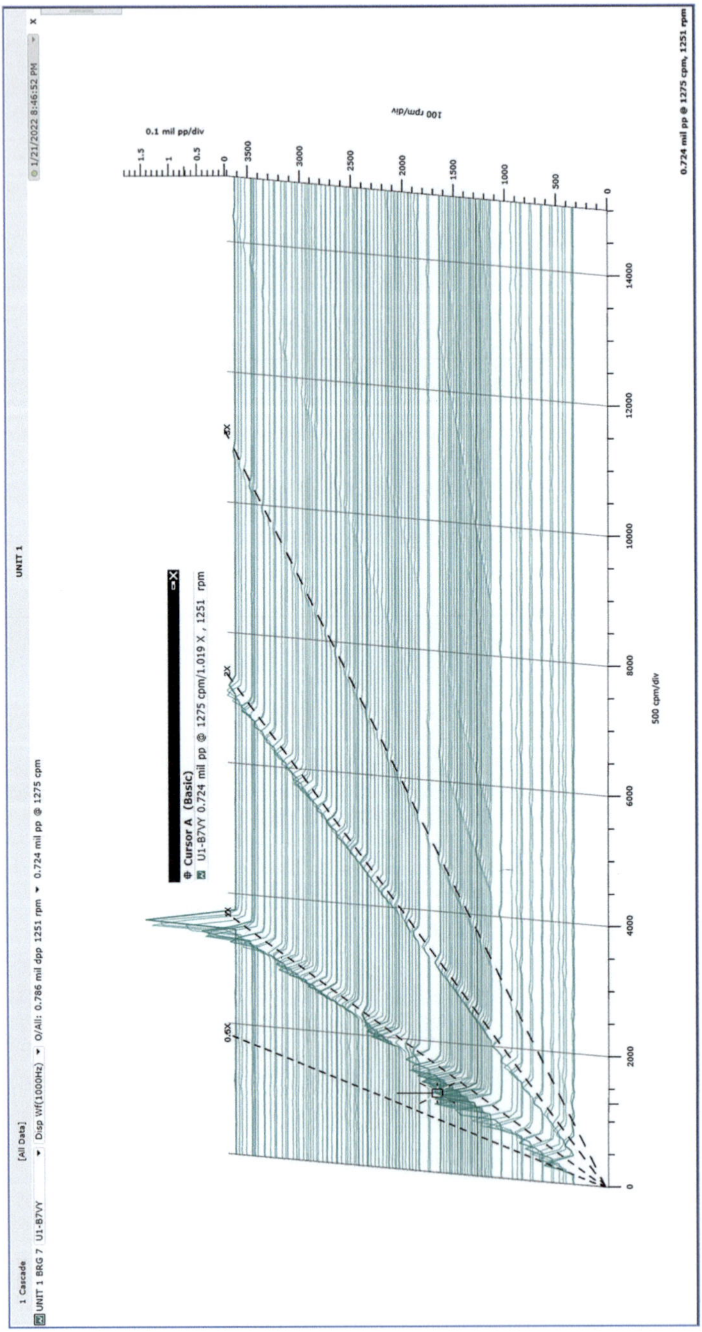

Fig. 14.14 Example cascade plot

14.5 Evaluation of Vibration Response

Any type of mechanical system responds to forces. Vibration of the system is one possible response, and the characteristics of that vibration provide information that can be used to evaluate the characteristics and health of the system and the nature of the forces. For a dynamic system such as a rotating machine, the vibration response is an important consideration. The forces that affect the vibration response include a combination of static and dynamic forces. Some examples of forces are listed in Table 14.3. The system characteristics influence how the system responds to the forces.

A dynamic system that is rotating is similar to other mechanical systems such as simple spring-mass-damper systems or structural beams. The system's characteristics determine the resonant frequency, the transient vibration response, and how the system responds to forces. A rotating system, such as a compressor rotor, introduces other effects such as gyroscopic forces.

A simplified view of the relationship among vibration response, forces, and system characteristics is that the vibration response is related to the forces and controlled by the dynamic stiffness, where the dynamic stiffness is a function of the system characteristics. The system characteristics are summarized as system mass, spring stiffness, and damping.

$$\text{vibration} = \frac{\text{force}}{\text{dynamic stiffness}} \tag{14.1}$$

$$\begin{aligned}\text{dynamic stiffness} &= \text{spring stiffness} + \text{mass stiffness}\\ &\quad + \text{damping stiffness}.\end{aligned} \tag{14.2}$$

The relationship described in Eqs. 14.1 and 14.2 can be used to evaluate changes in the vibration response. For example, if the vibration response experiences a step change, an obvious possible cause is a sudden change in the unbalance force with no change to the dynamic stiffness. Although less probable, the change could also be due to an instantaneous change in spring stiffness resulting from a crack.

Table 14.3 Examples of forces and responses for dynamic systems

Force	Type	Response
Unbalance	Dynamic	Forced vibration
Misalignment	Static	Modification of normal vibration
Impact due to rub	Dynamic	Transient vibration, modification of vibration response
Pipe strain	Static	Modification of normal vibration
Cross-coupled forces	Static/dynamic	Self-excited vibration

The relationships described in Eqs. 14.1 and 14.2 are simplified ways to characterize the rotordynamic behavior of a rotating machine such as a compressor. Chapter 12, Lateral Rotordynamics, provides additional details.

14.6 Challenges in Vibration Data Acquisition for Centrifugal Compressors

Centrifugal compressors introduce some challenges related to capturing data for analysis. These issues are not unique to compressors, but they should be given consideration when evaluating the machines.

Compressors are small machines in comparison to utility-scale turbine generators. Both machine types are classes of turbomachinery, and both are instrumented with similar instrumentation, including proximity probes, accelerometers, and piezo velocity transducers. However, the smaller size of compressors introduces complications for the installation of proximity probes and the evaluation of the data from the probes.

The linear range of a typical proximity probe (with a probe tip diameter of 8 mm [0.3 in.]) is often stated as about 80 mil or 0.080 in. (2 mm). This range implies that the full-scale range of vibration that could be measured using a proximity probe is ± 40 mil (±1 mm). Because the probe range should also include the anticipated shaft position change in the bearing, this offset should also be included.

For a typical compressor, bearings and journals normally are less than 10 in. (25.4 cm) in diameter, and diametrical clearances are normally less than 10 mil (254 μm). Assuming that the rotor runs in the center of the bearing, the change in position from stopped to full speed would then be only 5 mil (127 μm). The same machine would typically have vibration amplitudes of about 0.5 mil (12.7 μm) pk-pk. Therefore, the measurements would use about 5.25 mil (133 μm) of the 80-mil (2 mm) range.

For larger turbomachinery, bearing clearances could be 20 mil (508 μm) or more, and typical vibration amplitudes could be about 3.0 mil (76 μm) pk-pk. Therefore, the measurements would use about 11.5 mil (292 μm) of the 80-mil (2 mm) range.

Using an orbit plot as an example, the signal is often "AC-coupled." In other words, the DC offset is not considered. Because both machines are monitored with the same instrumentation, the noise level should be comparable. If a noise level of 0.25 mil (6.4 μm) pk-pk is assumed, the compressor would have a signal-to-noise ratio of 2.0 (0.5 mil/0.25 mil [12.8 μm/6.4 μm]), and the turbine generator would have a signal-to-noise ratio of 12.0 (3.0 mil/0.25 mil [76 μm/6.4 μm]). Therefore, the orbit plots from the larger machine would provide a much cleaner pattern than the plots from the smaller machine.

In some cases, this difference in plot readability due to noise is obvious on different measurements from the same machine. In the plots shown in Fig. 14.15, the orbit shown on the left has an amplitude of about 0.3 mil (7.6 μm) pk-pk, whereas the orbit shown on the right has a significantly higher amplitude, about 2.0 mil (50.8

μm) pk-pk. If the plots are placed on the same scale, the two measurements can be compared directly, as shown in Fig. 14.16.

The relatively small clearances in compressor bearings can also create difficulty when analyzing shaft centerline position plots. If the maximum expected shaft movement is only 0.005 in. (127 μm), then any minor change in the probe position due to thermal expansion or contraction of the machine can result in apparent position changes on the plot. Even changes to the rotor bow can result in changes in the average gap voltage that appear as shaft position changes on a shaft centerline plot. As with most analysis of vibration data, shaft centerline position data should not be analyzed without consulting other plots and relevant process data.

The plots provided in Fig. 14.17 were recorded from the startup of a compressor. The plot on the left (bearing 7) shows expected shaft centerline changes in position, whereas the plot on the right (bearing 8) shows highly abnormal movement. The plots show data from cylindrical journal bearings. For these bearing types, the typical shaft position change from low speed to operating speed is up and to the right with counterclockwise shaft rotation.

To fully evaluate the plot for bearing 8, the analyst would review other data such as shaft centerline data from a shutdown, bearing metal temperature data, and other process parameters. Often, shutdown data provide a more reliable view of the shaft centerline data during transient conditions because the thermal changes in the machine are less significant during the shutdown. Figure 14.18 provides plots for the same machine during a shutdown. For bearing 8, the data still show movement outside of the indicated bearing clearance, but the overall pattern of movement is normal and comparable to the movement on the other end of the rotor (bearing 7).

Often, a compressor train includes a gearbox used as a speed increaser. Speeds on the low-speed side of the gearbox may be in the range of 3600 rpm or less for a motor-driven machine. Compressor speeds could be between 5000 and 20,000 rpm. The author considers these gearboxes to be classified as "high-speed." Typical advice given for data collection is that the frequency range should be up to three times the gear mesh frequency, or 3XGMF.

Some example machine configurations encountered by the author are provided in Table 14.4.

The need to evaluate high-frequency data creates some difficulties for high-speed gearboxes. First, proximity probes measure displacement, and as shown in Fig. 14.1, displacement is attenuated at high frequencies (greater than ~ 1000 Hz). Additionally, gear faults tend to produce impacts, and the impacts simply do not result in significant displacement. Therefore, measurements on the gearbox housing are required to obtain the data needed to evaluate the condition of the gearbox. Gearbox shaft vibration measured with proximity probes is also required to evaluate faults such as misalignment, unbalance, whirl, and rubbing.

Figure 14.1 shows that acceleration units exaggerate vibration measurements at high frequencies. For the gearboxes listed in Table 14.2, measurements up to approximately 20,000 Hz would be required to capture frequencies up to three times the gear mesh frequency. For these high-frequency measurements, nonstandard accelerometers designed for high-frequency measurements may be required. Additionally,

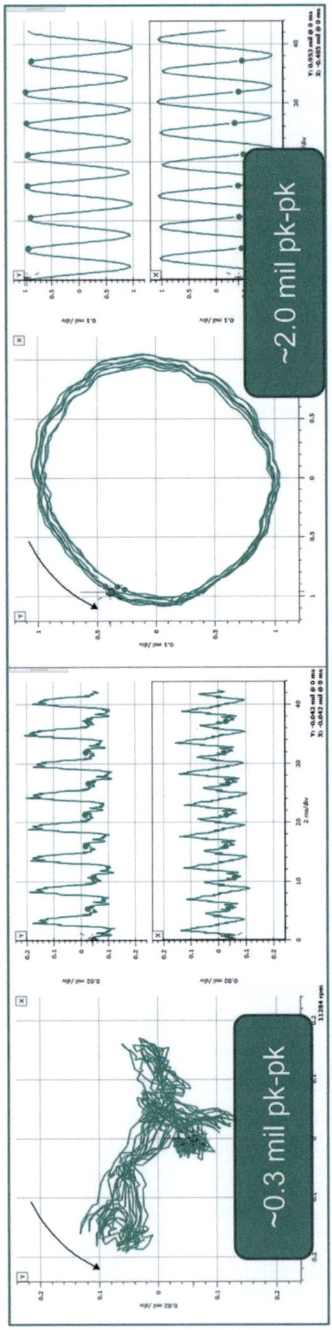

Fig. 14.15 Example of autoscaled orbit plots from the same compressor

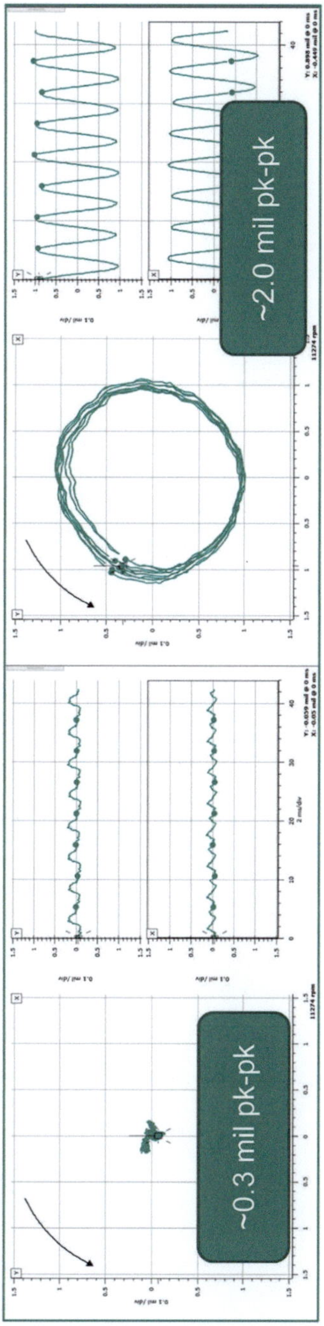

Fig. 14.16 Example of manually scaled orbit plots

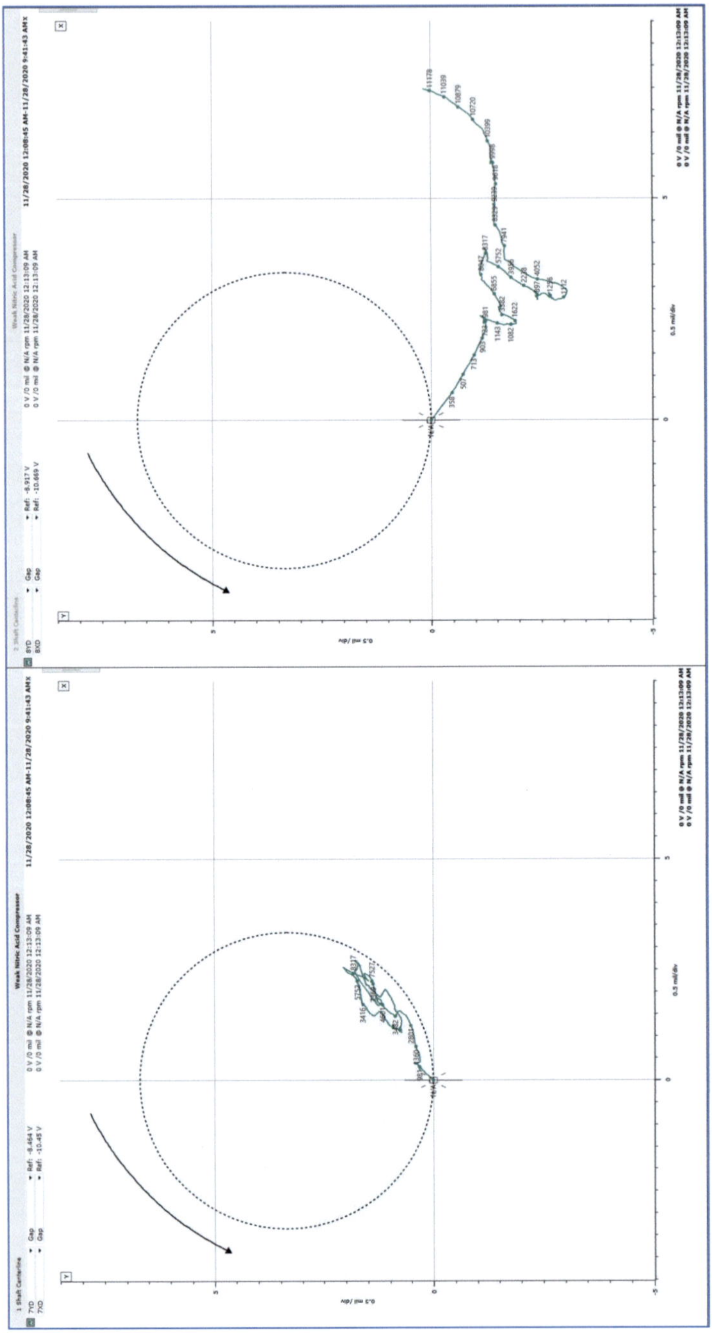

Fig. 14.17 Example shaft centerline plots from a startup

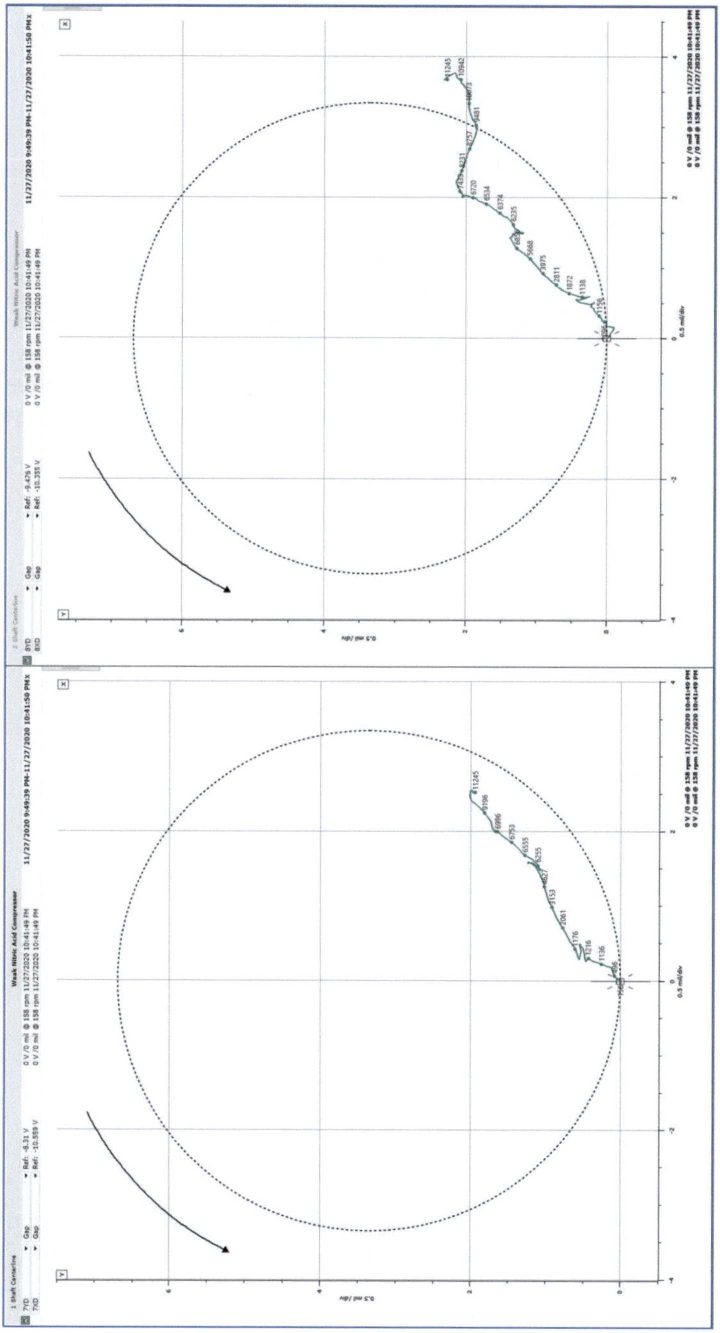

Fig. 14.18 Example shaft centerline plots from a shutdown

Table 14.4 Example gear box information for compressor trains

Machine No.	No. of gear teeth	No. of pinion teeth	Ratio	Driver speed (rpm)	Driven speed (rpm)	Gear mesh (cpm)	Gear mesh (Hz)	3XGMF (Hz)
Machine 1	95		3.276	1800	5897	171,000	2850	8550
Machine 2	223	47		1200	5693	267,600	4460	13,380
Machine 3	215	47	4.575	1766	8079	379,690	6328	18,984

permanent installation on a carefully prepared surface will be necessary to maximize the measurement range.

Another practical consideration that affects the evaluation of compressor data is that the machines are often a critical part of a process, and shutting down the compressor could require an expensive plant outage. As a result, the opportunities to collect valuable shutdown data can be very limited. Consider that a critical compressor shutdown may occur only once per year or once every three years or once every six years; the infrequent shutdowns can create considerable difficulty if data from hot shutdowns are needed to diagnose a problem.

14.7 Compressor Malfunctions and Vibration Symptoms

This section presents an overview of some common machinery malfunctions.

14.7.1 General Rotating Equipment Malfunctions: Unbalance, Misalignment

Compressors are subject to malfunctions typical of rotating equipment including the following:

- Unbalance.
- Rubs.
- Looseness.
- Misalignment.
- Shaft cracks.
- Fluid-induced instabilities.

Numerous references are available to review the expected vibration response to the listed malfunctions. Bently and Hatch [6] and Eisenmann and Eisenmann [7] are two references used by the author.

In this section, comments are provided regarding the common faults and how they relate to compressors. The discussion provided is generic in nature and based on the author's experience. An exhaustive research project was not undertaken to support the comments.

14.7.1.1 Unbalance

In general, compressors are balanced when the rotor is overhauled. Depending on the owner's preference, shop recommendations, and other advice, the rotors may be balanced in a low-speed balance machine or in a high-speed (at-speed) balance machine. Often, the process includes a "stack" balance, where individual impellers are balanced before they are attached to the shaft. Then, a final assembled balance is performed. Usually, the rotors operate in a well-balanced condition when placed in service.

If the rotor is in a fouling service, unbalance can occur during operation when fouling does not occur evenly. Wear and tear on the impellers can also cause unbalances to occur during operation.

Because the machines are not configured to support field balancing, options for addressing the unbalance without taking the machine apart are limited. In some cases, balancing on the coupling(s) may be attempted.

As with any rotating machine, monitoring for changes in the 1X amplitude and phase can be beneficial for identifying changes in the balance condition of the machine.

14.7.1.2 Rubs

Compressors are subject to rubs, but in the author's experience, rubs are less common in compressors than in other rotating equipment. Rubs that do occur on compressors are likely to occur in seals, as with most machines. Because compressors tend to operate at high speeds, tend to be relatively lightweight, and tend to operate near or above twice their first critical speed, the rubs may generate ½X and harmonics in spectral plots. However, the most common characteristic of a rub is a changing 1X amplitude and phase.

In some cases, rubs can result from fouling. Fouling material can build up in a location with tight clearances until a rub occurs. The rub then causes vibration to increase, resulting in a clearing of the fouling material. Vibration then drops back to lower levels, and the cycle continues. This behavior can also be described as being due to coking.

14.7.1.3 Misalignment

No machine is perfectly aligned. Compressors are surely misaligned as often as other machines. However, compressors are typically coupled with flexible couplings that minimize the effects of misalignment on the vibration response.

Some additional ideas on misalignment should also be mentioned. Often, claims are made that misalignment is easily identifiable using vibration data. However, in the author's experience, changes in vibration response are often poorly correlated with misalignment. The design of the coupling has a significant influence on the vibration response, and other machine design factors may also affect the response more than the misalignment. For additional information on detecting misalignment, the reader should consult the discussion provided by Piotrowski in Chap. 2 of his *Shaft Alignment Handbook* [2].

To evaluate the compressor data for signs of misalignment, the analyst should focus on proximity probe data when available. Comparisons between a known baseline with good alignment and the current conditions should be used. Possible symptoms could include changes in the orbit shapes, changes in the shaft centerline position, changes in the vibration frequency content, and changes in bearing metal temperatures.

The vibration response is related to the forces that drive the vibration, often unbalance, and the dynamic stiffness that controls the forces. Misalignment affects the dynamic stiffness by creating a static force that modifies the spring stiffness.

For turbomachinery with journal bearings, the shaft centerline and orbit plots can provide indications of possible misalignment, as shown in Fig. 14.19. The green points indicate a normal plot for a compressor with a tilting pad bearing. The motion from low speed to high speed is mostly vertical because of a lack of cross-coupling forces. The orbits are shown as open elliptical orbits, which could be typical for tilting pad bearings with a load on pivot configuration. For load between pivot bearings, the orbits could be more circular as a result of more symmetric bearing stiffness.

The overlay curves are shown in the gold color. For these curves, the rotor has moved to an abnormal position in the bearing, and the orbits have taken a more elliptical shape than shown in the baseline.

Figure 14.20 shows a possible situation for a rotor with two elliptical bearings. In this case, the normal shaft position is the bottom right quadrant, and the normal orbit shape is elliptical. Note that one of the plots shows the rotor operating slightly outside of the bearing clearance. This is often the case because of inaccuracies in the plot resulting from errors in the shaft clearance, thermal growth of the probe mounting device, and the axial location of the probe relative to the bearing.

It should be noted that the examples provided in Fig. 14.20 are generic examples of possible changes in vibration and position response due to misalignment as shown on shaft centerline plots obtained from hot shutdowns. Also, in both cases, bearing metal temperatures would be expected to change because of the misalignment.

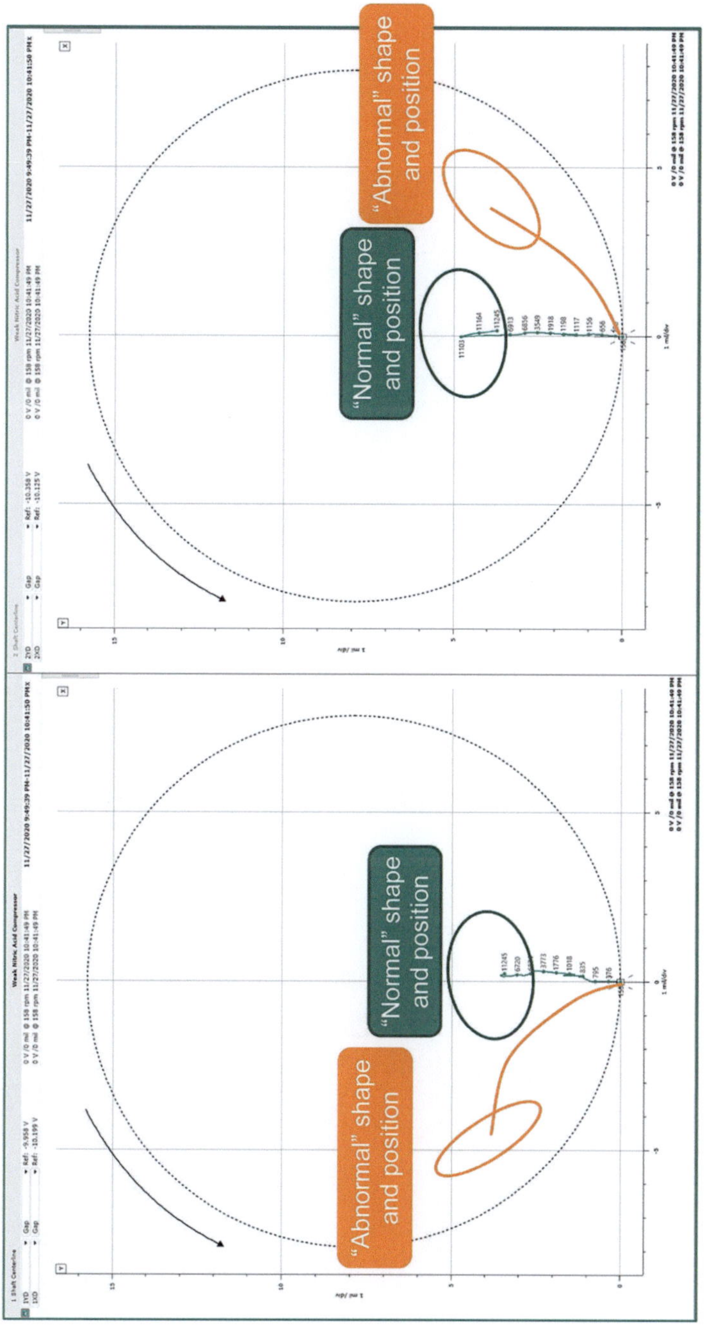

Fig. 14.19 Shaft centerline plot with orbit overlay showing baseline good conditions versus misaligned conditions for tilting pad bearings

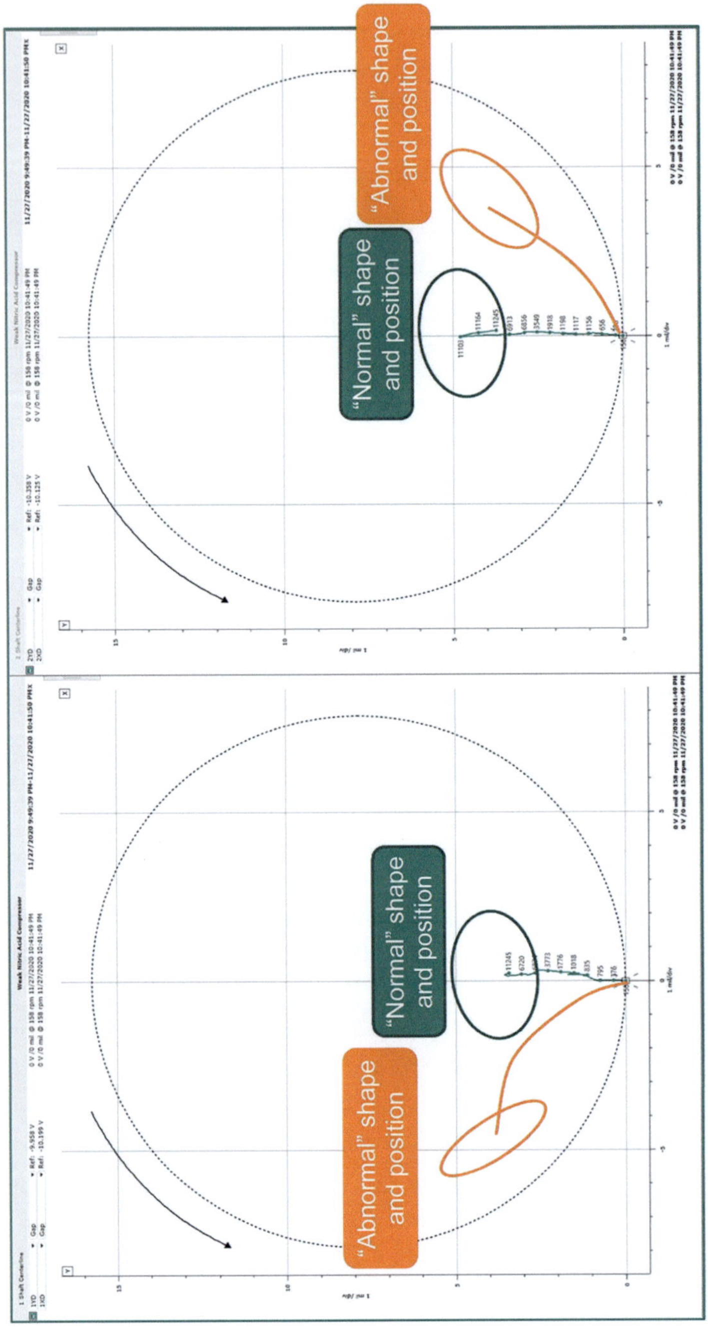

Fig. 14.20 Shaft centerline plot with orbit overlay showing baseline good conditions versus misaligned conditions for elliptical bearings

14.7.1.4 Shaft Cracks

Possible symptoms of shaft cracks include changes in the following:

- 1X amplitude and phase at steady-state conditions.
- 2X amplitude and phase at steady-state conditions.
- Slow-roll runout due to changing rotor bow.
- Critical speed response.
- Thermal sensitivity.

Shaft cracks in compressor rotors do not appear to be common, based on the author's experience. However, Eisenmann and Eisenmann discuss a shaft crack that occurred on a compressor rotor in their well-known text [7, pp 443–448].

Shaft cracks, although still rare, appear to occur in turbine generators and large vertical pumps more often than in compressor rotors. For example, there were a significant number of cracked shafts in reactor coolant and reactor recirculation pumps in the 1980s. Also, the author is aware of several cracked shafts on steam and gas turbine generators and no cracked shafts on compressors.

Because symptoms of shaft cracks can overlap with symptoms from other malfunctions, and because the diagnosis of a shaft crack is a "high-cost" diagnosis, the process is often one of eliminating other possibilities before determining that a shaft crack is the problem. On the other hand, failure of a machine while in operation due to a shaft crack can result in serious consequences.

14.7.1.5 Fluid-Induced Instabilities

Fluid-induced instability is a term used to describe instabilities that occur as a result of fluid forces resulting from the interaction of a fluid contained in a clearance and the rotor. Examples of fluid-induced instabilities include oil whirl and oil whip. These phenomena are also discussed in Chap. 12.

Truly unstable vibration would increase until the system fails. Instabilities in a rotating machine are technically limit-cycle vibrations. In other words, the vibration increases until the nonlinear characteristics of the system limit the vibration.

The force that drives the vibration in a fluid-induced instability is said to be related to the average circumferential flow of the lubricant in the cylinder. This idea is related to the understanding that the average flow velocity is typically slightly less than one-half of the surface speed of the rotor. The frequency of the whirl vibration then matches the average velocity of the fluid under certain conditions. However, the frequency of the whirl vibration is also a function of rotor rotative speed. Going from lower speeds to higher speeds, the whirl frequency will eventually match the critical speed of the rotor, typically the first critical speed. At that point, the instability is called whip, and the frequency of the vibration is locked onto the natural frequency of the unstable mode.

Bently and Hatch [6] related the average circumferential flow rate to a parameter they called lambda (λ). The use of the λ term and other aspects of Bently and

Hatch's methodology were not adopted by academia and industry. The recognition that the circumferential flow rate plays a part in driving the instability appears to have been accepted based on current texts, for example, pages 174–185 of Vance et al.'s *Machinery Vibration and Rotordynamics* [3], and the implementation of antiswirl technology in seals.

The circumferential flow results in a differential pressure around the bearing, giving rise to the follower force that is modeled by the cross-coupled terms in rotordynamic models. The follower force is in the direction of rotation, and it is a destabilizing force.

Although it is helpful to understand whirl and whip instabilities from the standpoint of a single bearing or seal as discussed to this point, the actual manifestation of the problem in a compressor is more complex and is typically evaluated using a rotordynamic model. Considering a simple compressor rotor with multiple impellers as an example, one can identify several machinery parts that contribute direct and cross-coupled stiffness and damping to the system. These parts include the bearings and seals. The longer seals such as the balance piston seal create stronger forces than the shorter seals. Dry gas seals contribute little to the stiffness and damping of the rotor and can be left out of the model. Ultimately, the combination of stiffness and damping throughout the rotor determines if the rotor modes are stable or not stable. The first mode is usually the only mode that is at risk of becoming unstable.

Factors that affect the probability that a rotor system will become unstable include the following:

- Rotor speed: Higher speeds create higher destabilizing forces and can lead to less stable operation.
- Rotor weight: Lighter rotors can be less stable than heavier rotors, although very heavy rotors can experience instabilities.
- Bearing design: Tilting pad bearings have very low cross-coupling and are often installed to correct a stability problem. Fixed-geometry bearings, especially cylindrical bearings, are prone to causing instabilities.
- Bearing loading: Improper alignment can result in bearings operating with lighter loading than the design loading, and this light loading can lead to an instability problem.

Because compressors tend to operate at relatively high speeds, they can be prone to instability problems. The machines are often designed or retrofitted with tilting pad bearings to improve stability characteristics.

Recognizing an instability in data from a machine in the field is sometimes straightforward, but there can be confusing factors. For example, rubs, looseness, and stall can produce similar frequencies. Also, a subsynchronous vibration may not be strong enough to dominate the spectrum, and the presence of other frequencies with significant amplitudes can complicate the motion of the rotor and the spectral data.

Another factor that creates confusion is that, often, reference material emphasizes details on how the problem occurs, and some of these aspects are not seen in all cases. This can lead an analyst to think that the full whirl-to-whip plots shown in textbooks will be observed when the machine has a stability problem. A more likely scenario is that the machine appears to be operating fine one minute, and then a sudden large subsynchronous vibration occurs as a result of the whip mode.

Consider that the whirl mode is difficult to excite in the field. For oil whirl to occur, the rotor would usually need to be operating toward the center of the bearing or, at a minimum, operating with a load that is lighter than anticipated. Sometimes, whirl can occur when the bearing has excess clearance. However, it is more likely that the rotor system will immediately go from stable operation into the whip mode. If the analyst is looking for whirl symptoms, the expected frequency would be slightly less than one-half, and the actual frequency could be quite different because the whip frequency is equal to the frequency of the rotor's first mode.

Figure 14.21 shows a cascade plot taken from the startup of a customized rotor kit that experienced oil whip conditions without showing evidence of oil whirl. As rotor speed increased from stopped to about 7000 rpm, the first critical peak in the vertical direction occurred at about 1700 rpm. The first critical peaks can be seen in the cascade plot at speeds near 1700 rpm. When the rotor reached higher speeds of approximately 6000 rpm and greater, a large subsynchronous peak appeared at about 1700 cpm, clearly matching the first critical peak.

Figure 14.22 shows typical orbit plots for both rotor kit bearings captured when the machine was operating under unstable conditions. The motion is mostly in the vertical direction because of the bearing characteristics. The orbits show Keyphasor™ blank and bright spots in random locations because the dominant subsynchronous vibration frequency is fixed relative to the rotor speed. Whirl vibration would show two groups of Keyphasor™ indications.

14.7.1.6 Compressor Malfunctions: Surge and Stall

One well-known compressor malfunction is surge. Chapter 4, Antisurge and Performance Control Systems, describes this issue. Surge results when the compressor flow is reduced below the surge line on the compressor map. Some of the effects of surge include rapid oscillations in flow and pressure, exponential rise in the gas temperature, and process instability.

The oscillations in process conditions during surge affect the machine vibration response. The typical surge symptoms include low-frequency vibration, especially in the axial direction. In some cases, the frequency is low enough for the changes in axial position to be observed on a trend plot, as shown in Fig. 14.23. During the same time, the compressor vibration response changed to some extent, but the amplitudes (< 0.6 mil [15 μm] pk-pk, direct and 1X) remained well below the alarm set points, as shown in Fig. 14.24.

Whereas Figs. 14.23 and 14.24 are provided as example plots of a compressor under surge conditions, the actual behavior depends on the characteristics of the compressor. In some cases, the radial vibration is higher in amplitude, and the frequency of the oscillations may be different. The frequency of the vibration and the changes in axial position corresponds to the flow-reversal frequency range of 0.3–3

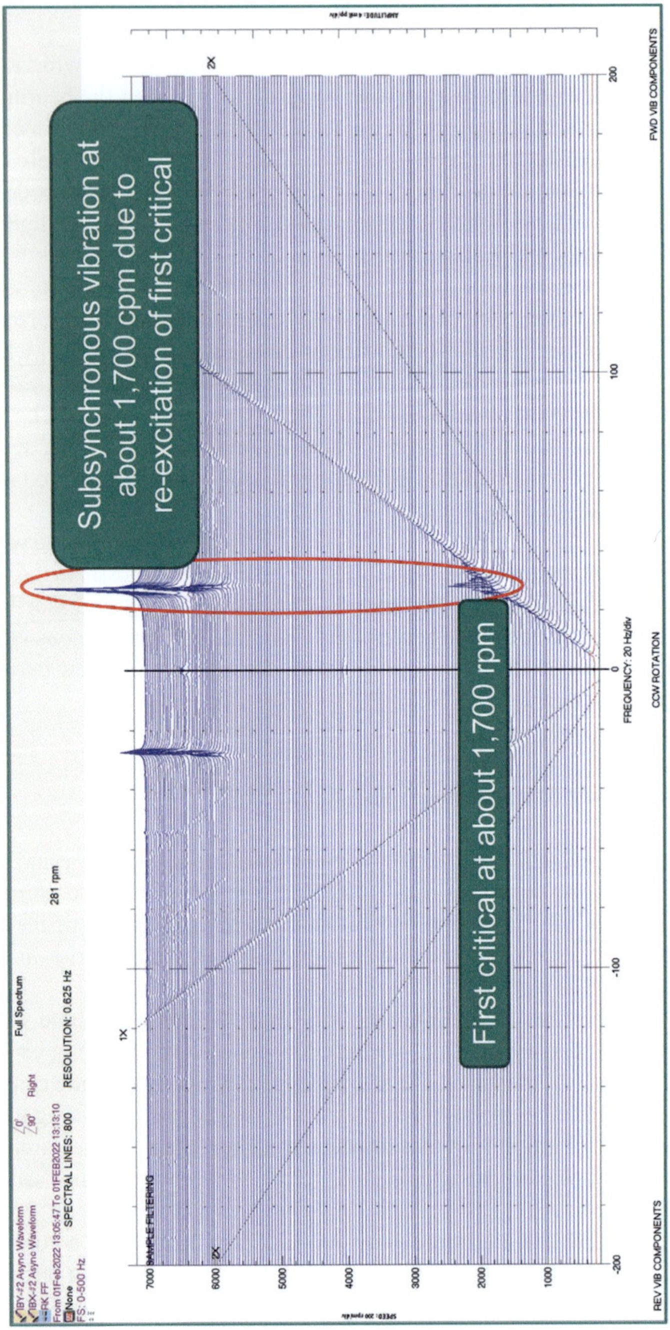

Fig. 14.21 Cascade plot showing fluid-induced instability resulting from oil whip

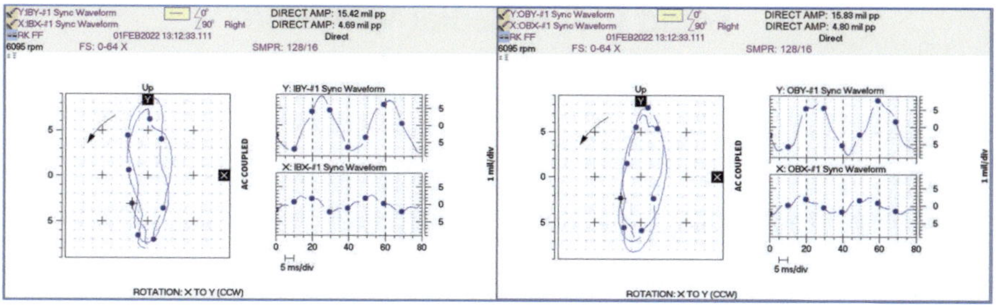

Fig. 14.22 Orbit plots showing fluid-induced instability resulting from oil whip

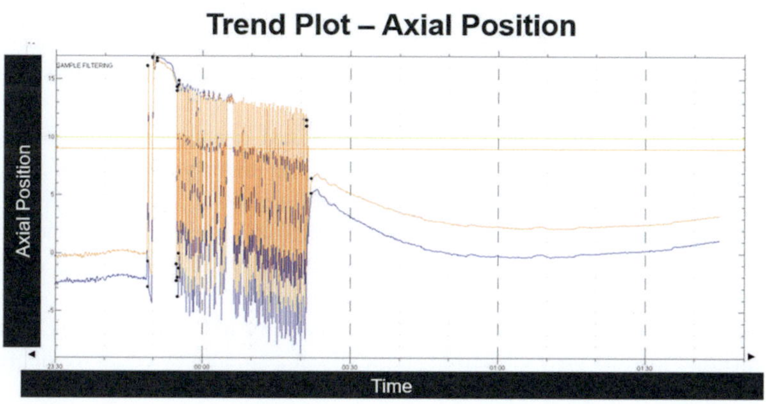

Fig. 14.23 Trend plot of the axial position for a compressor under surge conditions

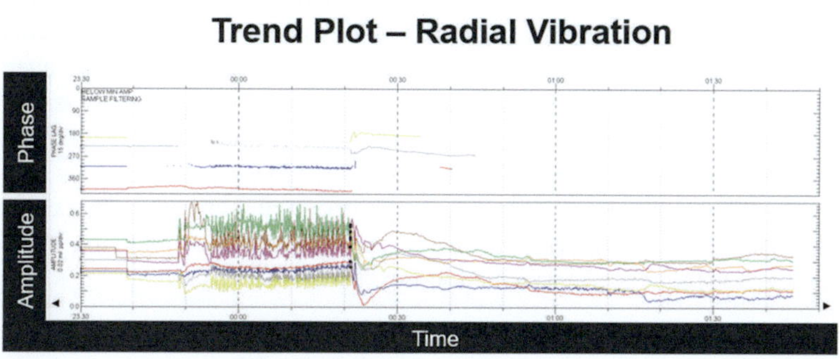

Fig. 14.24 Trend plot of the radial vibration for a compressor under surge conditions

Hz. Because the frequencies may be below typical vibration frequencies, the trend plots for the radial gap voltage and axial position should be considered in addition to the vibration plots.

The fact that surge can have a strong effect in the axial direction provides a basis for ensuring that the thrust probes are monitored for axial vibration. In the author's

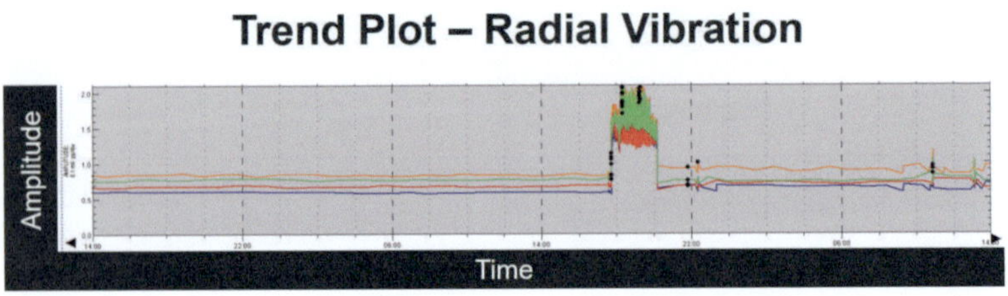

Fig. 14.25 Trend plot of the direct radial vibration during a short stall event

experience, this feature of a monitoring system is often overlooked because the primary purpose of axial measurements is to monitor the axial position of the rotor relative to the thrust bearing. In the example provided in Figs. 14.23 and 14.24, axial position measurements were sufficient. However, users should not assume that this will always be the case.

Stall is related to surge, but it is a more localized phenomenon. In his book on centrifugal compressors, Boyce refers to stall as the underlying cause of surge and provides a discussion of stall in the chapter on surge control [10, pp 225–233]. Stall occurs when the flow over a vane (impeller or diffuser) separates from the vane. In a compressor, the stall can be rotating (propagating) or stationary. Stall can cause subsynchronous vibrations and pressure pulsations [11].

Stationary stall cells can form in stationary parts of the compressor. These cells can contribute to vane pass vibrations, typically at a frequency equal to the number of blades times the rotor speed [11].

Rotating stall is a stall zone that propagates in the direction of rotation. The stall can occur in the impeller or diffuser. Often, rotating stall occurs near the surge line, but it has been reported to occur near design flow rates in some cases. The typical vibration resulting from stall is a subsynchronous peak with a forward-processing orbit. The vibration frequency may range from 0.08X to 0.80X, possibly even up to 1.0X, according to some sources, and the frequency tracks with speed. In general, stall located in the impeller is expected to occur at about 0.6X–0.8X, and stall occurring in the diffuser is expected at about 0.2X–0.3X [11, 12].

Figure 14.25 provides a trend plot showing direct radial vibration for a centrifugal compressor including a short stall event. A waterfall plot showing the 0.37X subsynchronous frequency is provided in Fig. 14.26.

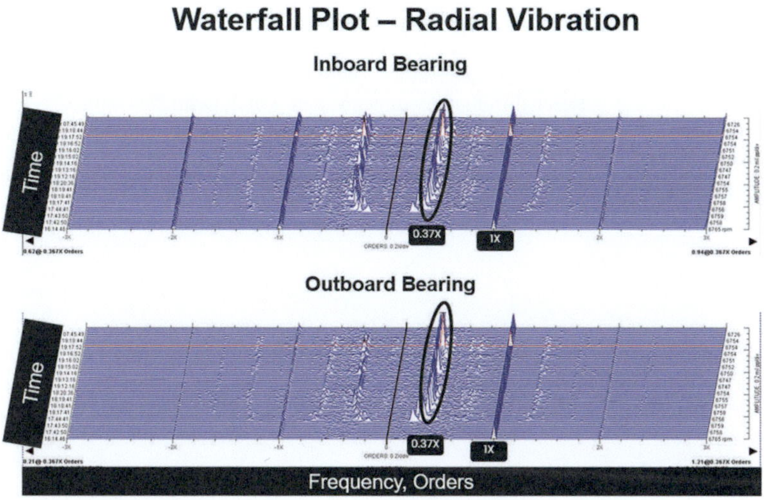

Fig. 14.26 Waterfall plot of a compressor in a short stall event

Acknowledgements All of the images/graphics/figures/tables in this chapter are courtesy of Bently Nevada, a division of Baker Hughes, Minden, NV. They were provided by Michael Titone, who is the contributing author of this chapter.

References

1. API Standard 670 (2025) Machinery protection systems, 6th edn. American Petroleum Institute, Washington, DC
2. Littrell N (2005) Understanding and mitigating shaft runout. Orbit 25(3):5–17
3. Maalouf MG (2007) Slow-speed vibration signal analysis. Orbit 27(2):4–16
4. Smith SW (1997) The scientist and engineer's guide to digital signal processing. California Technical Publishing, San Diego, CA
5. Cooley JW, Tukey JW (1965) An algorithm for the machine calculation of complex Fourier series. Math Comp 19(90):297–301
6. Bently DE, Hatch CT (2002) Fundamentals of rotating machinery diagnostics. Bently Pressurized Bearing Press, Minden, NV
7. Eisenmann RS Sr, Eisenmann RS Jr (1997) Machinery malfunction diagnosis and correction: vibration analysis and troubleshooting for the process industries, 1st edn. Prentice Hall PTR, Upper Saddle River, NJ
8. Piotrowski J (2006) Shaft alignment handbook, 3rd edn. CRC Press, Boca Raton, FL
9. Vance J, Zeidan F, Murphy B (2010) Machinery vibration and rotordynamics. Wiley, Hoboken, NJ
10. Boyce MP (2003) Centrifugal compressors: a basic guide. PennWell Corporation, Tulsa, OK
11. Marshall DF, Sorokes JM (2000) A review of aerodynamically induced forces acting on centrifugal compressors, and resulting vibration characteristics of rotors. In: Proceedings of the 29th turbomachinery symposium, Houston, TX, 2000. Turbomachinery Laboratories, Texas A&M University, College Station, TX, pp 263–280

12. Bently DE, Goldman P, Yuan J (2001) Rotor dynamics of centrifugal compressors in rotating stall. Orbit (2Q01):40–50
13. Sabin S, Chitwood R, Hamad A, Qureshi M (2006) Understanding discrepancies in vibration amplitude readings between different instruments, part 2 of 2. Orbit 26(1):40–57
14. Sabin S (2005) Understanding discrepancies in vibration amplitude readings between different instruments, part 1 of 2. Orbit 25(2):18–30
15. Sujatha C (2010) Vibration and acoustics: measurement and signal analysis. Tata McGraw Hill Education Private Limited, New Delhi

Chapter 15
Preservation and Packaging

Abstract This chapter outlines the minimum requirements for preserving and packaging compression equipment to maintain its integrity from the point of origin through transit and post-delivery. Packaging and preservation methods are determined by several factors, including the storage location at the jobsite—whether in a climate-controlled indoor facility or an outdoor area exposed or shaded—the site's ambient temperature range, and the expected duration of storage prior to installation. The preservation process aims to protect the equipment from contamination, corrosion, and physical damage caused by environmental elements such as dust, sand, humidity, sunlight, and rain. To facilitate efficient handling upon arrival, all items must be clearly tagged for identification. Additionally, shipping cards for each component should include the part number, description, purchase order number, line-item reference, quantity, and any other details required.

15.1 Preservation [1]

Immediately following the final shop assembly and post-testing, all external machined surfaces and internal components such as the casing, internals, and bearing housing must be treated with a removable rust-preventive coating. This is essential to safeguard against corrosion during shipment, storage, and field installation. Additionally, a base layer of rust-inhibiting paint should be applied to all external surfaces of the unit, excluding machined areas. This initial protective treatment typically remains effective for up to six months. Various steps implemented during preservation are described below.

- Rust Prevention and Preservation:

  The application of removable rust-preventive or rust-inhibiting materials involves a multi-step process:

  - Surface preparation: This includes thorough cleaning, degreasing, and drying.
  - Protective coating: Surfaces are treated with rust-preventive or inhibiting solvents.

N. Ghaisas, *Centrifugal Compressors in Hydrocarbon Processing Industries*,
https://doi.org/10.1007/978-3-032-03687-2_15

- Sealing: Equipment is wrapped and sealed to maintain protection during transit and storage.

Rust-preventive compounds may be oil-based, petroleum grease-based, or asphaltic. These materials must be chemically compatible with the process gases and elastomeric components used within the equipment.

- Surface Treatment and Cleaning:

  To eliminate rust and scale, mechanical methods such as shot-blasting, sand-blasting, and grinding are employed. Chemical treatments like pickling and passivation are also used:

  - Pickling/descaling or electrochemical cleaning removes grease, weld flux, slag, metal particles, abrasives, surface contaminants, and post-weld heat tint.
  - Passivation eliminates free iron and forms a protective oxide layer, enhancing corrosion resistance.

  Interior and exterior surfaces are cleared of dust and debris using clean, dry air. Residual fluids, ink markings, grease, and oil are removed with solvents. Rust-preventive coatings are applied uniformly until all designated surfaces are fully protected.

  Machined components—such as sliding surfaces, mounting pedestals, and alignment targets—are cleaned with specialized agents to eliminate handling marks.

- Protection of Openings and Components:

  All equipment openings and mounting plates are treated with volatile rust inhibitors:

  - Flanged openings: Sealed with steel covers or blind flanges and painted safety yellow (Fig. 15.1).
  - Threaded openings: Secured with long shank plugs.
  - Lube oil drain lines and compressor nozzles: Coated with vapor-phase corrosion-inhibiting oil-based protectants (Fig. 15.2).

  Weatherproof tags are used to identify areas treated with rust-preventive materials or containing desiccant bags. This tagging system supports inspection and ensures proper handling upon arrival at the jobsite. Figure 15.3 illustrates a centrifugal compressor string that has been preserved and is ready for packaging.

- Weld Bevel Protection:

  Weld bevels on equipment and auxiliary components must be free of contaminants such as dirt, oil, grease, scale, and rust. For carbon steel and ferritic alloy steel, cleaned weld bevels—both internal and external—should be coated with weldable primer to prevent corrosion using oil-based protectants (Fig. 15.2).

Spare compressor rotors are typically housed in nitrogen-purged containers engineered for either horizontal or vertical storage. Prior to shipment, these containers are pressurized and depressurized using low-pressure nitrogen—below 2 psi g (14 kPa g)—to lower the internal oxygen concentration to under 5%. To ensure proper preservation, the nitrogen purge pressure should be maintained in accordance with the

Fig. 15.1 Centrifugal compressor with a radially split casing prepared for preservation (Image courtesy of Mitsubishi Heavy Industries Compressor International, Houston, TX)

Fig. 15.2 Lube oil drain line and compressor nozzle cover prepared for preservation (Image courtesy of Mitsubishi Heavy Industries Compressor International, Houston, TX)

manufacturer's guidelines. Generally, the recommended purge pressure falls within the range of 1–5 psi g (7–35 kPa g).

Fig. 15.3 Centrifugal compressor unit prepared for preservation (Image courtesy of Mitsubishi Heavy Industries Compressor International, Houston, TX)

15.2 Packaging

Customer or purchaser-assigned inspectors typically perform both in-process and final inspections. Prior to shipping, the release inspection should verify the following:

- Dimensional and Mechanical Checks:
  - All dimensions align with the approved engineering drawings.
  - Mounting plate leveling screws are adequately protected against corrosion.
  - Tubing and utility header piping within the skid are visually inspected to confirm blind flanges are installed on customer connection points, preventing contamination from particulates, dust, or liquids.
  - Nitrogen pressure is maintained in the lube oil accumulator of the pressurized system, as well as in nitrogen purged containers for compressor spare rotors and dry gas seals.
  - Compressor nozzles are fitted with blind flanges or covers to block foreign material or liquid ingress.
- Electrical and Instrumentation Integrity:
  - Electrical housings on the skid are fully sealed, with desiccant placed inside.

  - All unused electrical entries are properly plugged or sealed, and instrument manifolds are securely closed.
  - Cable glands on instrument junction boxes are tightened, and unused entries are sealed. Internal components are protected from corrosion, especially in enclosed junction boxes (Fig. 15.4).

- Surface Protection and Identification:

  - Exposed machinery mounting pads, machined surfaces beneath the baseplate, and level targets are treated with rust-preventive coatings.
  - Equipment and instrument tags are correctly applied.
  - Any components restricted from movement are clearly identified with tags [2].

- Coating and Documentation Compliance:

  - Final paintwork meets the facility owner's specifications or adheres to the provided coating and painting standards.
  - Equipment nameplates and regulatory registration documents reflect the latest purchaser-approved revisions of certified final drawings. Design parameters such as pressure and temperature must match the approved datasheets. Examples of regulatory documents include ASME U-1A forms and Canadian Registration Number (CRN) letters.

- Tools, Spares and Shipping Readiness:

  - The special tools set is complete and properly tagged.
  - Spare parts are accurately counted and labeled with manufacturer part numbers and descriptions.
  - Shipping documentation is finalized, including the bill of lading, packing list, relevant Material Safety Data Sheets (MSDSs) for hazardous materials, and package markings and lift points as outlined in API Standard 617 [2], ensuring safe handling during transport and delivery.

Wrapping and sealing the equipment serves as a crucial final measure to extend the effectiveness of rust prevention (Fig. 15.5). Once wrapped, the package is placed into its shipping crate and firmly secured using bands or straps. To safeguard against moisture during transit, the interior surfaces of wooden crates should be lined with waterproof material or wax paper (Fig. 15.6). Additionally, any piping spools mounted on the unit must be tightly fastened to prevent movement or vibration [1].

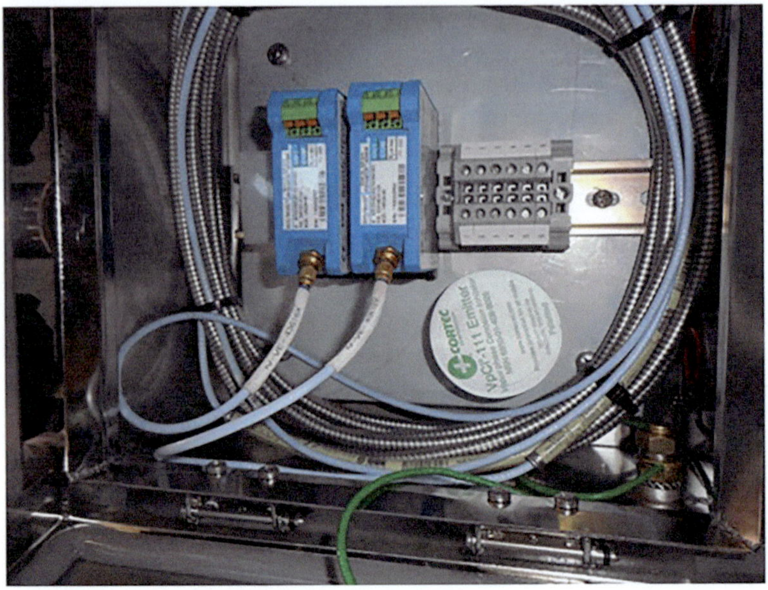

Fig. 15.4 Junction box prepared for corrosion protection (Image courtesy of Mitsubishi Heavy Industries Compressor International, Houston, TX)

Fig. 15.5 Shrink wrapping and crating of a compressor package (Image courtesy of Mitsubishi Heavy Industries Compressor International, Houston, TX)

Fig. 15.6 Lined interiors of a shipping crate for piping spools (Image courtesy of Mitsubishi Heavy Industries Compressor International, Houston, TX)

Acknowledgements All images/graphics/figures in this chapter were provided by Mitsubishi Heavy Industries Compressor International (MCO-I), Houston, TX.

References

1. Shop rust prevention and standard preservation and packing procedures provided by and described with permission from Mitsubishi Heavy Industries Compressor International (MCO-I), Houston, TX
2. API Standard 617 (2022) Axial and Centrifugal Compressors and Expander-compressors, 9th edn. American Petroleum Institute, Washington, DC

Chapter 16
Compressor Specification, Evaluation, and Design Coordination

Abstract Thorough and well-defined technical specifications for equipment inquiries and procurement are essential to avoid ambiguity, misinterpretation, and scope gaps. Incomplete or inaccurate purchase orders can lead to significant costs and complications during project execution. To mitigate these risks, experienced and knowledgeable machinery engineers should develop comprehensive technical requisitions that clearly outline relevant project specifications, applicable industry standards and codes, local regulatory requirements, inspection protocols, specialized analyses and testing procedures, spare parts provisions, and vendor documentation. This chapter illustrates the subject through three detailed scope narratives: one for a motor-gear-driven centrifugal compressor string, one focused on evaluating technical proposals, and another addressing design coordination with equipment manufacturers.

16.1 Technical Scope

The example presented in this chapter includes the scope of supply and performance tolerance requirements in a request for quotation (inquiry) document for a radially split type casing centrifugal compressor string in a refinery application.

16.1.1 Compressor

- Radially split forged cast steel casing with the following features:
  a. O-ring lubricants and elastomeric seals inside the casing, on the end walls (covers), and in the dry gas seal cartridges shall be suitable for the maximum allowable temperature and of one-piece molded construction.
  b. The casing axial anchor point and thrust bearing shall be located on opposite ends to avoid thermal expansion-induced force upon high-speed coupling.

N. Ghaisas, *Centrifugal Compressors in Hydrocarbon Processing Industries*, https://doi.org/10.1007/978-3-032-03687-2_16

- Vertically downward-oriented alloy steel inlet and discharge nozzles:

  a. Nozzle flanges shall be of American National Standards Institute/American Society of Mechanical Engineers (ANSI/ASME) B16.47 type with raised-face weld neck (RFWN).
  b. The accuracy on the projection of the casing nozzles shall be ± 1/8 in. (3.2 mm).
  c. The minimum allowable forces and moments on the compressor nozzles shall not be less than 3 times (allowable factor) the individual resultant forces and moments and combined resultant forces and moments derived from the equations in American Petroleum Institute (API) Standard 617 [1].

- The casing drain, stage drain, and balance piston cavity drain flanged nozzles shall have the same rating as the discharge nozzle. Drain valves shall be full port.
- All nozzle connections on the compressor casing that are nominal pipe size (NPS) 2 in. (equivalent to diameter nominal [nominal diameter, DN] 50) and smaller shall have welded bracings (two-plane gussets). The gusset material, size, shape, and attachment locations shall be selected for the calculated load (vibratory, thermal, static). This requirement applies to all gusseted joints specified in this requisition.
- Horizontally split diaphragms in fabricated steel construction.
- Vaneless diffusers.
- Backward lean, fully enclosed impellers in two-piece welded construction. The combination electron-beam welding and brazing method shall be used for narrow-tip-width impellers (0.300 in. [7.6 mm] and smaller). Narrow-width impellers require geometrical inspection with a laser scanner. Blades shall be milled out of forgings. All impellers shall be positively locked to the shaft by interference fit. The required amount of interference fit shall be based on the calculated maximum impeller stage torque.
- Forged steel one-piece shaft with a tapered end, suitable for hydraulic mounting of the coupling hub. The design of the shaft end-to-coupling hub juncture shall meet the following requirements:

  a. The shaft end taper shall be 1:24.
  b. The coupling, coupling-to-shaft juncture, and coupled shafts shall be able to transmit, without damage, 115% of the maximum transient torque from the torsional analysis, or the specified transient peak torque, whichever is greater (400% for an induction motor driver).
  c. The coupling hub advance on the taper shaft end shall be based on the interference rate of 0.001 in. per inch of shaft diameter (or 0.001 mm per millimeter of shaft diameter).
  d. O-rings used in hydraulically mounted hubs shall have a pressure backup (antiextrusion) ring.

- Renewable interstage shaft sleeves.
- Abradable seals at the impeller eye, balance piston, and interstage locations. Honeycomb or hole-pattern seals, if used as abradable seals, shall have retaining carrier rings to protect them against axial forces.

- The balance piston shall be axially restrained on the shaft and installed with an interference fit to prevent axial or radial movements relative to the shaft.
- Horizontally split, integral steel bearing housings.
- Tilting-pad radial bearings with steel-backed babbitted pads, hardened steel inserts, spherical seat, and center pivot.
- Self-equalizing, double-acting thrust bearing with the following features:

  a. Center pivot.
  b. Steel-backed babbitted pads with hardened steel inserts.
  c. Designed for directed lubrication.
  d. Hydraulically fitted replaceable thrust disk.

- Pressurized lubrication with oil supplied from a common lube oil system.

  a. Manifolded and RFWN flanged oil supply and oil return valve sets shall be supplied as shipped-loose assemblies to connect at the jobsite.
  b. The oil supply and oil return manifolds within the confines of the compressor baseplate shall be connected to the bearing housings of the served equipment (compressor, gear, and driver motor).

- The following scope requirements apply to solvent injection nozzles on the compressor casing:

  a. Check valves furnished on supply lines for each stage and located downstream of the supply manifold.
  b. Annular injection points for uniform distribution.
  c. Breakout piping spools to facilitate removal and insertion of injection lances.
  d. Welded bracings (two-plane gussets) on injection nozzles.

- A minimum of two (2) RFWN flanged casing drains (one balance piston cavity drain and one drain on the suction side of the outer casing), valved (globe-type valve) and manifolded. The purchaser's tie-in point shall have an RFWN flanged gate valve with a blind flange installed on its downstream.
- RFWN flanged individual stage drains, valved (globe-type valve) and manifolded. The purchaser's tie-in point shall have an RFWN flanged gate valve with a blind flange installed on its downstream.
- 100-Ω, 3-wire, duplex platinum resistance temperature detectors (RTDs) with transmitters:

  a. Two (2) in each radial bearing.
  b. Four (4) in the thrust bearing (two [2] on the active side, two [2] on the inactive side).

- Bently Nevada 3300XL series vibration and axial position probes with Proximitors® (Bently Nevada is a Baker Hughes business):

  a. One (1) *X–Y* pair of probes per radial bearing.
  b. Two (2) axial position probes.

- Dynamic simulation of the compressor–driver train. Dynamic simulation shall include all operating cases except the nitrogen case. The supplier shall confirm if a hot-gas bypass valve (HGBV) is required during an emergency shutdown (ESD). The following major scenarios shall be covered:
  a. 100% duty point steady-state operation.
  b. Startup (restart) of the compressor train from depressurized loop conditions.
  c. Startup of the compressor train from process conditions (normal start).
  d. Compressor train normal stop.
  e. Compressor train normal stop from operation at the surge control line point.
  f. Trip from operation at the surge control line point.
  g. Sudden opening of the recycle (antisurge) valve during compressor operation.
  h. Compressor train emergency stop from full-load, full-pressure, full-speed conditions.
- The compressor supplier shall analytically verify that the driver's electric motor can start, accelerate, and reach full load under the following concurrent conditions, all within the limits of the selected motor rating.
  a. Specified settle-out pressure.
  b. Minimum 10% margin between the load torque and motor speed–torque curves at 80% motor voltage.
- The supplier shall furnish scatter plots of head coefficient versus flow coefficient, mechanical tip speed versus flow coefficient, relative Mach number versus flow coefficient, stage power versus flow coefficient, and tip Mach number versus flow coefficient, each showing the referenced and proposed impellers and the blade lean.
- All instrument transmitters installed on the compressor shall be Ex-d–certified (explosion-proof; designed to withstand and prevent the spread of internal explosions) and installed in National Electrical Manufacturers Association (NEMA) 4X housings. All armored control and instrument cables shall be installed on channel trays terminated in Ex-e–certified (explosion-protective; designed to prevent the contents from creating any arcs, sparks, or hotspots) NEMA 4X junction boxes with cable glands.

16.1.2 Dry Gas Seals and Seal Support System

- To reduce the consumption of primary seal gas, the required method of seal gas control is flow control. The flow control system shall have the following features:
  a. Valve with a remote flow controller will compare the seal gas flow to each seal and adjust the flow control valve to maintain a minimum flow velocity of 16.4 ft/s (5 m/s) [2] across the inner labyrinth seal based on "low-select reference pressure" for each seal, measured downstream of the flow orifices.

b. Bypass line with manual block valves around the flow control valve.
c. Flow control valve with a carbon steel body with stainless-steel trim.
d. Seal gas supply line to each seal with a rotameter flow meter.

- The dry gas seal system shall be designed for all operating cases shown on the compressor data sheet, the settle-out pressure, and the specified process relief valve set point. The dry gas seal shall be of the "tandem with intermediate labyrinth" type.
- Process gas from the compressor discharge nozzle will be used as the primary seal gas. The associated requirements for the treatment of primary seal gas are as follows:

a. A seal gas electric heater with a thyristor control panel shall be furnished to maintain a dew-point margin of at least 20 K (36 °R) throughout the dry gas seal system [2] during startup after depressurization. This requires that a phase map computer simulation of the dry gas seal system from the primary seal gas supply point to the primary vent shall be completed to evaluate any potential for gas condensation under all specified operating conditions. The thyristor control panel shall be shipped loose. The supplier shall identify the recommended electric heat tracing areas on the primary seal gas supply line.
b. The seal gas heater and its control scheme shall include the following features:
   i. Bypass line with a gate valve.
   ii. Total of four (4) temperature-sensing elements, with three (3) for casting and the internal enclosure high limit and one (1) for cascade control (high- and low-temperature set points).
   iii. Total of four (4) flow transmitters, with two (2) on each end of the primary seal gas supply line to provide compressor autostart on L1 (NOT low flow) and seal gas heater cutout on L2 (low-low flow).
   iv. Unit encased in pyrogel insulation.
   v. Outer casing and casing hardware, seal gas flow coil, electrical enclosure, and mounting flange in austenitic stainless steel.
   vi. Heater casting in type 356 aluminum alloy.
   vii. National Association of Corrosion Engineers (NACE) MR0103-compliant pressure boundary components.
   viii. ASME Section VII, Division 1 Code construction.
c. The temperature of the seal gas shall be measured at the point of seal gas entry into the seal, not at the source of the seal gas supply. Two (2) temperature-sensing elements are required, with one on each primary seal gas entry point to provide signals to the thyristor control panel.

- Nitrogen is the source of the secondary/separation seal gas.

a. The secondary seal gas from the separation/secondary seal filters is supplied between the intermediate labyrinth and the secondary seal to provide a barrier and prevent primary seal leakage from reaching the secondary seal.
b. Indicating-type flow transmitters are required on the secondary seal gas supply to each end of the secondary seal.

c. Restriction orifices shall be installed downstream of the flow transmitters to regulate flow to the secondary seals.
d. Indicating-type pressure transmitters shall be installed on the secondary vent lines.
e. Secondary seal gas supply lines shall be separated from the separation gas supply lines, downstream of the pressure control valve on the nitrogen supply line.

- The material of the mating face of the dry gas seal shall be silicon nitride. Tungsten carbide is not permitted.

a. Polymer static seals shall be used.
b. O-rings shall have a minimum diameter of 0.125 in. (3.175 mm). O-rings with a diameter of 0.100 in. (2.54 mm) are acceptable in the areas of dry gas seal unit where the ring groove size cannot accommodate 0.125-in. (3.175-mm) diameter O-rings.
c. Dual O-rings shall be provided under the rotating seal sleeve.

- Primary vents from the dry gas seal system will be routed to the plant's flare. Check valves will be installed on primary vent lines to avoid reverse pressurization of the seal cavities.
- The dry gas seal system piping and vents shall be sized to prevent overpressurization of the bearing housings in the event of a seal failure. Outlet modules shall measure flow by differential pressure.
- Secondary seal vents from the dry gas seal system will be routed to the atmosphere and positioned at the jobsite to prevent rain or moisture ingress.
- The dry gas seal control system panel or frame shall be mounted on the compressor baseplate. It shall comply with the following requirements:

a. All connections between the panel or frame and the compressor shall be within the supplier's scope.
b. The control panel or frame and the gauge/instrument board frame shall be in austenitic stainless-steel material.

- Duplex full-flow, coalescing-type primary seal gas filters shall be in an austenitic stainless-steel shell. The filters shall have a minimum efficiency of 99.9% for particles larger than or equal to 1 μm (beta ratio, $\beta_1 \geq 1000$) [2]. The seal gas filters shall include the following features:

a. Liquid removal to 0.005 ppm (weight basis).
b. Filter shells shall have a minimum ½-in. raised-face (RF) flanged or ¼-in. national pipe tapered (NPT) vent and manual drain connections.
c. ASME Section VIII, Division 1 Code construction.
d. A pressure gauge shall be installed upstream of the seal gas filters, and a differential pressure transmitter shall be furnished across the filters.
e. Zero-leakage continuous-flow transfer valve of the single-body, six-port straight-plug type with resilient seats shall be used (a taper plug valve is not acceptable).

- A booster unit shall be furnished to supply the normal flow of seal gas at the differential pressure required (typically 30–40 psid [206–276 kPa differential]) to avoid contamination of the dry gas seal cavities under pressurized standstill conditions and during startup (rampup) of the compressor. The booster shall have the following features:

    a. The booster shall be of the centrifugal type.
    b. Hermetically sealed design. One (1) vibration sensor shall be provided on the casing.
    c. The control logic shall automatically start and stop the booster as described on the dry gas seal system schematic. Autostart of the booster driver motor shall be initiated upon detection of a low differential signal from the transmitters on the drive-end and nondrive-end primary seal gas supply lines. A pressure differential transmitter shall be provided across the booster to allow forced stop upon indication of a high differential pressure.
    d. The booster driver motor shall be of constant-speed NEMA MG1 alternating-current (ac) induction type with a totally enclosed fan-cooled (TEFC) enclosure and provided with a 120 V/1 Ph/60 Hz space heater.
    e. Society of Automotive Engineers (SAE) milled flat connections on the booster casing with an RFWN flanged inlet and discharge tie-in points.
    f. The casing shall be in austenitic type 316L stainless steel.
    g. The booster shall be installed on the dry gas seal system control panel or mounted separately on a rigid support within the confines of the compressor baseplate and connected to the dry gas seal support system.
    h. An RFWN flanged valve shall be provided in the inlet line upstream of the booster to facilitate online testing of the booster. The supplier shall tag it "Locked Close" prior to shipment and install a blind flange on its inlet. A drain valve shall be provided in the seal gas supply line to the booster.

- Duplex full-flow, coalescing-type secondary/separation gas filters in austenitic stainless-steel shell. The filters shall have a minimum efficiency of 99.9% for particles less than or equal to 10 μm (beta ratio, $\beta_{10} \geq 1000$) [2]. The separation/secondary gas filters shall include the following features:

    a. Liquid removal to 0.005 ppm (weight basis).
    b. Filter shells shall have a minimum ½-in. RF flanged or ¼-in. NPT vent and drain connections.
    c. A pressure gauge installed upstream of the separation gas filters and a differential pressure transmitter furnished across the filters.
    d. Zero-leakage continuous-flow transfer valve of the single-body, six-port straight-plug type with resilient seats (taper plug valve is not acceptable).

Note: Two three-way straight-plug valves permanently aligned and joined with a single operating lever are acceptable.

- The separation (barrier) seal shall be of the noncontacting bushing type and designed to minimize consumption of the separation gas under static and dynamic conditions.
- The supply of secondary seal gas/separation gas shall be controlled by a fixed-pressure control system (fixed pressure over the secondary vent). Dual check valves shall be provided on the secondary gas supply module to avoid reverse pressurization. The separation gas control system shall include the following:

  a. A pressure gauge and a pressure transmitter downstream of the pressure control valve on the nitrogen supply line to the secondary/separation seals. A valved bypass line is required around the pressure control valve.
  b. The pressure transmitter shall allow a start permissive for the lube oil pump after the separation gas supply is established.
  c. Indicating-type flow transmitters on the separation gas supply lines.
  d. Restriction orifices downstream of the transmitters to regulate the flow through the separation seals.

Note: The purchaser's secondary vent piping will be configured to limit the pressure drop to 5 psi (34 kPa).

- The dry gas seal supply and vent piping shall have RFWN flanged connections. The piping shall be American Society for Testing and Materials (ASTM) A312, type 316/316L, and the flange bolting shall be as in accordance with standard ASME B31.3. The bolts shall be ASTM A193, Grade B7, and the nuts shall be ASTM A194, Grade 2H. The piping for the separation supply gas shall be valved and manifolded for single purchaser connection points.
- The dry gas seal system piping shall be complete with fittings and integral flow and pressure instrumentation. It shall comply with the following requirements:

  a. Screwed or slip-on flanges are not permitted.
  b. Seal welding is not allowed on instruments or in locations where disassembly is required for maintenance.

- The dry gas seal control system shall have the range and controllability to maintain a minimum gas velocity of 16.4 ft/s (5 m/s) across the inner labyrinth seal (process-side seal), as well as at labyrinth clearances ranging from minimum to maximum and up to twice the maximum design clearance [2]. This is critical to avoid contamination of the seal and to maintain the required seal face temperature within the safe upper operating limit.
- The dry gas seal system shall be designed to maintain the seal chamber pressure greater than the primary vent system pressure throughout all operating conditions including startup, throttled suction pressure, and the specified settle-out pressure.
- All nozzle connections on the dry gas seal system that are NPS 2 in. (DN 50) and smaller shall have welded bracings (two-plane gussets). Breakout spools shall be provided on small-bore piping to facilitate maintenance.
- Threaded ¾-in. vent, drain, and minimum ½-in. instrument root valves including valve manifolds are required. RFWN flanged double block and bleed valves (inlet,

outlet, and vent/bleed valves) shall be installed on the process lines and process instruments that are on the dry gas seal control panel. Isolation valves shall be gate-type glove valves on bypass lines and ball valves on bleed (drain) and local vent lines.

   a. Double Block and Bleed valves
      i. Drain on 900# lines inside the gas seal module.
      ii. Instrument root valve on 900# lines.
      iii. Compressor seal gas cavity drains (900#).
   b. Double Block Valves
      i. Drain on 900# interconnecting piping between the gas seal module and compressor (if piping has low pockets).
      ii. Drain and vent of the primary seal gas filter (900#).
   c. Single Block Valves
      i. 150# lines (separation and secondary seal gas).
      ii. Isolation valve on the 900# primary seal gas line.

- Austenitic stainless-steel tube fittings (where tube connections are provided such as instrument takeoff) shall be Swagelok®.
- RFWN flanged manual low-point drain valves shall be provided on seal gas, separation gas, secondary seal gas, and primary and secondary seal cavities to allow for the removal of any accumulated lube oil that was carried over from the bearing housings during upset conditions and any accumulated liquids. Drain lines shall have sight flow glasses.
- The design pressure and the design temperature of the dry gas seal system shall not be less than the maximum allowable working pressure (MAWP) and the maximum allowable temperature, respectively, of the compressor casing, that is, the design pressure and the design temperature of the primary seal gas line, primary vent line up to the final element, and secondary line downstream of check valve will be the same as the pressure and temperature of the compressor casing.
- The MAWP or pressure rating of the following dry gas seal support subsystems shall be as specified in API 692 [2].
- The outboard or backup seal in the tandem seal shall be designed for full sealing pressure and temperature (same as for the inner or primary seal).
- The dry gas seal system shall include the two-out-of-three (2-o-o-3) voting function served by three (3) differential pressure transmitters for high-high primary vent flow trip.

   a. Primary vents shall have rotameters and transmitters.
   b. The control valves in this system shall be able to be manually opened for maintenance or forced open if their controller fails.

- Dry gas seal system phase map study and report.
- Seal gas flow velocity calculations report including analyzed situations such as with and without seal gas booster conditions in startup, coastdown, and all specified operating cases. The analysis shall include flow velocity versus compressor speed charts for each evaluated case.

16.1.3 Driver Motor

- Class F insulation, Class B temperature rise, totally enclosed air-to-air cooled (TEAAC) enclosure (corresponding NEMA MG1 Specification 1.26.8, minimum degree of ingress protection of IP54. Heat-exchanger tubes shall be in the material shown on the motor data sheet).
- Magnetic center indicator on the drive end.

  a. The motor rotor float shall exceed the coupling end float by at least 0.25 in. (6.35 mm).
  b. To facilitate field installation, the magnetic center shall be within 20% of the total end float from the center of the end float limit to provide clearances between the rotor journal shoulders and the bearing and end seal faces under all operating conditions with a limited end float coupling installed.

- American Iron and Steel Institute (AISI) 4000 series forged alloy steel one-piece shaft (shaft, spiders, and shaft end machined from one-piece heat-treated forging) with integral flanged shaft end.
- Two (2) frame space heaters (120 V/1 Ph/60 Hz).
- Two (2) 100-Ω, 3-wire duplex platinum RTDs with transmitters on a TEAAC enclosure to monitor the air temperature.
- Nine (9) 100-Ω, 3-wire single-element platinum RTDs. Three (3) RTDs per phase winding.
- Four (4) embedded 100-Ω, 3-wire duplex platinum RTDs with transmitters, two (2) per radial bearing.
- Four (4) Bently Nevada 3300XL series vibration probes and four (4) Proximitors®. One (1) *X*–*Y* pair of probes per radial bearing. (Bently Nevada is a Baker Hughes business.)
- Pressurized lubrication with oil supplied from the common lube oil system.
- Emergency STOP pushbutton inside a protective box, installed on the motor casing or a local gauge board.
- Babbitted radial bearings shall be of the hydrodynamic split sleeve type and shall have steel-backed babbitted replaceable shells.

  a. Antirotation pins shall be provided in the bottom-half shell.
  b. Bearings shall be axially secured.
  c. Both radial bearings shall be of the same size and electrically insulated.

- Oversized main terminal box with the following features:

  a. Bottom entry for the power cable.
  b. 120 V/1 Ph/60 Hz space heater.
  c. In addition to the main power feeds, the terminal box shall include current transformers, surge capacitors, and surge arrestors for motor differential protection.
  d. Partial discharge transducers (couplers) to facilitate online monitoring by others.

16.1.4 Speed Increasing Gear Unit

- Single-stage, high-speed, double-helical gear unit in fabricated steel casing.
- The mesh direction shall be downward, with the gear "downloaded" and the pinion "uploaded" for this application (upmesh gearset configuration).
- Hydrodynamic split-sleeve babbitted radial bearings on the gear shaft. The pinion shall have self-equalizing, tilting-pad radial bearings with steel pads, offset pivot, and load between pads.
- Kingsbury self-equalizing, double-acting thrust bearing of the leading-edge groove (LEG) type, offset pivot, and steel-backed babbitted pads. The thrust bearing shall be installed on the gear shaft. An equivalent of Kingsbury thrust bearing design is acceptable.
- Integral flanged shaft ends on the high-speed pinion shaft and the low-speed gear shaft. Integral flanged ends are required from considerations of the pinion speed, pinion weight, and any effect on rotordynamic stability due to concentrated weight at the shaft end.
- The pinion shall be integral with its shaft. The gear wheel shall be either integrally forged or fitted on its shaft with interference fit if the pitch line velocity is less than that specified in API 613 [3]. The gear rim shall be shrunk onto the gear wheel and subjected to final machining only after installation of the gear wheel on the shaft. Shafts shall be made from heat-treated forged alloy steel.
- Adapter flanges for high-speed and low-speed cylindrical coupling guards.
- Gear case vertical leveling jackscrews and horizontal adjustment jackscrews.
- Screwed and tack welded internal oil piping, including an oil spray manifold in 304 stainless steel.
- 2-in. 150# ANSI RFWN flanged and blinded vent connection in the gear casing.
- Pressurized lubrication, with oil supplied from common lube oil system.
- Minimum ¾-in. NPT purge connection on the gear casing.
- Labyrinth-type replaceable shaft end seals.
- Type 304 stainless-steel full-width shim pack under the gear casing.
- 100-Ω, 3-wire, duplex platinum RTDs with transmitters, two (2) per radial bearing, two (2) on the active side of the thrust bearing, and two (2) on the inactive side of the thrust bearing.
- Bently Nevada 3300XL series vibration, axial position, and phase reference probes with Proximitors® and gear-case-mounted accelerometers (Bently Nevada is a Baker Hughes business):

  a. One (1) pair of *X–Y* probes at each radial bearing.
  b. Two (2) axial position probes.
  c. Two (2) phase reference probes (one on the low-speed shaft and the other on the high-speed pinion).
  d. Two (2) Bently Nevada casing accelerometers with built-in charge amplifiers, extension cable, and interface module.
  e. Accelerometers shall be mounted on the lower half of the gear casing, one on the low-speed (bull gear) side and the other on the high-speed (pinion) side.

16.1.5 High-Speed and Low-Speed Couplings

- 24-in.-long coupling spacer tubes. The peak torque and momentary torque ratings of the couplings shall be as required by API 671 [4].
- Hub drill templates for integral flanged shaft ends (on the low-speed gear and high-speed pinion) and a set of matched plug and ring gauges and two-piece stop ring for the tapered-bore coupling hub on the compressor shaft.
- Component balance and assembly check balance of the coupling. (Method 3 in API 671 [4] shall be used for the high-speed coupling. Method 2 is applicable to the low-speed coupling.)
- Residual unbalance check of the assembled high-speed coupling after assembly balance.
- Balance repeatability check of the high-speed coupling.
- Surface protection coating on the coupling flexible elements.
- Trim-balance holes on the high-speed coupling. (They shall not be used during shop balancing of the coupling.)
- Oil-tight, fully enclosed, flanged cylindrical, spark-resistant coupling guards including minimum 2-in. ANSI/ASME B16.5 RF flanged vent.

16.1.6 Baseplate

- A common baseplate shall be furnished to support the compressor, gear, driver motor, dry gas seal panel, and local gauge board. The baseplate will be epoxy grouted onto its foundation. It shall have the following features:

    a. Two (2) diagonally opposite grounding pads or lugs with end covers with ¼-in. 20 Unified National Coarse (UNC) threads shall be furnished.
    b. All purchaser tie-in RF flanged connections shall be terminated at the edge of the baseplate and shall have a tolerance of $\pm$ 1/8 in. (3.2 mm) on projection.
    c. A local open-construction gauge board shall be provided on the compressor baseplate.
    d. Machined flat (within 0.001 in. per linear foot [0.083 mm per linear meter]) mounting pads shall be provided on the underside of the baseplate for a level support. At least six (6) machined target surfaces shall be provided on the baseplate top surface to facilitate setting the baseplate level in the field. These targets or pads shall be close to the machinery mounting pedestals.
    e. Machinery mounting surfaces shall meet the requirements specified in API 617 [1] for surface finish, flatness, parallelism, and difference among all mounting surfaces in the same horizontal plane.
    f. The sizes and sections of machinery mounting pedestals and baseplate structural members shall be selected to prevent sag or distortion of the mounting surfaces after the equipment is mounted on the baseplate.

g. The baseplate shall be designed to achieve level placement to within 0.003 in./ft (0.25 mm/m) at the site in the longitudinal and transverse directions with the equipment mounted on it.
h. Welded lugs shall be furnished for four-point lifting. Lifting lugs and their welds shall be suitable for the specified lowest ambient temperature.
   i. Lift load calculations shall be furnished for the purchaser's review.
   ii. Lifting lugs shall be full-penetration-welded if the main members are C-channels or I-beams or fillet-welded if the main members are square pipe sections. Continuous welding is required. Welding shall be in accordance with International Organization for Standardization (ISO) 15614 ANSI/American Welding Society (AWS) D1.1 standards. The design load capacity of the welded lifting lugs shall consider the type of weld joint. If the baseplate beam is constructed from square steel pipe and does not permit full-penetration welds, then fillet welds are acceptable.
   iii. Lifting lug welds shall be 100% tested by nondestructive examination (NDE) methods, namely, liquid penetrant (fluorescent particle) or magnetic particle.
i. Anchor-bolt holes shall have a noncumulative accuracy of ± 1/16 in (1.6 mm).

16.1.7 Lube Oil System

- Beam-type construction shall be used to ensure structural rigidity of the fabricated baseplate. Full-depth, full-width cross members (bracings) shall be provided under the oil reservoir, shell-and-tube exchangers, and pump and driver motor units. The baseplate will be perimeter epoxy grouted on its foundation mat. It shall include the following features:

  a. Vertical corners shall be rounded to a minimum 2-in. (50-mm) radius.
  b. Structural members shall be full-penetration, continuously welded.
  c. The baseplate top surface shall be covered with a sloped and full seal welded deck plate.
  d. The deck plate shall be coated with grit-embedded paint.
  e. Two (2) diagonally opposite grounding pads and end covers with ¼-in. 20 UNC threads shall be furnished.

- Lube oil reservoir in austenitic stainless-steel material, sized for 8-min retention and designed for a vacuum of 4 in. of water column (4 in. WC [10 mbar]) (considering an extraction-type mist eliminator installed on the reservoir).

  a. Insulation clips for 2-in. (50-mm) thick mineral wool insulation with unrestricted space around them for installation of the insulation material.
  b. Thermometer.
  c. Siphon breaker.
  d. Connection for an oil conditioner.

e. Reservoir pressure monitoring system.
f. Weld-pad reflex-type level glass.
g. Level indication by magnetic level transmitter, fitted with 2-in. isolation valves.
h. Two tapped grounding pads located diagonally opposite to each other.
i. Minimum 2-in. fill connection, ¾-in. oil sampling connection with an isolating valve, oil clarifier (conditioner) connection, valved drain connection, sloped bottom (¼ in./ft [21 mm/m]), and minimum 24 in. × 24 in. clean-out opening(s). These connections shall be RFWN flanged.
j. All atmospheric and pressurized oil return lines shall be separate, not combined.
k. Immersion-type electric heaters sized for a heating time of 6 h to achieve the supplier's required minimum startup oil temperature (considering that the site-specified minimum ambient temperature is lower than the required minimum startup oil temperature). This requirement shall be achieved with a temperature transmitter to provide start permissive for the compressor train.
   i. The heaters shall be designed to permit removal of the heating element online, such as for maintenance or repair when necessary.
   ii. Withdrawable ceramic core elements, austenitic stainless-steel plain surface core sheath, and a NEMA 4X terminal box in austenitic stainless steel, type 316L.
   iii. Each heater shall have a temperature transmitter, instead of a local thermostat, to allow remote temperature control.
l. Fan-assisted extraction oil mist eliminator, instead of a breather cap, installed on the oil reservoir, to reduce visible oil mist emissions or smoke. It shall include the following features:
   i. Designed to route trapped oil in the mist eliminator drain to the reservoir. The demister drain line shall have a sight glass, and a valved recycle line shall be furnished from the fan discharge to the fan suction to control the capacity of the demister.
   ii. Two (2) 100% coalescing filter cartridges with online isolation hardware.
   iii. Differential-pressure-indicating transmitter and remote alarm for the coalescing filter.
   iv. Two (2) constant-speed electric motor-driven 100%-capacity fans. Motors shall be of NEMA MG1 ac induction type with a TEFC enclosure with Ex-n certification and shall be provided with a 120 V/1 Ph/60 Hz space heater.
   v. Bypass valve(s) to maintain vacuum inside the reservoir to 2 in. WC (5 mbar) and facilitate maintenance of the mist eliminator during operation.
   vi. The mist eliminator cartridge efficiency shall be 99.97% for a droplet size of 0.3 μm (12 μin.) as required by API 614 [5] to help eliminate visible oil mist smoke.
m. Flame arresters installed on the mist eliminator vent (blower exhaust) lines.
n. Removable grating or checker plate on the top surface.

o. Safety-yellow-painted ladder with a cage made of galvanized carbon steel.

- Positive-displacement rotary screw-type oil pumps: one (1) 100%-capacity horizontal main oil pump and one (1) 100%-capacity horizontal auxiliary oil pump. Pumps and drivers shall be installed on a separate structurally rigid subbase. This subbase shall be mounted on the full-member base beams of the console base. Each lube oil pump unit shall include the following:

  a. The pumps shall be sized for continuous operation at the rated flow, at the pressure-limiting valve (PLV) set point and minimum viscosity [5]. The second requirement for pump sizing is that the pressure differential shall not be greater than 90% of the pump manufacturer's rating at the pump's rated capacity, with a minimum oil viscosity and subject to the PLV's set point.
  b. 20-mesh Y-type suction strainer.
  c. Carbon steel or alloy steel case, RF flanged inlet and outlet connections. Casing drain and vent connections.
  d. Cartridge-type single mechanical seal.
  e. External PLV and check valve on the discharge piping. Valved bypass line around the PLV to facilitate online maintenance.
  f. Disk pack coupling with a minimum 5-in. (125-mm) spacer tube and spark-resistant, fully secured coupling guard.
  g. NEMA MG1 ac electric-induction motor driver (TEFC enclosure, Ex-n certification, 120 V/1 Ph/60 Hz space heater). Motor sizing shall be based on the most demanding condition among the following:
     i. Power required for the pump's cold oil start temperature operation at the pump PLV setting plus accumulation.
     ii. 110% of the power required to pump the rated capacity at the normal oil reservoir temperature at the maximum pump discharge pressure.
  h. Start/stop/auto push buttons enclosed in protective boxes, installed on a local gauge board.

- Atmospheric vertical lube oil rundown tank to provide lubrication of the bearings in the served equipment string, during coastdown following a trip. It shall be sized for a minimum 3-min normal lube oil flow and furnished with atmospheric breather valve, and it shall include the following features:

  a. Shell constructed in austenitic stainless steel.
  b. Oil fill-in scheme described in API 614 [5].
  c. Low-level alarm and high-level permissive start functions, both served by a single level transmitter. Two block valves with drips shall be provided on each side of the fill-in orifice and located inside the bypass around the orifice.
  d. Overflow line sized for full-open quick-fill valve scenario to prevent oil spillage from the rundown tank vent should the quick-fill valve be inadvertently kept open instead of closed after filling of the rundown tank.
  e. Welded attachments for structural supports and 2-in. (50-mm) thick mineral wool installation.

- Pressure transmitters in the oil supply header downstream of the oil filters to monitor the oil supply pressure and initiate automatic start of the auxiliary oil pump on the predetermined low oil pressure.
- Austenitic stainless-steel tube fittings (where tube connections are provided) shall be Swagelok®.
- Two (2) 100%-duty Tubular Exchanger Manufacturers Association (TEMA) class C shell-and-tube lube oil heat exchangers with the following features:

  a. Thermal overpressure protection for the oil side (in shell) and coolant side (in tubes). The shell side of the exchanger shall include a fill equalizing line with an orifice to protect the standby exchanger from oil-side thermal overpressure. The orifice shall be sized for thermal relief. The valve on the equalizing line shall be locked in the open position before shipment.
  b. Spectacle blinds on both sides of the manual three-way valves outside the equalizing line to facilitate isolation.
  c. RF flanged vent and drain nozzles.
  d. Exchanger's oil-side drains manifolded together with filter clean-side oil drains.
  e. Vents on the oil side piped (tubed) back to the reservoir. Vent lines shall have flow indicators.
  f. Two-way, six-port, continuous-flow transfer valve of the single-body, six-port straight-plug type with resilient seats (taper plug valve is not acceptable).
  g. Removable tube bundles. AEW type is acceptable.
  h. ASME Section VIII, Division 1 Code construction.
  i. Tubes in Inhibited Admiralty brass and tube sheets in naval brass. Shells and channels in carbon steel.
  j. Oil heat exchanger's cooling water supply and return piping shall be valved and manifolded for single purchaser connection points.
  k. Each oil heat exchanger's supply and return cooling water lines within the confines of the lube oil system baseplate shall have a 2-in. RFWN flanged gate valve with a blind on its outlet.

- Three-way automatic temperature control valve (TCV) with a pneumatic controller. A bypass line with isolation valves shall be provided around the TCV.
- One (1) bladder-type lube oil accumulator with an austenitic stainless-steel body shall be furnished with a constant-pressure regulating system to maintain the lube oil pressure while the standby pump accelerates from idle conditions to its operating speed.

  a. The precharge pressure shall be just under the autostart pressure set point of the auxiliary oil pump. A block valve and a calibrated restriction orifice shall be provided to allow for controlled filling and draining of the accumulator. A bypass line with an isolation valve and a restriction orifice around the drain valve is acceptable.
  b. The lube oil accumulator shall be sized to provide normal oil flow to the entire system for a period of 5 s before the auxiliary oil pump comes to full speed.

c. The material of the bladder and secondary seals shall be compatible with the lube oil.
d. A connection shall be provided for charging and checking the precharge pressure in the accumulator bladder.
e. ASME Section VIII, Division 1 Code construction.

- Duplex full-flow 10 μm oil filters with an austenitic stainless-steel body. Their particle removal efficiency shall be as required by API 614 [5].

a. Filters shall have a pressure equalization line with a thermally sized orifice to provide thermal overprotection of the standby filter in addition to the primary function of pressure equalization for filling and switching filters. The valve on the equalizing line shall be locked in the open position before shipment.
b. Two-way, six-port, continuous-flow transfer valve of the single-body, six-port straight-plug type with resilient seats (taper plug valve is not acceptable).
c. Spectacle blinds on both sides of the manual valves outside of the equalizing line to facilitate isolation.
d. Replaceable synthetic polyester (poly-fiber) filter elements.
e. Minimum ¾-in. 300# ANSI RF flanged vent and drain shell-side connections.
f. ASME Section VIII, Division 1 Code construction.
g. Differential pressure (DP) indications (DP transmitter and DP gauge or indicating-type DP transmitter) shall be provided across the lube oil filters.
h. Filter vents routed to the reservoir, and each vent shall include a flow indicator with mechanical flow indication.

- The oil PLV shall be piped back to the reservoir through a flow indicator with a positive flow mechanical indication.
- A back-pressure control valve shall be provided to maintain the required oil supply pressure.

a. The back-pressure control valve shall be sized to properly control a wide range of oil flows (minimum to maximum) under varying pressure drop conditions (minimum to maximum) and for higher oil viscosity during cold oil circulation.
b. Drain valves shall be provided upstream and downstream of the back-pressure control valve inside the bypass around this valve.

- Pressure-reducing valve(s) shall be located downstream of the oil filters. If a pressure control bypass line is required, it shall be located at the discharge of the exchangers and upstream of the filters.
- Threaded ¾-in. vent, drain, and minimum ¾-in. instrument root valves. Plugged ¾-in. NPT low-point local drains and vents (see requirements for oil heat exchanger and oil filter vents and drains). Plugged bleed valves upstream and downstream of the check valves.
- Austenitic stainless-steel oil supply and return piping with RFWN flanges, fittings, and integral oil pressure instrumentation. Lube oil drain lines from the served equipment shall have flanged sight flow indicators and a temperature indicator.

Single tie-in points for oil supply and return shall be provided on the lube oil system baseplate. Flange bolting shall be in accordance with ASME B31.3. The bolts shall be ASTM A193, Grade B7, and nuts shall be ASTM A194, Grade 2H.

- Socket-welded fittings, socket-welded flanges, and slip-on flanges are not permitted in oil system piping. Seal welding is prohibited on instruments or in locations where disassembly is required for maintenance.
- All nozzle connections on the lube oil system that are NPS 2 in. (DN 50) and smaller shall have welded bracings (two-plane gussets). Breakout spools shall be provided where practical on all of this small-bore piping to facilitate maintenance.
- All control valves on the lube oil skid shall have a carbon steel body with austenitic stainless-steel trim. Block valves (hand valves) in the lube oil system shall have a carbon steel body with austenitic stainless-steel trim. Drain valves shall be furnished on the upstream and downstream of those control valves in this system that cannot be manually opened for maintenance.
- Baseplate-mounted open-construction gauge board for supervisory instrumentation.
- Sight glasses and associated hardware, including piping supports within the confines of the baseplate. All piping and fittings shall be prefabricated and fully assembled. All interconnecting flanges shall terminate at the edge of the baseplate.
- The lube oil system shall include three (3) pressure transmitters to allow two-out-of-three (2-o-o-3) voting on low-low lube oil supply pressure trip. Pressure transmitters shall be installed downstream of the lube oil filters.
- FluidLoc™ cable or equivalent proven sealing method, rather than an inverted gooseneck trap conduit, to prevent oil from leaking out of the equipment through cables.

16.1.8 Manufacturing Inspection and Records

- Table 16.1 describes the requirements for shop inspection and testing.

Table 16.1 Manufacturing inspection and nondestructive examination

No.	Activity description	Requirement(s)
Centrifugal compressor		
A	***Head (forging)***: end wall dry gas seal housing (separable housing)	
1.	Material inspection: chemical composition, physical properties	Material takeoff receipts
2.	Ultrasonic test and magnetic particle test after rough machining	Certified test reports (NDE records)
3.	Component part inspection after final machining	Inspection record
B	***Outer casing (forging) and nozzles***	
1.	Material inspection After heat treatment: chemical composition, physical properties, hardness test	MTR, heat treatment record
2.	Ultrasonic test and magnetic particle test after rough machining	Certified test reports (NDE records)
3.	Magnetic particle test of plates, edges, and weld preparation	Certified test report
	Welding	
4.	Magnetic particle test of all edges prepared for welding	NDE record
5.	Magnetic particle test of all weld back gouges, supports, and root pass during welding	NDE record
6.	Ultrasonic test of nozzle to casing welds (ultrasonic testing can be used only if all casing welds might not be completely accessible to conduct radiographic testing)	Certified test report
7.	Magnetic particle test of all welds before heat treatment	Certified test report
8.	Dimensional and visual inspection before heat treatment	Dimension sheet and data sheet
9.	Magnetic particle test of all welds after heat treatment	Certified test report
10.	Final weldment dimensional inspection	Dimension sheet and data sheet
11.	Hardness check nozzle-casing finish weld joint after PWHT, including chart to verify PWHT temperature and time	Certified test report, Brinell hardness ≤ 200
12.	Major repair welding map (requires purchaser's prior approval of the repair procedure, including description of the defect[s], root cause, pictures, WPS, and PQR[s]) and NDE record	Repair report
13.	Hydrostatic pressure test	Certified test report, QC document

(continued)

Table 16.1 (continued)

No.	Activity description	Requirement(s)
14.	Cast casings a. Weld repair maps for major repairs b. Casting inspection reports c. Heat treatment procedures, heat treatment cycle and charts	Documentation
15.	Inner casing (inner barrel)	MTR, dimension sheet, certified hardness test report
C	***Bearing housings***	
1.	Material inspection: chemical composition, physical properties	MTR
2.	Component part inspection: visual and dimensional inspection, including bearing retainer inner diameter	Inspection record
D	***Radial and thrust bearings***	
1.	Material inspection: chemical composition, physical properties	MTR
2.	Component part inspection: visual and dimensional inspection	Inspection record
3.	Ultrasonic test of each pad to verify 100% integrity of the bearing babbitt-to-pad bond	Test certificate
4.	Liquid penetrant (fluorescent particle) test	Test certificate
E	***Shaft***	
1.	Material inspection After heat treatment: chemical composition, physical properties, hardness test	MTR, heat treatment record
2.	Ultrasonic test after rough machining	Certified test report (NDE record)
3.	100% PMI after final machining	Certified test report
4.	Component part inspection: visual and dimensional inspection, including shaft end taper check and mechanical runout after final machining	Inspection record
5.	Magnetic particle test after final machining and residual magnetism check with calibrated Hall effect probe (limit: $\pm$ 2 Gauss)	Certified test report (NDE record) and inspection record for residual magnetism check
6.	Low-speed balance after final machining	Balance record
F	***Shaft sleeves, impeller spacers, thrust disk, and balance piston***	
1.	Material inspection After heat treatment: chemical composition, physical properties	MTR
2.	Magnetic particle test after finial machining	Certified test report (NDE record)
3.	Component part inspection: visual and dimensional inspection after final machining	Inspection record

(continued)

Table 16.1 (continued)

No.	Activity description	Requirement(s)
G	***Impellers***	
1.	Material inspection After heat treatment and stress relief: chemical composition, physical properties, hardness test	MTR, heat treatment record
2.	Ultrasonic test after rough machining and full-penetration welding	Certified test reports (NDE record)
3.	Radiography of fillet welds	Certified test report
4.	100% PMI after final machining	Certified test report
5.	Low-speed balance after final machining	Balance record
6.	Magnetic particle test after heat treatment and stress relief	Certified test report (NDE record)
7.	Component part inspection: visual and dimensional inspection before overspeed test	Inspection record
8.	Overspeed test	Test record
9.	Component part inspection: visual and dimensional inspection after overspeed test	Inspection record
10.	Magnetic particle test after overspeed test	Certified test report (NDE record)
11.	Liquid penetrant test after overspeed test	Certified test report (NDE record)
H	***Diaphragms, inlet wall/volute, discharge wall/volute***	
1.	Material inspection: chemical composition, physical properties	MTR
2.	Component part inspection: visual and dimensional inspection after final machining	Inspection record
I	***Rotor assembly***	
1.	Dimensional inspection	Dimension sheet
J	***Compressor assembly***	
1.	Bearing retainer inner diameter in bearing housings	Inspection record
2.	Bearing bore to diaphragm and end wall alignment (bore concentricity check)	Inspection record
3.	Rotor radial diameter	Inspection record
4.	Rotor float	Inspection record
5.	Radial bearing clearance, radial vibration probe gap (the probe gap shall be set at – 10 V DC [± 0.2 V DC] after 0 V DC has been verified [physical contact of probes with the observed surface])	Installation plan, QC document

(continued)

Table 16.1 (continued)

No.	Activity description	Requirement(s)
6.	Thrust bearing clearance, axial position probe gap (the probe gap shall be set at −10 V DC [± 0.2 V DC] after 0 V DC has been verified [physical contact of probes with the observed surface])	Installation plan, QC document
7.	Coupling hub pull-up (axial location) and contact check	Inspection record
8.	Residual magnetism check with calibrated Hall effect probe (limit: ± 2 Gauss)	Inspection record for residual magnetism check
9.	Combined electrical and mechanical runout check of surface areas observed by radial vibration and axial displacement proximity probes, measured at slow roll (~ 5 rpm) on balancing machine fitted with noncontacting *X*–*Y* pair of probes	Inspection record
10.	Low-speed and high-speed balance	Balancing record
11.	Unit surface inspection	Inspection record
K	***Dry gas seals and seal cartridges***	
1.	Material inspection: chemical composition, physical properties	MTR
2.	Component part inspection: visual and dimensional inspection after final machining	Inspection record
3.	Low-speed balance of seals	Test record
4.	Overspeed test of the mating ring	Test record
L	***Couplings and coupling guards***	
1.	Material inspection: chemical composition, physical properties	MTR
2.	Component part inspection: visual and dimensional inspection after final machining	Inspection record
3.	Magnetic particle test after final machining	Certified test report (NDE record)
4.	Coupling balance	Test record
5.	Surface protection coating inspection	Inspection record
6.	Assembly dimensional inspection	Inspection record
M	***Package preparation applicable to the compressor and all supplied auxiliaries: gear, main driver, lube oil system, and ship-loose systems***	

(continued)

Table 16.1 (continued)

No.	Activity description	Requirement(s)
1.	Final inspection: dimensional, piping assembly, anchor bolts, equipment initial alignment, tagging, painting, cleanliness, rust preventative, spare parts, special tools, calibration of instruments and controls, wiring continuity checks, local certification(s), and regulatory approvals Note that all items listed here might not be applicable to every package in the supplier's scope	In compliance with the quality and certification forms
N	***Positive material identification (PMI)***	
1.	Compressor pressure-containing parts: forging, weldments and nozzles	Material takeoff receipts, certified test report
2.	Impeller material	Material takeoff receipts, certified test report
Speed increasing gear unit		
A	***Gear and pinion material***	Raw material
1.	Mill test reports	Certified test report
2.	Chemical analysis	Certified test report
3.	Grain size	Material test certificate
4.	Hardness verification	Inspection record
5.	Mechanical properties	Certification
6.	Heat treatment (records of depth of case hardening, tooth flank hardness, chemical composition, material properties)	Certification
7.	Carbon content	Inspection record
8.	Effective case depth	Inspection record
9.	Core hardness	Inspection record
10.	Retained austenite	Inspection record
B	***Rotors***	
1.	Ultrasonic test after tooth finish turn	Test certificate
2.	Bearing radial runout	Inspection record
3.	Interference fit	Inspection record
4.	Residual magnetism check with calibrated Hall effect probe (limit: $\pm$ 2 Gauss)	Inspection record for residual magnetism check
5.	Combined electrical and mechanical runout check of surface areas observed by radial vibration and axial displacement proximity probes, measured at slow roll (~ 5 rpm) on balancing machine fitted with noncontacting X–Y pair of probes	Inspection record
6.	Low-speed balance of bull gear rotor	Balancing record

(continued)

Table 16.1 (continued)

No.	Activity description	Requirement(s)
7.	Multiplane dynamic balancing of pinion	Balancing record
C	***Inspection of rotors after teeth finishing***	
1.	(Ultrasonic test: see B.1 in the preceding table section)	
2.	Involute chart	Part drawings
3.	Lead chart	Inspection record
4.	Spacing error	Inspection record
5.	Pitch diameter	Inspection record
6.	Tooth surface finish	Inspection record
D	***Gear case inspection***	
1.	Weld inspection	Inspection record
2.	Magnetic particle test	Inspection record
3.	Bore alignment	Inspection record
4.	Bore diameter	Part drawings
E	***Bearing inspection***	
1.	Liquid penetrant (fluorescent particle) test	Test certificate
2.	Ultrasonic test of radial bearing shells and pads and thrust bearing babbitt-pad bond to verify bond is 100% integral	Test certificate
3.	Outside diameter	Part drawings
4.	Inside diameter	Part drawings
5.	Shell thickness	Part drawings
F	***Bull gear: pinion preassembly***	
1.	Tooth contact	Inspection record
2.	Contact on true centers	Inspection record
3.	Tape lifts	
G	***Bull gear: pinion final assembly***	
1.	Tape lifts	Inspection record
2.	Backlash	Inspection record
3.	Apex runout	Inspection record
4.	Seal clearance	Part drawings
5.	Radial bearing clearance, radial vibration probe gap (the probe gap shall be set at −10 V DC [± 0.2 V DC] after 0 V DC has been verified [physical contact of probes with the observed surface])	Installation plan, QC document

(continued)

Table 16.1 (continued)

No.	Activity description	Requirement(s)
6.	Thrust bearing clearance, axial position probe gap (the probe gap shall be set at −10 V DC [± 0.2 V DC] after 0 V DC has been verified [physical contact of probes with the observed surface])	Installation plan, QC document
7.	Shaft diameter	Installation plan
8.	Footprint	Installation plan
9.	Paint inspection	
H	***Base***	
1.	Weld inspection	Inspection report
2.	Dimensional inspection	Installation plan
3.	Flatness inspection	Inspection record
4.	Paint inspection	
I	***Electrical system***	
1.	Regulatory compliance verification	Wiring diagram
2.	Tagging inspection	Wiring diagram
Driver motor		
A	***Product identification and traceability***	(See B.1 in the following table section)
B	***Shaft and rotor assembly***	
1.	Material inspection After heat treatment: chemical composition, physical properties, hardness test	MTR, heat treatment record
2.	Ultrasonic test after rough machining	Certified test report (NDE record)
3.	100% PMI after final machining	Certified test report
4.	Component part inspection: visual and dimensional inspection, including shaft end taper check and mechanical runout after final machining	Inspection record
5.	Magnetic particle test after final machining	Certified test report (NDE record)
6.	Residual magnetism check with calibrated Hall effect probe (limit: ± 2 Gauss)	Inspection record for residual magnetism check
7.	Combined electrical and mechanical runout check of surface areas observed by radial vibration and axial displacement proximity probes, measured at slow roll (~ 5 rpm) on balancing machine fitted with noncontacting *X–Y* pair of probes	Inspection record
8.	Multiplane dynamic balancing after final machining	Balancing record
9.	Residual unbalance verification	Inspection record
C	***Bearing housings and bearings***	

(continued)

Table 16.1 (continued)

No.	Activity description	Requirement(s)
1.	Dimensional inspection	Dimension sheet
2.	Ultrasonic test of each split half of sleeve bearing to verify 100% integrity of bearing babbitt-to-shell bond	Test certificate
3.	Liquid penetrant (fluorescent particle) test	Test certificate
D	***Stator assembly***	
1.	Coils	Inspection record
2.	Stator endplates	Inspection record
3.	Stator core bars	Material takeoff receipts or mill certificate
4.	Stator core stack	Inspection record
5.	Wound stator	Inspection record
6.	Stator vapor phase inhibitor or desiccant for shipping purpose	Inspection record
7.	Frame fabrication	Inspection record
8.	Frame machining	Inspection record
9.	Core-to-frame assembly	Inspection record
E	***Final assembly***	
1.	Sleeve bearing assemblies, radial vibration probe gap [the probe gap shall be set at − 10 V DC (± 0.2 V DC) after 0 V DC has been verified (physical contact of probes with the observed surface)]	Quality control document
2.	Bearing bracket machining	Inspection record
3.	Enclosure	Inspection record
4.	Auxiliary conduit box assembly	Quality control document
5.	Main conduit box	Quality control document
6.	Final paint	Quality control document
7.	Nameplates	Quality control document
Base frames (compressor-driver and common lube oil system baseplates)		
1.	Material inspection: chemical composition, physical properties	MTR
2.	100% radiographic test or 100% liquid penetrant test or magnetic particle test of fillet welded lifting lug welds and full-penetration welded main members and full seal-welded cross bracings	Test report
3.	Postweld heat treatment (stress relief) after final welding	Quality control document
4.	Charpy V-notch impact test to qualify welded lifting lugs for − 43 °F (− 42 °C)	Certified test report

(continued)

Table 16.1 (continued)

No.	Activity description	Requirement(s)
5.	Painting and underside epoxy primer inspection	Inspection record
6.	Finish dimensional and visual examination: flatness, parallelism, machinery mounting pedestal surfaces, level pads, anchor bolt hole size and locations	Quality control document
Pressure vessels in lube oil system and dry gas seal system (lube oil filters, seal gas filters, seal gas booster, seal gas heater, lube oil shell-and-tube heat exchangers and lube oil accumulator)		
1.	Material inspection: chemical composition, physical properties	MTR
2.	Dimensional and visual inspection of finished welds (not applicable to forged body nonwelded and bolted construction if used for seal gas filters or lube oil filters)	Quality control document
3.	ASME U-1A data report	Quality control document
4.	Pressure equipment design registration as required by Alberta Boilers Safety Association (ABSA), Canadian Registration Number (CRN)	Design dossier, nameplate facsimile
Lube oil system		
	Lube oil reservoir and mist eliminator	
1.	Material inspection: chemical composition, physical properties	MTR
2.	Liquid penetrant test after welding	NDE report
3.	Reservoir water tightness test for 12 h duration	Inspection record
4.	Mist eliminator function test for a minimum of 30 min, including switching over of blowers	Certified test report
5.	Cleanliness, internal rust prevention, nameplate, and tagging verification	Inspection record
6.	Component part inspection: visual and dimensional inspection during assembly	Inspection record
	Mist eliminator driver motor	
1.	Dimensional inspection: junction box location, space heater leads, cable gland	Inspection record
2.	Insulation resistance check	Certified test report
3.	No-load running test	Certified test report
4.	Painting, nameplate, and tagging verification	Inspection record
	Lube oil pump	
1.	Material inspection: chemical composition, physical properties	MTR

(continued)

Table 16.1 (continued)

No.	Activity description	Requirement(s)
2.	Magnetic particle test or liquid penetrant test of casing and rotors	NDE report
3.	Casing hydrostatic test at 1.5× design pressure	Certified test report
4.	Operational test for 1 h, including noise and vibration measurements	Certified test report
5.	Cleanliness, internal rust prevention, painting, nameplate and tagging verification	Inspection record
6.	Component part inspection: visual and dimensional inspection during assembly	Inspection record
	Lube oil pump driver motor	
1.	Dimensional inspection: junction box location, space heater leads, shaft end, anchor bolt hole size, and locations, cable gland	Inspection record
2.	Insulation resistance check	Certified test report
3.	No-load running test	Certified test report
4.	Painting, nameplate, and tagging verification	Inspection record
	Shell-and-tube heat exchanger	
1.	100% radiography test for butt welds	NDE report
2.	PMI of lube oil shell-and-tube heat exchanger tubes and tube sheets	Certified test report
3.	Magnetic particle test or liquid penetrant test of pressure-part welded seams	NDE report
4.	Hydrostatic test at 1.5× design pressure	Certified test report
5.	Tube-to-tube sheet joint leak test with nitrogen	Certified test report
6.	Cleanliness, rust prevention, transfer valve action, nameplate, and tagging verification	Inspection record
7.	Component part inspection: visual and dimensional inspection during assembly	Inspection record
8.	Painting, nameplate, anchor-bolt hole size and locations, and tagging verification	Inspection record
	Lube oil filter and accumulator	
1.	PMI of pressure parts	Certified test report
2.	Magnetic particle test or liquid penetrant test of pressure-part welded seams	NDE report
3.	Hydrostatic test at 1.5× design pressure	Certified test report
4.	Leak test of accumulator bladder with nitrogen	Certified test report

(continued)

Table 16.1 (continued)

No.	Activity description	Requirement(s)
5.	Cleanliness, rust prevention, filter transfer valve action, nameplate, and tagging verification	Inspection record
6.	Component part inspection: visual and dimensional inspection during assembly	Inspection record
7.	Painting, nameplate, anchor-bolt hole size and locations, and tagging verification	Inspection record
	Lube oil rundown tank	
1.	Material inspection: chemical composition, physical properties	MTR
2.	Water tightness test for 12 h duration	Inspection record
3.	Cleanliness, internal rust prevention, painting, nameplate, and tagging verification	Inspection record
4.	Component part inspection: visual and dimensional inspection during assembly	Inspection record
Dry gas seal system		
	Primary seal gas and secondary/separation gas filters	
1.	PMI of pressure parts	Certified test report
2.	Magnetic particle test or liquid penetrant test of pressure-part welded seams	NDE report
3.	Hydrostatic test at 1.5× design pressure	Certified test report
4.	Cleanliness, rust prevention, transfer valve action, nameplate, and tagging verification	Inspection record
5.	Component part inspection: visual and dimensional inspection during assembly	Inspection record
	Seal gas heater	
1.	Magnetic particle test or liquid penetrant test of pressure-part welded seams	NDE report
2.	Hydrostatic test at 1.5× design pressure	Certified test report
3.	Electrical insulation resistance and wiring continuity tests: seal gas heater and thyristor control panel	Certified test report
4.	Thyristor control panel function test	Certified test report
5.	Cleanliness, rust prevention, nameplate, and tagging verification	Inspection record
6.	Component part inspection: visual and dimensional inspection during assembly	Inspection record
	Seal gas booster	
1.	Material inspection: chemical composition, physical properties	MTR

(continued)

Table 16.1 (continued)

No.	Activity description	Requirement(s)
2.	Component part inspection: visual and dimensional inspection during assembly	Inspection record
3.	Magnetic particle test or liquid penetrant test of casing and rotors	NDE report
4.	Casing integrity test with nitrogen at 1.1× design pressure	Certified test report
5.	Operational test for 1 h, including noise and vibration measurements	Certified test report
6.	Cleanliness, internal rust prevention, painting, nameplate, and tagging verification	Inspection record
	Seal gas booster driver motor	
1.	Dimensional inspection: junction box location, space heater leads, cable gland	Inspection record
2.	Insulation resistance check	Certified test report
3.	No-load running test	Certified test report
4.	Painting, nameplate, and tagging verification	Inspection record
Supplier-furnished lube oil system piping, accumulator, and dry gas seal system piping		
1.	Receipt of raw materials and 10% PMI	Material takeoff receipts, certified test report
2.	a. Finished process piping: 10% radiography test of each welder production (butt welds), test in accordance with ASME B31.3 code b. Finished utility piping: minimum 10% spot radiography test of each welder production (butt welds), test in accordance with ASME B31.3 code	Certified test reports
3.	Liquid penetrant test or magnetic particle test of weld joints (100%)	Certified test report
4.	Hydrostatic pressure test followed by dewatering, drying, and cleanliness inspection	Certified test report, QC document
Bolting		
	Alloy steel bolts including anchor bolts, nuts, washers: 2 in. and larger diameter	
1.	Material inspection: chemical composition, physical properties	MTR

Abbreviations ABSA Alberta Boilers Safety Association, *ASME* American Society of Mechanical Engineers, *CRN* Canadian Registration Number, *MTR* material test report or mill test report, *NDE* nondestructive evaluation, *PMI* positive material identification, *PQR* procedure qualification record, *PWHT* postweld heat treatment, *QC* quality control, *WPS* welding procedure specification

16.1.9 Testing and Engineering Analyses

16.1.9.1 Compressor

- Nonwitnessed dynamic balancing of the impellers.
- Nonwitnessed residual magnetism check with calibrated Hall effect probe (limit: $\pm$ 2 Gauss).
- Nonwitnessed combined electrical and mechanical runout verification of the rotor.
- Nonwitnessed low-speed balance check and high-speed dynamic balancing of the assembled rotor.
- Witnessed overspeed (spin) test of the impellers at 115% of the maximum continuous speed (N_{mc}) for a minimum duration of 1 min, including pre-test and post-test dimensional and magnetic particle inspection of the bore, eye seals, and outside diameters of the hub and cover in accordance with API 617 [1].
- Witnessed 4-h mechanical running test with the spare rotor; spare inner barrel assembly; spare high-speed coupling hub; and coupling, shop gearbox, contract shaft end seals, shop seal support system, and contract proximity probes and Proximitors®.

  a. The vacuum inside the compressor shall be maintained with the shop ejector system.
  b. The shop coupling hub can be installed on the driver side, considering that the shop driver is used for the mechanical running test.
  c. The flow in the primary vent line from each seal shall be measured for reference.
  d. The unbalance response test requires the compressor speed to be instantaneously dropped to zero by tripping the shop steam turbine driver. Polar plots shall be used to determine the first critical speed during the transient when the compressor speed is dropped from the maximum continuous speed to zero. Orbit plots, shaft centerline plots, Bodé plots, and polar plots will be checked during the 4-h running test.

- Witnessed 4-h mechanical running test with the installed (working) rotor, working inner barrel assembly, working high-speed coupling hub and coupling, shop gearbox, contract shaft end seals, shop seal support system, and contract proximity probes and Proximitors®.
- Witnessed high-speed coupling hub contact check during change-out of the installed rotor (if a spare rotor and inner barrel assembly have been purchased).

  a. Witnessed ASME Performance Test Code on Compressors and Exhausters (PTC-10) type 2, five-point performance test [6] with the spare rotor and spare inner barrel assembly. The surge point (slope of the line) shall be verified by running the compressor at 90 and 100% of the test speed.

- Noise-level survey during the performance test. At least one set of readings shall be collected. The method of measurement shall be included in the performance test procedure. The sound power level (PWL) and the overall weighted sound pressure level (SPL in dBA) shall be recorded. (The sound level (dBA), including the octave-band frequency (Hz), shall be measured for reference at a distance of 3 ft (0.9 m) from and near the centerline of the casing, and background noise shall be measured.)

- Witnessed post-test dismantling and reassembly inspection (complete inspection after the first mechanical running test and partial inspection after the second mechanical running test). This inspection shall be conducted before the gas leak test.
- Acquisition of vibration data to determine amplitudes at frequencies other than synchronous.
- Static gas leak test with nitrogen at the maximum sealing pressure in accordance with API 617 [1]. This test will be witnessed if conducted soon after the mechanical run test.
- The sequence of shop testing shall be as follows:

  a. Performance test with the spare rotor, spare inner barrel assembly, and spare coupling.
  b. Mechanical running test with the spare rotor, spare inner barrel assembly, and spare coupling. The running test includes the unbalance response verification test.
  c. Disassembly inspection of the spare rotor, spare inner barrel assembly, and spare coupling hub. Visual inspection of the bearings, seals, and internal parts.
  d. Spare coupling hub fit-up check (contact check).
  e. Mechanical running test with the job (working) rotor, job inner barrel assembly, and job coupling hub.
  f. Disassembly inspection of the thrust and radial bearings.
  g. Job coupling hub fit-up check (contact check).
  h. Static gas leakage test of the assembled compressor with the job rotor and job inner barrel assembly.

- Lateral critical speed, unbalance response, and stability analyses. The lateral critical speed analysis shall include two evaluations: considering the shop test bed support stiffness and the purchaser-provided foundation stiffnesses (static and dynamic stiffnesses in the vertical and horizontal directions).
- Train steady-state torsional natural frequency analysis.
- Train transient torsional analysis.

  a. This analysis shall include the short-circuit fault conditions "line-to-ground" and "two-phase-to-ground" to confirm that the maximum stresses in all shaft elements are below their corresponding shear yield strengths and that the peak torques in the low-speed and high-speed couplings are less than their peak torque ratings per API TR 684 [8].
  b. The peak torque at each interference fit shall be below the minimum torque required to initiate relative motion at that fit.

c. The peak torques at gear mesh shall be confirmed as acceptable by the gear manufacturer.
d. Line-to-line and three-phase short-circuit analyses shall be conducted if conditions occur under which the excitation frequency is within 10% of the first torsional mode.

- Compressor impeller stress and frequency analysis to confirm that interference does not exist at the operating speed under the effects of forcing frequencies (first four orders and blade passing frequency). Other excitation sources such as the return channel vanes and diffuser vanes (for vaned diffusers) shall be included in the analysis.

16.1.9.2 Gear Unit

- The following tests shall be conducted as described in API Standard 613 [3]:

a. Nonwitnessed gear accuracy check.
b. Nonwitnessed slow-roll check.
c. Nonwitnessed tooth surface finish check.
d. Nonwitnessed axial stability check.
e. Nonwitnessed radial runout check.
f. Witnessed mechanical run test [3].
g. Witnessed part-load (10% of the rated load if the test shop setup can provide this load), full-speed test.
h. Witnessed post-test dismantling and reassembly inspection.
i. Witnessed tooth contact check. (Tooth contact tape lift shall be supplied.) Nonwitnessed gear housing bore alignment check.

- Nonwitnessed residual magnetism check of gear and pinion rotors with a calibrated Hall effect probe (limit: $\pm$ 2 Gauss).
- Nonwitnessed combined electrical and mechanical runout verification of the rotors.
- Nonwitnessed multiplane balancing of the pinion to ISO 21940-11 [7] balance quality grade equivalent to API (4 W/N_{mc}) [1]. Low-speed balancing of bull gear.
- Undamped lateral critical speed analysis of gear and high-speed pinion and damped unbalance response analysis of the pinion including a speed map. Rotordynamic analysis shall include load cases corresponding to anticipated torque on the shop test bed during a no-load mechanical running test and for 10, 50, and 100% load cases. Test bed support stiffness shall be used in the analysis for the no-load case. The supplier shall also include in the lateral analysis report the bearing stiffness and damping characteristics at 10%, 50%, and 100% of rated power, established at the rated speed of the gear unit.
- Stability analysis for the part-load and rated load cases.
- Sound level survey during the mechanical run test.

16.1.9.3 Lube Oil System

- Witnessed 4-h operational test (console performance test) in accordance with API 614 [5] including witnessed functional test of instrumentation and controls.
- Nonwitnessed operational tests of lube oil pumps. Pump drivers and mist eliminator blower drivers.

16.1.9.4 Dry Gas Seal System

- Nonwitnessed at-speed dynamic balance (100% speed) of the dry gas seal cartridges to the same balance grade as the compressor rotor.
- Witnessed functional tests as specified in API 692 [2].
- The following seal gas heater tests:

  a. Bubble test of shroud and standoff tube welds.
  b. Hydrostatic test of pressure coils.
  c. Pressure test of electrical enclosure.

16.1.9.5 Driver Motor

- Nonwitnessed residual magnetism check of the rotor with a calibrated Hall effect probe (limit: $\pm$ 2 Gauss).
- Lateral critical speed analysis.
- Nonwitnessed combined electrical and mechanical runout verification of the rotor.
- Nonwitnessed multiplane balancing of the assembled rotor.
- Witnessed complete test.

16.1.9.6 Compressor Performance Tolerances

- Performance test tolerances on the head, capacity, and power shall be as specified in API 617 [1]. The test conditions shall be simulated to meet the volume ratio (q_i/q_d); capacity/speed ratio (q_i/N), which is proportional to the flow coefficient (Φ_d), machine Mach number (M_m), and Reynolds number (Re) within the limits specified in ASME PTC-10 Code [6].
- Additional tolerances for the test are as follows:

  a. $\pm$ 0% for flow at the surge point compared with the predicted surge flow.
  b. – 0% for discharge pressure.
  c. 5% for the head rise to surge.
  d. – 2% for flow at the end of the curve compared with the predicted end-of-curve flow.
  e. + 0, – 2% for the test operating speed compared with the calculated test speed.

16.2 Bidder Selection

- Bidder Selection and Evaluation Process:

Only qualified and capable bidders should be invited to submit proposals. Key evaluation criteria include the bidder's quality assurance systems; familiarity with the equipment's size, type, and construction materials; relevant installation references; manufacturing locations of major components; subvendor relationships; historical quality performance; material sourcing; and the availability and proximity of parts and service support infrastructure to the project site. Experienced owners and engineering contractors typically collaborate early in the project to conduct these assessments and identify suitable bidders.

- Importance of Prebid Meetings:

The value of prebid meetings cannot be overstated. Holding individual sessions with each bidder within two weeks of the inquiry allows for clear communication of specific expectations. These meetings should cover the scope of supply and services, equipment startup conditions (e.g., suction throttling or settle-out pressure), and operational constraints such as reduced-voltage starts, quality standards and certifications, shop inspections and testing protocols, specialized engineering analyses, optional features, documentation requirements, and drawing submissions. This process ensures bidders understand the project's needs rather than relying on assumptions. Additionally, prebid meetings help clarify site-specific details such as electrical area classifications, minimum design metal temperature (MDMT), and ambient temperature ranges.

- Shortlisting and Final Evaluation:

Shortlisting bidders is often necessary and should be based on factors such as commercial competitiveness, delivery timelines, adherence to commercial terms, total evaluated cost, logistical considerations (e.g., manufacturing location and shipping constraints), completeness of the technical proposal, and the nature and extent of any exceptions, deviations, or comments on the inquiry requisition. Final evaluations of shortlisted bidders—both commercial and technical—lead to a well-informed recommendation for purchase.

16.3 Technical Evaluation

The technical evaluation of shortlisted bidders may range from brief to highly detailed, depending on the complexity of the application. For critical machinery—such as centrifugal compressors, steam turbines, and gas turbines—a comprehensive comparison is conducted across multiple criteria to determine both acceptable and preferred vendors. These criteria typically include process performance, construction materials, design features of key components, auxiliary systems, utility requirements, overall weight and mounting configuration, as well as any exceptions, comments, or deviations from the defined scope, specifications, standards, and codes. A sample Technical Bid Evaluation form for an electric motor-gear-driven centrifugal compressor string is provided in Annex.

16.4 Design Coordination

- Design Coordination Meeting Timing and Scope:

The initial design coordination meeting at the manufacturer's facility should be scheduled immediately after the first round of reviews and comments on the machinery string's general arrangement drawing, foundation drawing, and schematics. This section specifically addresses the centrifugal compressor string outlined in Annex.

- Importance of Joint Reviews:

Collaborative reviews involving the vendor and their subvendors are essential. These sessions ensure that the technical requirements and scope defined in the purchase order—including any agreed-upon exceptions or deviations—are properly understood and accurately reflected in the vendor's engineering deliverables. They also serve as a critical checkpoint to identify and rectify any design flaws early in the process.

- Layout and Clash Detection:

During these coordination meetings, the machinery plot plan and its associated 3D model should be thoroughly examined to detect and resolve any potential layout clashes or interferences.

- Engineering Review Focus:

Engineering discussions should cover the topics outlined in the subsequent sections. Reference standard API 617 [1] provides detailed guidance on relevant drawings and documentation. Specific items requiring close attention are highlighted in this section. The description presented herein draws upon an example featuring an electric motor-gear-driven centrifugal compressor string as mentioned above.

16.4.1 General Arrangement Drawings for Compressor String and Lube Oil System

- Size, rating, and location of all customer connections.
- Overall weight and handling weight.
- Overall dimensions and maintenance and dismantling clearances.

- Shaft centerline height.
- Rotor weights: compressor, gear (bull gear and pinion), motor.
- ISO 21940 [7] rotor balance quality grades for rotors.
- Dimensions of the baseplate, train, or lube oil skids, complete with diameters.
- Number and locations of bolt holes and thicknesses of sections through which the bolts pass.
- Grounding (earthing) details.
- Maximum forces and moments allowed on compressor nozzles.
- Thermal movements at the compressor inlet, sideload, and discharge nozzles and the lube oil return nozzle.
- Centers of gravity of individual pieces of equipment and of the assembled skid and lifting points. Lift lug attachment details and load calculations.
- Shaft end separation and alignment data.
- Directions of rotation.
- Lifting of assembled machine skid, lube oil system skid, and major components (atmospheric rundown tank).
- Process gas interconnecting piping to allow installation of blind flanges at compressor inlet and outlet connections for positive isolation.

16.4.2 Customer (Purchaser) Connection List

- Quantity, description, size, rating, and facing of all tie-in points and plugged connections.
- Electrical and instrument connection points, including cable entry and thread data.

16.4.3 Foundation Details

- Table of dynamic loads at each anchor bolt location including electrical short-circuit loads, seismic forces, starting and operating torque loads, alignment forces, and thermal growth loads.
- Maximum allowable vibration levels on the compressor foundation in peak-to-peak mils.
- Anchor-bolt dry or lubricated torque, bolt prestretch, bolt coefficient of friction, and bolt ASTM material and grade.
- Dynamic unbalance force resulting from maximum permissible manufacturing tolerances or operating clearances.

16.4.4 Schematics and Bill of Material

16.4.4.1 Process Gas System

This system should be reviewed in conjunction with the piping three-dimensional model of the compressor area.

- Connection of drains from the suction and interstage knockout vessels, protection measures for high liquid levels and overpressure, isolation hardware.
- Distance between knockout vessels and compressor inlet lines; need for thermal insulation or heat tracing.
- Flow measurement devices in compressor inlet lines.
- Compressor suction and discharge piping loops; isolation (block valves, spectacle blinds).

16.4.4.2 Dry Gas Seal System

- Flows, temperatures, and pressures at each supply point.
- Control, alarm, and shutdown settings for pressures and temperatures.
- Utility requirements, including electrical and nitrogen.
- Pipe, valve, and orifice sizes.
- Instrumentation, safety devices, control schemes, and wiring diagrams.
- Filtration requirements: primary seal gas filter, separation/secondary gas filter.
- Seal gas heater details: heater arrangement drawing, recommended alarms, and shutdowns.
- Seal gas booster start and stop logic.
- Seal gas booster casing temperature monitoring versus vibration monitoring.
- Double block and bleed valves as described in Sect. 16.1.2.

16.4.4.3 Lube Oil System

- Oil flows, temperatures, and pressures at each supply point.
- Control alarm shutdown settings for pressures and temperature.
- Total heat loads.
- Utility requirements, including electrical, water, and air.
- Pipe, valve, and orifice sizes.
- Instrumentation, safety devices, control schemes, and wiring diagrams.

16.4.4.4 Machinery Monitoring System

- Bearing temperature, vibration, axial position, and Keyphasor™ measurements.
- Alarms and shutdowns, voting logic.

16.4.5 Compressor and Auxiliary System Data Sheets

- Compressor: thermodynamic performance, Minimum Design Metal Temperature (MDMT), casing maximum allowable working pressure and temperature, maximum capacity, other construction features, and tests and inspections.
- Auxiliary system data sheets: construction features, tests, and inspections.

16.4.6 Materials of Construction

- Compressor shaft, impellers, diaphragms, balance drum (piston) seal, interstage seals, casing, inner barrel.
- Gear casing.
- Motor shaft.
- Bull gear.
- Pinion.

16.4.7 Compressor Predicted Performance Curves for Each Specified Case

Request from the compressor manufacturer the ratio of the Mach number at the impeller inlet to that at the impeller tip for each stage.

16.4.8 Compressor Sectional Drawing

- Inner barrel roller supports and their tracks.
- Location of casing nozzles (if downward nozzle openings interfere with roller tracks).

16.4.9 Rotor Assembly Details and Parts List

- Impeller-to-diffuser lineup with axial tolerances. How are rotor-to-stator radial clearances determined and measured at the top, bottom, and sides—especially at interstage seals, balance piston seal, separation seals, and in radial bearings—and are dynamic effects like rotor vibration, thermal growth, and bearing pivot deflection taken into account?
- Rotor float.

16.4.10 Compressor Thrust Bearing Assembly Details

- Thrust bearing clearance.
- Rotor impeller–diffuser lineup, including tolerances.
- Load at rated condition, maximum allowable load.
- Pad pivot: center or offset (%).
- Rotor axial float setting.

16.4.11 Compressor Radial (Journal) Bearing Assembly Details

- Load at rated condition, maximum allowable load.
- Load direction: on pads or between pads.
- Pad pivot: center or offset (%).
- Bearing journal surface speed at N_{mc}.

16.4.12 Dry Gas Seal Assembly Drawing and Parts List

16.4.13 Maximum and Maximum Allowable Dry Gas Seal Cavity Pressures

16.4.14 Balance Piston Leakage Rates and Thrust Forces

- At the design point.
- At two times the design radial clearance.

16.4.15 Balance Piston Differential Pressure Versus Thrust Load Curve

16.4.16 Lateral Critical Speed Analyses [8] (Chap. 12)

- Rotor model (sketch) with station numbers.
- Bearing data; source of bearing coefficients; identity of the code used for the calculation of the bearing coefficients.
- Bearing pedestal data; method of determining the pedestal stiffness: finite-element analysis (FEA) or modal test.

- Two evaluations should be included in the analysis; one considering the shop test bed or assumed support stiffness, and one considering the purchaser-provided foundation stiffness values (static and dynamic stiffnesses in the vertical and horizontal directions).
- If the compressor driver is an electric motor, check how the effect of the magnetic stiffness of its core is modeled.
- Undamped critical speed map and undamped mode shapes; check natural frequencies related to the rigid-body modes at low bearing stiffness values ($< 10^5$ lbf/in. [$< 17.5 \times 10^4$ N/mm]).
- Predicted unbalance response, amounts of unbalances that are applied 4 W/N_{mc} for rotors and 40 W/N_{mc} for couplings, number of cases analyzed: maximum unbalance at midspan, half unbalances out-of-phase at shaft ends, maximum and minimum bearing clearance cases (maximum stiffness and minimum stiffness cases). Check if unbalance is applied to rotor elements that have the highest amplitudes in the mode shapes, for example, couplings and the impeller. (It is impractical to apply a trial weight on the centrifugal compressor impeller for the unbalance response verification test.)
- Check if peaks occur at speeds higher than the undamped critical speeds (indicative of the presence of significant cross-coupling coefficients).
- Verify that the Bodé plots and amplification factors are associated with minimum, normal, and maximum bearing clearances; probe response locations; and unbalance distribution(s) per API TR 684 [8].
- Stability analysis. (Deswirl features such as honeycomb or hole-pattern division wall seal in a single-casing two-section configuration and swirl cancellers at the balance piston or inlet/interstage labyrinth seals, if provided, must be described with supporting rationale.)

16.4.17 Train Torsional Analysis [8] (Chap. 13)

- Torsional model.
- Coupling details: continuous torque rating, peak torque rating, and one-time torque rating (i.e., momentary torque rating for short-circuit analysis of synchronous motors).
- Torsional natural frequencies and mode shapes: coupling mode, component body mode.
- Campbell diagram.
- Need for steady-state response analysis if the interference point does not meet the separation margin and does not involve the 1X excitation and a coupling mode.

16.4.18 Mechanical Design and Constructional Data: Compressor

It is important to note that certain checks outlined in this section may necessitate access to the design calculations, which would be contingent upon a signed nondisclosure agreement between the vendor and the purchaser.

- Scatter plots of referenced and job impellers (described in Sect. 16.1.1).
- Stresses at the blade fillet welds for each fabricated impeller at N_{mc}, overspeed, trip, and momentary speed overshoot (if the compressor is driven by a turbine and trip occurs upon a sudden loss of load). Margin of the stress at each speed from the minimum material yield strength of the weld [the stress evaluation at speeds greater than N_{mc} is not applicable to the example cited in Sect. 16.4).
- Manufacturing details for the casing, shaft, and impellers (refer to details in Table 16.1).
- Compressor casing fixed and sliding support details, including vertical and side clearances at the sliding interface.
- Casing movement direction under steady-state operating conditions.
- Casing-to-mounting pedestal design (refer to description in Chap. 1).
- Casing bolt stress at the maximum differential pressure and operating temperature, maximum allowable bolt stress.
- Casing depressurization rate based on the sealed pressure and operating temperature and considering situations such as emergency shutdown from full-load, full-pressure, full-speed operation, and sudden opening of a recycle valve during normal operation.
- Securing or positive locking arrangements for casing internal fasteners.
- Diffusers in each stage: vaned or vaneless; flow coefficients of each; locations of stage volutes, whether over the impeller tip or offset.
- Diaphragm fixation method. Are the diaphragms installed in a recess to allow transfer of the axial thrust to the casing?
- Differential pressure across each diaphragm (considering the highest value resulting from the specified operating cases and end-of-curve flows, near the surge control line, and near stonewall).
- Diaphragm material of construction: cast iron or nodular iron or fabricated, where the choice of material is governed by the differential pressure (see the preceding point).
- Maximum allowable differential pressure across diaphragms.
- Rated and maximum polytropic head per stage.
- Interference fits at the impeller-to-shaft, interstage shaft sleeves, and hydraulically fitted coupling hub; peak torque at each interference fit; and minimum torque required to initiate relative motion at these locations (refer to Chap. 2).
- Stage torque (transmitted through the impeller interference fit) at 100% speed and N_{mc}; maximum allowable stage torque at 100% speed.
- Compressor shaft end diameter sizing basis; torque transmitted through the coupling-to-shaft juncture at 100% speed and N_{mc}.

- Thrust bearing lubrication: directed or flooded.
- Relative positions and orientations of the lube oil supply and return headers on the compressor baseplate and of the return nozzle on the lube oil reservoir.
- Maximum allowable pressure drop in the secondary vent line of the dry gas seal (if the compressor is installed in a building or enclosure and secondary vents are routed to the exterior of building).

16.4.19 Phase Map Analysis (Chap. 6)

The objective of the phase map study is to determine whether the seal gas heater is required during startup following depressurization across the defined operating scenarios. This evaluation ensures a minimum safety margin or superheating of 35 °F (20 °C) above the dew-point line, thereby preventing liquid condensation within the seal gas system. This concept is illustrated in Fig. 16.1.

16.4.20 Seal Gas Booster (Chap. 6)

The seal gas booster ensures adequate seal gas flow at the necessary differential pressure during conditions such as pressurized standstill, startup, coastdown, and recycling. Its primary function is to prevent contamination from process gas migrating through the inner labyrinth into the dry gas seal cavities. Figure 16.2 illustrates, based on the simulation results, that the booster is required to be active during coastdown,

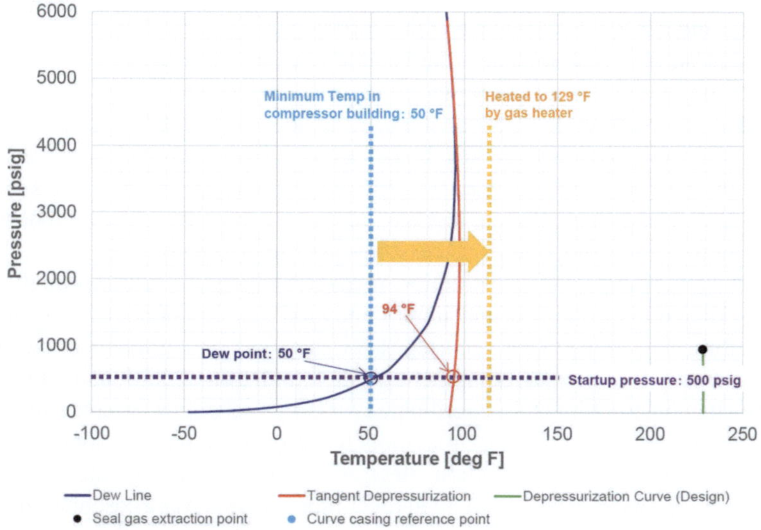

Fig. 16.1 Need for a seal gas heater

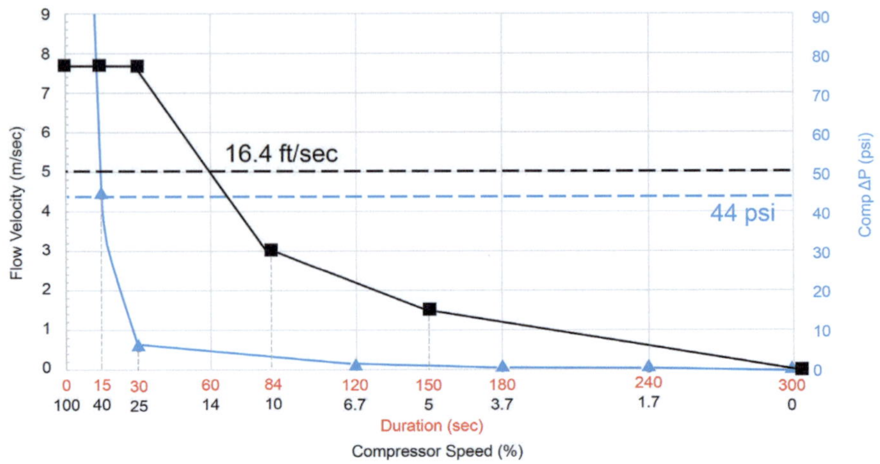

Fig. 16.2 Need for a seal gas booster

since seal gas velocity falls below the minimum threshold of 16.4 ft/s (5 m/s) within 60 s after the compressor shuts down or trips.

16.5 Installation

Engineering firms develop Construction Work Packages (CWPs) that provide a comprehensive narrative outlining the scope of work required for field installation. Alongside this narrative, each CWP includes relevant technical documents, drawings, and a bill of materials. For machinery packages, the CWP is designed to supplement—rather than replace—the vendor's installation manual.

Given the importance of coordination between engineering teams, vendors, and subvendors, it is advisable to conduct a joint review of the CWP's scope to ensure alignment on the sequence of key installation activities. A well-prepared CWP enables the site construction team to effectively plan and execute tasks such as equipment assembly, installation, leveling, alignment, and the integration of piping, instrumentation, subcomponents, accessories, and related hardware.

CWPs also prompt site construction management to perform thorough inventory checks and inspections of machinery and auxiliary components. This includes documenting any deficiencies, shortages, or damages and implementing corrective measures before placing equipment on its foundation. Additionally, the CWPs help the site construction team coordinate workforce schedules, prepare lifting and rigging strategies, and supply essential tools for leveling, alignment, and installation hardware for piping, instrumentation, electrical systems, and associated auxiliaries.

Sample scope narratives are provided for components such as a baseplate-mounted centrifugal compressor, gear assembly, driver motor, dry gas seal support system, and a separately mounted pressurized lube oil system. It should be noted that jobsite receiving inspections fall outside the scope of these CWPs.

Note: The term "supplier" includes "seller" or "vendor."

16.5.1 Compressor String

Baseplate Preparation and Installation:

- Clean the underside of the baseplate and inspect the epoxy primer for any damage or irregularities.
- Verify anchor bolt positions using the anchor-bolt template. Ensure:
  - Bolt locations, projections, and diameters align with the baseplate holes.
  - Exposed threads are protected, and sleeves are filled with a pliable material.
  - The concrete pour line elevation matches the foundation drawing specifications.
- Mount the assembled baseplate onto the foundation. It will support the compressor, gear, motor, seal gas modules, gauge board, and associated piping.

Baseplate Positioning and Grouting:

- Align the baseplate per the general arrangement and foundation drawings, adjusting elevation, coordinates, and level. Use machined targets on the top surface for leveling.
- Apply epoxy grout beneath the baseplate.
- Before the grout sets, confirm the baseplate's elevation and level.
- Once cured, torque the anchor bolts and recheck the level.
- Inspect the grout for soft spots using a magnetic dial indicator. Reference the concrete foundation to ensure movement does not exceed 0.001 in. (0.025 mm) when bolts are loosened and retightened.

Compressor and Rotor Assembly:

- Remove rotor-locking devices from the compressor, gear, and motor.
- Center the compressor rotor and support it at both ends.
- Install dry gas seals on both ends of the compressor.
- Fit and adjust the radial and thrust bearing assemblies.
- Set rotor axial position and float by modifying shim thickness on both thrust-bearing sides. Install bearing end covers on both drive and nondrive ends.

Coupling Installation and Alignment:

- Mount the hydraulically fitted coupling hub onto the compressor shaft's tapered end. Refer to coupling drawings for DBSE, hub pull-up, and spacer prestretch values.

- Conduct a soft foot check on each machinery foot during initial alignment. Final alignment should be done without process piping connected. Treat the gear casing as fixed, with the compressor and motor as movable components.

Instrumentation, Piping, and Auxiliaries:

- Install proximity probes for radial vibration and axial displacement. Set each probe gap to (– 10 V DC [± 0.2 V DC] after 0 V DC verification).
- Connect process piping to compressor nozzles. Ensure flange parallelism and perform strain checks to maintain alignment.
- Insert a temporary strainer in the compressor inlet piping.
- Recheck machinery string alignment.
- Install both low-speed and high-speed couplings and their guards.

Insulation and Electrical Systems:

- Apply thermal insulation to compressor inlet and discharge piping. Adjust pipe supports and hangers accordingly.
- Install electric heat tracing on the primary seal gas supply line as per the dry gas seal schematic (for cold environments).
- Attach shipped-loose components, such as the relief valve on the secondary seal gas line, per the schematic.

Final Checks and Connections:

- Reconfirm probe gap settings on gear and motor proximity probes (– 10 V DC [± 0.2 V DC] after 0 V DC verification).
- Connect supplier-provided instruments (RTDs, proximity probes, phase reference probes, accelerometers) to local junction boxes.
- Integrate third-party systems like the plant's machinery monitoring system or Distributed Control System (DCS).
- Complete wiring between driver and driven units and connect power and grounding to motor junction boxes.

For compressors installed on soleplates, the site installation process involves positioning, securing, and grouting both the soleplates and sub-soleplates, followed by grout curing. After the grout has set, compressor support pedestals are placed, leveled, and fixed in position. Subsequent installation steps mirror those used for compressors mounted on baseplates.

16.5.2 Lube Oil System

To facilitate streamlined integration of interconnections, the lube oil system and atmospheric lube oil rundown tank should be installed prior to compressor installation.

- Clean the underside of the baseplate thoroughly.

- Mount the assembled baseplate onto the foundation.
- Align the baseplate according to the general arrangement and foundation drawings, ensuring correct elevation, coordinates, and level.
- Apply cementitious grout beneath the baseplate.
- Verify the baseplate's elevation and level prior to grout setting.
- Once the grout has cured and the level is rechecked, torque-tighten the anchor bolts.
- Inspect the cured grout for any soft spots. Use a magnetic dial indicator on the baseplate and reference the concrete foundation to ensure that surface movement does not exceed 0.001 in. (0.025 mm) when each anchor bolt is loosened and retightened.
- Conduct a soft foot check on each oil pump and driver motor foot during initial alignment. Final alignment must be performed without oil piping connected to the pumps.
- Connect the suction and discharge piping to the lube oil pumps.
- Install couplings and coupling guards between the lube oil pumps and their drivers.
- If shipped separately, install the electric heaters into the oil reservoir.
- Mount the lube oil mist eliminator and flame arrester as per the general arrangement drawing.
- Position and install the atmospheric lube oil rundown tank.
- Install the prefabricated interconnecting piping.
- Apply mineral wool thermal insulation to the lube oil reservoir, rundown tank, and oil piping, following the lube oil system schematic.
- Connect electrical power and grounding to the designated termination points of the supplier's equipment, including the mist eliminator blower motor and lube oil pump driver motor junction boxes.

Annex

Attachment					
Technical bid evaluation form: motor—gear driven centrifugal compressor					
		Service:		Bidder:	
		Tag(s):		Proposal No.:	
Item No.	Description	Units	Requirements		
1	**General**				
1.1	Compressor tag number				
1.2	Compressor model				
1.3	Number of casings				

(continued)

(continued)

Attachment					
Technical bid evaluation form: motor—gear driven centrifugal compressor					
		Service:		Bidder:	
		Tag(s):		Proposal No.:	
Item No.	Description	Units	Requirements		
1.4	Compressor casing split				
1.5	Number of sections				
1.6	Number of impeller stages in each section				
1.7	Governing industry standard				
1.8	Quality certification				
1.9	Equipment delivery				
1.10	Delivery point				
1.11	Hazardous area classification standard				
1.12	Zone				
1.13	Group				
1.14	Class				
2	**Process conditions**				
				As specified	As quoted
2.1	Section				
2.1.1	Normal or alternative condition		Yes/No		
2.1.2	Guarantee point		Yes/No		
2.1.3	Weight flow basis—dry/wet	lb/h			
2.1.4	Gas handled				
2.1.5	Molecular weight—dry/wet				
2.1.6	**Inlet conditions**				
2.1.7	Pressure	psia			
2.1.8	Temperature	°F			
2.1.9	$K = C_p/C_v$				
2.1.10	Compressibility (Z_s)				
2.1.11	Actual inlet volumetric flow	icfm (acfm)			
2.1.12	**Discharge conditions**				
2.1.13	Pressure	psia			
2.1.14	Temperature	°F			

(continued)

(continued)

Attachment					
Technical bid evaluation form: motor—gear driven centrifugal compressor					
		Service:		Bidder:	
		Tag(s):		Proposal No.:	
Item No.	Description	Units	Requirements		
2.1.15	Maximum discharge temperature @ surge point	°F			
2.1.16	Ratio of specific heats ($K = C_p/C_v$)				
2.1.17	Compressibility (Z_d)				
2.1.18	Stage pressure ratio				
2.1.19	Discharge flange flow	icfm (acfm)			
2.1.20	Brake horsepower	hp			
2.1.21	Seal/bearing losses	hp			
2.1.22	Speed	rpm			
2.1.23	Estimated surge	icfm (acfm)			
2.1.24	Polytropic head total	ft-lb/lbm			
2.1.25	Polytropic head per stage	ft-lb/lbm			
2.1.26	Head rise to surge	%			
2.1.27	Discharge pressure rise to surge	%			
2.1.28	Polytropic efficiency	%			
2.1.29	Percent inlet volume @ surge	%			
2.1.30	Turndown	%			
2.1.31	End of curve % design head	%			
2.1.32	End of curve % design flow	%			
2.1.33	Inlet Mach number @ operating point/end of curve				
3	**Materials of construction**				
3.1	Compressor outer casing				
3.2	Compressor inner casing/barrel				
3.3	Diaphragms				
3.4	Shaft				
3.5	Impellers				
3.6	Impeller eye labyrinth seals				
3.7	Interstage labyrinth seals				
3.8	Balance piston labyrinth seals				
3.9	Balance piston				

(continued)

(continued)

Attachment					
Technical bid evaluation form: motor—gear driven centrifugal compressor					
		Service:		Bidder:	
		Tag(s):		Proposal No.:	
Item No.	Description	Units	Requirements		
3.10	Shaft sleeves				
3.11	Bearing housings				
3.12	Radial bearings—pad base material				
3.13	Thrust bearing—pad base material				
3.14	Dry gas seal piping				
3.15	Lube oil system piping				
3.16	Lube oil reservoir				
3.17	Lube oil rundown tank				
3.18	Speed increaser gear unit casing				
3.19	Gear wheel				
3.20	Gear rim				
3.21	Pinion				
3.22	Driver shaft material				
4	**Material inspection and testing**				
4.1	**Compressor**				
4.2	Material certificates				
4.3	Case radiography				
4.4	Seal housing				
4.5	– Magnetic particle test				
4.6	Outer casing				
4.7	– Magnetic particle test				
4.8	– Ultrasonic test				
4.9	– Hardness check				
4.10	– Hydrostatic test				
4.11	Shaft				
4.12	– Ultrasonic test				
4.13	– Magnetic particle test of forging				
4.14	– Fluorescent penetrant test				
4.15	Impeller				
4.16	– Ultrasonic test				
4.17	– Magnetic particle test				
4.18	**Gear unit**				

(continued)

(continued)

Attachment					
Technical bid evaluation form: motor—gear driven centrifugal compressor					
		Service:		Bidder:	
		Tag(s):		Proposal No.:	
Item No.	Description	Units	Requirements		
4.19	Material certificates				
4.20	Heat treatment records				
4.21	Gear and pinion material				
4.22	– Hardness verification				
4.23	– Heat treatment				
4.24	– Grain size				
4.25	– Carbon content				
4.26	– Case depth				
4.27	– Retained austenite				
4.28	Rotor				
4.29	– Ultrasonic test				
4.30	– Magnetic particle test				
4.31	Gear housing				
4.32	– Magnetic particle test				
4.33	Bearings				
4.34	– Dye penetrant test				
4.35	**Driver electric motor**				
4.36	Material certificates				
4.37	Ultrasonic test				
4.38	Magnetic particle test of forging				
4.39	**Lube oil system and dry gas seal system**				
4.40	ASME code stamp				
4.41	Spot radiography				
4.42	PMI				
4.43	Hydrostatic test				
5	**Technical features**				
	SPEEDS				
5.1	Maximum continuous speed	rpm			
5.2	Maximum design limiting speed	rpm			
5.3	Trip speed	rpm			
5.4	Maximum tip speed @ 100% speed	ft/s			

(continued)

(continued)

Attachment

Technical bid evaluation form: motor—gear driven centrifugal compressor

		Service:		Bidder:	
		Tag(s):		Proposal No.:	
Item No.	Description	Units	Requirements		
5.6	Estimated damped first and second lateral critical speeds	rpm			
5.7	Amplification factor at first critical speed				
5.8	CASING				
5.9	Casing construction cast or fabricated				
5.10	Casing maximum allowable working pressure	psig			
5.11	Casing hydrotest pressure	psig			
5.12	Casing maximum design pressure	psig			
5.13	Casing minimum design pressure	psig			
5.14	Minimum suction pressure at which dry gas seal is suitable to operate continuously	psig			
5.15	Casing corrosion allowance	in.			
5.16	Casing maximum capacity	icfm (acfm)			
5.17	Casing joint design				
5.18	Casing support				
5.19	Location of casing sliding support		Thrust end or non-thrust end		
5.20	DIFFUSERS				
5.21	Diffusers		Vaned/ Vaneless		
5.22	Differential pressure across diaphragms	psi			
5.23	Maximum allowable differential pressure across diaphragms	psi			
5.24	TEMPERATURES				
5.25	Highest discharge temperature at surge and maximum continuous speed	°F			
5.26	Maximum allowable discharge temperature for casing/internal seals	°F			

(continued)

(continued)

Attachment					
Technical bid evaluation form: motor—gear driven centrifugal compressor					
		Service:		Bidder:	
		Tag(s):		Proposal No.:	
Item No.	Description	Units	Requirements		
5.27	Casing max./min. operating temperature	°F			
5.28	Casing minimum allowable inlet temperature	°F			
5.29	IMPELLERS				
5.30	Number of impellers				
5.31	Maximum number of impellers in casing				
5.32	Impeller/diffuser stage that has single stage test data				
5.33	Impeller user record				
5.34	Number of impellers in each section				
5.35	Impeller diameter—quoted	in.			
5.36	Maximum impeller diameter	in.			
5.37	Number of vanes on each impeller				
5.38	Impeller tip width	in.			
5.39	Impeller eye diameter	in.			
5.40	Impeller inlet eye flow area	in.2			
5.41	Impeller blade tip exit angle	°			
5.42	Impeller diameter ratio				
5.43	Impeller nominal polytropic efficiency	%			
5.44	Impeller type (enclosed/open)				
5.45	Impeller fabrication (welded/milled)				
5.46	Impeller stress @ rated speed	psi			
5.47	Impeller stress @ MCS	psi			
5.48	Impeller stage torque at rated speed	in.-lb			
5.49	Maximum torque allowable by impeller fit	in.-lb			
5.50	Impeller material maximum yield strength	psi			
5.51	Impeller material Brinell hardness (maximum and minimum)				
5.52	Impeller fixation method				

(continued)

(continued)

Attachment					
Technical bid evaluation form: motor—gear driven centrifugal compressor					
		Service:		Bidder:	
		Tag(s):		Proposal No.:	
Item No.	Description	Units	Requirements		
5.53	Maximum impeller head @ 100% speed	ft-lb/lbm			
5.54	Impeller specific speed				
5.55	Impeller rated flow coefficient				
5.56	Impeller flow coefficient @ 120% of rated flow				
5.57	Impeller surge flow coefficient				
5.58	Impeller rated head coefficient				
5.59	Impeller head coefficient @ end of curve flow				
5.60	Impeller inlet Mach number at rated flow				
5.61	Impeller inlet Mach number @ 120% of rated flow				
5.62	Impeller exit Mach number at rated flow				
5.64	SHAFT				
5.65	Shaft diameter @ impellers	in.			
5.66	Shaft diameter @ coupling end	in.			
5.67	Shaft diameter @ bearings	in.			
5.68	Bearing span	in.			
5.69	Shaft end				
5.70	Shaft maximum yield strength	psi			
5.71	Shaft torque @ rated speed and rated point	ft-lb			
5.72	Shaft maximum torque capability	ft-lb			
5.73	Shaft stress @ coupling end	psi			
5.74	BEARINGS				
5.75	Radial bearing number of pads				
5.76	Radial bearing load—on pads/between pads				
5.77	Radial bearing pivot—center/offset				
5.78	Radial bearing *L*/*D* ratio				
5.79	Radial bearing pad material				

(continued)

(continued)

Attachment					
Technical bid evaluation form: motor—gear driven centrifugal compressor					
		Service:		Bidder:	
		Tag(s):		Proposal No.:	
Item No.	Description	Units	Requirements		
5.80	Radial bearing load—actual/ allowable	psi			
5.81	Thrust bearing pivot—center/offset				
5.82	Thrust bearing load—maximum/ ultimate	psi			
5.83	Thrust bearing pad material				
5.84	Thrust bearing lubrication—directed/flooded				
5.85	LABYRINTHS				
5.86	Impeller eye labyrinth type				
5.87	Interstage labyrinth type				
5.88	Balance piston labyrinth type				
5.89	DRY GAS SEAL				
5.90	Dry gas seal design pressure	psig			
5.91	LUBE OIL				
5.92	Lube oil ISO viscosity grade				
5.93	NOZZLES				
5.94	**Process inlet connections**				
5.95	Inlet flange size	in.			
5.96	Inlet flange rating				
5.97	Inlet flange facing				
5.98	Inlet nozzle orientation				
5.103	Inlet nozzle gas velocity @ rated point	ft/s			
5.104	Inlet nozzle max. allowable gas velocity	ft/s			
5.105	**Process discharge connections**				
5.106	Discharge flange size	in.			
5.107	Discharge flange rating				
5.108	Discharge flange facing				
5.109	Discharge nozzle orientation				
5.110	Discharge nozzle gas velocity @ rated point	ft/s			

(continued)

(continued)

Attachment					
Technical bid evaluation form: motor—gear driven centrifugal compressor					
		Service:		Bidder:	
		Tag(s):		Proposal No.:	
Item No.	Description	Units	Requirements		
5.112	Casing drain flange rating				
5.113	Casing drain flange facing				
5.114	COUPLINGS				
5.115	Coupling service factor (low-speed/ high-speed)				
5.116	Coupling rating				
5.117	– Low-speed	hp/ 100 rpm			
5.118	– High-speed	hp/ 100 rpm			
5.119	Coupling Normal Torque				
5.120	– Low-speed	in.-lb			
5.121	– High-speed	in.-lb			
5.122	Coupling maximum steady torque				
5.123	– Low-speed	in.-lb			
5.124	– High-speed	in.-lb			
5.125	Coupling slip torque				
5.126	– Low-speed	in.-lb			
5.127	– High-speed	in.-lb			
6	**Utility summary**				
6.1	Main driver rated power	hp			
6.2	Main driver space heater	kW			
6.3	Main lube pump driver rated power	hp			
6.4	Auxiliary lube pump driver rated power	hp			
6.5	Lube oil pump driver space heater	kW			
6.6	Lube oil cooler heat load	Btu/h			
6.7	Lube oil reservoir heater	Btu/h			
6.8	Lube oil reservoir extraction demister motor	hp			
6.10	Seal gas heater	kW	If furnished		

(continued)

(continued)

Attachment					
Technical bid evaluation form: motor—gear driven centrifugal compressor					
		Service:		Bidder:	
		Tag(s):		Proposal No.:	
Item No.	Description	Units	Requirements		
6.11	Nitrogen buffer flow—max. for dry gas secondary seals and separation seals (standard volume flow)	scfm			
6.12	Nitrogen for primary seal gas booster pump (Standard volume flow)	scfm	If furnished		
6.14	Instrument air—field instruments (Standard volume flow)	scfm			
6.17	Cooling water flow (S&T lube oil heat exchanger)	gpm			
7	**Weight/mounting data**				
7.1	Driver-driven train	lb			
7.2	Lube oil console	lb			
7.3	Heaviest maintenance weight	lb			
7.4	Space requirement for complete unit	ft			
7.5	Space requirement for lube oil console	ft			
8	**Option items**				
8.1	Capital spare parts for compressor				
8.2	Start-up and commissioning spare parts for compressor				
8.3	Two-year operational spare parts including maintenance spares for compressor				
8.4	Start-up and commissioning spare parts for lube oil system				
8.5	Start-up and commissioning spare parts for dry gas seal system				
8.6	Start-up and commissioning spare parts for driver motor				
8.7	Start-up and commissioning spare parts for gear				
9	**Major sub-suppliers**				
9.1	Gear unit				
9.2	Compressor driver				

(continued)

(continued)

Attachment

Technical bid evaluation form: motor—gear driven centrifugal compressor

		Service:		Bidder:	
		Tag(s):		Proposal No.:	
Item No.	Description	Units	Requirements		
9.3	Lube oil console				
9.4	Heat exchangers				
9.5	Dry gas seal system				
9.6	Baseplates—compressor and lube oil system				
10	**Gear**				
10.1	Model number				
10.2	Gear rated speeds—input/output	rpm			
10.3	Speed ratio				
10.4	Gear service factor				
10.5	Gear and pinion shaft end configuration				
10.6	Gear and pinion arrangement				
10.7	Gear and pinion load direction				
10.8	Ratio of gear weight to pinion weight				
10.9	Gear unit rated power	hp			
10.10	Gear unit mechanical rating	hp			
10.11	Gear pitch line velocity	ft/s			
10.15	Gear and pinion attachment to shaft				
10.16	Gear shaft end torque @ maximum continuous speed	ft-lb			
10.17	Pinion shaft end torque @ maximum continuous speed	ft-lb			
10.20	Gear teeth/pinion manufacturing process				
10.21	Radial bearing load—at rated point	psi			
10.22	Radial bearing load—allowable	psi			
10.23	Thrust bearing load—at rated point	psi			
10.24	Thrust bearing load—allowable	psi			
10.25	Gear/pinion hardening method				
11	**Driver motor**				
11.1	**Electrical characteristics**				

(continued)

(continued)

Attachment					
Technical bid evaluation form: motor—gear driven centrifugal compressor					
		Service:		Bidder:	
		Tag(s):		Proposal No.:	
Item No.	Description	Units	Requirements		
11.2	Type of enclosure				
11.3	Motor frame designation				
11.4	Synchronous speed	rpm			
11.5	Service factor/insulation class				
11.6	Temperature rise above 104 °F (40 °C) @ 1.0 SF	°F			
11.7	Starting method				
11.8	**Performance**				
11.9	Full load current	A			
11.10	Maximum locked rotor current as a percentage of full load current	%			
11.11	**Technical features**				
11.12	Type of radial bearings				
11.13	Bearing *L*/*D* ratio—coupling end/outboard bearing				
11.14	Bearing loading—rated/maximum	psig			
11.15	Space heaters for main conduit box				
11.16	Current transformers for differential protection				
11.17	Shaft end configuration				
11.18	Grounding pads				
12	**Heat release**				
12.1	Compressor casing	Btu/h			
12.2	Gear casing	Btu/h			
12.3	Driver motor casing/heat exchanger	Btu/h			

References

1. API Standard 617 (2022) Axial and centrifugal compressors and expander-compressors, 9th edn. American Petroleum Institute, Washington, DC
2. API Standard 692 (2018) Dry gas sealing systems for axial, centrifugal, rotary screw compressors and expanders, 1st edn. American Petroleum Institute, Washington, DC

3. API Standard 613 (2021) Special purpose gear units for petroleum, chemical, and gas industry services, 6th edn. American Petroleum Institute, Washington, DC
4. API Standard 671 (2020) Special-purpose couplings for petroleum, chemical, and gas industry services, 5th edn. American Petroleum Institute, Washington, DC
5. API Standard 614 (2022) Lubrication, shaft-sealing, and oil-control systems and auxiliaries, 6th edn. American Petroleum Institute, Washington, DC
6. ASME PTC-10 1997 (R2014) (2014) Performance test code on compressors and exhausters. American Society of Mechanical Engineers, New York
7. ISO 21940-11:216 (2016, confirmed 2021) Mechanical vibration—rotor balancing—part 11: procedures and tolerances for rotors with rigid behaviour, 1st edn. International Organization for Standardization, Geneva, Switzerland
8. API TR 684-1 (2019) API Standard paragraphs rotordynamic tutorial: lateral critical speeds, unbalance response, stability, train torsionals, and rotor balancing, 1st edn. American Petroleum Institute, Washington, DC

Chapter 17
CO_2 Compressors

Abstract Anthropogenic CO_2 emissions from the combustion of fossil fuels and various industrial activities are a primary driver of global warming and its related climate impacts. To address this challenge, carbon capture, utilization, and storage (CCUS) technologies play a vital role in global decarbonization strategies. A range of commercially available carbon capture solutions exist, each adaptable to specific CO_2 recovery applications. Once captured and purified, CO_2 must be compressed for efficient transport, further processing, and long-term storage. This chapter provides a comprehensive overview of key CCUS considerations, including the context of global decarbonization initiatives, CO_2 product requirements, available capture technologies, and the unique thermodynamic challenges associated with CO_2 compression. It also explores the mechanical design aspects, common construction materials, compressor train configurations, shaft sealing techniques, capacity control methods, and modular design strategies for the two predominant types of CO_2 compressors: centrifugal and integrally geared.

17.1 CO_2 Process

Carbon capture, utilization, and storage (CCUS) refers to the process of capturing carbon dioxide (CO_2) generated from the combustion of fossil fuels and various industrial activities, preventing it from being released into the atmosphere. Once captured, the CO_2 is transported to a designated site for long-term storage—either by injecting it into geological formations or converting it chemically for use in alternative applications.

CCUS technologies are essential for reducing emissions from existing industrial infrastructure, especially in sectors where decarbonization is particularly challenging. Beyond addressing current emissions, carbon dioxide removal (CDR) methods—such as direct air capture (DAC) and bioenergy with carbon capture

N. Ghaisas, *Centrifugal Compressors in Hydrocarbon Processing Industries*,
https://doi.org/10.1007/978-3-032-03687-2_17

and storage (BECCS)—target legacy emissions by extracting CO_2 directly from the atmosphere.

However, the processes involved in capturing, compressing, transporting, and storing CO_2 come with distinct technical complexities that differ from conventional industrial operations. This section explores key design considerations specific to CCUS, with a focus on the unique requirements of CO_2 compression systems.

17.1.1 Global Warming and Decarbonization

Since the onset of the industrial era, human-driven carbon emissions have risen steadily, largely fueled by the widespread use of carbon-based energy sources—an engine of global economic growth and development. In 2022, annual CO_2 emissions from energy production and industrial activities reached 40.6 billion tons (36.8 gigatons), a sharp increase from 6.3 billion tons (5.7 Gt) in 1950 and under 2.2 billion tons (< 2 Gt) in 1900 [1].

These anthropogenic emissions are a primary contributor to global warming and the resulting climate change. The consequences are already evident: more frequent and intense extreme weather events are damaging infrastructure, agriculture, and ecosystems, while also posing serious risks to human health. Vulnerable and low-income communities are expected to bear the brunt of these impacts.

To avoid escalating climate-related threats, it is imperative to rapidly reduce carbon emissions to levels comparable to those before industrialization.

A wide array of approaches can support decarbonization efforts, including:

- Transitioning to electrification and adopting energy sources with minimal or zero emissions.
- Deploying carbon capture, utilization, and storage (CCUS) technologies.
- Transforming agricultural and industrial methodologies.
- Enhancing the capacity of natural carbon sinks for both immediate and long-term sequestrations.
- Advancing energy efficiency across sectors.

Achieving meaningful decarbonization will demand profound shifts in existing infrastructures, societal norms, and behavioral patterns—changes that will carry significant economic and social consequences. Success hinges on a unified, comprehensive strategy that leverages all available pathways, with CCUS expected to play a pivotal role in the transition.

17.1.2 Mandates, Policies, and Markets

While the impact of human-induced CO_2 emissions on global warming and climate change has been acknowledged and studied since the mid-twentieth century, formal international efforts took shape later. The Intergovernmental Panel on Climate Change (IPCC) was established in 1988, followed by the enactment of the United Nations Framework Convention on Climate Change (UNFCCC) in 1994. The UNFCCC serves as the foundation for the annual Conference of the Parties (COP) meetings, which culminated in the landmark Paris Agreement at COP21.

Adopted as the first legally binding global treaty on climate change, the Paris Agreement aims to limit the rise in global average temperatures to below 2.7–3.6 °F (1.5–2 °C) above preindustrial levels. To achieve the more ambitious 2.7 °F (1.5 °C) target, greenhouse gas emissions must peak by 2025, fall by 43% by 2030, and reach net zero by 2050 [2].

The Paris Agreement provides the framework for member states to develop climate action plans and submit nationally determined contributions (NDCs), which outline their specific emissions reduction goals. Approaches to reducing emissions differ across governments and jurisdictions, but they can generally be grouped into the following categories:

Approaches to reducing greenhouse gas emissions differ across governments and jurisdictions, but they generally fall into the following key categories:

- Carbon Pricing Mechanisms: These include carbon taxes and cap-and-trade systems designed to internalize the environmental cost of emissions. Notable examples are carbon tax frameworks in Canada, Sweden, Denmark, and Japan, as well as cap-and-trade programs in the European Union, China, and South Korea.
- Incentives for Renewable Energy Adoption: Governments promote low- and zero-carbon energy solutions through financial mechanisms such as tax credits, subsidies, and grants. Prominent initiatives include the Inflation Reduction Act (IRA) in the USA, India's Production-Linked Incentive (PLI) scheme, and Canada's Investment Tax Credits (ITCs).
- Emissions Regulations and Standards: These policies set limits on emissions from specific products or industrial sectors. Examples include vehicle emissions standards and the Clean Air Act in the USA.
- Energy Efficiency Initiatives: Programs aimed at improving energy use include the Energy Star certification in the US, the EU Energy Label in Europe, and Australia's Green Star rating system.
- Reforestation and Ecosystem Conservation: Efforts to preserve and expand natural carbon sinks play a vital role in long-term carbon sequestration.
- Waste Reduction and Recycling Programs: Strategies that minimize landfill waste and promote recycling help curb emissions associated with waste decomposition and management.

As part of broader emissions reduction strategies, carbon markets serve as platforms for trading credits that represent the reduction, removal, or avoidance of greenhouse gas emissions. Compliance markets operate within jurisdictions that have established carbon pricing frameworks, assigning a monetary value to carbon emissions. These markets are typically confined to the boundaries of their respective regulatory systems, with limited cross-border credit trading.

Table 17.1 CO_2 emissions by sector (2022) [3]

Sector	Associated 2022 emissions (billion tons [Gt])
Power	16.15 (14.65)
Industry	10.1 (9.15)
Transportation	8.80 (7.98)
Buildings	3.27 (2.97)

Alongside these, voluntary carbon markets offer a more flexible alternative. Though less regulated, they enable individuals and organizations to purchase carbon offsets as a means of mitigating their own emissions and contributing to climate goals.

Both carbon pricing and market mechanisms are critical in assessing the feasibility and economic viability of carbon capture, utilization, and storage (CCUS) initiatives, helping to shape investment decisions and policy development.

17.1.3 Emission Sources and CCUS Siting

Table 17.1 provides a breakdown of CO_2 emissions by sector, along with their respective contributions for the year 2022.

Carbon capture, utilization, and storage (CCUS) plays a vital role across a wide range of power and industrial sectors. Its relevance spans gas- and coal-fired power generation, natural gas and oil refining, chemical production, pulp and paper processing, steel manufacturing, and cement production. This broad applicability positions CCUS as a key technology with the potential to significantly reduce global greenhouse gas emissions.

The concentration of CO_2 in feed gases destined for capture varies depending on the source. For instance:

- Flue gas from natural gas-fired boilers typically contains 8–10% CO_2 by volume.
- Flue gas from natural gas-fired power plants contains a lower concentration, around 3–4%.
- Chemical process streams can yield CO_2 concentrations exceeding 15–20%.
- In contrast, ambient atmospheric CO_2 levels are approximately 400–450 ppmv.

The CO_2 concentration of the emission source is a critical factor in determining the feasibility of CCUS projects. Streams with higher CO_2 concentrations—such as syngas or tail gas from steam methane reforming (SMR)—are generally more attractive for capture due to lower energy requirements and greater economic viability compared to sources with lower concentrations, such as direct air capture (DAC).

Key considerations for siting CCUS facilities include:

- The scale and intensity of CO_2 emissions (preferably large point sources).
- Presence of oxygen and other contaminants in the gas stream.

- Availability of physical space for capture equipment.
- Integration with existing utilities and infrastructure.
- Proximity to CO_2 disposal sites or markets for CO_2-derived products.

17.1.4 CO_2 Storage and Utilization Applications

To effectively mitigate the climate impact of CO_2 emissions captured through carbon capture initiatives, long-term storage is essential. This can be accomplished either by storing the recovered CO_2 underground in geological formations or by converting it chemically into useful products for alternative applications.

Geological CO_2 storage entails injecting CO_2 into deep, porous rock formations. Suitable geological sites must exhibit:

- Permeability, allowing CO_2 to be injected and retained.
- Permanence, ensured by a cap rock that prevents upward migration or leakage.

Potential storage formations include depleted oil and gas reservoirs, unmineable coal seams, and deep saline aquifers. The capacity for geological sequestration is immense—according to the Global CCS Institute's 2022 estimate, global storage potential exceeds 15.4 trillion tons (14,000 Gt) of CO_2 [4].

Carbon utilization, on the other hand, focuses on transforming CO_2 into products that enable reuse and storage. The US Department of Energy's Carbon Dioxide Conversion Program identifies three primary pathways:

- Biological uptake, such as algae cultivation and bioproduct generation.
- Catalytic conversion, producing fuels and chemicals.
- Mineralization, forming stable inorganic compounds.

Unlike storage-focused projects, utilization efforts offer the added benefit of generating revenue through value-added products, helping to offset the costs of CO_2 capture. This field encompasses a wide array of technologies and remains a vibrant area of ongoing research and innovation.

In practice, large-scale emissions reduction strategies typically prioritize sequestration and storage for permanent carbon removal. Meanwhile, utilization projects are often tailored to specific industrial contexts, where they can be more easily integrated and economically optimized.

17.1.5 Carbon Capture Technology

A diverse range of technologies is available for CO_2 capture and recovery. When selecting among these options, project developers must evaluate several critical factors. Foremost is the alignment between the CO_2 source—its concentration and purity—and the specifications required for the intended CO_2 product. Key attributes of the emissions source, such as pressure, CO_2 concentration, and the presence of oxygen or other impurities, play a significant role in determining the most suitable technology.

In addition to capturing CO_2, the chosen technology must also manage associated waste streams and residual emissions. Many of the same considerations that influence site selection—outlined in Sect. 17.1.3—also affect technology choice. Constraints related to utility availability or limited plant space can further narrow the range of viable options.

The implementation of CO_2 capture technologies typically involves three distinct stages:

1. Pretreatment—Preparing the gas stream by removing contaminants and conditioning it for efficient capture.
2. CO_2 Recovery—Extracting CO_2 from the treated stream using physical or chemical processes.
3. Post-conditioning—Refining the recovered CO_2 to meet specifications for transport, storage, or utilization.

17.1.5.1 Pretreatment

For many CO_2 recovery technologies, it is essential to pretreat the emission source to properly condition the feed gas before initiating CO_2 capture. Common pretreatment steps include:

- Eliminating fly ash and particulates from coal-fired power plant emissions.
- Removing sulfur oxides (SO_x) ahead of chemical absorption processes.
- Extracting water and mercury prior to cryogenic separation.

The main purpose of pretreatment is to safeguard the performance and longevity of the CO_2 recovery system by minimizing exposure to harmful contaminants present in the source gas.

17.1.5.2 CO_2 Recovery

The effectiveness of CO_2 recovery technologies is primarily influenced by the targeted recovery rate and the desired purity of the CO_2 product. As previously discussed in this chapter, the suitability of a given recovery method depends heavily on the specific attributes of the feed gas—such as its pressure, CO_2 concentration, and the presence of oxygen or other impurities. Broadly speaking, CO_2 recovery technologies can be classified into the following categories:

- **Chemical Absorption**: This method involves circulating a chemical solvent that reacts with CO_2 in the feed gas, effectively capturing it. The solvent, now rich in CO_2, is regenerated by applying heat, which releases the CO_2 for collection. A similar approach is used in fixed-bed chemical absorbents, where the absorbent is anchored to a porous or structured substrate with a high surface area to maximize CO_2 binding sites. Metal–organic frameworks (MOFs) are a notable example of this design. Feed gas flows through the fixed bed, and CO_2 is absorbed. Regeneration occurs separately through thermal treatment to release the captured CO_2.

- **Physical Adsorption**: This technique employs either a physical solvent or a fixed-bed adsorbent to extract CO_2 from the feed gas. Unlike chemical absorption, it relies on CO_2 solubility or weak van der Waals interactions, which significantly lowers the energy required for CO_2 release. Regeneration is typically achieved by reducing pressure, sometimes combined with heat application.
- **Cryogenic Separation**: This process separates CO_2 based on differences in dew points among the feed-gas components. Operating at extremely low temperatures, it demands careful preconditioning and specialized materials. The result is a liquid CO_2 product, which can be efficiently transported using pumps.
- **Membrane Separation**: This method utilizes a semipermeable membrane to separate CO_2 based on differential mass transfer rates. Driven by partial pressure differences between the feed gas and the permeate side, membrane systems require high feed-gas pressure. On their own, they may not achieve high CO_2 recovery rates unless integrated with a secondary recovery technology.

Beyond the established processes, a range of emerging CO_2 recovery technologies are being developed for specialized industrial integration and coprocessing applications. It's also worth noting that the field is rapidly evolving, with many of these innovations progressing swiftly toward higher levels of technological maturity and deployment readiness.

17.1.5.3 Post-conditioning

Product quality requirements often necessitate additional processing steps to ensure compliance with established specifications. Examples of such post-conditioning measures include:

- Dehydration to meet CO_2 moisture standards, typically achieved through glycol-based systems or molecular sieve units, depending on the intended end use of the CO_2.
- Oxygen removal, commonly implemented via catalytic oxidation units to reduce residual oxygen levels.
- Liquefaction, involving the installation of a dedicated plant for converting CO_2 into liquid form for transport, especially if not integrated within the primary recovery system.

In many cases, these post-treatment steps are incorporated alongside CO_2 compression to streamline processing and improve system efficiency.

17.1.6 CO_2 Transportation

CO_2 capture facilities are typically situated near the source of emissions, necessitating the transport of captured CO_2 to sites for further processing or geological storage. While small quantities of CO_2 can be moved via truck or rail, large-scale operations generally depend on dedicated pipelines for land transport or specialized vessels for offshore delivery.

Many CO_2 transport models aim to create shared infrastructure that enables multiple producers to utilize a common network. This collaborative approach helps reduce transportation expenses, which rank as the second-highest cost in the carbon capture, utilization, and storage (CCUS) chain—just behind the cost of capture itself.

CO_2 pipelines usually operate at very high pressures—ranging from 1450 to 2180 psi (10–15 MPa)—exceeding the critical pressure of CO_2. This allows the gas to be transported in a supercritical state, which is more efficient due to its higher density and reduced volume compared to lower-pressure gas transport. For sequestration projects located near pipeline networks, wellhead conditions also demand high pressures, influencing the required delivery pressure of the pipeline system.

Offshore shipping systems, on the other hand, must carefully balance capital investment with operational costs. Transporting CO_2 in liquid form is feasible at pressures between its triple point and critical pressure. Operating at higher pressures tends to reduce energy consumption and refrigeration needs while increasing transport efficiency due to higher density and mass flow per unit volume. However, these benefits come with increased costs for pressure-rated storage and shipping equipment. Currently, CO_2 shipping pressures range from 218 to 290 psi (1500–2000 kPag), though both lower and higher-pressure strategies are being explored to optimize cost-performance trade-offs based on specific shipping requirements.

17.1.7 CO_2 Physical Properties

Understanding the thermodynamic characteristics and related physical properties of CO_2 is essential for designing its transportation infrastructure and specifying appropriate compression equipment. This section outlines the fundamental properties of pure CO_2 to establish a reference point for transportation planning.

Table 17.2 presents key CO_2 properties, while Fig. 17.1 illustrates its phase diagram. Representative charts for enthalpy and density are shown in Figs. 17.2 and 17.3, respectively.

Many CO_2 recovery processes yield a low-pressure product that must be compressed to meet final delivery specifications. As CO_2 transitions from a low-pressure gas to a supercritical fluid, its density increases rapidly due to enhanced compressibility. Consequently, compression and transportation systems must be engineered with the nonideal behavior of CO_2 at elevated pressures in mind.

Table 17.2 Key physical properties of CO_2 [5]

Parameter	Units	Value
Molecular weight	g/mol	44.02
Critical-point temperature	°R (K)	547.6 (304.2)
Critical-point pressure	psia (kPa)	1070.4 (7,380)
Triple-point temperature	°R (K)	389.9 (216.6)
Triple-point pressure	psia (kPa)	75.1 (518)

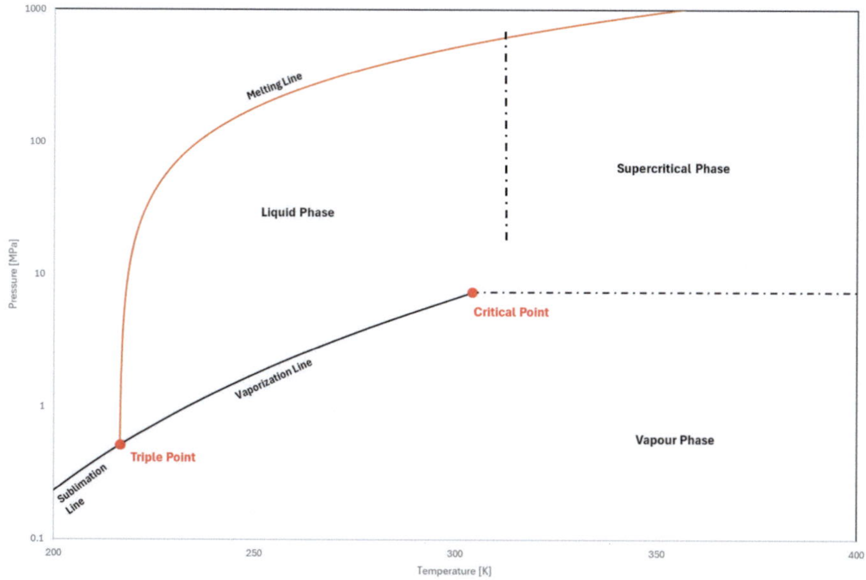

Fig. 17.1 CO_2 phase diagram

Supercritical CO_2 exhibits significantly higher compressibility and thermal expansion compared to typical gases transported in hydrocarbon pipelines. Above its critical temperature, CO_2 does not undergo a distinct phase change; instead, its density changes continuously, with pronounced sensitivity near the critical point (refer to Fig. 17.3).

These thermodynamic complexities are further influenced by the presence of impurities in the CO_2 stream. Accurate specification of compression equipment and reliable prediction of both steady-state and dynamic system performance depend on the careful selection and application of appropriate thermodynamic property models.

As illustrated in Fig. 17.1, CO_2 does not exist in liquid form at atmospheric pressure (14.6 psia [101 kPaA]). Instead, it sublimates directly from solid to gas at 351.4°R (195.2 K), equivalent to – 458.7 °F [– 272.6 °C]. To prevent dry ice formation during liquefaction or liquid separation, CO_2 must be maintained above

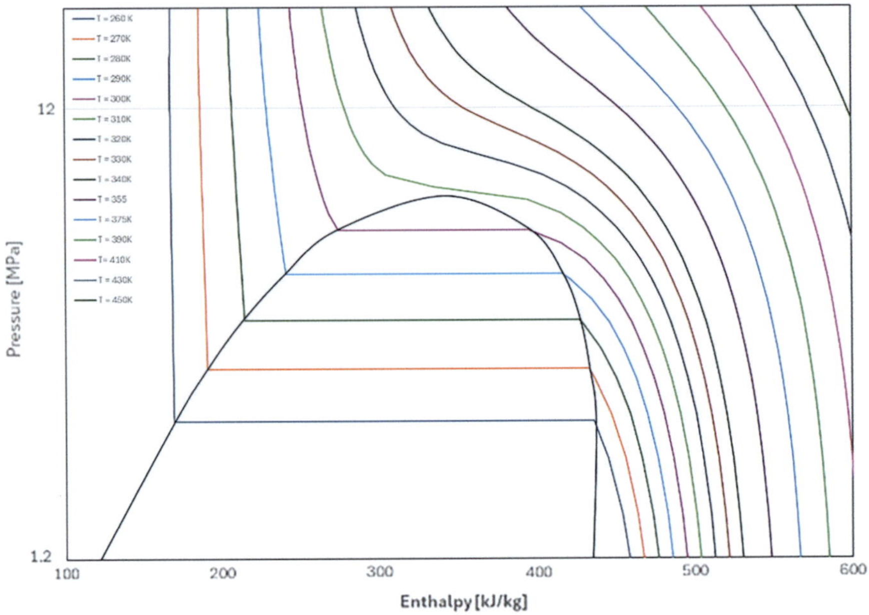

Fig. 17.2 CO_2 pressure–enthalpy diagram

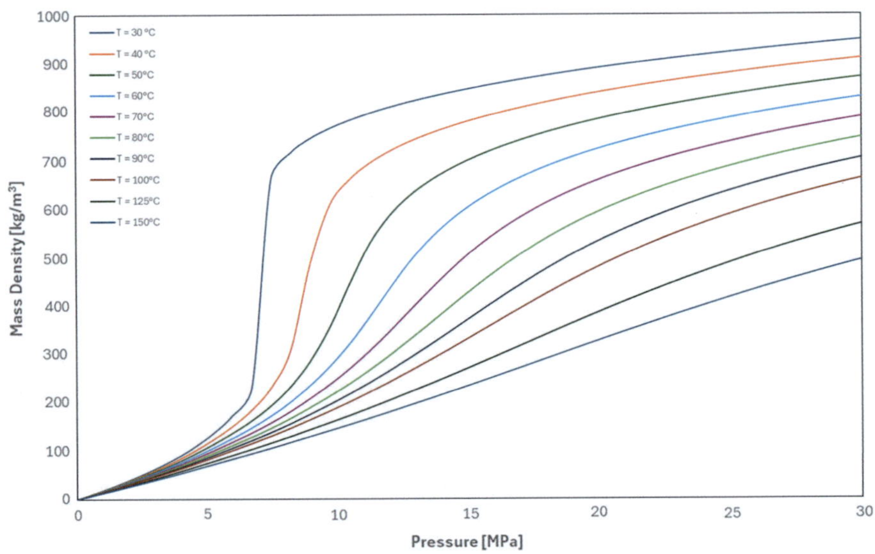

Fig. 17.3 CO_2 pressure–density diagram

389.9°R (216.6 K) and at pressures exceeding the vapor pressure corresponding to the system's operating temperature.

In systems designed to transport both liquid and supercritical CO_2, rapid depressurization can trigger Joule–Thomson cooling, causing the fluid to approach the sublimation temperature of 351.4°R (195.2 K) as it nears atmospheric pressure. Beyond managing cryogenic conditions, engineers must also account for the potential formation of dry ice and hydrates due to residual moisture during such pressure drops.

As shown in Fig. 17.2, CO_2 compressor power consumption is minimized when the system operates in a directionally isothermal manner at low temperatures—ideally approaching the temperature of the available cooling utility. While this approach limits opportunities for heat integration due to the rejection of compression heat as low-grade energy, it enhances compressor efficiency and reduces overall power demand.

In regions where electricity is both low-carbon and relatively inexpensive, but fuel costs are high, it may be advantageous to adjust the compression profile toward an isotherm with a higher cooling temperature. This shift elevates the grade of heat rejected during compression, making it more suitable for integration with other thermal processes. As a result, it can help offset utility heating requirements associated with carbon capture, though it comes at the cost of increased compressor power consumption.

17.1.8 CO_2 Product Specifications

Product specifications at a carbon capture facility are determined by both the intended destination of the CO_2 and the characteristics of the transportation system. For CO_2 streams intended for end-use applications, specifications are shaped by the feed quality requirements of the downstream process. In contrast, streams designated for geological sequestration must meet criteria based on the porosity and storage capacity of the target formation. Across both pathways, additional factors must be addressed—including corrosion control within the transport system, the risk of impurity condensation or precipitation, hydrate formation, and the presence of noncondensable gases in liquid-phase transport.

Maintaining dryness in CO_2 streams is essential to prevent the formation of carbonic acid, which can compromise pipeline integrity. To mitigate carbonic acid corrosion, stainless steel should be used throughout the entire compression system. The water dew point in CO_2 reaches its lowest at approximately 870 psig (6000 kPag), and water solubility increases as CO_2 is compressed into the supercritical phase.

If CO_2 dehydration is required as a post-conditioning step, it is typically implemented at an intermediate compression stage. This approach enables the removal of free water during interstage cooling in the early compression phases, thereby reducing the burden on the dehydration system and allowing for more efficient sizing and pressure rating of the equipment.

Beyond the CO_2 product compressor, transportation pipelines—commonly constructed from carbon steel—must be properly isolated and depressurized during commissioning, startup, and upset conditions to prevent carbonic acid formation.

Contaminants in the CO_2 product, such as oxygen, sulfur compounds, nitrous oxides, and hydrogen sulfide, can lead to the formation of strong acids or trigger alternative corrosion mechanisms. To preserve the integrity of carbon steel pipelines, these impurities must be tightly controlled.

Organic fluid concentrations must also be limited to maintain accurate dew-point control. For supercritical CO_2 pipeline systems, ambient temperature (for above-ground routing) or ground temperature (for below-ground routing) will influence thermal design parameters to prevent the emergence of a secondary liquid phase.

Carryover of organic substances—such as amine-based solvents from CO_2 recovery or glycol from dehydration—can significantly affect the dew point and must be restricted according to product specifications. Depending on the carbon source and upstream recovery process, the presence of liquid hydrocarbons is also possible and should be considered in system design.

Hydrate formation poses a significant challenge, as hydrates can naturally develop in CO_2 that hasn't been properly dehydrated—even under moderately cool ambient conditions. These hydrates may accumulate within pipelines, leading to blockages and potentially severe operational disruptions. To mitigate this, the water content specification of the CO_2 must incorporate a sufficient design margin, considering wintertime ambient and subsurface temperatures along the pipeline route and burial depth.

In the case of liquid CO_2 transport—such as via ships for offshore injection or geological sequestration—the presence of noncondensable gases (e.g., hydrogen, methane, argon, and nitrogen) directly influences the bubble point of the liquid CO_2. Higher concentrations of these gases elevate the pressure required to maintain CO_2 in its liquid state at a given temperature. This necessitates equipment with higher-pressure ratings and thicker walls, making it essential to minimize noncondensable gas content. While their impact on pipeline transport is less pronounced than in liquid systems, these gases still contribute to operational inefficiencies due to increased volume requirements and must be carefully controlled.

Contaminants in the CO_2 stream can also affect corrosion behavior, phase stability, and solubility dynamics. Many CO_2 transport projects rely on integrated networks that gather CO_2 from diverse emission sources and utilize various capture technologies. The interactions among these contaminants are complex and remain an active area of research and development. As the industry gains more operational experience, CO_2 product specifications will continue to evolve. These specifications will also differ across transportation networks, depending on the characteristics of the CO_2 sources and the intended end-use or storage destination.

17.1.9 *Additional Considerations*

17.1.9.1 Off-Specification CO_2 Product Handling

Maintaining dry CO_2 is essential to prevent the formation of carbonic acid, which can compromise pipeline integrity and operation. Any prolonged loss of dehydration must be carefully evaluated for its impact on downstream transport infrastructure, with appropriate safeguards such as system redundancy or alarms put in place. Equal consideration should be given to other off-spec CO_2 scenarios resulting from upstream process disruptions or failures in CO_2 recovery and post-conditioning systems. At a minimum, systems must be in place to safely divert nonspec CO_2 to a vent or disposal route until product quality is restored or upstream operations can be safely halted.

17.1.9.2 CO_2 Venting and Monitoring Systems for Carbon Capture Facilities

A dedicated CO_2 vent system and stack are essential for managing compressor train startup and shutdown, handling off-spec CO_2 product, and accommodating overpressure relief from carbon capture equipment—particularly if routing excess gas to a flare system is not feasible or required. This vent system must be exclusively designed for CO_2 disposal, with the stack positioned at a safe elevation and location to ensure proper dispersion of discharged gas.

The vent piping should be engineered to withstand the cryogenic temperatures resulting from depressurizing supercritical CO_2. Additionally, the stack must include a water boot to remove any accumulated moisture within the system. Design considerations should address sonic expansion of the vent stream, incorporate dispersion modeling to determine optimal stack height and exit velocity, and account for noise control and winterization to ensure safe and reliable operation under all conditions.

Beyond venting infrastructure, robust CO_2 leak detection is critical in CO_2 recovery and compression facilities. Point detectors should be strategically placed near potential release points, such as compressor buildings, pipeline interfaces, and metering skids. To detect larger releases, open-path CO_2 detectors should be installed around the plant perimeter to identify drifting gas plumes effectively.

17.2 Centrifugal Compressors

17.2.1 Design Challenges and Best Practices

Since the late 1960s, centrifugal compressors have been employed to compress carbon dioxide in urea production facilities. Over time, this application has led to a deep understanding of CO_2's unique characteristics—particularly its significant volume change during compression and its high density in elevated pressure zones.

Due to CO_2's high molecular weight, it's possible to achieve elevated Mach numbers—along with substantial pressure and volume ratios—while maintaining tip speeds well below mechanical limits. This enables exceptionally high-pressure ratios in CO_2 systems; for example, urea plants can reach overall pressure ratios of 150 or more using a single compressor train.

Such high-pressure ratios produce compressor performance curves that appear nearly vertical when charting the full compression process from initial suction to final discharge. Figure 17.4 illustrates a typical set of performance curves for a compressor driven at fixed speed, with flow regulated by inlet guide vanes (IGVs).

The near-vertical shape of the performance curves for each inlet guide vane (IGV) angle stems from the behavior of suction flow relative to the design point. When suction flow exceeds the design value, the pressure ratio of the first-stage drops. This reduction not only allows more flow into the compressor inlet but also amplifies the suction flow into the next stage due to the diminished pressure ratio from the preceding stage. This creates a compounding—or snowball—effect.

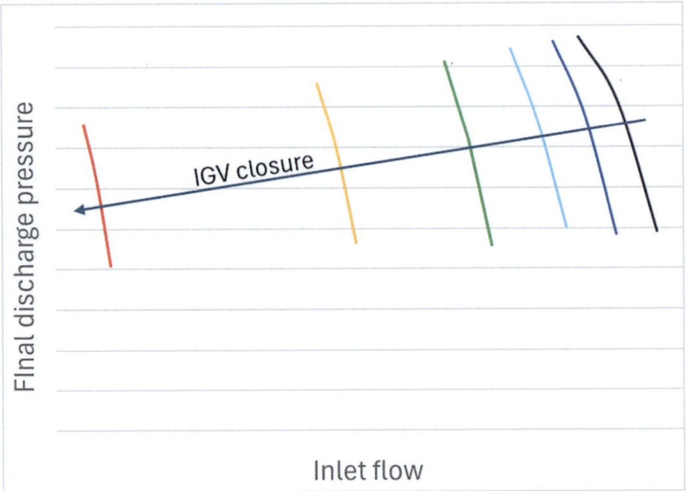

Fig. 17.4 Performance curve as a function of inlet guide vane angle (Image courtesy of Nuovo Pignone Tecnologie srl [part of Baker Hughes Company], Firenze, Italy)

Conversely, when suction flow falls below the nominal value, the first-stage pressure ratio increases. As a result, the suction flow into the second stage decreases—not just because of the reduced inlet flow, but also due to the higher-pressure ratio delivered by the first stage. This feedback loop intensifies with an increasing number of stages and a higher overall pressure ratio, as is typical in CO_2 compression systems used in urea production.

This cascading effect significantly influences overall performance predictions, which rely on the combination of individual sectional performance curves. If any section deviates from expected behavior, the curves shown in Fig. 17.4 may contract—both in terms of flow range (horizontal span) and pressure ratio range (vertical span). This leads to more complex compressor control and may compromise the ability to achieve the required discharge conditions.

In urea plants, the standard configuration for inline CO_2 compressors involves two machines operating in series within a single compression string, typically driven by a shared power source (refer to Figs. 17.5 and 17.6). One of these compressors may function as a pass-through unit, mechanically linked to other turbomachinery on both ends—for instance, connected to the driver on one side and another compressor on the other.

The defining characteristics of CO_2 compressors used in urea plants include:

- Low-Pressure Compressor: Comprises two stages housed in a horizontally split casing (for pass-through configurations). The first stage features a high inlet flow coefficient and elevated Mach number, utilizing three-dimensional (3D) impellers to reduce overall compressor size. As the flow volume decreases in the second stage, smaller impellers are employed, resulting in a lower peripheral Mach number.
- High-Pressure Compressor: Also consists of two stages, enclosed within a barrel-type casing. It typically uses two-dimensional (2D) impellers designed for low volume flow and operates at medium to low Mach numbers. The use of 2D impellers contributes to a more compact flow path, which enhances rotor stiffness.

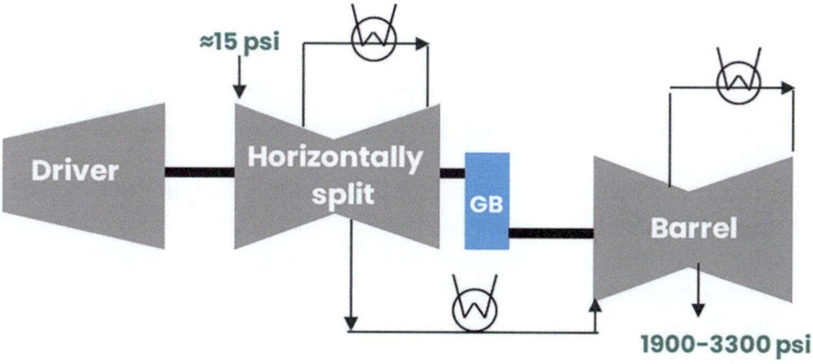

Fig. 17.5 CO_2 compression string with a gearbox (GB) between the low- and high-pressure compressors (Image courtesy of Nuovo Pignone Tecnologie srl [part of Baker Hughes Company], Firenze, Italy)

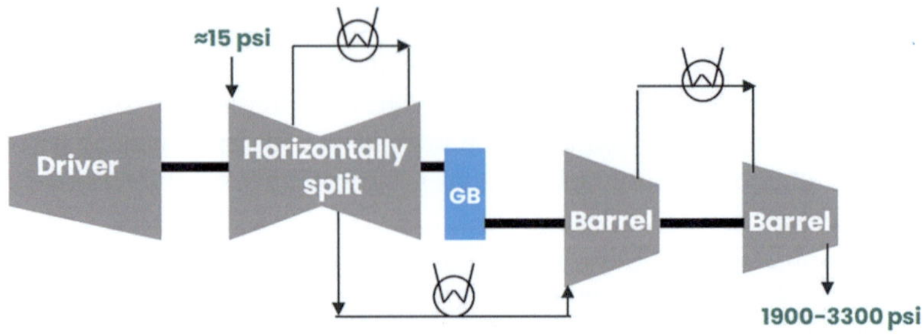

Fig. 17.6 CO_2 compression string with a driver, a gearbox (GB), and three compressors (Image courtesy of Nuovo Pignone Tecnologie srl [part of Baker Hughes Company], Firenze, Italy)

This compressor configuration, originally developed for urea production facilities, has been effectively adapted for carbon capture applications where pressure conditions are comparable. In these cases, the CO_2 stream typically enters at near-atmospheric suction pressure, while discharge pressures range between 1900 and 3300 psi (13–23 MPa), depending on the specific scenario. For enhanced oil recovery, operating conditions can vary significantly. For instance, offshore injection compressors may need to compress CO_2-rich streams to pressures as high as 8000 psi (55 MPa). These compressors often utilize 2D impellers operating at low tip speeds, which results in a lower peripheral Mach number. This design choice enhances flow coefficients across stages, thereby improving overall efficiency.

Compression train configurations are highly flexible and depend on both the type of driver and the process requirements. In early urea plants, a common setup—illustrated in Fig. 17.5—involved a low-pressure compressor directly coupled to the driver, with a speed-increasing gearbox used to optimize the high-pressure casing speed. This approach improved efficiency and reduced compressor size and cost. In some arrangements, the gearbox was placed between the driver and a horizontally split casing, if the driver (such as a four-pole fixed-speed electric motor) couldn't meet the speed demands of the low-pressure unit.

The horizontally split casing design for pass-through machines offers maintenance advantages: the top half can be lifted to access the low-pressure section, and if process connections are located on the bottom half, maintenance can proceed without disconnecting piping. The high-pressure section is serviced by axially removing internal components. To reduce shaft power consumption, each compression stage is intercooled.

For applications involving low-volume flow, the configuration illustrated in Fig. 17.5 can be modified by separating the final two sections into distinct units, as shown in Fig. 17.6. This adjustment enables the use of a number of small-diameter impellers in the high-pressure stages, which enhances the average flow coefficient and overall efficiency—surpassing what is achievable with a single unit containing two back-to-back sections. However, incorporating this additional unit may necessitate removing the machine from the compression train to extract the bundle from the pass-through barrel casing, which in turn requires disassembly of the process

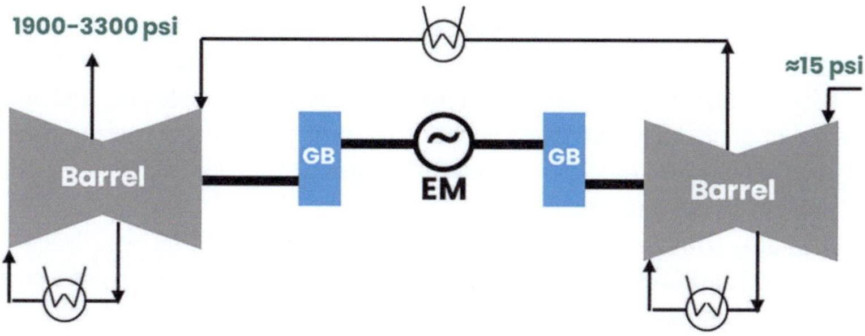

Fig. 17.7 CO_2 compression string with its driver in a double-ended shaft (Image courtesy of Nuovo Pignone Tecnologie srl [part of Baker Hughes Company], Firenze, Italy)

piping. This design was commonly employed in early, smaller-scale urea plants, but is rarely adopted in larger-capacity systems.

An alternative configuration involves a pass-through driver—typically an electric motor (EM)—paired with two speed-increasing gearboxes (GBs), as illustrated in Fig. 17.7. This setup enables each compressor unit to operate at its optimal speed, striking an effective balance between efficiency and rotordynamic performance. Both compressors may feature barrel casings, with internal components extracted axially from the nondrive end. However, if the process connections are oriented downward, the low-pressure casing can be designed with a horizontal split, allowing internal access by removing the upper half of the casing.

Some of the large compressors currently in operation are engineered for large-scale carbon capture initiatives, handling inlet flow rates as high as 82,000 ft^3/min (140,000 m^3/h) and achieving discharge pressures nearing 2900 psi (20 MPa). These machines demand substantial power input—approximately 30,000 hp (22,000 kW)—representing the shaft power needed for compressors driven by either electric motors or gas turbines.

17.2.2 Materials of Construction

The choice of materials for the key components of CO_2 centrifugal compressors [6] is influenced by a combination of critical factors:

- Mechanical performance, including the applied loads and resulting stress distribution across the components.
- Operating conditions and environmental exposure, particularly scenarios that may lead to corrosion.
- Relevant standards and regulatory requirements.
- Manufacturing techniques, such as welding.
- Economic considerations.

When evaluating the risk of CO_2 corrosion and acid gas corrosion more broadly, the material selection process must begin with an assessment of potential water condensation within the compressor. This depends on the pressure and temperature profiles at each stage and must account for both standard operating conditions and possible deviations—such as upset, off-design, or transient states—especially if water is present in the gas stream.

CO_2 is notably soluble in water, forming carbonic acid (H_2CO_3), which accelerates the corrosion of carbon steel by facilitating iron dissolution. The overall corrosion reaction can be expressed as:

$$Fe + 2H_2CO_3 \rightarrow Fe^{2+} + 2HCO_3^- + H_2$$

If a water phase saturated with CO_2 is present, the corrosion rate of carbon steel (CS) or low-alloy steel (LAS) can be estimated using established industry corrosion models. These models help predict the total metal loss over the compressor's operational life. If the projected loss remains within the corrosion allowances defined during design, CS or LAS may be considered viable options. Final grade selection should then be based on factors such as tensile strength, manufacturing process, and the material's performance at low temperatures and under impact.

Initial material selection should prioritize CS and LAS due to their lower cost and widespread availability. However, CO_2 sourced from combustion or industrial processes typically contains substantial moisture, increasing the likelihood of a liquid water phase. Given the high CO_2 concentration and elevated operating pressures, the resulting water phase is expected to have a low pH, leading to corrosion rates that exceed acceptable limits for CS and LAS. In such cases, switching to a corrosion-resistant alloy (CRA) becomes necessary. The appropriate alloy family or grade should be chosen based on its mechanical strength and specific corrosion resistance.

General guidelines for CRA selection include:

- Martensitic and Precipitation-Hardened Stainless Steels: These have a long history of successful use in CO_2 compressors, even under high partial pressures of CO_2 where pH levels may drop to 3 or lower during transient settling-out conditions. However, their application can be restricted by contaminants like H_2S and chlorides.
- Austenitic Stainless Steels: These offer full resistance to wet CO_2 and better tolerance to contaminants compared to martensitic grades. Their lower mechanical strength, however, confines their use to stator internals.
- Highly Alloyed Stainless Steels and Nickel-Based Alloys: These materials provide superior combinations of corrosion resistance and mechanical strength, though their adoption is often constrained by cost considerations.

To reduce manufacturing costs while ensuring compatibility with environments containing contaminants, equipment manufacturers have developed several strategies that utilize carbon or low-alloy steel as the base material, with surface protection methods to prevent corrosion. The primary approaches include:

- Weld Overlay: A corrosion-resistant alloy is applied in multiple welded layers onto the base material, forming a protective surface with a minimum undiluted thickness of approximately 0.04 in. (1 mm). Common overlay materials include austenitic stainless steels (e.g., 316 L) or nickel-based alloys (e.g., alloy 625), selected based on the specific gas composition. This method requires extensive welding time and may introduce typical welding challenges—such as casing deformation, especially if the original casing wall is thin. Therefore, simulation studies and potential redesign of the base thickness are often necessary. Weld overlays are commonly used for compressor casings and head flanges.
- Lining: This method involves welding stainless steel sheets or plates directly onto the base component to achieve corrosion resistance. While generally simpler than weld overlay, lining is best suited for flat or uncomplicated surfaces and is often used in conjunction with overlay techniques. Its application is typically limited to compressor casings.
- Electroless Nickel Plating (ENP): ENP is a chemical deposition process that applies a uniform layer of nickel–phosphorus alloy, known for its excellent corrosion resistance—often exceeding the operational lifespan of compressors. Its adaptability to complex geometries makes ENP a preferred solution for protecting various compressor components, including casings, head flanges, stator internals, impellers, and even fully assembled rotors.

In urea production facilities, humid CO_2 typically lacks significant contaminants. However, with the advent of enhanced oil recovery (EOR) technologies, the CO_2-rich streams intended for compression began to include impurities such as hydrogen sulfide (H_2S), which influence both material selection and seal configuration [7].

Moreover, anthropogenic CO_2—generated during emission and capture processes—often carries contaminant species and concentrations that differ markedly from those found in natural gas [8]. These may include sulfur oxides (SO_x), nitrogen oxides (NO_x), ammonia (NH_3), and various amines. Several of these compounds are currently under scientific scrutiny due to their potential to form highly corrosive acids (a phenomenon known as acid dropout), even in conditions that are nominally dry. Such acidic phases pose a serious corrosion threat to both carbon steel/low-alloy steel (CS/LAS) and corrosion-resistant alloy (CRA) materials.

Although gas treatment is expected to reduce these contaminants to trace levels, there is no universally accepted threshold that guarantees effective mitigation of associated risks. Therefore, when selecting materials for compressors in carbon capture, utilization, and storage (CCUS) applications, it is essential to thoroughly evaluate the potential impact of these impurities—even in minimal concentrations.

17.2.3 Rotordynamics

Higher gas density associated with CO_2 can result in increased destabilizing seal coefficients. However, these effects are readily mitigated through well-established countermeasures that have consistently shown strong positive impact:

- **Shunt Holes**: Gas with elevated static pressure—typically sourced from the scroll or diffuser exit—is redirected upstream of the seal with zero or negative preswirl. While some users have noted the potential for clogging, this risk is generally minimal. Shunt holes are usually designed as multiple large openings, and clogging tends to occur only under highly specific, unanticipated conditions during compressor design—such as the presence of unusually high concentrations of elemental sulfur that may solidify due to pressure drops within passages and seals.
- **Swirl Brakes**: These are commonly integrated into the seal design at the impeller eye to suppress swirl in the leakage flow entering the cavity shroud. Swirl brakes have proven effective in reducing preswirl at the seal inlet. Computational fluid dynamics (CFD) simulations are used to assess the flow characteristics post-brake, enabling accurate calculation of the seals' dynamic coefficients and rotor logarithmic decrement for the first mode. These predictions have been successfully validated through experimental measurements, as detailed in Sect. 17.2.8.
- **Honeycomb and Hole-Pattern Seals**: The stator portion of the seal is machined with a pattern of hexagonal or circular cells, providing robust direct damping. This configuration is especially suitable for high-pressure applications (discharge pressures approaching 8,000 psi [55 MPa]) and ensures stable operation even with dense gases. To further reduce leakage, these seals are often paired with a labyrinth seal featuring a tighter clearance.
- **Pocket Damper Seals**: Similar in function to hole-pattern seals, pocket damper seals are engineered to restrict circumferential gas motion while delivering substantial direct damping performance.

Seal Design Considerations in Multisection Compressors Include:

- As differential pressure across seals increases, fluid forces intensify—making it essential to carefully consider seal design.
- In two-section high-pressure compressors, the interstage balance drum typically experiences the highest differential pressure. Therefore, the damping seal here plays a critical role in system stability.
- In contrast, the external balance drum, which usually sees the pressure difference between the suction sides of the two sections, is less critical. As a result, low-clearance labyrinth seals are commonly employed in this area.

Thermal and Phase Behavior in External Balance Drums:

- The external balance drum in a two-section machine generally links the suction side of the high-pressure section to that of the low-pressure section.

- These sections are often intercooled to reduce power consumption, meaning the gas passing through the drum is at the cooler's outlet temperature.
- The pressure drop from high-pressure suction to low-pressure suction can cause CO_2 condensation or even solid CO_2 formation.
- This risk must be evaluated early in the design process. If condensation is likely, an alternative balancing strategy—such as routing gas at the compressor discharge temperature to the seal—should be considered.

Seal Configuration in Low-Pressure Machines:

- In low-pressure compressors, the interstage balance drum or piston may not require a damper seal and is often replaced with a labyrinth seal.
- Additionally, if the thrust bearing can handle the residual axial load, the external balance drum may be omitted entirely.
- This streamlined configuration reduces leakage across the balance drum, resulting in enhanced performance and improved efficiency.

17.2.4 Sealing Options

Since their introduction in the mid-1990s, dry gas seals have become the preferred sealing solution for high-pressure machinery, primarily due to their effectiveness in minimizing process gas leakage. However, these seals demand strict gas quality standards, particularly regarding the absence of solid or liquid contaminants. To meet these requirements, a gas conditioning system is employed to inject filtered and, if necessary, overheated seal gas upstream of the seals. Overheating may be essential depending on the gas composition and operating pressure and temperature, as it prevents liquefaction or solidification during pressure reduction across the seal—an effect driven by the Joule–Thomson phenomenon.

As illustrated in Fig. 17.8, injecting seal gas directly in front of the dry gas seal ring ensures that untreated process gas does not pass through the seal. In oil and gas applications, the standard configuration is a tandem seal arrangement, featuring two rings in series. The inboard (primary) seal separates the process gas from the primary vent and is exposed to the full differential pressure between the process and near-atmospheric conditions. This vent may discharge to the atmosphere in safe zones or be routed to a flare system if handling flammable gases.

The outboard (secondary) seal serves as a backup, preventing uncontrolled leakage in the event of primary seal failure. On the atmospheric side, a barrier separation seal is installed, accompanied by the injection of a separation gas. This prevents process gas from migrating toward the bearings, which could compromise the lubrication system. The separation gas—typically instrument air or nitrogen (used if air poses an explosion risk due to gas composition)—may also be injected between the inboard and outboard seals to dilute any toxic process gas in the secondary vent.

In industrial settings where the compressed gas is inert or nontoxic, brief uncontrolled emissions may not pose significant environmental or safety risks. In such

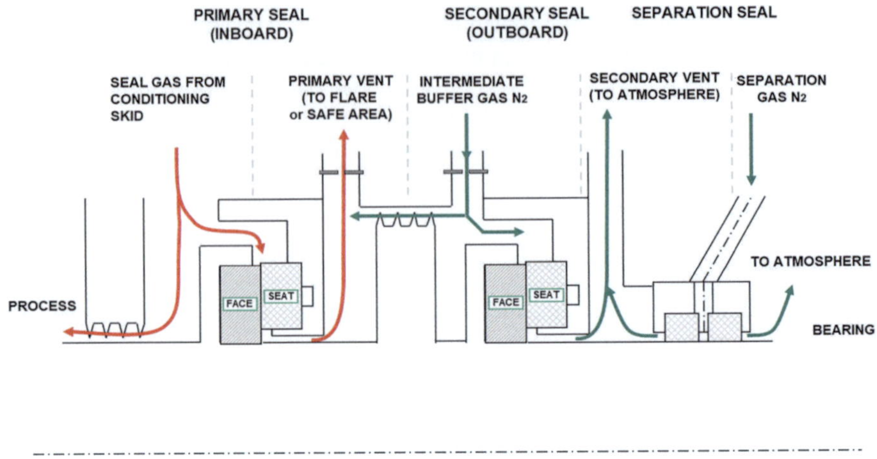

Fig. 17.8 Tandem dry gas seal (Image courtesy of Nuovo Pignone Tecnologie srl [part of Baker Hughes Company], Firenze, Italy)

cases, the tandem seal configuration is often omitted in favor of a single dry gas seal ring. This is common in CO_2 compression applications where no harmful contaminants are present. Even with a single-ring setup, a seal gas conditioning system and separation gas system remain necessary. Despite the simpler design, single-ring dry gas seals still deliver high sealing performance with minimal atmospheric leakage—making them ideal for reducing CO_2 emissions in compression processes.

- Dry Gas Seal Considerations:

  When a continuous flow of seal gas is maintained to prevent contamination, dry gas seals—supported by a seal gas booster—can theoretically sustain pressurized conditions indefinitely. However, material selection becomes critical during extended pressurized standstills to prevent unexpected corrosion. Additionally, initiating startup under pressure increases torque demand, which may exceed the driver's capacity. In such cases, partial depressurization may be necessary to enable the compression train to restart. One of the key advantages of dry gas sealing systems is their ability to minimize or eliminate the need for full or partial venting of process gas, though this benefit comes with the added expense of a gas treatment system.
- Alternative Sealing for Inert Gases:

  For machines handling gases that permit greater leakage at the shaft ends—such as inert gases that are nonflammable and nontoxic—simpler and more robust sealing options like labyrinth seals or carbon ring seals may be used. These alternatives do not require filtration or heating systems, making the sealing arrangement more straightforward. Typically, the setup includes two sealing stages with process gas flowing directly through the seal. The larger clearances allow small particles or liquid droplets to pass through the sealing faces, and one or two connections redirect the process gas to lower-pressure zones.

- Barrier Gas Management:

  In this configuration, the barrier gas is introduced after the secondary seal. Depending on the required purity of the process gas, this barrier gas may need to be vented to the atmosphere. Figure 17.9 illustrates a typical carbon ring seal arrangement, with labyrinth seal configurations following a similar design.

In systems utilizing less restrictive seals—such as labyrinth seals or carbon rings—pressure tends to drop following shutdown due to leakage at the shaft ends. Simultaneously, the gas temperature within the machinery gradually decreases through heat exchange with the surrounding environment. As the CO_2 stream expands within the shaft-end seal, this cooling effect can lead to the formation of liquid or ice on the sealing surfaces. While these seals may tolerate minor liquid accumulation, it is strongly recommended that compressors not be restarted if the seals are inundated with liquid or ice. To mitigate this risk, the compression train should be depressurized promptly after shutdown.

When it comes to sealing stator components (e.g., between the casing and end head), standard limitations on depressurization rates must be observed. CO_2 is not known to contribute to explosive decompression effects, so the primary concern with depressurizing CO_2 loops lies in the cooling that occurs due to the Joule–Thomson effect.

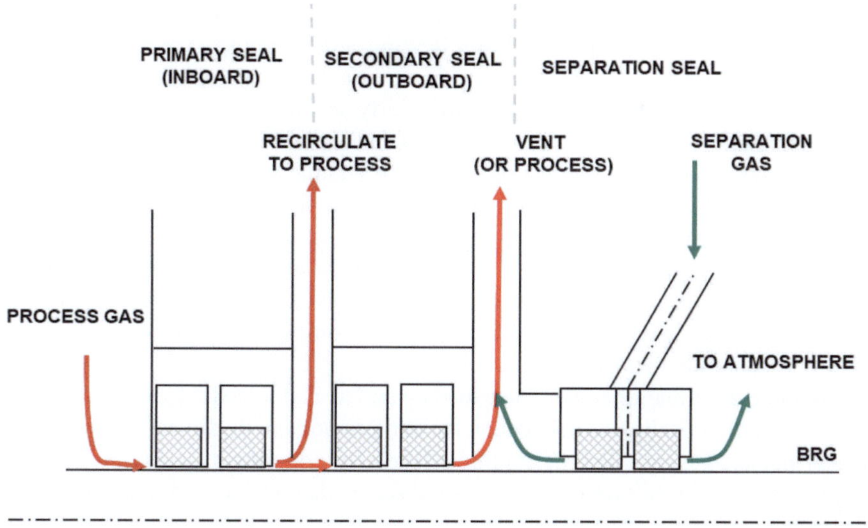

Fig. 17.9 Carbon ring seals (Image courtesy of Nuovo Pignone Tecnologie srl [part of Baker Hughes Company], Firenze, Italy)

17.2.5 Considerations for Availability

Maintenance tasks are generally categorized into shaft-end procedures and internal inspections. Shaft ends are designed for quick disassembly, allowing for efficient preventive maintenance and minimal downtime during repairs. Compared to integrally geared compressors (IGCs), this configuration uses fewer seals—components that are typically more susceptible to failure or wear—resulting in maximized train availability.

17.2.6 Process Connections

During compressor operation, process piping connections can impose substantial loads on the compressor nozzles. Although originally designed for steam turbines, the National Electrical Manufacturers Association (NEMA) standard SM 23 [9] is commonly used as a reference for evaluating these loads in centrifugal compressors. As a result, customer specifications often require load limits expressed as a multiple of the NEMA base values—such as "15 times NEMA"—which may apply to a single directional load, the total load on a nozzle, or the combined load across the entire machine.

Piping loads typically arise from the weight of the piping itself and the influence of process gas pressure and temperature. In addition to these, acceleration loads must also be considered. These may stem from dynamic conditions in offshore environments, such as those experienced on floating production storage and offloading (FPSO) vessels, or from seismic activity in earthquake-prone regions. In areas with high seismic risk, these loads can be particularly severe. Local regulations usually specify the required acceleration values based on the plant's geographic location and the intended design life of the equipment. Since seismic load predictions are probabilistic, longer equipment lifespans correlate with a higher likelihood of encountering a high-energy earthquake.

The resulting piping loads can increase stress on the compressor flanges or casing. However, as design pressure rises, casing thickness typically increases as well, making these loads less critical to casing integrity. Consequently, this risk is generally less significant than those discussed below.

Baseplate deformation under these loads—and to a lesser extent, casing deformation—can cause misalignment between the compressor and its driver. This misalignment is managed using a flexible coupling. While the coupling may operate within its fatigue limits, it can still transmit substantial forces that lead to elevated compressor vibration.

If nozzle forces act in a way that lifts the compressor off the baseplate, the resulting loss of contact can cause further positional deviation. This, in turn, amplifies the forces transmitted through the coupling, exacerbating vibration and mechanical stress.

If a 3D finite-element analysis (FEA) is conducted, the resulting effects become clearly observable. However, in cases where design decisions are based on practical experience—such as when a compressor end-user requests a configuration suitable for a specific NEMA multiplier—the manufacturer may rely on precedent or reference analyses to validate the design without performing a dedicated simulation. Despite this, fundamental considerations—like the potential lifting or unloading of the compressor feet—must be thoroughly evaluated to prevent unexpected shaft-line misalignment.

Inline compressor units are generally very robust and capable of enduring substantial external forces. A common worst-case scenario involves a pipeline compressor with horizontal process flanges, which can produce a global NEMA multiplier exceeding 50. Under such conditions, the barrel casing typically undergoes moderate additional stress. However, these forces can also cause the feet to lose contact with the mounting surface.

Figure 17.10 presents the results of an FEA performed on a barrel casing, illustrating the impact on the feet contact surfaces from a bottom-up perspective. The analysis reveals that external loading causes the feet to shift slightly, which may eventually lead to shaft-line misalignment. In both visualizations, the color scale ranges from low values (blue) to high values (red). The contact pressure plot indicates that although full contact across the entire foot surface is not maintained, none of the feet is entirely lifted—ensuring that the compressor casing remains adequately constrained. Stress data are omitted, as the levels induced by external loading are negligible compared to those generated by internal pressure.

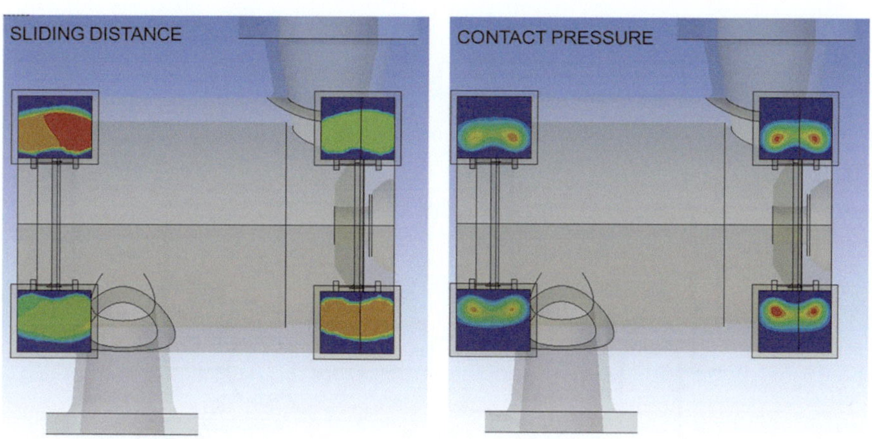

Fig. 17.10 Sliding distance and contact pressure on the casing due to the preload and nozzle loads (Image courtesy of Nuovo Pignone Tecnologie srl [part of Baker Hughes Company], Firenze, Italy)

17.2.7 Compressor Startup

Integrating a compressor unit into a process loop requires thorough evaluation to ensure safe operation under both steady-state conditions and anticipated transient or upset scenarios—such as startup, fluctuations in process flow, or emergency shutdowns. Simulation models play a key role in predicting system behavior and guiding the development of process design and control strategies. These models help assess the effects of modifying, removing, or improperly sizing equipment within the loop.

The remainder of this section presents examples of process simulations to illustrate these concepts. Figure 17.11 provides a simplified process flow diagram (PFD) to highlight the fundamental principles. A critical component for compressor protection is the antisurge valve (ASV) loop, which enables flow recirculation. When the gas flow through the system drops below the compressor's minimum required flow, the ASV loop redirects gas back to the compressor inlet, maintaining adequate flow and preventing surge conditions.

Figure 17.12 illustrates the compressor's operating point on its performance map during a ramp-up sequence. Initially, the compressor is at rest, with zero inlet flow and zero head. Startup typically begins with the recycle antisurge loop engaged—meaning the antisurge valve is fully open to allow maximum recirculation. Under these conditions, the operating point shifts along the right-hand side of the performance curve, transitioning across speed lines as the driver accelerates.

The duration and nature of the startup sequence depend on the driver's characteristics. For instance, a steam turbine may need to idle for a period to reach thermal equilibrium. During this phase, when the compressor operates below its minimum

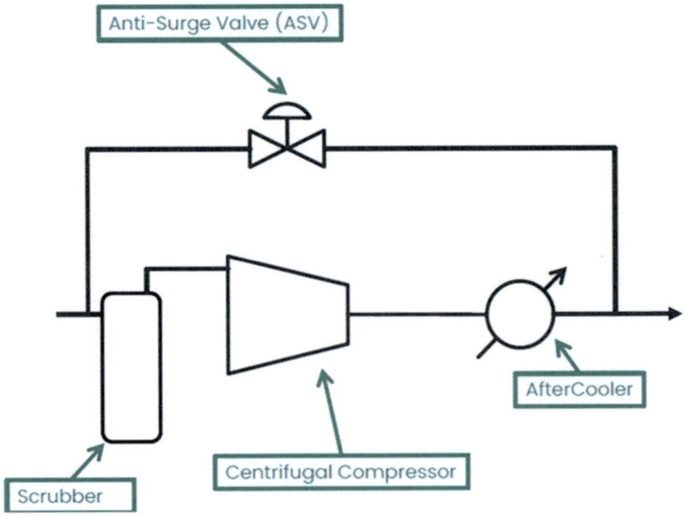

Fig. 17.11 Basic flow diagram for compressor section (Image courtesy of Nuovo Pignone Tecnologie srl [part of Baker Hughes Company], Firenze, Italy)

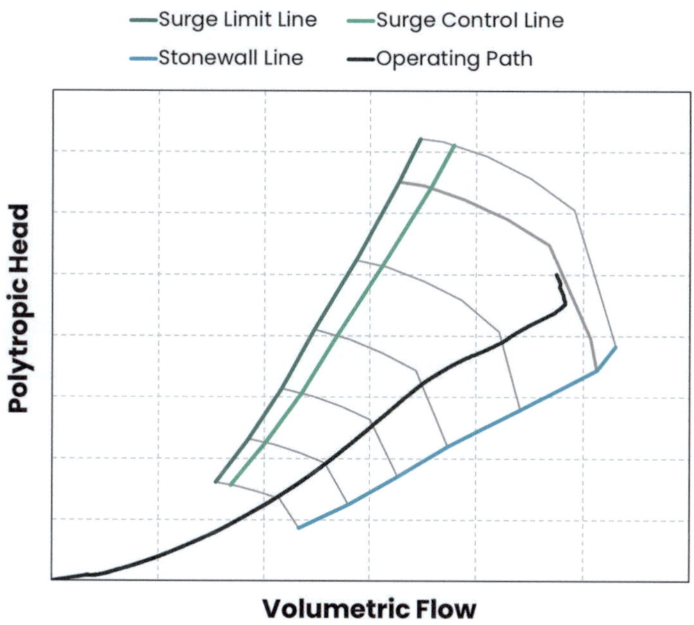

Fig. 17.12 Compressor envelope and operating point position during startup ramp (Image courtesy of Nuovo Pignone Tecnologie srl [part of Baker Hughes Company], Firenze, Italy)

speed, it's crucial to avoid lingering near lateral critical speeds to prevent excessive vibration. Additionally, the operating point must remain within the compressor map, and surge conditions should be avoided—even though lower speeds generally involve less energy and milder surge effects.

Once the compressor reaches its minimum operating speed, the full performance map becomes usable. At this stage, two control loops govern operation:

- Antisurge control monitors the proximity of the operating point to the surge line and opens the antisurge valve if necessary.
- Performance control adjusts the compressor speed to meet the plant's operational requirements.

The compressor operates within a defined envelope on the performance map, bounded by minimum and maximum continuous speed lines and corresponding flow limits. Flow cannot exceed the "stonewall line," which represents a sonic limit in high-Mach-number applications. Crossing this line can lead to unstable behavior and intense aerodynamic forces. Maximum flow is also constrained by plant design factors, such as pressure drops in the piping system.

On the low-flow side, operation must stay above the surge control line. This line is derived from the surge limit—based on expected performance curves or validated through testing—and includes a safety margin. If the process flow drops below this

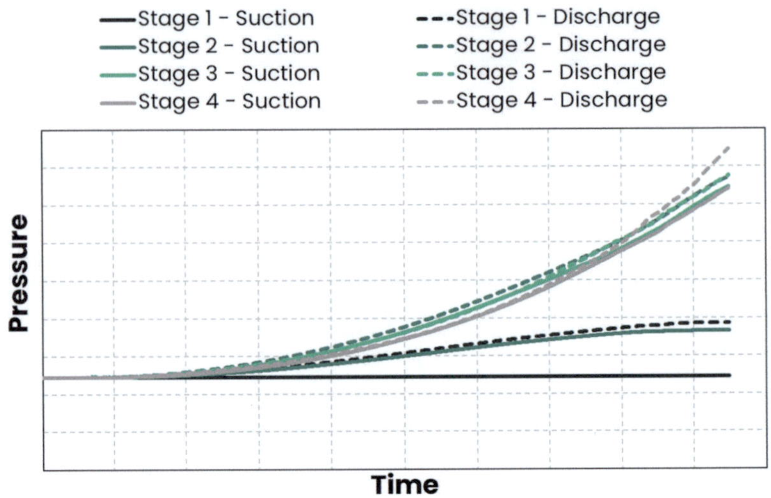

Fig. 17.13 Pressure versus time profile during startup for a four-section compressor (Image courtesy of Nuovo Pignone Tecnologie srl [part of Baker Hughes Company], Firenze, Italy)

threshold, the antisurge controller responds by opening the valve to boost inlet flow and maintain stability.

Figure 17.13 shows that as speed ramps up, the pressure ratio across each process section increases. This progression enables the compression train to reach its target pressure ratio and become fully operational within the process.

Emergency shutdowns are critical scenarios that require thorough evaluation. Unlike planned shutdowns, which follow established procedures to maintain compressor operation within safe parameters, emergency shutdowns can lead to adverse conditions—most notably, surge cycles. Accurately predicting these phenomena demands detailed modeling of the process fluid diagram, including system volume.

As illustrated in Fig. 17.14, an emergency shutdown can cause the compressor's operating point to cross the surge limit line, initiating a surge cycle. Although the antisurge valve typically opens fully in response, this action alone may not suffice to prevent the surge, as demonstrated in the figure.

The compressor manufacturer must assess the severity of such events by analyzing the frequency and velocity of the surge cycles, which directly relate to the energy involved. If high-speed, repeated cycles are anticipated, a review of the system design is essential to explore potential countermeasures that could mitigate or prevent this condition.

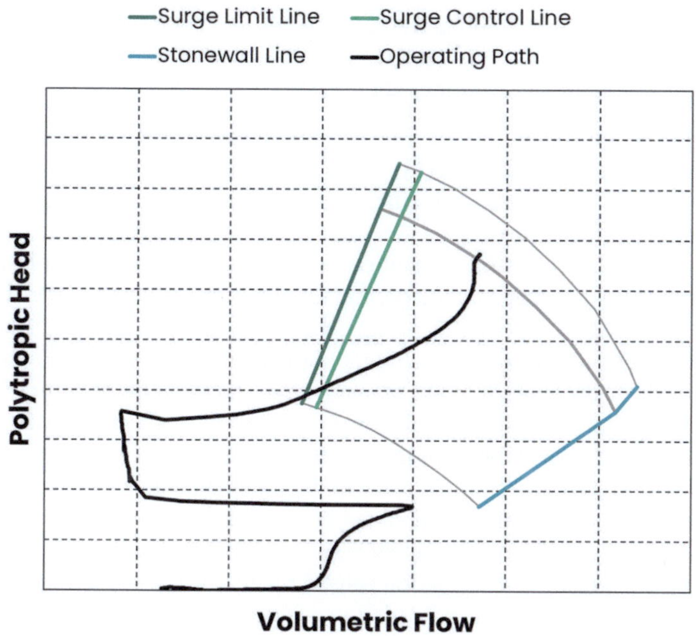

Fig. 17.14 Compressor operating point during emergency shutdown (Image courtesy of Nuovo Pignone Tecnologie srl [part of Baker Hughes Company], Firenze, Italy)

17.2.8 Special Engineering Analyses and Shop Tests

To ensure a compressor unit is fit for operation in its designated environment, a series of tests are conducted at manufacturing and assembly facilities. Performing these evaluations in advance helps avoid extensive rework at the installation site, which can be challenging due to constraints such as limited personnel, tool availability, or site-specific restrictions. For special engineering analyses, refer to Sect. 17.2.6.

Typical testing procedures for assembled compressor units include:

- Mechanical Running Test: Mandatory if American Petroleum Institute (API) Standard 617 [10] is invoked; typically conducted using the shop driver and auxiliary systems.
- ASME PTC-10 Type 2 Performance Test: Performed at reduced power, pressure, and speed in accordance with the American Society of Mechanical Engineers (ASME) Performance Test Code on Compressors and Exhausters [11], using the shop driver and auxiliaries.
- ASME PTC-10 Type 1 Performance Test: Executed under site-specific power and pressure conditions, also using the shop driver and auxiliaries.
- Complete Mechanical Running Test: Conducted with the job driver and shop auxiliaries to validate the unit's mechanical integrity.

- String Test: Involves the job driver and auxiliaries, potentially under full-speed and full-load conditions, to assess the integrated performance of the entire package in simulated operating conditions (Fig. 17.15).

The commonly monitored parameters are those explicitly outlined in the acceptance criteria of API 617 [10]—such as probe-measured vibration, bearing pad temperatures, and head and efficiency relative to inlet flow, all compared against agreed performance targets and tolerances. Performance evaluations can be conducted at standard operating conditions as well as at the extremes of the performance curve (e.g., surge and choke points) to more precisely define the surge limit line.

Comprehensive unit testing—including string testing—extends validation to the entire shaft line. This enables the implementation of a torsional vibration measurement system to determine whether the driver's characteristics might induce significant torsional excitation in the shaft line. These measurements, typically taken at the coupling section, can be performed using either a permanent torquemeter or temporary strain gauge instrumentation mounted on the coupling spacer. The torquemeter generally captures relative displacement between the ends of the spacer, while strain gauges measure localized deformation caused by both static and dynamic torsional loads. In either case, the data can be used to evaluate torsional vibrations and assess whether the resulting stress levels could compromise continuous operation.

For electric motors with an accessible shaft end (i.e., single-sided drive), direct measurement of motor torsional oscillation is also feasible. These data can be utilized by an adjustable-frequency drive (AFD) control system to actively dampen torsional vibrations by responding to the oscillation itself. This method is particularly effective if targeting torsional natural frequencies. However, the mode shape plays a critical role—if the torsional mode exhibits minimal or no amplitude at the motor location,

Fig. 17.15 String test arrangement for an electric motor-driven centrifugal compressor (Image courtesy of Nuovo Pignone Tecnologie srl [part of Baker Hughes Company], Firenze, Italy)

the driver will have limited influence on the vibration, making it ineffective for both excitation and damping when vibrations are triggered by other sources.

It's also important to note that torsional natural frequencies typically lie outside the range of most excitation sources, except for driver-induced pulsating torques. Fluid dynamic forces, for instance, usually occur at significantly higher frequencies.

The critical speed of the system is determined, and the rotordynamic model can be verified through an unbalance response test. Additionally, it is possible to estimate the natural frequencies and associated damping of the compressor rotor. While the conventional method for this is experimental modal analysis (EMA)—which involves installing a magnetic exciter to apply a frequency sweep and measure the rotor's response—a more efficient alternative is operational modal analysis (OMA) [12]. OMA utilizes the inherent fluid forces within the system as excitation, eliminating the need for a magnetic exciter. This avoids modifications such as rotor elongation, which could adversely affect dynamic behavior. OMA can be implemented without altering the compressor design and has shown good correlation with EMA results in published studies.

Table 17.3 presents a comparison between predicted and measured logarithmic decrements (denoted as "log dec") for a prototype unit operating at various speeds and pressure levels. The unit was designed to achieve a logarithmic decrement below 0.1 under operating conditions, allowing for a broad assessment of damping characteristics [13].

The data show strong agreement at the first two test points. However, the third test point reveals a more noticeable discrepancy. This can be attributed to the limitations of applying a single-degree-of-freedom damping model: When the predicted logarithmic decrement reaches 2.3, the corresponding amplification factor is approximately 1.5. Although this mode falls within the acceptable operating range defined by API Standard 617, its minimal vibration amplitude makes it difficult to detect. Consequently, precise measurement of damping at such high levels is both challenging and of limited practical relevance and is therefore not pursued further in this analysis.

Table 17.3 Logarithmic decrement of the first mode under different operating conditions

Test point	Percentage of rated speed (%)	Suction (psi)	Discharge (psi)	Predicted log dec	Measured log dec
1	100	87	595	0.05	0.02
2	86	1247	2407	0.3	0.33–0.34
3	51	2393	5525	2.3	> 1.5

17.3 Integrally Geared Compressors

17.3.1 Design Challenges and Best Practices

One of the primary challenges in compressing CO_2 lies in the substantial variation in specific volume, which can result in low-flow coefficients and, consequently, reduced thermodynamic efficiency. Integrally geared compressor (IGC) units address this issue by enabling individual speed adjustments for each pinion. This flexibility allows for the selection of impellers with optimal flow coefficients—while still adhering to rotordynamic stability constraints. Leading gear manufacturers have successfully implemented configurations with up to five pinions per unit, accommodating as many as ten compression stages within a single integrally geared system.

Another notable benefit of IGC units is the potential to incorporate a cooler after each compression stage. This approach aims to minimize power consumption and lower operating costs. However, installing intercoolers at every stage can significantly increase capital expenditure and expand the system's physical footprint. Therefore, the number of coolers is typically determined through a cost–benefit analysis that weighs upfront investment against long-term operational savings. This evaluation—often conducted by the end user or contractor—is based on the total cost of ownership over the projected life of the project. Figure 17.16 illustrates a four-stage integrally geared compressor configuration.

Fig. 17.16 Four-stage integrally geared compressor (Image courtesy of Nuovo Pignone Tecnologie srl [part of Baker Hughes Company], Firenze, Italy)

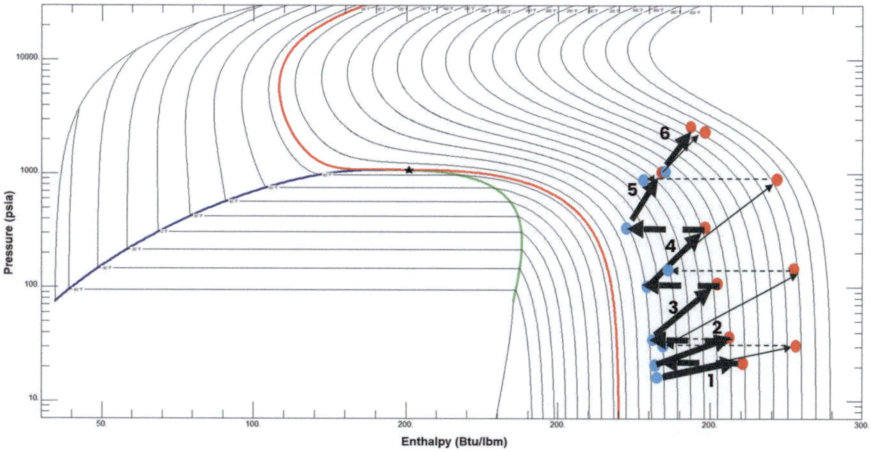

Fig. 17.17 CO_2 compression pressure–enthalpy diagram (Image courtesy of Nuovo Pignone Tecnologie srl [part of Baker Hughes Company], Firenze, Italy)

Figure 17.17 illustrates a representative pressure–enthalpy diagram for CO_2 compression within a urea production facility. The bold black lines trace the transformation of the gas through an inline centrifugal compressor train, which consists of four compression stages split between two compressors. Blue dots indicate the inlet conditions for each stage, while red dots on the right mark the corresponding outlet conditions. Dashed lines show the cooling process that occurs after each stage.

Compression begins at atmospheric pressure and increases to a range of 500–700 psi (3.4–4.8 MPa), remaining below CO_2's critical pressure of 1,070 psi (7.4 MPa). This process involves two intercoolers between stages. A third intercooler is then used to cool the gas to a precisely controlled temperature. This temperature must be selected with great care, as operating near the critical pressure causes the isothermal lines to flatten significantly. In this region, even minor pressure changes can result in dramatic shifts in gas density.

With an integrally geared compressor, increasing the number of stages enables enhanced intercooling capabilities. Figure 17.17 illustrates the results for a configuration using six compression stages—considered a moderate setup, given that such systems can accommodate up to ten stages. In this scenario, the number of intercoolers increases from three to four, though additional coolers could be incorporated in designs with more stages.

Notably, in this case, there an opportunity to cool the gas stream between stages 5 and 6. However, since the gas was nearing its critical pressure, further cooling toward its critical temperature of 87.8 °F (31 °C) risked reducing compressibility and inducing a dense phase. As a result, Fig. 17.17 shows identical suction conditions for stage 6 and discharge conditions for stage 5.

17.3.2 Rotordynamics

- The rotordynamics of Integrally Geared Compressors (IGCs) exhibit two key characteristics that distinguish them significantly from inline machines.
  - Rotor Geometry and Critical Speed Behavior: IGCs feature overhung impeller stages, which create a unique rotor configuration. This results in an undamped critical speed map that differs markedly from that of inline compressors. For a typical pinion equipped with one impeller on each end, the first critical modes—commonly referred to as "conical" and "cylindrical"—fall below the operating speed range. Consequently, the machine typically operates between its second and third critical speeds.
  - Thrust Transfer and Bearing Load Dynamics: Unlike inline machines, IGCs generally lack a thrust bearing on the high-speed pinion. Instead, axial thrust from the impellers is redirected to the low-speed shaft via collars mounted on the pinions that engage the bull gear. This thrust is counterbalanced at the gear mesh—the point of torque transmission—where axial forces are localized.

    This configuration introduces a second major distinction: the influence of fluid operating conditions extends beyond seal effects. It also imposes additional loads on the radial bearings due to residual axial thrust and torque transmission. Specifically, the gear mesh generates tangential forces on the pinion teeth, which are absorbed—both in force and moment—by the journal bearings.
- Load Variation and Rotordynamic Sensitivity:

These dynamic forces can exceed the static rotor weight by an order of magnitude. As a result, journal bearing loads can vary dramatically—from near-zero during mechanical running tests to substantial levels under full operating conditions. Moreover, different process conditions lead to different bearing loads.

Since bearing performance is highly load-dependent, the rotordynamic behavior of the machine is also sensitive to operating conditions. This variability affects both the critical speed values and amplification factors. To ensure stable rotordynamics across varying conditions, IGCs are typically operated at a fixed speed. Fixed-speed electric motors are the most common drivers, though fixed-speed turbines are also frequently used.

Figure 17.18 illustrates an undamped critical speed diagram for an Integrally Geared Compressor (IGC). In this diagram, the solid lines depict how critical speeds vary with bearing stiffness. On the left side of the graph, bearing stiffness is extremely low, and the first two modes—near-zero cycles per minute—correspond to rigid-body translation. Conversely, the right side reflects very high stiffness values, where the bearings behave almost like fixed supports.

The horizontal line in the figure marks the rated operating speed. For machines equipped with one impeller on each side, this typically falls between the second and third vibrational modes. Dotted lines on the left indicate bearing characteristics during mechanical running tests, where the compressor operates in a vacuum without process

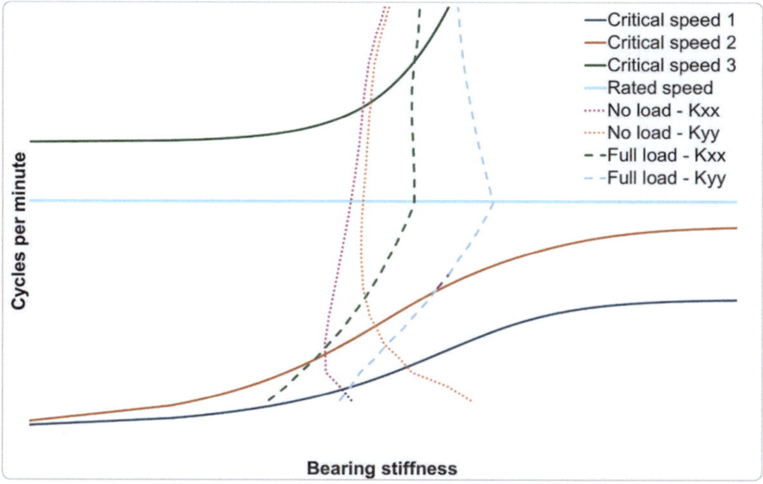

Fig. 17.18 Undamped critical speed map, bearing stiffness as a function of load (Image courtesy of Nuovo Pignone Tecnologie srl [part of Baker Hughes Company], Firenze, Italy)

gas load—only the rotor's weight contributes to the load. Dashed lines represent bearing behavior under actual operating conditions, where the load varies with speed based on the compressor's power absorption curve.

Beyond 100% speed, the power input is assumed constant, as it is limited by the motor's maximum capacity. Notably, the points where the load curves intersect the critical speed lines shift depending on the load. This effect becomes especially pronounced in lightweight, high-speed rotors transmitting substantial power. Therefore, it is essential to thoroughly evaluate bearing loading scenarios to ensure that each critical speed remains safely separated from the rated operating speed.

17.3.3 Compressor Startup

As stated in the preceding section, Integrally Geared Compressors (IGCs) typically operate at a fixed speed. If driven by an electric motor, a common method of initiating operation is the direct online (DOL) startup, where the motor accelerates to its nominal speed within a few seconds—usually around 10 s. This rapid acceleration imposes significant demands on system design, as the sudden surge in current can cause a noticeable voltage drop at the motor terminals. Simultaneously, the motor must generate enough torque to overcome several opposing forces: the inertial load from acceleration, the resistance from compressing process gas, and mechanical losses along the shaft line (such as friction and internal leakage). The torque required for gas compression is particularly sensitive to the pressure within the process circuit.

Given these factors, pressurized startup conditions are especially challenging and must be rigorously validated through simulations that account for both mechanical and process dynamics. A key concern is extended startup duration, which can result in repeated cycles of pulsating torque acting on the coupling. This may lead to component failure or motor overheating, potentially causing the startup sequence to abort.

Figure 17.19 illustrates the impact of excessive load torque. Operating near the first torsional natural frequency for an extended period significantly amplifies the pulsating torque on the load coupling, which adds to the average torque. The total peak torque is therefore the sum of the average torque—determined by compressor load resistance—and the pulsating torque, which intensifies when the system remains close to its torsional resonance. If this combined torque exceeds the slippage threshold between the coupling hub and compressor shaft, or surpasses the coupling's peak torque capacity, mechanical failure may occur. To prevent such outcomes, startup conditions must be thoroughly assessed using detailed train design data.

An adjustable-frequency drive (AFD) enables a controlled startup ramp, where slower acceleration minimizes the inertial component of resistant torque. This allows for higher process pressure levels during startup and may eliminate the need for compressor loop depressurization after shutdown—a common requirement in fixed-speed direct-on-line (DOL) systems to reduce startup torque.

If full motor power isn't necessary for the process, the AFD can be undersized accordingly and used to sequentially start multiple compressor trains. This staggered approach reduces the cost impact of the AFD on each train compared to continuous AFD operation across all units.

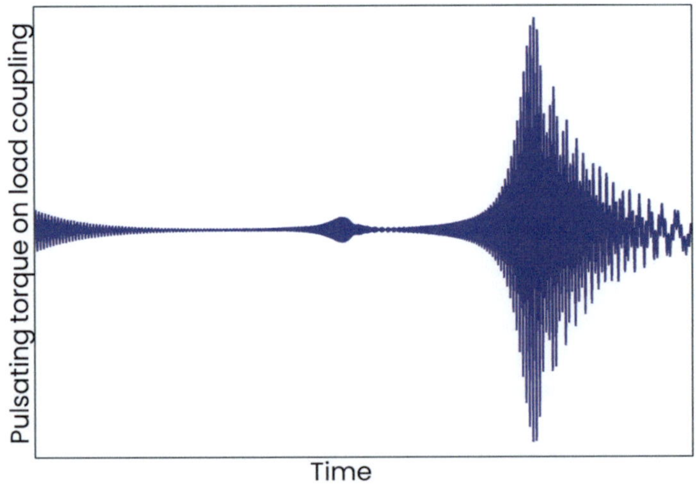

Fig. 17.19 Pulsating torque on the load coupling as a function of time (Image courtesy of Nuovo Pignone Tecnologie srl [part of Baker Hughes Company], Firenze, Italy)

17.3.4 Capacity Control Through Variable Inlet Guide Vanes

Variable inlet guide vanes (IGVs) are the most widely used method for regulating compressor performance to meet specific process demands. These vanes are controlled by a single actuator that adjusts the angle of incoming flow at the impeller inlet, effectively shifting the compressor's performance curve toward lower inlet flow rates and pressure heads.

Under normal operating conditions, each stage equipped with variable IGVs introduces a degree of flexibility, allowing the stage curve to be modified and one process parameter to be fine-tuned. For example, this setup can help maintain a constant compressor discharge pressure even when variables such as gas composition, flow rate, temperature, or inlet pressure fluctuate.

In multistage compressors, each IGV can only control one parameter—either the final discharge pressure or the discharge pressure of a specific stage. If both the final discharge pressure and the interstage pressure must remain stable, an additional IGV with independent control is necessary. The term "independent control" is important here because IGV angles can also be adjusted according to a fixed relationship across stages—such as keeping all vane angles uniform. While this approach can accommodate a broader range of operating conditions, it doesn't add another independent control variable.

Additionally, variable speed offers another layer of control over operating conditions. One example of this configuration was implemented by a compressor manufacturer in a steam turbine-driven syngas compressor system, which featured two sections: makeup and recycle. In that setup, the primary control variable was the train speed, while the IGVs in the recycle section were adjusted to maintain a consistent pressure in the injection stream between the two sections.

17.3.5 Sealing Options

The sealing principles outlined in Sect. 17.2.4 for beam compressors are equally applicable to integrally geared compressors (IGCs). However, the distinct procedures involved in seal inspection and replacement—covered in the following section—have led to a preference for labyrinth seals (typically used in low-pressure or less critical applications like air compression) and carbon ring seals.

A notable difference lies in the pressure dynamics: while inline machines feature a balancing line that equalizes pressure across both shaft-end seals, IGCs inherently expose each seal to the pressure differential between its respective compression stage and atmospheric pressure. This makes the sealing arrangement more complex, especially if using dry gas seals. Furthermore, due to the increased number of seals in IGCs, even with dry gas sealing, leakage tends to be higher compared to beam compressors.

17.3.6 Considerations for Availability

- Maintenance Considerations for Integrally Geared Compressors (IGCs):

  Maintenance operations for integrally geared units generally involve longer downtimes compared to inline compressors. This is primarily due to the more challenging accessibility of key components—such as bearings and seals—that are typically inspected during routine preventive maintenance.

  However, design enhancements can help mitigate these challenges. For instance, incorporating split bearings into an IGC allows for pad inspection or replacement without needing to dismantle the compressor casing. Seal maintenance, on the other hand, presents greater complexity. While split-design labyrinth seals can be serviced with minimal disruption, high-performance sealing solutions like dry gas seals—often housed in barrel cartridges—require axial removal and installation. This process demands the disassembly of several major components, including the suction diaphragm and impeller, as well as the detachment of process connections, resulting in extended equipment downtime.
- Optimizing Maintenance Intervals:

  Extending the intervals between maintenance activities is key to enhancing system availability. For CO_2 compressors, carbon ring seals have gained popularity due to their resilience against solid and liquid contaminants. They offer a favorable balance between sealing efficiency and ease of maintenance, making them a practical choice in applications where reliability and uptime are critical.

17.3.7 Process Connections

Integrally geared compressor units can accommodate up to 10 stages within a single gear assembly, resulting in as many as 20 process connections linked to one unit. This configuration subjects the gear to the combined forces of all connected stages, effectively treating it as a single compressor with process loads distributed across up to 20 points.

Each compressor casing is generally rigid, and a critical internal check involves verifying the bolted connection between the suction diaphragm and the outer casing—both of which are pressure-containing components. These bolts serve as pressure-retaining elements, and it is essential to confirm their integrity under process loads to prevent any risk of gas leakage.

Next, the bolts securing the compressor casing to the gear must also be assessed. These fasteners bear the loads from the suction and discharge flanges of each stage, as well as the weight of the stator components.

The most crucial structural evaluation, however, is of the gear itself. It must be verified to withstand all forces transmitted from the compressor casings. Gear casing strength can be enhanced by increasing wall thickness and adding structural ribs. Alternatively, the design of the process piping can be optimized to reduce the mechanical stress imposed on the gear. Introducing flexibility into the piping—either through strategic routing (e.g., adding bends or segments to increase compliance) or by using non-standard connections instead of conventional bolted ASME B16.5 flanges [14] to significantly mitigate these loads.

Flexible piping connections are common in low-pressure industrial systems. They accommodate misalignment using gaskets capable of absorbing substantial relative displacement. However, in high-pressure systems involving toxic or flammable gases, such flexibility may pose a leakage risk due to the potential for gasket failure under extreme conditions.

17.3.8 Compressor Unit Installation Options

The dimensions and configurations of CO_2 compression strings and their drivers can vary significantly depending on the project. Drivers may include gas turbines, steam turbines, or electric motors, and can be single- or double-ended, offering multiple installation possibilities.

For smaller systems powered by electric motors or steam turbines, a single base-plate design can streamline site installation. In contrast, gas turbines or specific project requirements may necessitate mounting the driver on a separate skid.

Regardless of the driver type or system size, the most common practice is to package the equipment at the factory or a dedicated packaging facility prior to shipment. Once delivered, the compression system—including all primary process piping—is assembled on-site, often involving complex procedures and hot work in a production environment.

To address challenges posed by strict local regulations or remote/offshore locations, modularized solutions offer a major advantage. These systems can be pre-assembled at the factory, reducing on-site interfaces and minimizing installation work. Pipe welding and assembly are completed in a controlled environment, and depending on transport constraints, modules can be shipped either fully assembled or in submodules tailored to transport limitations. These submodules are then reassembled on-site without the need for hot work.

Factory assembly provides a critical benefit: it enables thorough testing and verification of piping alignment with design specifications, significantly reducing the likelihood of on-site modifications.

Additionally, modularization supports the implementation of multilevel compression systems, where the compressor shaft line is positioned above the process coolers. This configuration reduces the overall footprint and space requirements, while also minimizing on-site activities—ultimately leading to substantial cost savings for the compression system.

Acknowledgements Matthew Gutscher is the contributing author of the section on the CO_2 process (Sect. 17.1). Stefano Falomi, Massimo Camatti, and Guido Peano contributed sections on centrifugal and integrally geared compressors (Sects. 17.2 and 17.3). All images in this chapter except Figs. 17.1, 17.2, and 17.3 were provided by Nuovo Pignone Tecnologie srl (part of Baker Hughes Company), Firenze, Italy.

References

1. IEA (2023) Global CO_2 emissions from energy combustion and industrial processes, 1900–2022. IEA, Paris. https://www.iea.org/data-and-statistics/charts/global-co2-emissions-from-energy-combustion-and-industrial-processes-1900-2022. License CC-BY-4.0. Accessed 3 Mar 2025
2. UN Climate Change (2025) The Paris Agreement. UN Climate Change, Bonn, Germany. https://unfccc.int/process-and-meetings/the-paris-agreement. Accessed 6 Mar 2025
3. IEA (2022) Global CO_2 emissions by sector, 2019–2022. IEA, Paris. https://www.iea.org/data-and-statistics/charts/global-co2-emissions-by-sector-2019-2022. License CC-BY-4.0. Accessed 6 Mar 2025
4. Oil and Gas Climate Initiative (OGCI) (2025) CO_2 storage resource catalogue. Pale Blue Dot Energy, Banchory, Aberdeenshire, Scotland. https://co2storageresourcecatalogue.com. Accessed 6 Mar 2025
5. National Institute of Standards and Technology (2025) NIST chemistry webbook, SRD 69, carbon dioxide. National Institute of Standards and Technology, Gaithersburg, MD. https://webbook.nist.gov/cgi/cbook.cgi?Name=carbon+dioxide&Units=SI&cTG=on&cTC=on&cTP=on#Thermo-Phase. Accessed 6 Mar 2025
6. Association for Materials Protection and Performance (AMPP) (2023) Guideline for materials selection and corrosion control for CO_2 transport and injection, AMPP Guide 21532-2023. AMPP, Houston, TX
7. Cappuccini F et al (2015) Materials selection and development for centrifugal compressors operating in extreme sour & acid gas service. Paper presented at the 3rd Middle East Turbomachinery Symposium, Doha, Qatar, 15–18 Feb 2015
8. Smith L, Billingham M (2010) Corrosion and selection of materials for carbon capture and storage. IEAGHG, Cheltenham, UK. https://ieaghg.org/publications/corrosion-and-selection-of-materials-for-carbon-capture-and-storage. Accessed Sept 2023
9. NEMA Standards Publication No. SM 23-1991 (R1997, R2002) Steam turbines for mechanical drive service. National Electrical Manufacturers Association, Rosslyn, VA
10. API Standard 617 (2022) Axial and centrifugal compressors and expander-compressors, 9th edn. American Petroleum Institute, Washington, DC
11. ASME PTC-10 1997 (R2014) (2014) Performance test code on compressors and exhausters. American Society of Mechanical Engineers, New York
12. Baldassarre L et al (2015) Operational modal analysis (OMA) application for the measure of logarithmic decrement in centrifugal compressor. Presented at the 44th Turbomachinery Symposium, Houston, TX, 14–17 Sept 2015
13. Vannini G et al (2017) Rotordynamic test results from a high flexibility ratio—high pressure fully instrumented centrifugal compressor test vehicle. Presented at the 46th Turbomachinery Symposium, Houston, TX, 11–14 Dec 2017
14. ASME B16.5 (2020) Pipe flanges and flanged fittings: NPS 1/2 through NPS 24, metric/inch standard. American Society of Mechanical Engineers, New York

Correction to: Centrifugal Compressors in Hydrocarbon Processing Industries

Correction to:
N. Ghaisas, *Centrifugal Compressors in Hydrocarbon Processing Industries*,
https://doi.org/10.1007/978-3-032-03687-2

The original version of this book was updated with a few typographical corrections in Chapters 1, 5, 6, 7, 9 and 16. The book has been updated with the changes.

The updated version of this book can be found at
https://doi.org/10.1007/978-3-032-03687-2

N. Ghaisas, *Centrifugal Compressors in Hydrocarbon Processing Industries*,
https://doi.org/10.1007/978-3-032-03687-2_18

Bibliography

1. Abdelhamid AN (1980) Analysis of rotating stall in vaneless diffusers of centrifugal compressors. In: Proceedings of the ASME 1980 international gas turbine conference and products show. Volume 1B: general, New Orleans, LA, 10–13 Mar 1980. ASME paper no. 80-GT-184
2. ACI Committee 351 (2018) ACI 351 3R-18: report on foundations for dynamic equipment. American Concrete Institute, Farmington Hills, MI
3. ANSI/NACE MR0103-2015/ISO 17945:2015 (R2023) Petroleum, petrochemical and natural gas industries — Metallic materials resistant to sulfide stress cracking in corrosive petroleum refining environments. NACE International, Houston, TX
4. API TR 684-1 (2019) API Standard Paragraphs Rotordynamic Tutorial: Lateral Critical Speeds, Unbalance Response, Stability, Train Torsionals, and Rotor Balancing. 1st edn. American Petroleum Institute, Washington, DC
5. API RP 686 (2009) (R2024) (E2025) Recommended practice for machinery installation and installation design. American Petroleum Institute, Washington, DC
6. API Standard 520 (2020) Sizing, selection, and installation of pressure-relieving devices part I—sizing and selection, 10th edn. American Petroleum Institute, Washington, DC
7. API Standard 521 (2020) Pressure-relieving and depressuring systems, 7th edn. American Petroleum Institute, Washington, DC
8. API Standard 526 (2023) Flanged steel pressure-relief valves, 8th edn. American Petroleum Institute, Washington, DC
9. API Standard 611 (2022) General-purpose steam turbines for petroleum, chemical, and gas industry services, 6th edn. American Petroleum Institute, Washington, DC
10. API Standard 613 (2021) Special purpose gear units for petroleum, chemical, and gas industry services, 6th edn. American Petroleum Institute, Washington, DC
11. API Standard 614 (2022) Lubrication, shaft-sealing, and oil-control systems and auxiliaries, 6th edn. American Petroleum Institute, Washington, DC
12. API Standard 617 (2022) Axial and centrifugal compressors and expander-compressors, 9th edn. American Petroleum Institute, Washington, DC
13. API Standard 670 (2025) Machinery protection systems, 6th edn. American Petroleum Institute, Washington, DC
14. API Standard 671 (2020) Special-purpose couplings for petroleum, chemical, and gas industry services, 5th edn. American Petroleum Institute, Washington, DC
15. API Standard 676 (2022) Positive displacement pumps—rotary, 4th edn. American Petroleum Institute, Washington, DC
16. API Standard 692 (2018) Dry gas sealing systems for axial, centrifugal, rotary screw compressors and expanders, 1st edn. American Petroleum Institute, Washington, DC

N. Ghaisas, *Centrifugal Compressors in Hydrocarbon Processing Industries*, https://doi.org/10.1007/978-3-032-03687-2

17. API TR 684-1 (2019) Technical report: API Standard paragraphs rotordynamic tutorial: lateral critical speeds, unbalance response, stability, train torsionals, and rotor balancing, 1st edn. American Petroleum Institute, Washington, DC
18. ASME B16.5 (2020) Pipe flanges and flanged fittings: NPS 1/2 through NPS 24, metric/inch standard. American Society of Mechanical Engineers, New York
19. ASME PTC 10-1997 (R2014) (2014) Performance test code on compressors and exhausters. American Society of Mechanical Engineers, New York
20. Association for Materials Protection and Performance (AMPP) (2023) Guideline for materials selection and corrosion control for CO_2 transport and injection, AMPP guide 21532-2023. AMPP, Houston, TX
21. Baldassarre L et al (2015) Operational modal analysis (OMA) application for the measure of logarithmic decrement in centrifugal compressor. Presented at the 44th turbomachinery symposium, Houston, TX, 14–17 Sept 2015
22. Bansal PN, Kirk RG (1975) Stability and damped critical speeds of rotor-bearing systems. ASME J Eng Ind 97(4):1325–1332
23. Bently DE, Hatch CT (2002) Fundamentals of rotating machinery diagnostics. Bently Pressurized Bearing Press, Minden, NV
24. Bently DE, Muszynska A, Littrell NB (1996) Margin of stability. Orbit 15(2):4–6
25. Bently DE, Goldman P, Yuan J (2001) Rotor dynamics of centrifugal compressors in rotating stall. Orbit (2Q01):40–50
26. Berot F, Dourlens H (1999) On instability of overhung centrifugal compressors. In: Proceedings of the ASME international gas turbine & aeroengine congress & exhibition, Indianapolis, IN, 7–10 June 1999. ASME paper no. 99-GT-202
27. Boyce MP (2003) Centrifugal compressors: a basic guide. PennWell Corporation, Tulsa, OK
28. Brun K, McBain M, Kurz R (2008) Application guideline for centrifugal compressor surge control systems, release version 4.3, Gas Machinery Research Council, Southwest Research Institute, San Antonio, TX, Apr 2008
29. Budynas RG, Nisbett JK (2010) Shigley's mechanical engineering design, 9th edn. McGraw-Hill series in mechanical engineering. McGraw-Hill, New York
30. Cameron JA, Danowski FM Jr (1973) Some metallurgical considerations in centrifugal compressors. In: Proceedings of the second turbomachinery symposium, gas Turbine Laboratories, Texas A&M University, College Station, TX, Oct 1973
31. Cappuccini F et al (2015) Materials selection and development for centrifugal compressors operating in extreme sour & acid gas service. Paper presented at the 3rd Middle East turbomachinery symposium, Doha, Qatar, 15–18 Feb 2015
32. Childs D (1993) Rotordynamic introduction to hydrodynamic bearings and squeeze-film dampers, chap 3. In: Turbomachinery rotordynamics: phenomena, modeling, and analysis. Wiley, New York, pp 132–226
33. Chockfast® Foundation Systems (1991) Electronic machinery grouting manual. ITW Philadelphia Resins, Montgomeryville, PA
34. Cooley JW, Tukey JW (1965) An algorithm for the machine calculation of complex Fourier series. Math Comput 19(90):297–301
35. Corbo MA, Melanoski SB (1996) Practical design against torsional vibration, tutorial. In: Proceedings of the twenty-fifth turbomachinery symposium. Texas A&M University, College Station, TX, pp 189–222
36. Dashefsky GJ (1930) The elimination of torsional vibration. Presented at the third national oil and gas power meeting, The Pennsylvania State College, University Park, PA, 12–14 June 1930
37. de Jongh FM, Morton PG (1994) The synchronous instability of a compressor rotor due to bearing journal differential heating. In: Proceedings of the ASME 1994 international gas turbine and aeroengine congress and exposition, The Hague, The Netherlands, 13–16 June 1994. ASME paper no. 94-GT-35
38. Dickau R, Perera L (2000) Mechanical vibration problems with variable frequency drives. In: Proceedings of the international pipeline conference, vol 2, Calgary, Alberta, Canada, 1–5 Oct 2000. ASME, New York, pp 1415–1423

39. Educational notes RTO-EN-AVT-143, paper 10. RTO, Neuilly-sur-Seine, France
40. Eisenmann RS Sr, Eisenmann RS Jr (1997) Machinery malfunction diagnosis and correction: vibration analysis and troubleshooting for the process industries, 1st edn. Prentice Hall PTR, Upper Saddle River, NJ
41. Eldridge TM, Soulas TA (2005) Mechanism and impact of damper seal clearance divergence on the rotordynamics of centrifugal compressors. In: Proceedings of ASME turbo expo 2005: power for land, sea, and air. Volume 4: turbo expo 2005, Reno, NV, 6–9 June 2005, pp 907–915
42. European Forum for Reciprocating Compressors (2017) Guidelines for vibrations in reciprocating compressor systems. Available at https://www.recip.org/wp-content/uploads/2020/04/EFRC-Guidelines-for-Vibrations-in-Recipr.-Compr.-Systems_4th-edition.pdf
43. European Forum for Reciprocating Compressors (2017) Summary of international guidelines, standards, and best practices of foundations, anchor bolts, and grouting of reciprocating compressor systems. Available at https://www.recip.org/wp-content/uploads/2020/04/EFRC-RD-Report-Foundation-Anchor-Bolts-Grouting-October-19th-2017.pdf
44. Frigne P, Van den Braembussche R (1984) Distinctions between different types of impeller and diffuser rotating stall in a centrifugal compressor with vaneless diffuser. ASME J Eng Gas Turbines Power 106(2):468–474
45. Gigatec SmartRock™ literature
46. Gunter EJ (1966) Dynamic stability of rotor-bearing systems. Technical report NASA SP-113. National Aeronautics and Space Administration, US Government Printing Office, Washington, DC
47. Harrison DM (2013) The grouting handbook: a step-by-step guide for foundation design and machinery installation, 2nd edn. Elsevier, Waltham, MA, p 6
48. Holdrege J, Subler W, Frasier W (1983) AC induction motor torsional vibration consideration—a case study. IEEE Trans Ind Appl 19(1):68–73
49. How disc couplings work. Form #10036E. Regal Rexnord Corporation, Milwaukee, WI
50. Hu Q, Zhu M, Yang J (2020) Study on thermal unstable vibration of rotor under journal whirl with large amplitude in journal bearing. Int J Rotat Mach 2020:1980759. https://doi.org/10.1155/2020/1980759
51. Huetten V, Zurowski RM, Hilscher M (2008) Torsional interharmonic interaction study of 75 MW direct-driven VSDS motor compressor trains for LNG duty. In: Proceedings of the thirty-seventh turbomachinery symposium. Turbomachinery Laboratory, Texas A&M University, College Station, TX, pp 57–66.
52. IEA (2022) Global CO_2 emissions by sector, 2019–2022. IEA, Paris. https://www.iea.org/data-and-statistics/charts/global-co2-emissions-by-sector-2019-2022. License CC-BY-4.0. Accessed 6 Mar 2025
53. IEA (2023) Global CO_2 emissions from energy combustion and industrial processes, 1900–2022. IEA, Paris. https://www.iea.org/data-and-statistics/charts/global-co2-emissions-from-energy-combustion-and-industrial-processes-1900-2022. License CC-BY-4.0. Accessed 3 Mar 2025
54. ISO 21940-11:216 (2016, confirmed 2021) Mechanical vibration—rotor balancing—part 11: procedures and tolerances for rotors with rigid behaviour, 1st edn. International Organization for Standardization, Geneva, Switzerland
55. ISO 2631-1:1997 (1997, confirmed 2021) Mechanical vibration and shock—evaluation of human exposure to whole-body vibration. Part 1: general requirements, 2nd edn. International Organization for Standardization, Geneva, Switzerland
56. Jackson C, Leader ME (1983) Design and commissioning of two synchronous motor–gear–air compressor trains. Presented at the 13th Texas A&M turbomachinery symposium, Houston, TX, 13–15 Nov 1983
57. Jeffcott HH (1919) XXVII. The lateral vibration of loaded shafts in the neighbourhood of a whirling speed. The effect of want of balance. Philos Mag Ser 6 37(219):304–314
58. Kaiser TF, Osman RH, Dickau RO (2008) Analysis guide for variable frequency drive operated centrifugal pumps. In: Proceedings of the twenty-fourth international pump users symposium. Turbomachinery Laboratory, Texas A&M University, College Station, TX, pp 81–106

59. Keogh PS, Morton PG (1994) The dynamic nature of rotor thermal bending due to unsteady lubricant shearing within bearing. Proc R Soc Lond A 445:273–290
60. Ker Wilson W (1959) Practical solution of torsional vibration problems, vols 1–5. Wiley, New York
61. Kirk RG (1980) Stability and damped critical speeds: how to calculate and interpret the results. CAGI Tech Dig 12(2):1–14
62. Kirk RG, Reedy SW (1988) Evaluation of pivot stiffness for typical tilting-pad journal bearing designs. ASME J Vib Acoust Stress Reliab Des 110(2):165–171
63. Kirk RG, Guo Z, Balbahadur AC (2003) Synchronous thermal instability prediction for overhung rotors. In: Proceedings of the thirty-second turbomachinery symposium, Turbomachinery Laboratory, Texas A&M University, College Station, TX, 8–11 Sept 2003, pp 121–135
64. Kuly J (2010) Best practices in compressor mounting. Presented at the 7th conference of the EFRC, Florence, Italy, 20–22 Oct 2010
65. Kuly J, Adkins J, Krich R (2014) GMRC short course: best practices in compressor and gas-engine mounting. Presented at the gas machinery conference, Nashville, TN, 6–8 Oct 2014
66. Kuzdzal MJ, Hustak JF (1996) Squeeze film damper bearing experimental vs. analytical results for various damper configurations. In: Proceedings of the twenty-fifth turbomachinery symposium, Turbomachinery Laboratory, Department of Mechanical Engineering, Texas A&M University, College Station, TX, 17–19 Sept 1996
67. Kuzdzal MJ, Hustak JF, Sorokes JM (1994) Identification and resolution of aerodynamically induced subsynchronous vibration during hydrocarbon testing of a 34,000 HP centrifugal compressor. In: Proceedings of the 4th international conference on rotordynamics, IFToMM, Chicago, IL, 7–9 Sept 1994
68. Leary J, O'Malley R (2004) Grouting large skid mounted compressor packages. Presented at the 2004 gas machinery conference, Albuquerque, NM, 4–7 Oct 2004
69. Lee YI, Barkey ME, Kang HT (2012) Metal fatigue analysis handbook. Elsevier, Amsterdam, The Netherlands
70. Littrell N (2005) Understanding and mitigating shaft runout. Orbit 25(3):5–17
71. Lund JW (1964) Spring and damping coefficients for the tilting-pad journal bearing. ASLE Trans 42(4):342–352
72. Lund JW (1974) Stability and damped critical speeds of a flexible rotor in fluid bearings. Trans ASME Ser B 96(2):509–517
73. Lund JW (1980) Review of analytical methods in rotor-bearing dynamics. Tribol Int 13(5):233–236
74. Lund JW, Orcutt FK (1967) Calculations and experiments on the unbalance response of a flexible rotor. ASME J Eng Ind 89(4):785–796
75. Maalouf MG (2007) Slow-speed vibration signal analysis. Orbit 27(2):4–16
76. Marshall DF, Sorokes JM (2000) A review of aerodynamically induced forces acting on centrifugal compressors, and resulting vibration characteristics of rotors. In: Proceedings of the 29th turbomachinery symposium, Houston, TX. Turbomachinery Laboratories, Texas A&M University, College Station, TX, pp 263–280
77. Marshall DF, Hustak JF, Memmott EA (1993) Elimination of subsynchronous vibration problems in a centrifugal compressor by the application of damper bearings, tilting pad seals, and shunt holes. Paper presented at the rotating machinery conference and exposition, Somerset, NJ, 10–12 Nov 1993. Paper no. NJIT-ASME-HI-STLE
78. Memmott EA (1987) Damper bearings and stability of a CO_2 compressor. Presented at the seventh annual rotating machinery & controls industrial research conference, San Diego, CA, 21–25 June 1987
79. Memmott EA (1992) Stability of centrifugal compressors by applications of tilt pad seals, damper bearings, and shunt holes. In: Proceedings of the 5th international conference on vibrations in rotating machinery, Bath, England, 7–10 Sept 1992, pp 99–106

80. Memmott EA (1994) Stability of a high-pressure centrifugal compressor through application of shunt holes and a honeycomb labyrinth. In: 13th machinery dynamics seminar, CMVA, Toronto, Canada, 12–13 Sept 1994, pp 211–233
81. Memmott EA (2000) Empirical estimation of a load related cross-coupled stiffness and the lateral stability of centrifugal compressors. In: Proceedings of the 18th machinery dynamics seminar, CMVA, Halifax, Nova Scotia, Canada, 26–28 Apr 2000, pp 9–20
82. Mirsky S, Jacobson W, Tiscornia D, McWhirter J, Zaghloul M (2012) Development and design of antisurge and performance control systems for centrifugal compressors. In: Proceedings of the forty-second turbomachinery symposium, Turbomachinery Laboratory, Texas A&M Engineering Experiment Station, Oct 2012
83. Moore JJ, Hill DL (2000) Design of swirl brakes for high pressure centrifugal compressors using CFD techniques. In: Proceedings of the 8th international symposium on transport phenomena and dynamics of rotating machinery, ISROMAC-8, vol II, Honolulu, HI, 26–30 Mar 2000, pp 1124–1132
84. Moore JJ, Soulas TA (2003) Damper seal comparison in a high-pressure re-injection centrifugal compressor during full-load, full-pressure factory testing using direct rotordynamic stability measurement. In: Proceedings of the ASME 2003 international design engineering technical conference and computers and information in engineering conference. Volume 5: 19th biennial conference on mechanical vibration and noise, parts A, B, and C, Chicago, IL, 2–6 Sept 2003, pp 1319–1326
85. NACE MR0175-2021/ISO 15156:2020 Petroleum and natural gas industries—Materials for use in H2S-containing Environments in oil and gas production—Part 1: General principles for selection of cracking-resistant materials. NACE International/ISO, Houston, TX
86. National Institute of Standards and Technology (2025) NIST chemistry webbook, SRD 69, carbon dioxide. National Institute of Standards and Technology, Gaithersburg, MD. https://webbook.nist.gov/cgi/cbook.cgi?Name=carbon+dioxide&Units=SI&cTG=on&cTC=on&cTP=on#Thermo-Phase. Accessed 6 Mar 2025
87. Nelson HD (1980) A finite rotating shaft element using Timoshenko beam theory. ASME J Mech Des 102(10):793–803
88. Nelson HD (1985) Rotordynamics equations in complex form. ASME J Vib Acoust Stress Reliab Des 107(4):460–461
89. Nelson HD, McVaugh JM (1976) The dynamics of rotor-bearing systems using finite elements. ASME J Eng Ind 98(2):593–600
90. NEMA Standards Publication No. SM 23-1991 (R1997, R2002) Steam turbines for mechanical drive service. National Electrical Manufacturers Association, Rosslyn, VA
91. Nestorides EJ (1958) A handbook on torsional vibration. Cambridge University Press, Cambridge, UK
92. Newkirk BL, Taylor HD (1924) Shaft whipping due to oil action in journal bearings. Gen Electr Rev 28(8):559–568
93. Nicholas JC (1994) Tilting pad bearing design. In: Proceedings of the twenty-third turbomachinery symposium, Turbomachinery Laboratory, Texas A&M University, College Station, TX, pp 179–194
94. Nicholas JC (1996) Hydrodynamic journal bearings—types, characteristics and applications. Mini course notes. In: 20th annual meeting. The Vibration Institute, Willowbrook, IL, p 76
95. Nicholas JC, Barrett LE (1984) The effect of bearing support flexibility on critical speed prediction. ASLE Trans 29(3):329–338
96. Oil and Gas Climate Initiative (OGCI) (2025) CO_2 storage resource catalogue. Pale Blue Dot Energy, Banchory, Aberdeenshire, Scotland. https://co2storageresourcecatalogue.com. Accessed 6 Mar 2025
97. Pash N, Hayles GC Jr, Hirata D, Kiuchi D (2012) A comparison of manufacturing methods, accuracy, quality control and testing methods as they relate to high head low flow impeller efficiency and overall compressor performance. Paper presented at the 41st turbomachinery symposium, Houston, TX, 24–27 Sept 2012

98. Pilkey WD, Pilkey DF (2008) Peterson's stress concentration factors, 3rd edn. Wiley, New York
99. Piotrowski J (2006) Shaft alignment handbook, 3rd edn. CRC Press, Boca Raton, FL
100. Process Industry Practices (2017) PIP STC01015, structural design criteria. Process Industry Practices, Austin, TX
101. Raimondi AA, Boyd J (1958) A solution for the finite journal bearing and its application to analysis and design: I. Trans ASLE 1(1):159–174
102. Raimondi AA, Boyd J (1958) A solution for the finite journal bearing and its application to analysis and design: II. Trans ASLE 1(1):175–193
103. Raimondi AA, Boyd J (1958) A solution for the finite journal bearing and its application to analysis and design: III. Trans ASLE 1(1):194–209
104. Ramesh K (2002) A state-of-the-art rotor dynamic analysis program. In: Proceedings of the 9th of international symposium on transport phenomena and dynamics of rotating machinery, Honolulu, HI, 10–14 Feb 2002
105. Ramesh K, Kirk RG (1995) Design procedure for evaluating stability of turbomachinery supported on squeeze film dampers. ASME J Tribol 117(4):742–744
106. Ramesh K, Kirk RG (1999) Nonlinear response of rotors supported on squeeze film dampers. In: Proceedings of the ASME 1999 design engineering technical conferences. Volume 7A: 17th biennial conference on mechanical vibration and noise, Las Vegas, NV, 12–16 Sept 1999, pp 1079–1085
107. Renfro EM (1985) Machinery foundations & grouting. Adhesive Services, Houston, TX
108. Rice RC (ed) (1988) SAE fatigue design handbook, AE-10, 2nd edn. SAE International, Warrendale, PA
109. Rice RC (ed) (1997) SAE fatigue design handbook, AE-22, 3rd edn. SAE International, Warrendale, PA
110. Riverhawk (Indikon) torque metering system literature. Regal Beloit America, Inc. (now Regal Rexnord), Milwaukee, WI
111. Ruddy AV, Summers-Smith D (1980) An introduction to the influence of the bearings on the dynamics of rotating machinery. Tribol Int 13(5):199–203
112. Ruhl RL, Booker JF (1972) A finite element method for distributed parameter turborotor system. ASME J Eng Ind 94(1):128–132
113. Sabin S (2005) Understanding discrepancies in vibration amplitude readings between different instruments, part 1 of 2. Orbit 25(2):18–30
114. Sabin S, Chitwood R, Hamad A, Qureshi M (2006) Understanding discrepancies in vibration amplitude readings between different instruments, part 2 of 2. Orbit 26(1):40–57
115. San Andrés L (2006) Hydrodynamic fluid film bearings and their effect on the stability of rotating machinery. Paper presented during the AVT-143 RTO AVT/VKI lecture series, Rhode St. Genèse, Belgium, 20–23 Mar 2006
116. Shade N (2012) Foundation and skid design considerations for large, high-speed reciprocating compressors. CompressorTECH^2 Magazine, Oct 2012. Reprint available at https://vdn.woodplc.com/assets/pdfs/Technical_Articles/CT2-2012_Foundations-and-Skids_large-high-speed-recips.pdf
117. Simmons JEL, Advani SD (1987) Michell and the development of tilting pad bearing. In: Dowson D, Taylor CM, Godet M, Berthe D (eds) Fluid film lubrication—Osborne Reynolds Centenary: proceedings of the 13th Leeds–Lyon symposium on tribology. Tribology series, vol 11. Elsevier Science, Amsterdam, pp 49–56
118. Simmons H, Smalley A (1984) Lateral gear shaft dynamics control torsional stresses in turbine driven compressor train. Presented at the 29th international gas turbine conference and exhibit, Amsterdam, The Netherlands, 3–7 June 1984. ASME paper 84-GT-28
119. Simmons H, Smalley A (1989) Effective tools for diagnosing elusive turbomachinery dynamics problems in the field. Presented at the ASME 1989 international gas turbine and aeroengine congress and exhibition, Toronto, Ontario, Canada, 4–8 June 1989. ASME paper 89-GT-71

120. Smalley AJ, Pantermuehl PJ (1997) Foundation guidelines. GMRC technical report no. TR-97-2. Gas Machinery Research Council, Dallas, TX
121. Smith SW (1997) The scientist and engineer's guide to digital signal processing. California Technical Publishing, San Diego, CA
122. Smith L, Billingham M (2010) Corrosion and selection of materials for carbon capture and storage. IEAGHG, Cheltenham, UK. https://ieaghg.org/publications/corrosion-and-selection-of-materials-for-carbon-capture-and-storage. Accessed Sept 2023
123. Sorokes JM, Kuzdzal MJ, Marshall DF (2016) A review of aerodynamically induced forces acting on centrifugal compressors, and resulting vibration characteristics of rotors. Paper presented at the 45th turbomachinery and 32nd pump symposia, Houston, TX, 12–15 Sept 2016. https://doi.org/10.21423/R16K5W
124. Soulas TA, Colby G, Desai AR, Griffin T (2011) CO_2 compression for capture and injection in today's environmental world. In: Proceedings of the first Middle East turbomachinery symposium, Doha, Qatar, 13–16 Feb 2011
125. Sujatha C (2010) Vibration and acoustics: measurement and signal analysis. Tata McGraw Hill Education Private Limited, New Delhi
126. Tanaka K, Nemoto H, Takahashi N, Fukushima Y, Akita Y, Tobise M (2010) Measurement and simulation of forced torsional vibration with inter-harmonic frequencies in variable speed drive motor driven compressor. In: Proceedings of the eighth IFToMM international conference on rotor dynamics, Seoul, Korea, 12–15 Sept 2010. IFToMM, Fredericton, New Brunswick, Canada, pp 844–851
127. Torquemeters Limited technical catalogues (various). Regal Beloit America, Inc. (now Regal Rexnord), Milwaukee, WI
128. UN Climate Change (2025) The Paris Agreement. UN Climate Change, Bonn, Germany. https://unfccc.int/process-and-meetings/the-paris-agreement. Accessed 6 Mar 2025
129. Vance JM (1988) Rotordynamics of turbomachinery. Wiley, Hoboken, NJ
130. Vance J, French R (1984) Measurement of torsional vibration in rotating machinery. Presented at the 7th annual conference on design engineering, Birmingham, UK, 25–27 Sept 1984. ASME paper 84-DET-55
131. Vance J, Zeidan F, Murphy B (2010) Machinery vibration and rotordynamics. Wiley, Hoboken, NJ
132. Vannini G et al (2017) Rotordynamic test results from a high flexibility ratio—high pressure fully instrumented centrifugal compressor test vehicle. Presented at the 46th turbomachinery symposium, Houston, TX, 11–14 Dec 2017
133. Verma AK, Sarangi S, Mahesh K, Banerjee S (2012) Oil whip detection using stator current monitoring. In: 2012 IEEE symposium on computers and informatics, ISCI 2012, Penang, Malaysia, 18–20 Mar 2012. IEEE Press, Piscataway, NJ, pp 119–124. https://doi.org/10.1109/ISCI.2012.6222678
134. Wachel JC (1983) Case histories of rotor dynamic instabilities in the field. Short course in rotordynamics of turbomachinery. Turbomachinery Laboratory, Texas A&M University, College Station, TX
135. Walker DN (2003) Torsional vibration of turbomachinery. McGraw-Hill, New York
136. Young WC, Budynas RG, Sadegh AM (2012) Roark's formulas for stress and strain, 8th edn. McGraw Hill, New York

Index

N. Ghaisas, *Centrifugal Compressors in Hydrocarbon Processing Industries*, https://doi.org/10.1007/978-3-032-03687-2

T

U

V

MIX
Papier aus verantwortungsvollen Quellen
Paper from responsible sources
FSC® C105338

If you have any concerns about our products,
you can contact us on
ProductSafety@springernature.com

In case Publisher is established outside the EU,
the EU authorized representative is:
Springer Nature Customer Service Center GmbH
Europaplatz 3, 69115 Heidelberg, Germany

Printed by Libri Plureos GmbH
in Hamburg, Germany